Feature Review Papers in Electrocatalysis

Guest Editor

Vincenzo Baglio

MDPI

Basel • Beijing • Wuhan • Barcelona • Belgrade • Novi Sad • Cluj • Manchester

Guest Editor
Vincenzo Baglio
CNR-ITAE Institute for
Advanced Energy
Technologies "N. Giordano"
Messina
Italy

Editorial Office
MDPI AG
Grosspeteranlage 5
4052 Basel, Switzerland

This is a reprint of the Special Issue, published open access by the journal *Catalysts* (ISSN 2073-4344), freely accessible at: https://www.mdpi.com/journal/catalysts/special_issues/474PX3F749.

For citation purposes, cite each article independently as indicated on the article page online and as indicated below:

Lastname, A.A.; Lastname, B.B. Article Title. *Journal Name* **Year**, *Volume Number*, Page Range.

ISBN 978-3-7258-7252-7 (Hbk)
ISBN 978-3-7258-7253-4 (PDF)
https://doi.org/10.3390/books978-3-7258-7253-4

Contents

About the Editor

Vincenzo Baglio

Vincenzo Baglio is a Research Director at the CNR – Institute of Advanced Technologies for Energy "Nicola Giordano" in Messina, Italy. He obtained a B.Sc. degree in Chemistry from the University of Messina (Italy) and a Ph.D. in "Materials for Environment and Energy" from the University of Rome "Tor Vergata" (Italy). His current research focuses on energy conversion and storage, particularly in the fields of fuel cells, electrolysers, and metal–air batteries. Vincenzo Baglio has published more than 250 articles in international journals, 10 book chapters, 1 book, and 1 international patent. He has also contributed to approximately 400 national and international conferences, including more than 30 invited or keynote presentations.

catalysts

MDPI

Editorial

Feature Review Papers in Electrocatalysis

Vincenzo Baglio

Institute of Advanced Energy Technologies "Nicola Giordano", Consiglio Nazionale delle Ricerche, Via Salita Santa Lucia sopra Contesse, 5, 98126 Messina, Italy; vincenzo.baglio@itae.cnr.it

Electrocatalysis lies at the heart of many of the technologies required for a sustainable energy future, enabling the efficient interconversion of chemical and electrical energy under mild conditions and with direct compatibility with renewable power sources. From hydrogen production and fuel cells to carbon dioxide capture, environmental remediation, and electrochemical sensing, advances in electrocatalyst design are critical to overcoming kinetic limitations, reducing energy losses, and improving durability and scalability. In recent years, remarkable progress has been witnessed, driven by deeper mechanistic understanding, precise control of active sites at the atomic and molecular level, and the emergence of advanced materials architectures.

This Special Issue, entitled "*Feature Review Papers in Electrocatalysis*", brings together nine comprehensive reviews that collectively reflect the breadth, maturity, and forward momentum of the field. The contributions span fundamental concepts, materials chemistry, theoretical descriptors, and device-level considerations, highlighting how rational electrocatalyst design can address key challenges in energy conversion, storage, and sustainability.

Several contributions focus on *advanced materials platforms and atomic-scale design principles*. Single-atom catalysts (SACs), reviewed in Contribution 6, represent an advancement in electrocatalysis due to their maximized metal utilization and well-defined active sites. This work provides a thorough analysis of substrates, metal–support interactions, coordination environments, and synthesis strategies while critically discussing how structural and electronic regulation governs activity, selectivity, and stability across reactions such as ORR, OER, HER, CO_2RR, and NRR. Complementarily, Contribution 2 highlights asymmetric atomic-structure catalysts for the oxygen reduction reaction, demonstrating how symmetry breaking through heteroatom doping, defect engineering, or cluster formation can disrupt linear scaling relationships and significantly enhance catalytic performance. Together, these reviews underscore the importance of precise active-site engineering beyond conventional structure–activity paradigms.

The role of *carbon-based and metal–nitrogen–carbon (M–N–C) systems* in fuel cell technologies is addressed in Contributions 3 and 9. Contribution 3 critically reviews recent advances in Fe–N–C catalysts as platinum-group metal-free alternatives for PEM fuel cells, emphasizing synthesis strategies such as MOF-derived templates and cascade anchoring that improve active site density and durability in acidic media. Contribution 9 broadens the perspective by examining structured carbon components in electrocatalyst layers, revealing key correlations between composition, degradation resistance, and fuel cell performance under harsh operating conditions. Together, these reviews provide valuable insights into reducing cost while extending the lifetime of fuel cell devices.

Electrocatalysis for *small-molecule activation and sustainable chemical transformations* is another central theme of this issue. Contribution 1 explores electrochemical CO_2 capture using redox-active metal–organic frameworks, positioning these materials at the intersection of coordination chemistry, electrochemistry, and materials engineering. By defining what constitutes true redox activity in MOFs and proposing cross-cutting performance metrics, this review advances the rational design of energy-efficient, cyclic CO_2 capture systems. Contribution 7 focuses on phosphorous-based catalysts for the electrocatalytic oxidation of urea, highlighting their dual role in wastewater treatment and energy generation. This work demonstrates how compositional tuning and surface engineering can unlock efficient and durable waste-to-energy pathways.

Water electrolysis and hydrogen production are addressed from both materials and theoretical perspectives. Contribution 5 reviews self-supported transition metal-based electrocatalysts fabricated via electrospinning, emphasizing how this versatile technique enables high surface area, efficient mass transport, and robust electrode architectures for HER and OER. Contribution 8 complements this experimental focus by reviewing energy-based theoretical descriptors for alkaline HER, illustrating how parameters such as water dissociation barriers and hydroxyl adsorption free energies can guide catalyst design and identify true active sites.

Finally, Contribution 4 expands the scope of electrocatalysis into *electrochemical sensing*, reviewing advanced alloy-based modifiers for carbon paste electrodes. By showcasing stainless steel, high-entropy alloys, and shape-memory alloys for the detection of biologically and environmentally relevant analytes, this contribution highlights how electrocatalytic principles extend beyond energy applications and can reshape analytical electrochemistry.

Taken together, the reviews in this Special Issue illustrate the dynamic and interdisciplinary nature of electrocatalysis, where advances in theory, materials chemistry, and device engineering converge. We hope that this collection will serve as a valuable reference for both established researchers and newcomers to the field, stimulating further innovation toward efficient, durable, and scalable electrochemical technologies for a sustainable future.

Acknowledgments: The Guest Editor sincerely thanks all the authors and reviewers whose valuable contributions and dedicated efforts made this Special Issue possible. Special appreciation is also extended to the journal Special Issue Editor for her invaluable support in managing the Special Issue.

Conflicts of Interest: The author declares no conflicts of interest.

List of Contributions

1. Castro-Castillo, C.; Suazo-Hernández, J.; Espinoza-González, R.; García, G. Smart Materials for Carbon Neutrality: Redox-Active MOFs for Atmospheric CO_2 Capture by Electrochemical Methods. *Catalysts* **2025**, *15*, 1134. https://doi.org/10.3390/catal15121134.
2. Qiu, H.; Wen, S.; Fu, Q.; Zhao, X. Oxygen Reduction Reactions of Catalysts with Asymmetric Atomic Structures: Mechanisms, Applications, and Challenges. *Catalysts* **2025**, *15*, 615. https://doi.org/10.3390/catal15070615.
3. Helsel, N.; Choudhury, P. Non-Platinum Group Metal Oxygen Reduction Catalysts for a Hydrogen Fuel Cell Cathode: A Mini-Review. *Catalysts* **2025**, *15*, 588. https://doi.org/10.3390/catal15060588.
4. Rajendrachari, S.; Chalageri, G.; Mahale, R.; Altas, E.; Chapke, Y.; Adimule, V. Investigation of Electrocatalytic Applications of Various Advanced Nanostructured Alloys—An Overview. *Catalysts* **2025**, *15*, 259. https://doi.org/10.3390/catal15030259.
5. Che, S.; Jia, Y.; Li, Y. Design and Synthesis of Self-Supported Water-Splitting Transition Metal-Based Electrocatalysts via Electrospinning. *Catalysts* **2025**, *15*, 205. https://doi.org/10.3390/catal15030205.
6. Ajmal, S.; Huang, J.; Guo, J.; Tabish, M.; Mushtaq, M.; Alam, M.; Yasin, G. Substrate Engineering of Single Atom Catalysts Enabled Next-Generation Electrocatalysis to Power a More Sustainable Future. *Catalysts* **2025**, *15*, 137. https://doi.org/10.3390/catal15020137.

7. Abd El-Lateef, H.; Khalaf, M.; Mohamed, I. Advances in Phosphorus-Based Catalysts for Urea Electrooxidation: A Pathway to Sustainable Waste to Energy Conversion Through Electrocatalysis. *Catalysts* **2024**, *14*, 937. https://doi.org/10.3390/catal14120937.
8. Ogunkunle, S.; Mortier, F.; Bouzid, A.; Hinsch, J.; Zhang, L.; Wu, Z.; Bernard, S.; Zhu, Y.; Wang, Y. Navigating Alkaline Hydrogen Evolution Reaction Descriptors for Electrocatalyst Design. *Catalysts* **2024**, *14*, 608. https://doi.org/10.3390/catal14090608.
9. Zasypkina, A.; Ivanova, N.; Spasov, D.; Mensharapov, R.; Sinyakov, M.; Grigoriev, S. Recent Advances in the Development of Nanocarbon-Based Electrocatalytic/Electrode Materials for Proton Exchange Membrane Fuel Cells: A Review. *Catalysts* **2024**, *14*, 303. https://doi.org/10.3390/catal14050303.

https://doi.org/10.3390/catal16030207

catalysts

MDPI

Review

Smart Materials for Carbon Neutrality: Redox-Active MOFs for Atmospheric CO_2 Capture by Electrochemical Methods

Carmen Castro-Castillo [1,*], Jonathan Suazo-Hernández [2,3], Rodrigo Espinoza-González [1] and Gonzalo Garcia [4,*]

[1] Department of Chemical Engineering, Biotechnology and Materials, FCFM, Universidad de Chile, Avenida Beauchef 851, Santiago 8320000, Chile; roespino@ing.uchile.cl

[2] Center of Plant, Soil Interaction and Natural Resources Biotechnology, Scientific and Biotechnological Bioresource Nucleus (BIOREN-UFRO), Universidad de La Frontera, Avenida Francisco Salazar, Temuco 01145, Chile; jsuazo@udla.cl

[3] Facultad de Medicina Veterinaria y Agronomía, Universidad de Las Américas, Sede Concepción, Concepción 4030000, Chile

[4] Instituto Universitario de Materiales y Nanotecnología, Departamento de Química, Universidad de La Laguna, P.O. Box 456 Santa Cruz de Tenerife, Spain

* Correspondence: carmen.castro.c@uchile.cl (C.C.-C.); ggarcia@ull.edu.es (G.G.)

Abstract: The electrochemical capture and transformation of carbon dioxide (CO_2) (ECC) has recently emerged as a transformative alternative to conventional sorbent-based processes, enabling fully reversible operation under mild conditions and direct compatibility with renewable energy sources. This review focuses on redox-active metal–organic frameworks (MOFs) as electrosorbent materials for the electrochemical capture of CO_2. Rather than encompassing all electrochemical CO_2 capture technologies, we use molecular, polymeric, and COF-based systems as a framework to define what makes a MOF truly "redox-active" for CO_2 electrosorption and how its performance can be assessed. This includes capacitive versus faradic electrosorption mechanisms and design strategies based on the redox chemistry associated with metal nodes, π-conjugated ligands, and strongly redox-active units such as tetrathiafulvalene, viologen, and ferrocene. The way in which defects affect hybrid MOF composites was highlighted, and in situ and operando spectroscopic techniques have improved the understanding of the reaction mechanism in carbon dioxide capture and release under controlled potential. Research comparing carbonaceous materials, redox polymers, and hybrid structures has highlighted both the opportunities and limitations of MOFs, particularly in terms of energy efficiency, scalability, structural robustness, and reproducibility. From a broader perspective, redox-active MOFs occupy a unique position at the intersection of coordination chemistry, electrochemistry, and materials engineering for large-scale applications. In this review, we analyze how redox activity in MOFs—at the metal nodes, ligands, and extended structures—can be harnessed to design energy-efficient, cyclic electrochemical CO_2 capture systems. Furthermore, we propose cross-cutting metrics and design rules that enable meaningful comparisons between materials and device architecture.

Keywords: redox-active MOFs; electrochemical CO_2 capture; conducting structures; electrosorption mechanism; ligand-centered redox activity; defect engineering; energy-efficient carbon capture

1. Introduction

The greatest scientific and technological challenges of this century have been the reduction in climate change. One of the key strategies proposed to systematically reduce carbon dioxide (CO_2) emissions is carbon capture and storage (CCS), and carbon capture, utilization, and storage (CCUS) have become necessary instruments in global environmental decarbonization policies [1,2]. CO_2 capture processes are based on chemical absorption using solvents, physical adsorption in microporous materials, or selective separation using membranes. However, these approaches have several disadvantages, such as high energy requirements for sorbent regeneration, solvent degradation, limited molecular selectivity, and low scalability under industrial conditions [3]. Under these circumstances, a new paradigm has emerged: electrochemical CO_2 capture (ECC), an option that can be carried out under ambient conditions, with high reversibility and compatibility with current challenges, such as renewable energy sources. Unlike thermal processes, electrochemical methods regulate the electrochemical potential of CO_2, allowing capture and electrochemical conversion to be combined in a single device [4,5]. This approach represents a promising step toward closed-loop carbon technologies, where electrochemical CO_2 capture and transformation are integrated into energy-efficient and potentially scalable systems [6–8].

The success of ECC depends exclusively on the electrosorbent material in which the reaction is generated. Early studies focused mainly on heteroatom-doped carbons, redox-active polymers, and hybrid compounds [9–11]. These studied materials promise rapid kinetics and excellent reversibility, but they present limited molecular selectivity and insufficient control of host-guest interactions. In contrast, metal–organic frameworks (MOFs) have emerged as a highly efficient, adsorptive, and versatile class of materials thanks to their large active surface area, modularity and structural defects, in addition to their ability to incorporate specific active sites by designing metal nodes and organic bonds [12]. Despite their success in CO_2 capture, conventional MOFs show low intrinsic electrical conductivity, which limits their direct use and development in conventional electrochemical systems [13]. Due to these limitations, the development of a subclass of conducting or redox-active MOFs has been encouraged. This subclass has developed electron transport facilitated by metal nodes, π-conjugated ligands, or structural defect engineering [14–16].

Two-dimensional (2D) MOFs such as $Cu_3(HHTP)_2$ and $Ni_3(HITP)_2$ have been shown in several studies to have reversible CO_2 adsorption modulated by the applied electrochemical potential, while maintaining structural stability over multiple voltammetric cycles [17,18]. This is attributed to the charge redistribution of the electron density, induced by oxidation/reduction processes that change to a higher affinity of the active sites towards CO_2 molecules (Scheme 1) [19]. However, the scientific community remains divided regarding the true potential of these MOF scaffold systems. Several researchers highlighted the tunable selectivity and molecular-level control offered by porous MOFs compared to widely studied traditional materials [20,21], while others emphasize practical limitations such as instability under humid conditions, scalability in synthetic processes, and energy efficiency [22,23]. In this sense, emerging approaches such as surface defect engineering and post-synthetic functionalization of the material seek to simultaneously improve electrical conductivity, selectivity and increase the density of active sites [24–26].

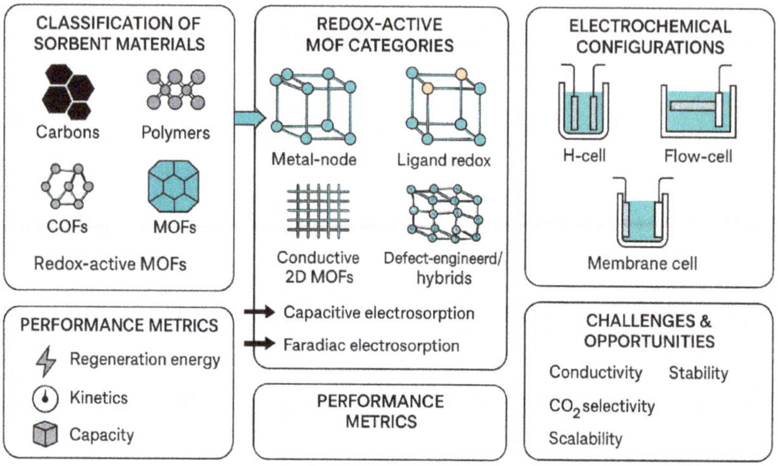

Scheme 1. Schematic overview of this review on redox-active MOFs for electrochemical CO_2 capture. The figure first classifies the main families of electrosorbent materials (carbons, polymers, COFs, MOFs and redox-active MOFs), then highlights different categories of redox-active MOFs (metal-node, ligand-redox, conductive 2D frameworks and defect-engineered/hybrid systems) and their capacitive vs. faradaic electrosorption behavior, and finally links these materials to typical electrochemical cell configurations (H-cell, flow cell and membrane cell), key performance metrics (regeneration energy, kinetics, capacity, stability) and current challenges and opportunities (conductivity, CO_2 selectivity and scalability).

The convergence of coordination chemistry, electrochemistry, materials engineering, and sustainability research positions redox-active MOFs as a promising and prominent platform in the field of smart carbon management. However, a comprehensive review critically analyzing their role in electrochemical CO_2 capture, unifying mechanistic, molecular, and industrial scale applied perspectives, is still lacking. This review aims to address this deficiency by providing a systematic overview encompassing the fundamentals of capacitive and faradaic electrosorption; recent advances in redox-active MOF design, material conductivity, and metal-ligand behavior; and technological challenges, such as the prospects for large-scale implementation and integration with electrochemical conversion processes, over the past 10 years of research in this field. Furthermore, this review aims to demonstrate how redox activity in MOFs, whether in metal nodes, ligands, or extended structures, can be rationally harnessed to design electrochemical CO_2 capture systems that contribute to carbon neutrality, and to propose cross-cutting metrics and rules for the design of these materials that allow for meaningful comparison between materials and devices. Figure 1 summarizes the conceptual framework adopted in this work, which structures the following sections into five interconnected dimensions: molecular design, redox control, CO_2 capture kinetics, benchmarking and limitations, and device integration for carbon-neutral processes.

The scope of this review is specific. We do not intend to provide a comprehensive analysis of all electrochemical approaches to CO_2 capture with redox-active MOFs. Instead, we first describe the general principles and representative examples of electrochemically mediated CO_2 capture using molecular, polymeric, and COF-based redox carriers, as a point of reference. Next, we focus on redox-active metal–organic frameworks (MOFs) and hybrid systems based on MOFs, analyzing how redox activity at the metal nodes and/or ligands controls the thermodynamics, kinetics, and selectivity of CO_2 binding. To conclude this review, we propose cross-cutting performance metrics and design rules that can be used to compare MOFs with other electrosorbents and to guide the development of next-generation redox-active MOFs for integrated CO_2 capture and conversion technologies.

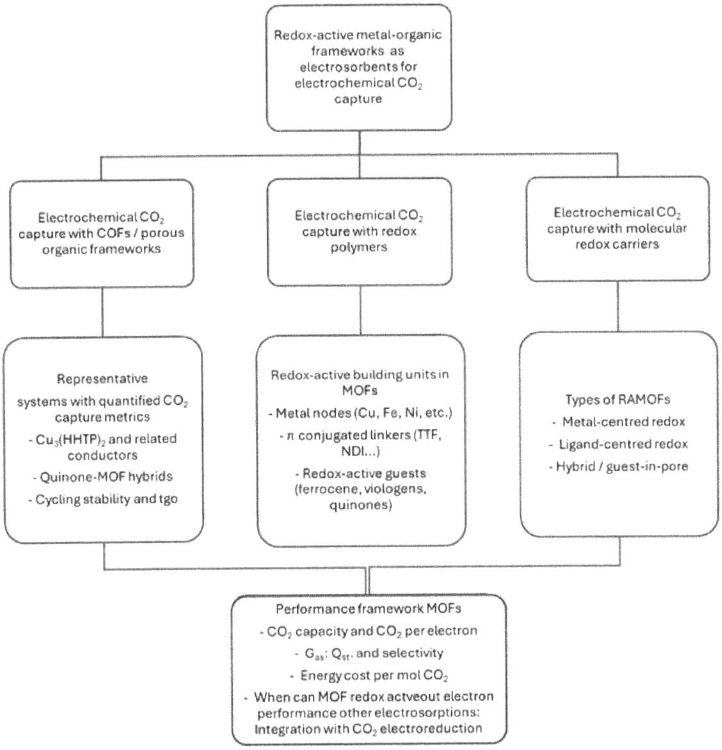

Figure 1. Conceptual framework of this review, highlighting redox-active metal–organic frameworks as electrosorbents for electrochemical CO_2 capture.

2. Fundamental and General Principles of CO_2 Electrosorption in MOFs

In this direction, the electrochemical capture and transformation of CO_2 was presented as a promising alternative, unlike conventional methods (thermal absorption or adsorption), since it allowed modulating the gas/solid interaction by applying a controlled electrochemical potential, this was classified as electrosorption, which allowed the reversible adjustment of the electron density and surface charge of the material used, thus determining the affinity of the adsorbent for the CO_2 molecules studied [27,28].

Electrosorption formed the conceptual core of electrochemical CO_2 capture and transformation, introducing a level of dynamic and reversible control that is very difficult to achieve in other, more conventional chemical systems. Depending on the charge storage process, two main mechanisms can be distinguished: capacitive electrosorption and Faradaic or redox-active electrosorption. While both mechanisms can occur simultaneously, they vary according to their physical and chemical nature [29,30].

Storage in the electrical double layer: this capacitive mechanism prevailed in porous materials such as graphene aerogels, activated carbons, and heteroatom-doped carbons. In these cases, adsorption was generated by physical interactions, driven by an electric field, formed at the electrode/electrolyte interface when a voltage was applied. The formation of an electrical double layer caused a redistribution of ions that modified the attraction of polar molecules, temporarily retaining them within the pores of the MOF material [31,32]. This process was characterized by rapid kinetics and high reversibility, as the potential change was sufficient to release the adsorbed gas. The absence of chemical transformations supported excellent stability. However, its molecular selectivity was low, and the capture capacity directly depended on the surface area and porosity of the particles [33–35].

However, due to its conceptual simplicity, capacitive electrosorption proved insufficient for achieving more selective CO_2 capture. Its future development lies in hybrid configurations, where it could serve as a stable and conductive junction for redox-active compounds.

Faradaic electrosorption: Redox processes and molecular control involve oxidation-reduction processes at functional sites of the studied material, modifying the electron density and directly modulating the interaction with the analyzed gas [35,36] This mechanism has been studied in redox-active materials such as electroactive polymers; for example: quinones or viologens; metal oxides and redox-active MOFs.

During the reduction study, the redox sites gained electrons, generating transient anionic species that stabilized the CO_2 molecule through dipolar interactions. During oxidation, the process was reversed, promoting desorption. This electronic switch provided more precise molecular control than capacitive electrosorption, resulting in higher selectivity and lower energy consumption of the system [37,38].

However, this mechanism introduced new challenges: redox stability often degraded after multiple cycles, especially in the presence of impurities; and balancing conductivity, active site density, and porosity remained a structural challenge [37].

Faradaic electrosorption promised a selective and efficient route, but its success depended on material design. Without electronic design, the redox advantage was lost due to structural degradation resulting from the material's low durability and resistive losses.

Critical comparison of both electrosorption methods: Capacitive electrosorption stood out for its simplicity and robustness, but lacked molecular discrimination, which required structural modifications; Faradaic electrosorption introduced chemical control and high selectivity, although it required materials with advanced electronic conductivity, modularity, and stability, which necessitated changes in the design architecture. Consequently, molecular hybrid systems combining porous conductive components with redox-active sites lead us to a more promising system, integrating molecular redox modulation with capacitive stability [39,40].

Redox-active MOFs emerged as promising and unique candidates to integrate both mechanisms into a single material, combining high porosity, large surface area, programmable electronic (switch) functionality, and a well-ordered, crystalline architecture.

2.1. Redox-Active MOFs as a Platform for Faradaic Electrosorption

MOFs, unlike doped carbons or electroactive polymers, their lattice structure allows for parallel tuning of conductivity, active site density, and CO_2 selectivity through compositional modification, doping, and structural defects. MOFs provide a molecular and defect engineering platform that enables redox binding to both metal nodes and organic ligands [15,28].

At redox-active metal nodes, multivalent metal centers such as Cu, Fe, Ni, and Co alternated their oxidation states, acting as switchable CO_2 capture sites. In systems derived from HKUST-1 or Fe carboxylates, node reduction increased the local electron density, favoring transient coordinative and electrostatic interactions with CO_2 [8,10,41,42].

These materials demonstrated the concept of switched trapping, their poor and limited electronic conductivity prevented a homogeneous response of the material, highlighting the need to improve the coupling between the redox-active metal nodes and the conductive network.

On the other hand, π-conjugated ligands and intrinsic conductivity, the use of π-conjugated electroactive ligands, such as HHTP and HITP, introduced continuous electronic channels that exhibit distributed redox processes and direct charge modulation within the pores and defects of the material.

MOFs such as $Cu_3(HHTP)_2$ and $Ni_3(HITP)_2$ [10,15,43] showed high structural stability and electron mobility, demonstrating the possibility of tuning their affinity toward CO_2 by applying an electric potential [17,44–46].

Consequently, these MOF-type materials represent the current frontier between adsorbents and functional semiconductors. While their synthesis remained complex, delicate, and expensive, a full understanding of the relationship between selectivity and molecular durability, as well as their electronic structure, which required further study, was still needed.

Ni-bis(diimine)-based MOFs have also been studied. The most recent study by Liu et al. [43] marked a milestone by demonstrating solid-state CO_2 capture and release, confirming the cycling stability and redox durability of Ni-bis(diimine)-based MOFs. These studies validated the feasibility of solid-state devices with applied potential-controlled response.

The study of defect engineering and postsynthetic functionalization has been fundamental. The insertion of structural defects, whether metal vacancies, heteroatomic substitution, and absence or presence of ligands, altered the density of electronic states and created high-energy adsorption centers, improving the redox response of the MOF [28,47]. Similarly, postsynthetic functionalization with viologenic groups increased the density of active sites, which improved and enhanced the reversible electrochemical capture of CO_2 [28].

Clearly, defect engineering offered an effective approach to fine-tuning electronic properties; long-term stability for industrial application and synthetic reproducibility remained problematic to date; standardized protocols for peer validation were required.

In a comparative evaluation of MOFs, between redox polymers or capacitive carbons, redox-active MOFs stood out for their tunable selectivity through structural design, high reversibility (crystalline rigidity), and functional integration capability in electrochemical CO_2 capture and conversion processes. Limitations of these MOFs included high synthesis costs and low scalability, as well as their fragility to contaminants and moisture [43,44].

In general, redox-active MOFs represent a highly versatile platform for the electrochemical capture and transformation of CO_2. However, their transition from laboratory to industrial applications depends on advances in intrinsic conductivity, environmental stability, and macroscale scalable synthesis methodologies.

2.2. Electrochemical Techniques: Characterization of Redox Activity and Adsorption Dynamics for MOFs in CO_2 Capture

Electrosorption mechanisms in redox-active MOFs are fundamental to study and understand, requiring the correlation of redox processes with gas adsorption and structural changes induced by the applied electrochemical potential. In this regard, cyclic voltammetry (CV) and electrochemical impedance spectroscopy (EIS) were fundamental tools to evaluate the electronic response of the material and the interaction mechanism with CO_2 molecules [46].

Cyclic voltammetry (CV) and square wave voltammetry (SWV) techniques have allowed to know the redox couples associated with metallic nodes or π-conjugated ligands, for example, $Cu^{2+} \rightarrow Cu^{+}$ or $Fe^{3+} \rightarrow Fe^{2+}$, and the correlation of these events with variations in the amount of adsorbed CO_2. In systems such as $Cu_3(HHTP)_2$ or Ni-bis(diimine) units in MOFs, the processes associated with anodic and cathodic currents were directly linked to the capture and release of CO_2, which corroborates the faradaic and reversible nature of the process [10,43].

EIS provided information on charge transfer, interfacial capacitance, and internal resistances of the material. This technique showed a decrease in charge transfer resistance during CO_2 adsorption. Nyquist diagrams indicated an increase in local electron density, resulting in higher conductivity within the porous and defective channels of the two-dimensional MOFs studied [48,49].

The combination of CV and EIS techniques allowed us to differentiate between capacitive and faradaic behaviors, revealing the efficient coupling between electrochemical processes and gas adsorption. However, comparative analyses remained scarce and lacked standardized protocols that would allow for an adequate interpretation of what happened at the surface with CO_2 in a quantitative and reproducible manner across different laboratories.

2.3. Spectroscopic Techniques: Structural Correlation and Mechanistic Evidence of CO_2 in MOFs

On the other hand, electrochemical techniques were aided by in situ and operational spectroscopic techniques, providing direct evidence of the structural and chemical changes that occur during CO_2 electrosorption. Techniques such as infrared spectroscopy (FTIR), X-ray absorption spectroscopy (XAS), and in situ Raman spectroscopy allowed the identification of intermediate species, changes in the oxidation of metal nodes, and modifications in the vibrations of π-conjugated ligands, respectively [38]. These studies confirmed the formation of surface carbonates and bidentate species, which disappeared when the potential applied to the CO_2 electrodesorption was reversed, thus corroborating the reversibility of the process.

Electrochemical and spectroscopic characterization was essential to understanding the synergy of conducting MOFs between redox processes and molecular adsorption. However, this field still required reproducible protocols and universal parameters to verify capture efficiency in different materials and compare them across systems. Establishing experimental standards analogous to those used in electrochemical energy storage was considered a crucial step to consolidate the industrial application of redox-active MOFs in CO_2 capture.

3. Redox-Active MOFs as Electrosorbents for CO_2: Archetypes and Mechanisms

In the context established above for the electrochemical capture of CO_2 from redox-active MOFs, this section focuses on how these electrosorbent materials, highlight structural systems in which metal-centered and/or ligand-centered redox processes directly control the adsorption and desorption of CO_2 under an applied potential.

3.1. Classic Redox MOFs: HKUST-1 and Fe Carboxylates

Classical systems already studied, such as HKUST-1 (Cu-BTC) and Fe carboxylates describe the first attempts to employ redox-active metal centers for CO_2 capture and absorption [50].

In one of the most classic and studied MOFs, Cu-BTC, the Cu^{2+}/Cu^{+} reduction increased the electron density at the metal nodes, which improved the adsorption, although its instability in the presence of moisture restricted its applicability [5,38,51]. Another widely investigated MOF is Fe-based MOFs, which demonstrated that Fe^{3+}/Fe^{2+} states could actively participate in the adsorption processes, naturally linking CO_2 capture and electrochemical conversion within the same material, which marked the next steps to follow [49,52].

Understanding these types of MOFs was conceptually fundamental to future advances, but they had a major limitation: low conductivity and a tendency toward degradation, making them historical models, although they retained their value for structural comparison and mechanistic analysis. Table 1 shows representative electrochemical CO_2 capture systems based on redox carriers.

Table 1. Representative electrochemical CO_2 capture systems based on redox carriers.

Active System/Material	Type of Material/Mechanism	CO_2 Medium	Reported Capture Metrics	Key Comment	References
Polyanthraquinone–CNT (PAQ–CNT) + PVFc–CNT (Faradaic electro-swing)	Redox polymer (quinone) supported on CNT; reactive electro-swing	Solid-state cell, electrolyte [Bmim][Tf_2N]; gas 0.6–10% v/v CO_2 in N_2	Faradaic efficiency ≈ 490%; 40–90 kJ·mol^{-1} CO_2 work; capacity virtually independent of inlet concentration; loss < 30% after 7000 cycles	Iconic demonstration of reactive electro-swing with quinones; high energy efficiency and excellent cyclability	[53]
Molecular redox-active amine	Dissolved redox organic molecule (electrochemically mediated capture)	Electrochemical cell with amine redox in organic electrolyte; simulated CO_2 currents	Electronic utilization (mol CO_2/mol e^-) up to 1.25; reversible capture and release over multiple cycles	It shows that a molecular carrier can overcome the 1 CO_2/e^- limit through chemical design of the amine	[54]
1,8-ESP (phenazine bis-sulfonate)	Phenazine redox molecule; pH-swing cell coupled to electrical storage	Highly soluble aqueous solution (>1.35 M) over a wide pH range; captures from concentrated and dilute streams	Capacity 0.86–1.41 mol CO_2·L^{-1}; energy cost 36–55 kJ·mol^{-1} CO_2; capacity degradation < 0.01% per day	Complete CO_2 capture with electrochemical storage; key reference in pH-swing cells	[55]
BDT-Q (benzodithiophene-quinone)	Redox heterocyclic quinone; stable electrochemical capture in the presence of O_2	Cells with BDT-Q in electrolytes containing O_2; CO_2 currents with O_2	High stability in the presence of O_2; maintains virtually unchanged capture capacity after prolonged cycling (values detailed in the article)	Overcomes one of the bottlenecks of quinones: O_2 deactivation	[56]
Quinones grafted on porous carbon (f-CMK-3 and others)	Mesoporous carbons covalently functionalized with anthraquinone	Functionalized carbon electrodes in organic/ionic media; CO_2 in flow	Significant increase in electrochemical capture capacity compared to non-functionalized carbon; simultaneous improvement in charge storage (specific capacities and faradaic efficiencies reported)	First systematic family of quinone-functionalized carbons for ECC; basis for optimizing structure-performance	[57]
Quinone-functionalized carbons (structure-performance study)	A series of porous carbons with varying porosities/surface areas, functionalized with anthraquinone.	Electrodes in different electrolytes; CO_2 atmosphere.	All materials exhibit reversible electrochemical capture; capacity and kinetics are correlated with the pore environment; capture energies and kinetic parameters are detailed in the paper.	Clear relationships are established between electrode structure and capture performance.	[58]
Anthraquinone COF (COF-AQ)	Conductive organic covalent framework with anthraquinone units	COF electrodes in ionic liquid electrolyte and aqueous media; CO_2 in flow	Electrochemical capacity > 2.6 mmol CO_2 g^{-1} COF (≈50% of theoretical capacity); energy ~31 kJ·mol^{-1} CO_2; stable capture for 500 cycles with 99.6% coulombic efficiency	First demonstration of electrochemical CO_2 capture with a COF; very high capacity and good stability	[59]

3.2. *Electrosorption in a Two-Dimensional MOF: $Cu_3(HHTP)_2$*

Another important MOF has been $Cu_3(HHTP)_2$, a two-dimensional molecule that marked the first clear experimental evidence of reversible CO_2 capture and release upon applying electrochemical potential (a fixed potential was used for capture and the reverse potential for release). Vetik et al. and Zhao et al. [10,60] demonstrated that redox modulation of HHTP ligands transformed the material's affinity for CO_2, which was validated by CV, differential electrochemical mass spectrometry (DEMS) and density functional theory (DFT). Applying reduction potentials increased the electron density and reached a capacity of approximately 2 mmol g^{-1}, while changing the potential completely released the gas, confirming its Faradaic reversibility in the CO_2 reaction study.

This MOF represented a functional model that combined conductivity, structural stability, and π-delocalization. However, its response was highly dependent on the pressure and temperature conditions used, as well as the electrolyte composition, limiting its scalability beyond controlled laboratory environments [61].

Figure 2 shows the structural diversity of redox-active MOFs investigated to date. MOFs based on redox-active metal centers, either Cu, Fe, or Ni, offered nodes with multiple electrochemical oxidation states, while ligand-centered systems, such as TTF, viologens, and ferrocene, provided charge transport and electronic flexibility. Together, these architectures explained that electronic modulation could be implemented from both metal nodes and organic ligands, opening new avenues to control selectivity, charge density, and adsorption kinetics.

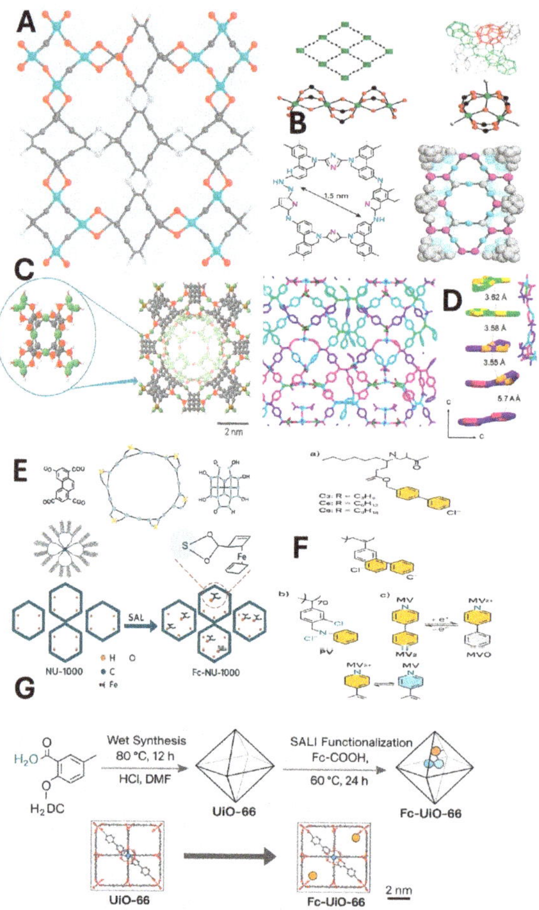

Figure 2. Representative MOFs redox-active for electrochemical CO_2 capture, highlighting archetypal examples such as (**A**) HKUST-1 (Cu(BTC)) [62], (**B**) Fe-carboxylate MOFs [63,64], (**C**) 2D conductive MOFs ($Ni_3(HHTP)_2$) [15], and (**D**) ligand-centered systems incorporating TTF [65,66], (**E**) ferrocene [67] (**F**) Viologen-based redox-active ligands [68]. (**a**) Schematic structures of viologen-containing polypeptides, in which the viologen units are connected to the backbone through alkyl spacers of three (C3), six (C6), or nine (C9) methylene groups. (**b**) Chemical structure of the reference viologen polymer (PV), where the viologen moieties are attached to a non-peptidic backbone. (**c**) Simplified redox scheme of methyl viologen, illustrating the interconversion between the dication

(MV^{2+}), the radical cation ($MV^{\cdot+}$), and the neutral species (MV^0). And (**G**) ferrocene functionalities [69]. All subfigures are reproduced or adapted from the cited references with permission from the respective publishers.

3.3. Metallic Conductivity in Two-Dimensional Structures: $Ni_3(HITP)_2$

A 2D MOF such as $Ni_3(HITP)_2$, initially conceived as a supercapacitor material [15,70,71], stood out for its metallic conductivity and the presence of electrochemically accessible Ni redox states. To date, no systematic studies on its electrochemical interaction with CO_2 have been performed; its highly ordered two-dimensional structure and broad electron mobility have positioned it as a structural reference for the design of next-generation redox-active MOFs.

This $Ni_3(HITP)_2$ MOF illustrated how conducting MOFs could act as electronic platforms before being functionalized for CO_2 capture. The key was to combine its excellent conductivity with coordinating groups that facilitated chemical interaction with the CO_2 molecule.

3.4. MOF with Ni-bis(Diimine) Units

As mentioned above, Liu et al. [43] achieved a fundamental breakthrough by demonstrating reversible solid-state CO_2 capture and release using a Ni-bis(diimine) MOF. This study validated that the metal nodes acted as stable redox switches under laboratory conditions. However, the structural robustness and high faradaic efficiency of this system laid the groundwork for solid-state capture devices with controlled response by the application of an electrochemical potential.

These studies validated redox shifting as a mechanism of operation but also highlighted the need to investigate the degradation of nickel sites under extended cycles of more than 10^4, an aspect that has not yet been adequately documented in recent publications, and which would pave the way for the implementation of these materials at an industrial level.

3.5. Emerging Strategies: Defect Engineering and Doped ZIFs

Research on these redox-active materials has led to the discovery of new strategies focused on the introduction of defects and metallic dopants to improve electrochemical stability, increase the density of redox sites, and optimize the molecular matrix of these systems. Studies by Zhang et al. [47] demonstrated that defect engineering facilitated electronic transport and altered the charge distribution over the MOF structural ligands, transforming conventional ZIFs and MOFs into partially conducting platforms.

Defect engineering was considered one of the most promising approaches, although risky, since an excess of vacancies could collapse the crystal lattice and alter selectivity. The challenge lay in controlling defects without compromising the integrity of the structure.

3.6. Other Redox-Active Families: TTF MOFs, Viologens and Ferrocenes

MOFs based on organic ligands with high redox activity expanded the spectrum of functional design:

Tetrathiafulvalene (TTF): has demonstrated reversible redox states and high conductivity. Its MOFs have been explored in molecular electronics and could be tuned for electrosorption due to their electronic tunability [65,66,72,73]

Viologens: These ligands were known for their reversible radical dication behavior and provided potentials for electroswitchable matrices, where adsorption and release directly depended on the potential applied to the system [74–76].

Ferrocene MOFs: Iron (Fe^{2+}/Fe^{3+}) redox pathways provided chemical stability and accessible redox behavior, with a long history in energy storage applications [77–80]. However, their direct application in electrosorption was not yet tested, they represented a promising frontier based on established coordination chemistry.

These MOF systems based on redox-active organic ligands offered the greatest design flexibility, but their main challenge lay in maintaining an effective electronic conduction path along their entire crystal lattice.

3.7. Thin Films and Hybrids Are Integrated with MOFs

The current use of redox-active MOFs depends on their methodology and integration into functional architectures. The most relevant strategies were:

(1) **Electrosynthesized thin films:** this type of system on surfaces allows MOFs to be anchored directly on conductive substrates, for example, ITO, graphene, among others, optimizing electrical contact and reducing interfacial resistance [81–83].

(2) **MOF-COF type molecular hybrids:** they combined the crystalline porosity of MOFs with the extended π conductivity of COFs, improving mechanical stability and charge mobility [84].

(3) **MOF-carbon based composites:** dispersion of MOF particles in carbon nanotube or graphene matrices increased conductivity and increased accessibility to active sites [85].

These strategies shifted the focus from chemistry to engineering. It was not enough to design a redox-active MOF; it was essential to process it into functional configurations that would allow its integration into industrial and real-world electrochemical systems.

Table 2 summarizes the fundamental differences between carbons, redox polymers, and redox-active MOFs. While the former differs in stability and low cost, and the latter in chemical modularity, MOFs offer a unique property that combines molecular selectivity, electronic control, and structural design.

Table 2. Comparison of families of materials for electrochemical CO_2 capture.

Materials	Dominant Mechanism	Main Advantages	Critical Limitations	References
Porous carbons	Capacitive electrosorption (double layer)	High cyclic stability; low cost; wide availability	Limited selectivity; low gravimetric capacity	[4,49]
Redox polymers	Faradaic electrosorption on quinone, viologen, etc. groups.	Chemical modularity; high reversibility; selectivity potential	Long-term chemical degradation; limited conductivity	[86,87]
Redox-active MOFs	Electron redistribution in metallic nodes and ligands	Molecular selectivity; structural tunability; possibility of integrating capture and conversion	Structural stability; expensive synthesis; limited scalability	[15]

Based on this background, redox-active MOFs were considered the most versatile platform for the development of electrochemical CO_2 capture technologies. Their advancement depends on overcoming obstacles related to conductivity, stability, and scalability, as well as establishing comparable metrics that similarly assess capture efficiency across different families of MOF-type materials.

Following the mechanistic framework proposed by Vetik et al. [10], Figure 3 shows an adapted scheme that summarizes the main electrochemical CO_2 capture mechanisms in redox-active MOFs, namely capacitive, faradaic, and hybrid pathways.

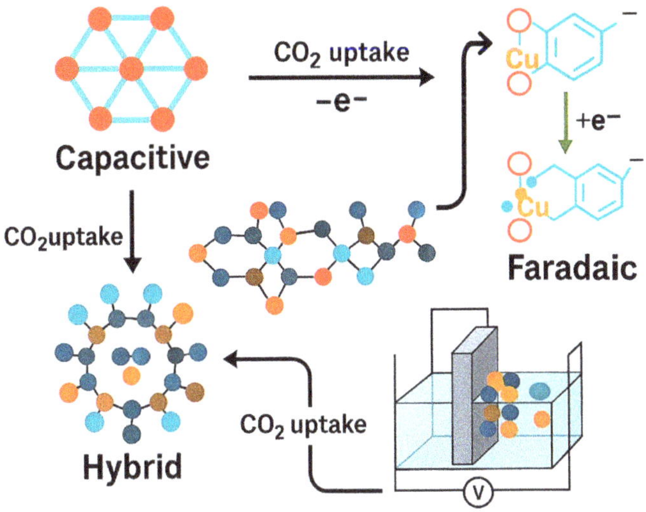

Figure 3. Schematic representation of the main electrochemical CO_2 capture mechanisms in redox-active MOFs, including capacitive, faradaic, and hybrid pathways. Adapted with modifications from I. Vetik et al., "Advancing Electrochemical CO_2 Capture with Redox-Active Metal-Organic Frameworks," arXiv:2411.16444 (2025) [10].

4. Redox-Active MOFs: Proposed Mechanisms of Electrochemical Trapping

The electrochemical trapping of CO_2 in redox-active MOFs resulted from the synergistic combination of physical interactions and localized redox processes that altered the electron density at the metal nodes and organic ligands [88].

Unlike purely porous adsorbents, conductive MOFs allowed the energetics of the gas–solid interaction to be modulated in real time by applying an electrochemical potential, enabling dynamic control of the adsorption and desorption of CO_2 molecules (Figure 4).

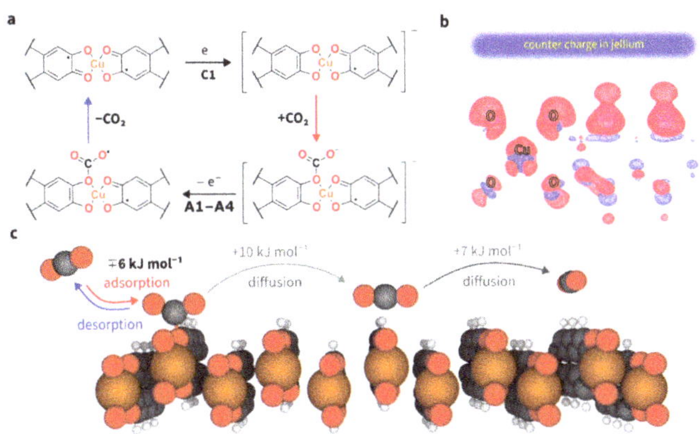

Figure 4. Schematic representation of electron redistribution in redox-active ligands under applied potential. In 2D conducting MOFs such as $Cu_3(HHTP)_2$, the conjugated HHTP ligands act as electron

reservoirs, delocalizing charge along the molecule; (**a**) mechanism; (**b**) absorption energy; (**c**) counter exchange in jellium [10]. In TTF-MOFs, reversible oxidation of tetrathiafulvalene regulates the electron density for CO_2 binding. Viologen-based ligands undergo redox dication/radical transitions, generating localized charge and enhanced polarization. Ferrocene-functionalized structures, e.g., NU-1000-Fc, provide stable Fe^{2+}/Fe^{3+} pairs, which act as redox markers and tunable binding sites. Simultaneously, these systems illustrate how ligand-centric redox processes can be harnessed to reversibly and controllably modulate CO_2 affinity.

4.1. *Modulation of Metallic Nodes in MOFs*

Redox-active metal nodes constitute the main interaction sites in classical MOFs. Reversible transitions such as Cu^{2+}/Cu^{+} in HKUST-1 or Fe^{3+}/Fe^{2+} in carboxylate MOFs regulated the local electron distribution and generated the intensity of Lewis-type interactions between the reduced metal center and the CO_2 carbon atom.

With the application of reducing potential, the nodes increased their electron density, generating a transient coordinative stabilization in the adsorbed CO_2 [89]. This redox modulation showed us the changes in the in situ IR and XPS spectra, where the carbonyl band (C=O) shifted to a lower energy during the reduction, indicating a reinforced metal and CO_2 (M-CO_2) interaction.

In this sense, modulation of metallic nodes provided direct control over the molecule's affinity for the gas but often carried the risk of structural degradation and partial dissolution of the metal under intense polarization, resulting in an unstable material. The main challenge was to design metallic nodes with operating redox processes that balanced the material's activity and stability.

4.2. *Electron Redistribution in π-Conjugated Ligands in MOF-Type Systems*

Two-dimensional MOFs containing HHTP ligands, such as $Cu_3(HHTP)_2$ act as electron reservoirs, generating electron movement and charge redistribution toward the ligand pores upon application of a reducing potential. This phenomenon, occurring in 2D MOFs, generated locally charged environments that brought CO* molecules closer together through electrostatic interaction and polarization induced by an external potential, explaining the Faradaic reversibility observed experimentally in these materials [10,15,90]. In extended π-conjugated systems, electron delocalization reduced the charge transfer resistance, which benefited the coupling between gas trapping and the redox reaction.

The π-conjugated ligands not only transferred electrons but also functioned as electronic switches, allowing precise molecular control of CO_2 adsorption and desorption. Their role in selectivity toward other molecules such as H_2O, N_2, and CH_4 had not yet been thoroughly studied.

4.3. *Strongly Redox Ligands: TTF and Viologen*

On the other hand, MOFs with redox-active organic ligands represented a new avenue for electronic modulation, as charge transfer in the ligands used was deciphered. In MOFs with TTF ligand, the reversible oxidation of tetrathiafulvalene alternated between low- and high-electron-density states, modulating the polarity of the material's pores or defects and, consequently, its specific affinity for CO_2 [65,66,91,92].

Furthermore, MOFs with Viologen ligands utilize the cation-dication radical transition in their mechanism, producing localized charges that induce electrostatic attraction toward the molecules. These processes are used as molecular switches controlled by the applied potential, facilitating the design of programmable materials that respond to partial pressure and gas composition [93].

These Viologen TTF ligands have represented a major advance in molecular electronics and gas adsorption. However, a quantitative correlation between redox potentials and adsorption free energy (ΔG_{ads}), a key parameter for assessing the feasibility of their technological application, was still needed.

Taken together, these examples illustrate different design strategies for making MOFs redox-active for electrochemical CO_2 capture:

(i) Housing a molecular carrier within a porous structure.
(ii) Incorporating redox-active bonds or nodes so that the structure itself acts as a sorbent.
(iii) Leveraging MOF environments that pre-concentrate and activate CO_2 at electrochemically polarized interfaces.

While the underlying chemical reactions are diverse, they share the common characteristic that CO_2 binding can be modulated by an applied potential. To provide a concise comparison of these systems, Table 3 summarizes representative redox-active MOFs and MOF-based materials, highlighting their electrochemical configurations and reported CO_2 capture metrics (including Q_{st}, the isosteric heat of CO_2 adsorption), as well as the specific role the MOF plays in each case.

Table 3. Redox-active MOFs and MOF-based materials for electrochemical CO_2 capture electrochemical configurations, CO_2 capture metrics and role of the MOF.

MOF/Material	Metal Anode/Redox Ligand	Electrochemical Configuration	CO_2 Capture Metrics	Comment on the Role of the MOF	References
$Cu_3(HHTP)_2$ (conducting MOF)	Redox-active Cu(II) node; aromatic 2,3,6,7,10,11-hexahydroxytriphenylene (HHTP) ligand	$Cu_3(HHTP)_2$ electrodes in aqueous electrolyte; dissolved and gas-phase CO_2; CV, GCD, and DEMS	Capacity of ≈2 mmol CO_2 g^{-1}; ΔH_ads ≈ −20 kJ·mol^{-1}; reversible capture/release in redox cycles in water	First MOF to switch its CO_2 capture capacity in aqueous electrolytes; The combination of Cu redox + conjugated π-ligand governs electrosorption	[10]
Hybrid MOF/quinone (Wenger–D'Alessandro)	Stable MOF framework (Mg-MOF-74 type or other robust analog) mechanochemically impregnated with a redox quinone molecule	MOF + quinone cathode electrodes; potential-dependent controlled capture, studied by spectroelectrochemistry	Reversible electrochemical capture of CO_2 is demonstrated; faradaic currents associated with the quinone/CO_2 reaction and in situ spectroscopic changes are reported (capacities and energies detailed in the paper)	First explicit demonstration of a redox-active MOF-based adsorbent for electroswing; the MOF acts as a porous scaffold and stabilizes the redox host	[94]
Zn(NDC) (DPMBI) and its reduced forms	Zn(II) node; redox-active benzenenotetracarboxidiimide ligand (DPMBI)	Controlled chemical/electrochemical reduction (Na^+ introduction) of the framework; static CO_2 adsorption measurements	Increased CO_2 capacity and increased reduction degree up to a certain limit; also improved CO_2/N_2 selectivity compared to the neutral MOF	Q_{st}	[95]
Ni_2(dobdc) and functionalized derivatives (e.g., pip-Ni_2 (dobdc))	Ni(II) node; dobdc ligand with open sites and grafted amines	Adsorption cells where oxidation/reduction in the framework (or ligands) modulates CO_2 affinity; primarily studied by adsorption + T/P stimulation and redox	Higher CO_2/N_2 selectivity and even higher after functionalization; changes in affinity have been explored by adjusting the metal's redox state and the density of basic sites	Q_{st}	[96]

Table 3. *Cont.*

MOF/Material	Metal Anode/Redox Ligand	Electrochemical Configuration	CO_2 Capture Metrics	Comment on the Role of the MOF	References
Various redox-active MOFs (TTF, NDI, etc.) with CO_2 response	Frameworks with tetrathiafulvalene (TTF), naphthalenediimide (NDI), or other π-redox systems ligands	SURMOF-type thin films and single crystals; spectroelectrochemical studies under CO_2 atmospheres	Changes in UV-Vis-NIR spectra and adsorption capacity upon switching between redox states; in some cases, an increase in Q_{st} and CO_2 adsorbed charge	These studies demonstrate how redox ligands on MOFs can serve as a platform for gas electrosorption, although they are not always quantified as complete ECC cells.	[97]
Cu- and Ni-MOFs functionalized for CO_2 capture/activation	Different nodes (Cu, Ni, Co) and ligands with basic/redox sites	MOF or MOF-derivative electrodes in electrochemical cell configurations (sometimes coupled to CO_2 reduction)	CO_2 adsorbed onto the MOF prior to reduction; some studies report enhanced capture capabilities under polarization and correlation with electrocatalytic activity	Bridge examples between capture and conversion; useful for your integrated capture+reduction section	[98]

Qst: isosteric heat of CO_2 adsorption obtained from multi-temperature adsorption isotherms (kJ·mol^{-1}).

4.4. Ferrocene-Functionalized MOFs as a Redox Tracer for CO_2 Capture

On the other hand, ferrocene in MOFs provided a stable, reproducible, and easily tunable Fe^{2+}/Fe^{3+} redox signal. A MOF that gave good results in this matter is NU-1000-Fc [69,99]: it was shown that the variation of the applied potential transforms the electron density of the Fc ligand structure, thus modifying the selectivity towards CO_2 and other molecules with similar dipoles [93].

When ferrocene is functionalized in MOFs, they behave as chemically robust redox tracers, enabling precise monitoring of charge transfer processes and high redox activity, making them a suitable platform for experimental validation of CO_2 mechanisms. Ferrocene-containing MOFs represent a suitable platform for correlating in situ spectroscopy with quantifiable redox behavior, but their efficiency in direct electrosorption remains to be demonstrated (Figure 5). Their properties under operating conditions will be essential to confirm their role in electrochemical processes scalable to the real world.

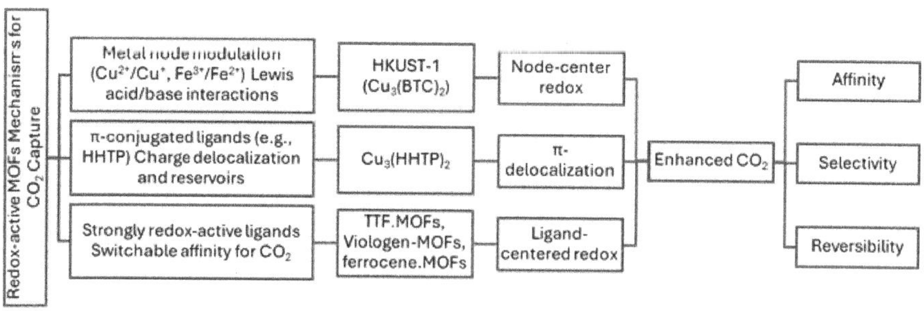

Figure 5. Schematic classification of redox-active mechanisms in MOFs for electrochemical CO_2 capture. Metal-centric processes, such as HKUST-1, π-conjugated ligand systems $Cu_3(HHTP)_2$, and MOFs with highly redox-active ligands such as TTF, viologen, and ferrocene, provide complementary routes to tune affinity, reversibility, and selectivity in electrosorptive CO_2 capture.

The current challenge lies not only in the synthesis of new MOFs, but also in understanding how electrons interact with the thermodynamics of adsorption during charge transfer. The development of quantitative models linking electron density, applied potential, and ΔG_{ads} is expected to transform electrochemical CO_2 adsorption from an experimental approach to a predictive and industrially scalable technology.

4.5. Formation of Intermediate Species in MOFs: Spectroscopic Evidence

Understanding the electrochemical trapping mechanisms in redox-active MOFs comes from the use of in situ and operando spectroscopic techniques, especially IR, Raman and XAS, which allowed monitoring changes in the structures during adsorption/desorption cycles.

These studies showed the transient formation of surface carbonate or bicarbonate species, as well as coordinative adducts between the metal and CO_2 (M-CO_2), whose appearance was closely related to the oxidation/reduction state of the material [38,43]. By reducing potentials, the high electron density at metal nodes or π-conjugated ligands facilitated the formation of semi-coordinated species and/or bidentate carbonates, while upon reversal of the potential, these bands rapidly disappeared from the IR and Raman spectra, indicating the reversibility of the process. In XAS studies, it was shown that the absorption changes confirmed periodic oscillations between the metal oxidation states of the MOF, which were synchronized with the CO_2 adsorption and desorption processes [38].

Intermediate species analysis provided one of the strongest evidences of the coupling between electronic dynamics and transient chemical interactions between redox-active MOFs and CO_2. However, spectroscopic studies performed under operating conditions, such as pressure, complex electrolytes, or the presence of H_2O, were limited, limiting representative validation of mechanistic models for CO_2 and MOFs.

4.6. The Role of Defect Engineering and Hybrid Materials in MOFs

Controlling local electronic environments through defect engineering and structural hybridization has emerged as a strategy to optimize the redox response and energy efficiency of MOFs. The deliberate inclusion of metal or ligand vacancies, as well as heteroatomic doping (N, S, or P), has made it possible to modify the electron density of states (DOS) and create high-energy sites capable of interacting more intensely with the CO_2 molecule [47,99].

Hybrid materials, such as MOF-COFs or MOF-carbon composites, described a new generation of functional systems in which extended electronic connectivity improved charge transport, reducing resistive losses and increasing faradaic efficiency. On the other hand, hybrid MOF-COFs combined the crystallinity and modularity of MOF ligands with the intrinsic conductivity of COFs. These studies contributed to describing the electron transfer between two materials toward the same goal: CO_2 capture. While carbon-based MOFs facilitated the dispersion of the active site and improved structural stability during long-term oxidation-reduction cycles [84].

Overall, structural hybridization and structural defect engineering represented the bridge to understanding molecular design. However, quantitative control of defect density remained a challenge: excessive vacancies could induce structural collapse of the MOF, while low defect densities limited the material's activity. The balance between connectivity, stability, and redox site density was crucial for the development of scalable, durable, and stable MOFs.

Commonly, the proposed mechanisms for electrochemical trapping in redox-active MOFs could be classified according to the origin of the electronic modulation: (1) Metal-centered (nodal): direct control of the charge density, but limited stability. (2) π-conjugated ligands: higher reversibility and extended electron mobility. (3) Strongly redox ligands

(TTF, viologen, ferrocene): design flexibility and fine-tuning of polarity. (4) Defects and hybridization: higher accessibility to the active sites and enhanced electron transport.

The key to progress in this field lies in integrating an understanding of the reaction mechanism with multiscale structural engineering. After that, it will be possible to transform laboratory-derived redox-active MOFs into predictive and energy-efficient systems for the electrochemical capture and conversion of CO_2.

4.7. Coordination, Electric Field and Trap Dynamics

The metallic nodes, whether Cu^{2+}, Ni^{2+}, Fe^{3+} or Co^{2+} and their coordination geometries (octahedral, square-planar, or tetrahedral), determined the density of electronic states (DOS) and, consequently, the MOFs ability to stabilize reduced species during CO_2 capture. Some examples are: (1) In HKUST-1 type MOFs, the $Cu^{2+} \rightarrow Cu^{+}$ reduction generated partially vacant d orbitals that facilitated the formation of coordination adducts with CO_2 [12,100]. (2) In Fe-carboxylate type MOFs, the $Fe^{3+} \rightarrow Fe^{2+}$ transition leads to reversible Lewis-type interactions with the carbon atom of CO_2 gas [100].

The metal-ligand field also played a decisive role. Carboxylate ligands generated moderate ligand fields that promoted redox accessibility, although they sacrificed structural stability [49,96]. However, the use of phosphonates and π-conjugated ligands offered delocalization channels that stabilized the transferred electrons in the reduction reaction [38,73,101].

Elucidating the mechanism depends not only on the metal used, but also on how its structure and electronic boundary modulate the coordination environment of CO_2 and the MOF. Designing hybrid systems that facilitate electron transfer between the metal and the ligand, balancing stability and selectivity, was considered one of the most promising avenues for next-generation MOFs.

4.7.1. Connection with Other MOF Redox Materials

In comparative studies, it was shown that redox polymers exhibited structural heterogeneity that made it difficult to control the active molecular sites present in the MOF structures, while the metal oxides in these molecules showed irreversible phase transformations after multiple cycles in CO_2 capture.

On the other hand, this family of MOFs offered precise and periodic redox modulation within a more stable molecular architecture, although their structural rigidity could become a limitation under extreme CO_2 capture conditions [15,43,102]. The advantage and capability of MOFs lies in their reproducible coordination symmetry, which allowed the analysis of electron transfer phenomena and the adsorption of intermediates with quantum resolution, a feature impossible to achieve in amorphous matrices.

MOFs described a balance between the electronic plasticity of polymers and the structural robustness of oxides. Therefore, their technological maturity depends on the scope of their properties, such as the crystalline scale of the material.

4.7.2. Theoretical Approach and Energy Metrics in MOFs

Theoretical DFT calculations began to reveal how the application of potential transformed the adsorption free energy (ΔG_{ads}) and the local electron density (DOS/PDOS). In $Cu_3(HHTP)_2$ MOFs, studied by Vetik et al. [10], it was shown that electron redistribution under polarization directly controlled the trapping reversibility, while Coudert and Poloni et al. [103,104] described MOFs as dynamic materials, highly sensitive to pressure, charge, and the presence of adsorbates in the reaction.

Unfortunately, a systematic framework correlating applied potential (E), DOS/PDOS, and ΔG_{ads} was still lacking. This correlation would have enabled the transition from qualitative studies to predictive materials design. At the same time, the scientific community needed to agree on universal metrics, such as electrical energy per mole of CO_2 captured on an electroactive and adsorbent surface, Coulombic efficiency, and characteristic time t_{90}, to provide electrosorption technologies with a common and comparable thermodynamic basis [38,49].

Without the development of a common metric language, the field risks fragmentation and unprecedented results. The future of these capture systems depends on the integration of theoretical modeling, spectroscopic validation, and experimental standardization to translate molecular thermodynamics into technological and performance terms.

4.7.3. Mechanistic Gaps and Exploration Opportunities in MOF

Despite significant progress, structural and methodological gaps persisted. There is a lack of in situ and operando spectroscopies that correlate coordination structure and redox response in real time [15,43,45]. The current need is to generate standardized electrochemical protocols and parameters, including electrolyte composition, working electrochemical windows, DEMS, and reference systems [49]. A limited search for alternative families of redox-active ligands could have offered new avenues to optimize selectivity, reversibility, and reproducibility [67,68,75,105].

The greatest challenge was not the synthesis and generation of new MOFs, but rather the quantitative understanding of how they electrochemically captured and released CO_2. The integration of electrochemical, spectroscopic, and theoretical data into unified mechanistic models was expected to transform this field into a predictive discipline comparable to electrocatalysis.

Table 4 summarizes how different metal nodes and ligands of redox-active MOFs modulate CO_2 interaction and capture through characteristic electron transfer mechanisms and intermediates. However, the main challenge was not to identify additional systems, but rather to elucidate the mechanistic diversity using numerical performance criteria that would allow an objective assessment of the stability of these systems against this reduction reaction.

Table 4. Several representative examples of redox-active MOFs applicable to electrochemical CO_2 capture are discussed.

MOF	Main Redox Unit	Characteristic	References
HKUST-1 (Cu)	Cu^{2+}/Cu^{+} at carboxylate nodes	Lewis affinity modulation with CO_2	[106]
Fe-carboxylates	Fe^{3+}/Fe^{2+}	Redox states of more than one redox state; strong interaction with CO_2	[63]
$Cu_3(HHTP)_2$	π-conjugated ligands (HHTP)	Intrinsic conductivity; reversible 2D MOF pores	[15]
TTF-MOFs	Tetrathiafulvalene ($TTF/TTF^{+}\bullet$)	Reversible oxidation; electron density control	[107]
Viologen-MOFs	Viologeno (dication/radical catión)	Selective adsorption by modulated polarizability	[67,108]
Ferrocene-MOFs	Fe^{2+}/Fe^{3+} in ligand	Stable redox signal; switchable selectivity	[68,69]

The subsequent Section 5 focuses on linking coordination chemistry with the kinetic metrics and tunable parameters of redox-active MOFs. Only through this convergence will be possible to assess which MOFs are truly competitive for industrially scalable electrochemical CO_2 capture.

5. Prospects and Challenges of Redox-Active MOFs in CO_2 Capture

This section discusses the prospects and challenges of redox-active MOFs as platforms for electrochemical CO_2 capture. Building on the archetypes and mechanisms described in Sections 3 and 4, we now explore under what conditions redox-active MOFs can realistically outperform other electrosorbents and conventional thermal technologies, and what design principles are essential for achieving competitive performance.

Recent results, as reviewed in previous sections, have demonstrated the capacity of these structures for reversible and selective toward CO_2 adsorption onto diverse materials in electrochemical applications. However, the transition from model systems to functional systems requires addressing challenges in structural, energetic, and scalable technologies.

5.1. Stability and Scalability in MOF for CO_2 Capture

The stability of MOFs under real-world conditions, such as pressure and temperature challenges, as well as humidity, corrosive electrolytes, and operation over long analysis times (stability), remained the main obstacle to the implementation of these systems [103,109,110].

Future developments required stable, hydrotolerant MOFs, such as Zr, Ti, and Al carboxylates, capable of maintaining their structural integrity after thousands of redox cycles. Hybrid compounds such as carbon MOFs, COF-MOFs, and others, reduce dissolution and improve electronic connectivity. Long-term durability testing ($\geq 10^4$ cycles) was a standard requirement to validate the industrial potential of these adsorbent composites [49,111].

Chemical stability should not have been evaluated as a static and immutable parameter, but as a functional property under continuous electrochemical operation, where electron transport and hydration control the actual lifetime of the material for CO_2 capture.

5.2. Synergy Between Capture and Electrochemical Conversion in MOFs for CO_2

The most promising frontier was the direct integration of electrochemical capture and catalytic CO_2RR. Redox-active MOFs, particularly those based on $Cu_3(HHTP)_2$, could function as bifunctional electrodes, where the applied potential would determine whether CO_2 was stored or converted to CO or hydrocarbons, either gaseous (CH_4, C_2H_6, C_2H_4, among others) or aqueous (CH_3OH, C_2H_5OH, among others) [112–114].

Furthermore, electroactive ligands such as TTF, ferrocene, or viologen could simultaneously modulate CO_2 adsorption and desorption, as well as charge transfer, making the concept of reactive capture and entrapment of substances of interest a reality.

Beyond this background, the true advantage of redox-active MOFs lies in their ability to combine selectivity and reactivity, uniting capture, separation, and conversion in a single electrochemical interface.

5.3. Standardization, Advanced Design, and Predictive Modeling of MOFs in CO_2 Capture

Overall, these MOF-like molecular systems advanced the field of electrochemical CO_2 capture. This field was based on a common concept of characterization and evaluation that allowed for the comparison of different materials and structural architectures in CO_2 electrosorption. Unlike electrocatalysis, where established metrics existed, in CO_2 electrosorption, the methodology limited the selectivity and reproducibility of these systems.

It was necessary to systematically incorporate different parameters such as free energy of adsorption (ΔG_{ads}), energy efficiency (kJ mol^{-1} of captured CO_2) as a comparative metric with the thermal processes of the reaction, and adsorption/desorption kinetics and efficiency as indicators of reversibility [115]. In addition, in situ and operational techniques, such as IR, Raman, XAS, and DEMS, among others, were employed to link the active redox response with CO_2 capture.

Simultaneously, the expansion toward new electroactive ligands has been studied, as was previously depicted. This has allowed for the expansion of electrochemical potential windows and the tuning of electron affinity without compromising the material's porosity or structural defects, facilitating charge transfer and the formation of reaction intermediates of interest [74]. The integration of these units through electrosynthesis or electrodeposition, in a simple and environmentally friendly manner, has achieved subsequent synthetic functionalization or further defect engineering, showing us a realistic path toward more stable, tunable, and sensitizable materials [24,42,60,90].

On the other hand, computational modeling was crucial in advancing the design of these molecules in a more precise and scalable way for the experimental phase. The combination of DFT calculations and molecular dynamics allowed us to predict the structural and electronic properties of the MOFs, connecting microscopic parameters (DOS, PDOS, ΔG_{ads}) with macroscopic metrics such as capacitance, energy, and reaction kinetics [116].

The major qualitative leap in this field was expected to come not from experiments, but from predictions. It involved a transition from experimental observations to practical designs based on electronic principles and coordination chemistry. In this sense, the bottleneck was no longer experimental, but purely conceptual. Integrated theories and models that would translate the electronic language into the operational performance of these systems were still lacking.

5.4. *Cross-Sectional Comparison and Sustainability of MOFs for CO_2 Capture*

On the other hand, doped carbons, redox polymers, metal oxides, and redox-active MOFs were compared. The latter stood out for their molecular selectivity toward the CO_2 molecule and their redox control, although they had not yet demonstrated quantitative advantages in industrial parameters such as the target cost (<50 USD per ton of captured CO_2), apparent density (>2 mmol cm^{-3}), and operational stability (>10^4 cycles) in humid environments [91–93].

The path to scalability of these materials involved hybridization with engineered platforms, such as hierarchical electrodes, polymer-MOF membranes, and optimized electrochemical flow cells [68,94–97].

Furthermore, sustainability required prioritizing the most abundant transition metals, such as Fe, Cu, Ni, and Co, and developing green and recyclable synthesis pathways, such as electrosynthesis, aligned with carbon neutrality goals.

The field of MOFs did not need more new materials, but rather more demanding criteria. Progress was expected to be measured by reproducibility and industrial scalability, not by the novelty of the MOF.

Figure 6 presents a conceptual summary of the main performance factors governing electrochemical CO_2 capture systems. The pie chart illustrates the relative contribution of each element to the overall process performance: energy efficiency (35%), operational stability (25%), energy and operating costs (20%), electrochemical kinetics and response (15%), and sustainability and scalability factors (5%).

Together, these summarize the evolution from material-level durability to process energetics and operational robustness in next-generation electrochemical CO_2 capture systems.

Beyond individual cases, progress in this field depends on defining a common evaluation framework and structure. Table 5 proposes a set of critical metrics, including adsorption of free energy, energy efficiency, capture kinetics, cycling stability, specific cost, and volumetric density. These were expected to become standard parameters for the objective comparison of redox-active MOF materials for electrochemical CO_2 capture.

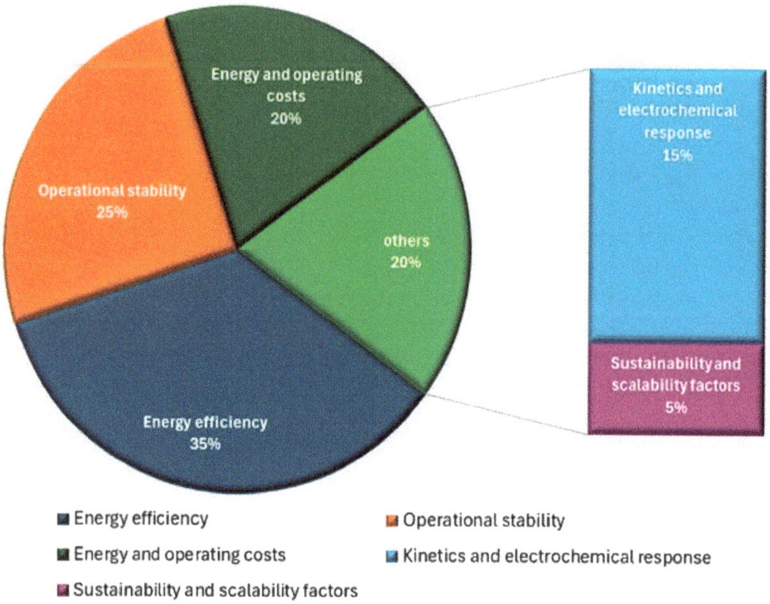

Figure 6. This section presents a conceptual summary of the main performance factors governing electrochemical CO_2 capture systems. These proportions reflect the balance between the thermodynamic, kinetic, and technoeconomic parameters reported in state-of-the-art electrochemical CO_2 capture systems. The chart integrates information from experimental trends (adsorption capacity and cyclic stability), energy analyses (EMAR vs. thermal routes), and device-level testing, providing a holistic perspective on performance parameters and design priorities for redox-active materials and hybrid systems for reversible CO_2 electrosorption.

Table 5. Critical parameters for evaluating redox-active MOFs in electrochemical CO_2 capture.

Parameter	Description/Importance	Indicative Value	References
ΔG_{ads} (kJ mol^{-1})	Free energy of adsorption under applied potential	-20 to -40 kJ mol^{-1} CO_2 for selectivity without irreversibility	[3,99]
Energy efficiency	Electrical energy consumed per mole of captured CO_2	<100 kJ mol^{-1} CO_2	[38,49]
Adsorption kinetics (t_{90})	Time to reach 90% capacity	<60 s under applied potential	[34]
Cyclical stability	Number of adsorption/desorption cycles without significant loss	$\geq 10^4$ cycles	[34]
Specific cost	Estimated cost per ton of captured CO_2	<50 USD/ton CO_2	[100]
Volumetric capture density	Normalized capacity per electrode volume	>2 mmol cm^{-3}	[49]

The synthesis and analysis presented in Table 5 demonstrate that the field of redox-active MOFs for electrochemical CO_2 capture is at a scientific and technological turning point. There are clear guidelines on critical metrics and parameters that need to be standardized, from adsorption free energy to volumetric capture density, as well as promising device integration strategies. However, there remains a significant limitation between the performance achieved under laboratory experimental conditions, such as stability, manufacturing cost, and the energy efficiency required for large-scale industrial implementation.

The advances achieved to date in this field position redox-active MOFs as a strategic and scalable platform for the development of smart electrochemical CO_2 capture and conversion technologies. The immediate step is to transform molecular knowledge into device engineering for immediate application, establishing a quantitative and reproducible evaluation framework that links the structural architecture, energy transfer, and operational performance of these systems.

Ultimately, the future of redox-active MOFs depends not only on the synthesis of new materials, which is clearly no longer relevant, but also on the coherent integration of fundamental science and engineering to create predictive and technologically scalable models.

Only through this convergence and integration of these materials into technological systems will be possible to evolve from an academic commitment to effective industrial tools, concretely contributing to the global goals of decarbonization, sustainability, and renewable energy.

In summary, the main message of this review is that redox-active MOFs should be considered not just as another class of porous adsorbents, but as a distinct platform for the electrochemical capture and transformation of CO_2, where coordination chemistry, electronic structure, and device-level metrics can be rationally linked.

By placing redox-active MOFs within the broader context of electrochemical CO_2 capture, alongside molecular redox carriers, whether polymeric or COF-based, the conditions under which MOFs can offer real advantages in terms of selectivity, reversibility, and energy efficiency can be identified. Looking to the near future, we maintain that the most pressing needs are:

- Standardized electrochemical and spectroscopic protocols that quantify CO_2 capture in a manner comparable to electrocatalysis and thermal CCS.
- Design strategies that integrate nodes (monometallic or multimetallic) and redox-active ligands with stable and hydrotolerant structures.
- Predictive models that decipher the electronic descriptors in CO_2 capture performance.

6. Conclusions

In this review, we have presented redox-active metal–organic frameworks (MOFs) as a distinct platform for electrochemical CO_2 capture, rather than as isolated case studies. By comparing electrochemically mediated CO_2 capture with molecular, polymeric, and COF-based redox carriers, and then analyzing archetypal redox-active MOFs, their redox-active structural units, and representative systems with quantified CO_2 capture metrics, we offer a performance-oriented perspective on how coordination chemistry, electronic structure, and device architecture are related. The cross-cutting metrics proposed in this work aim to provide a common language for comparing redox-active MOFs with other electrosorbents and with conventional thermal capture routes.

The advancement of redox-active MOFs for the electrochemical capture and transformation of CO_2 has transcended the scope of laboratory-based academic studies and has emerged as a new approach for applied coordination chemistry. These redox-active MOF materials have demonstrated extraordinary electrosorption capabilities, while integrating structural modularity, molecular selectivity for value-added compounds for CO_2 reduction, and reversible electronic control of adsorption and desorption in a single system.

However, this promising system has not yet passed the validation phase. Most advances are limited to controlled laboratory conditions, with little evidence of long-term stability and results difficult to compare using common metrics. This field presents several critical shortcomings. The lack of standardized protocols impedes objective assessment of energy efficiency and reversibility; low structural stability, especially under corrosive conditions; and a poor connection between theory and experimental practice, as computational modeling rarely predicts the most common design and synthesis problems.

Current advances in this field do not require further synthesis and studies of new materials, but rather more demanding and competitive criteria when introducing ligands and advancing the molecular electronics of these systems.

The evolution of these systems requires the integration of in situ and operational characterization, multiscale modeling, and process and materials engineering, moving from qualitative to quantitative descriptions with common metrics.

Redox-active MOFs could become true bridges between electrocatalysis and structural and molecular architecture: an intelligent design capable of capturing, storing, and transforming CO_2 into a single operational product.

Finally, the success of redox-active MOFs will not be measured by their structural novelty, but by their functional and applied impact. When these materials demonstrate stability, selectivity, reproducibility, and high energy efficiency under scalable industrial conditions, they will have surpassed their academic promise of becoming true catalysts comparable to noble metals. This is one of the most urgent challenges for mitigating climate change and global warming.

Author Contributions: Conceptualization, validation and research C.C.-C. and J.S.-H.; resources, G.G. and R.E.-G.; writing (original draft), C.C.-C. and G.G.; revision and editing, C.C.-C. and G.G.; supervision, G.G. and R.E.-G. All authors have read and agreed to the published version of the manuscript.

Funding: This research received no external funding.

Data Availability Statement: No new data were created or analyzed in this study. Data sharing is not applicable to this article.

Acknowledgments: The authors wish to express their gratitude to: C.C.-C. for Fondecyt Postdoctoral Grant No. 3240375, and R.E.-G. for FONDECYT No. 1231474. J.S.-H Fondecyt Postdoctoral Grant No 3230179. GG thanks to NANOtec, INTech, the Tenerife Island Council, and the University of La Laguna (ULL) for providing laboratory facilities.

Conflicts of Interest: The authors declare no conflicts of interest.

References

1. Becker, H.; Dickinson, E.J.F.; Lu, X.; Bexell, U.; Proch, S.; Moffatt, C.; Stenström, M.; Smith, G.; Hinds, G. Assessing potential profiles in water electrolysers to minimise titanium use. *Energy Environ. Sci.* **2022**, *15*, 2508–2518. [CrossRef]
2. Alsaç, E.P.; Zhou, K.; Rong, W.; Salamon, S.; Landers, J.; Wende, H.; Smith, R.D.L. Identification of non-traditional coordination environments for iron ions in nickel hydroxide lattices. *Energy Environ. Sci.* **2022**, *15*, 2638–2652. [CrossRef]
3. Sanz-Pérez, E.S.; Murdock, C.R.; Didas, S.A.; Jones, C.W. Direct Capture of CO_2 from Ambient Air. *Chem. Rev.* **2016**, *116*, 11840–11876. [CrossRef]
4. Sharifian, R.; Wagterveld, R.M.; Digdaya, I.A.; Xiang, C.; Vermaas, D.A. Electrochemical carbon dioxide capture to close the carbon cycle. *Energy Environ. Sci.* **2021**, *14*, 781–814. [CrossRef]
5. Renfrew, S.E.; Starr, D.E.; Strasser, P. Electrochemical Approaches toward CO_2 Capture and Concentration. *ACS Catal.* **2020**, *10*, 13058–13074. [CrossRef]

6. Wang, Y.; Hou, P.; Wang, Z.; Kang, P. Zinc Imidazolate Metal–Organic Frameworks (ZIF-8) for Electrochemical Reduction of CO_2 to CO. *Chemphyschem* **2017**, *18*, 3142–3147. [CrossRef]
7. Albo, J.; Vallejo, D.; Beobide, G.; Castillo, O.; Castaño, P.; Irabien, A. Copper-Based Metal–Organic Porous Materials for CO_2 Electrocatalytic Reduction to Alcohols. *ChemSusChem* **2017**, *10*, 1100–1109. [CrossRef]
8. Dong, B.-X.; Qian, S.-L.; Bu, F.-Y.; Wu, Y.-C.; Feng, L.-G.; Teng, Y.-L.; Liu, W.-L.; Li, Z.-W. Electrochemical Reduction of CO_2 to CO by a Heterogeneous Catalyst of Fe–Porphyrin-Based Metal–Organic Framework. *ACS Appl. Energy Mater.* **2018**, *1*, 4662–4669. [CrossRef]
9. Rohland, P.; Schröter, E.; Nolte, O.; Newkome, G.R.; Hager, M.D.; Schubert, U.S. Redox-active polymers: The magic key towards energy storage—A polymer design guideline progress in polymer science. *Prog. Polym. Sci.* **2022**, *125*, 101474. [CrossRef]
10. Vetik, I.; Žoglo, N.; Kosimov, A.; Cepitis, R.; Krasnenko, V.; Qing, H.; Chandra, P.; Mirica, K.; Rizo, R.; Herrero, E.; et al. Advancing Electrochemical CO_2 Capture with Redox-Active Metal-Organic Frameworks. *arXiv* **2025**, arXiv:2411.16444. [CrossRef]
11. Schubert, U.S.; Winter, A.; Newkome, G.R. *An Introduction to Redox Polymers for Energy-Storage Applications*; John Wiley & Sons: Hoboken, NJ, USA, 2023; Available online: https://www.wiley.com/en-us/An+Introduction+to+Redox+Polymers+for+Energy-Storage+Applications-p-9783527350902 (accessed on 25 November 2025).
12. Furukawa, H.; Cordova, K.E.; O'Keeffe, M.; Yaghi, O.M. The chemistry and applications of metal-organic frameworks. *Science* **2013**, *341*, 1230444. [CrossRef]
13. Talin, A.A.; Centrone, A.; Ford, A.C.; Foster, M.E.; Stavila, V.; Haney, P.; Kinney, R.A.; Szalai, V.; El Gabaly, F.; Yoon, H.P.; et al. Tunable Electrical Conductivity in Metal-Organic Framework Thin-Film Devices. *Science* **2014**, *343*, 66–69. [CrossRef] [PubMed]
14. Latimer, A.A.; Kulkarni, A.R.; Aljama, H.; Montoya, J.H.; Yoo, J.S.; Tsai, C.; Abild-Pedersen, F.; Studt, F.; Nørskov, J.K. Understanding trends in C–H bond activation in heterogeneous catalysis. *Nat. Mater.* **2017**, *16*, 225–229. [CrossRef] [PubMed]
15. Sheberla, D.; Bachman, J.C.; Elias, J.S.; Sun, C.-J.; Shao-Horn, Y.; Dincă, M. Conductive MOF electrodes for stable supercapacitors with high areal capacitance. *Nat. Mater.* **2016**, *16*, 220–224. [CrossRef]
16. Lu, Z.; Godfrey, H.G.W.; da Silva, I.; Cheng, Y.; Savage, M.; Tuna, F.; McInnes, E.J.L.; Teat, S.J.; Gagnon, K.J.; Frogley, M.D.; et al. Modulating supramolecular binding of carbon dioxide in a redox-active porous metal-organic framework. *Nat. Commun.* **2017**, *8*, 14212. [CrossRef] [PubMed]
17. Nwosu, U.; Siahrostami, S. Copper-based metal–organic frameworks for CO_2reduction: Selectivity trends, design paradigms, and perspectives. *Catal. Sci. Technol.* **2023**, *13*, 3740–3761. [CrossRef]
18. Doheny, P.W.; Babarao, R.; Kepert, C.J.; D'aLessandro, D.M. Tuneable CO_2 binding enthalpies by redox modulation of an electroactive MOF-74 framework. *Mater. Adv.* **2021**, *2*, 2112–2119. [CrossRef]
19. Kim, J.; Ling, J.; Lai, Y.; Milner, P.J. Redox-Active Organic Materials: From Energy Storage to Redox Catalysis. *ACS Mater. Au* **2024**, *4*, 258–273. [CrossRef]
20. Kang, D.-Y.; Lee, J.S. Challenges in Developing MOF-Based Membranes for Gas Separation. *Langmuir* **2023**, *39*, 2871–2880. [CrossRef]
21. Razavi, S.A.A.; Chen, W.; Zhou, H.-C.; Morsali, A. Tuning redox activity in metal–organic frameworks: From structure to application. *Coord. Chem. Rev.* **2024**, *517*, 216004. [CrossRef]
22. Deffo, G.; Tamo, A.K.; Fotsop, C.G.; Tchoumi, H.H.B.; Talla, D.E.N.; Wabo, C.G.; Deussi, M.C.N.; Temgoua, R.C.T.; Doungmo, G.; Njanja, E.; et al. Metal-organic framework-based materials: From synthesis and characterization routes to electrochemical sensing applications. *Coord. Chem. Rev.* **2025**, *536*, 216680. [CrossRef]
23. Fu, Y.; Qiu, W.; Zhou, S.-H.; Huang, H.; Luo, Y.; Lin, X.; Zhu, Q.-L. First-principles calculation studies of metal-organic frameworks and their derivatives for electrochemical energy conversion and storage. *Coord. Chem. Rev.* **2025**, *544*, 216982. [CrossRef]
24. Patil, P.D.; Gargate, N.; Tiwari, M.S.; Nadar, S.S. Defect metal-organic frameworks (D-MOFs): An engineered nanomaterial for enzyme immobilization. *Coord. Chem. Rev.* **2025**, *531*, 216519. [CrossRef]
25. An, D.; Chen, L.; Liang, Y.; Hou, J.; Chen, J. Defect-containing metal–organic framework materials for sensor applications. *J. Mater. Chem. A* **2023**, *12*, 38–58. [CrossRef]
26. Muthukumaran, M.K.; Govindaraj, M.; Kogularasu, S.; Sriram, B.; Raja, B.K.; Wang, S.-F.; Chang-Chien, G.-P.; J, A.S. Recent advances in metal-organic frameworks for electrochemical sensing applications. *Talanta Open* **2025**, *11*, 100396. [CrossRef]
27. Sullivan, I.; Goryachev, A.; Digdaya, I.A.; Li, X.; Atwater, H.A.; Vermaas, D.A.; Xiang, C. Coupling electrochemical CO_2 conversion with CO_2 capture. *Nat. Catal.* **2021**, *4*, 952–958. [CrossRef]
28. Xia, Q.; Zhang, K.; Zheng, T.; An, L.; Xia, C.; Zhang, X. Integration of CO_2 Capture and Electrochemical Conversion. *ACS Energy Lett.* **2023**, *8*, 2840–2857. [CrossRef]
29. Wang, C.; Jiang, K.; Yu, H.; Li, S.; Zhao, Y.; Zheng, Z.; Liu, H.; Xia, X.; Zhao, P.; Li, Y.; et al. Review of electrochemical carbon dioxide capture towards practical application. *Next Mater.* **2025**, *8*, 100660. [CrossRef]

30. Xu, Z.; Mapstone, G.; Coady, Z.; Wang, M.; Spreng, T.L.; Liu, X.; Molino, D.; Forse, A.C. Enhancing electrochemical carbon dioxide capture with supercapacitors. *Nat. Commun.* **2024**, *15*, 7851. [CrossRef] [PubMed]
31. Dissanayake, K.; Kularatna-Abeywardana, D. A review of supercapacitors: Materials, technology, challenges, and renewable energy applications. *J. Energy Storage* **2024**, *96*, 112563. [CrossRef]
32. Holze, R.F.; Béguin, E. Frąckowiak (eds): Supercapacitors—Materials, Systems, and Applications. *J. Solid State Electrochem.* **2015**, *19*, 1253. [CrossRef]
33. Zito, A.M.; Clarke, L.E.; Barlow, J.M.; Bím, D.; Zhang, Z.; Ripley, K.M.; Li, C.J.; Kummeth, A.; Leonard, M.E.; Alexandrova, A.N.; et al. Electrochemical Carbon Dioxide Capture and Concentration. *Chem. Rev.* **2023**, *123*, 8069–8098. [CrossRef]
34. Rheinhardt, J.H.; Singh, P.; Tarakeshwar, P.; Buttry, D.A. Electrochemical capture and release of carbon dioxide. *ACS Energy Lett.* **2017**, *2*, 454–461. [CrossRef]
35. Seo, H.; Rahimi, M.; Hatton, T.A. Electrochemical Carbon Dioxide Capture and Release with a Redox-Active Amine. *J. Am. Chem. Soc.* **2022**, *144*, 2164–2170. [CrossRef] [PubMed]
36. Zhang, J.; Cao, Y.; Ou, P.; Lee, G.; Zhao, Y.; Liu, S.; Shirzadi, E.; Dorakhan, R.; Xie, K.; Tian, C.; et al. A redox-active polymeric network facilitates electrified reactive-capture electrosynthesis to multi-carbon products from dilute CO_2-containing streams. *Nat. Commun.* **2025**, *16*, 3553. [CrossRef]
37. Iijima, G.; Naruse, J.; Shingai, H.; Usami, K.; Kajino, T.; Yoto, H.; Morimoto, Y.; Nakajima, R.; Inomata, T.; Masuda, H. Mechanism of CO_2 Capture and Release on Redox-Active Organic Electrodes. *Energy Fuels* **2023**, *37*, 2164–2177. [CrossRef]
38. Diederichsen, K.M.; Sharifian, R.; Kang, J.S.; Liu, Y.; Kim, S.; Gallant, B.M.; Vermaas, D.; Hatton, T.A. Electrochemical methods for carbon dioxide separations. *Nat. Rev. Methods Prim.* **2022**, *2*, 68. [CrossRef]
39. Li, J.; Ott, S. The Molecular Nature of Redox-Conductive Metal–Organic Frameworks. *Accounts Chem. Res.* **2024**, *57*, 2836–2846. [CrossRef]
40. Sahayaraj, A.F.; Prabu, H.J.; Maniraj, J.; Kannan, M.; Bharathi, M.; Diwahar, P.; Salamon, J. Metal–Organic Frameworks (MOFs): The Next Generation of Materials for Catalysis, Gas Storage, and Separation. *J. Inorg. Organomet. Polym. Mater.* **2023**, *33*, 1757–1781. [CrossRef]
41. Rozaini, M.T.; Grekov, D.I.; Bustam, M.A.; Pré, P. Shaping of HKUST-1 via Extrusion for the Separation of CO_2/CH_4 in Biogas. *Separations* **2023**, *10*, 487. [CrossRef]
42. Yañez-Aulestia, A.; Trejos, V.M.; Esparza-Schulz, J.M.; Ibarra, I.A.; Sánchez-González, E. Chemically Modified HKUST-1(Cu) for Gas Adsorption and Separation: Mixed-Metal and Hierarchical Porosity. *ACS Appl. Mater. Interfaces* **2024**, *16*, 65581–65591. [CrossRef]
43. Liu, J.; Yang, M.; Zhou, X.; Meng, Z. Solid-State Electrochemical Carbon Dioxide Capture by Conductive Metal–Organic Framework Incorporating Nickel Bis(diimine) Units. *J. Am. Chem. Soc.* **2024**, *146*, 33093–33103. [CrossRef]
44. Huang, J.-M.; Zhang, X.-D.; Huang, J.-Y.; Zheng, D.-S.; Xu, M.; Gu, Z.-Y. MOF-based materials for electrochemical reduction of carbon dioxide. *Coord. Chem. Rev.* **2023**, *494*, 215333. [CrossRef]
45. Zhou, S.; Liu, T.; Strømme, M.; Xu, C. Electrochemical Doping and Structural Modulation of Conductive Metal-Organic Frameworks. *Angew. Chem. Int. Ed. Engl.* **2024**, *63*, e202318387. [CrossRef]
46. Liu, K.-K.; Meng, Z.; Fang, Y.; Jiang, H.-L. Conductive MOFs for electrocatalysis and electrochemical sensor. *eScience* **2023**, *3*, 100133. [CrossRef]
47. Zhang, Y.; Li, Q.; Zhang, G.; Lv, T.; Geng, P.; Chen, Y.; Pang, H. Recent advances in the type, synthesis and electrochemical application of defective metal-organic frameworks. *Energy Mater.* **2023**, *3*, 300022. [CrossRef]
48. Mukherjee, D.; Saha, A.; Moni, S.; Volkmer, D.; Das, M.C. Anhydrous Solid-State Proton Conduction in Crystalline MOFs, COFs, HOFs, and POMs. *J. Am. Chem. Soc.* **2025**, *147*, 5515–5553. [CrossRef]
49. Rahimi, M.; Khurram, A.; Hatton, T.A.; Gallant, B. Electrochemical carbon capture processes for mitigation of CO_2 emissions. *Chem. Soc. Rev.* **2022**, *51*, 8676–8695. [CrossRef] [PubMed]
50. Kumar, R.S.; Kumar, S.S.; Kulandainathan, M.A. Highly selective electrochemical reduction of carbon dioxide using Cu based metal organic framework as an electrocatalyst. *Electrochem. Commun.* **2012**, *25*, 70–73. [CrossRef]
51. Gurkan, B.; Su, X.; Klemm, A.; Kim, Y.; Sharada, S.M.; Rodriguez-Katakura, A.; Kron, K.J. Perspective and challenges in electrochemical approaches for reactive CO_2 separations. *iScience* **2021**, *24*, 103422. [CrossRef] [PubMed]
52. Hinogami, R.; Yotsuhashi, S.; Deguchi, M.; Zenitani, Y.; Hashiba, H.; Yamada, Y. Electrochemical reduction of carbon dioxide using a copper rubeanate metal organic framework. *ECS Electrochem. Lett.* **2012**, *1*, H17–H19. [CrossRef]
53. Voskian, S.; Hatton, T.A. Faradaic electro-swing reactive adsorption for CO_2 capture. *Energy Environ. Sci.* **2019**, *12*, 3530–3547. [CrossRef]
54. Hemmatifar, A.; Kang, J.S.; Ozbek, N.; Tan, K.-J.; Hatton, T.A. Electrochemically Mediated Direct CO_2 Capture by a Stackable Bipolar Cell. *ChemSusChem* **2022**, *15*, e202102533. [CrossRef] [PubMed]

55. Pang, S.; Jin, S.; Yang, F.; Alberts, M.; Li, L.; Xi, D.; Gordon, R.G.; Wang, P.; Aziz, M.J.; Ji, Y. A phenazine-based high-capacity and high-stability electrochemical CO_2 capture cell with coupled electricity storage. *Nat. Energy* **2023**, *8*, 1126–1136. [CrossRef]
56. Abdinejad, M.; Massen-Hane, M.; Seo, H.; Hatton, T.A. Oxygen-Stable Electrochemical CO_2 Capture using Redox--Active Heterocyclic Benzodithiophene Quinone. *Angew. Chem. Int. Ed. Engl.* **2024**, *63*, e202412229. [CrossRef] [PubMed]
57. Hartley, N.A.; Pugh, S.M.; Xu, Z.; Leong, D.C.Y.; Jaffe, A.; Forse, A.C. Quinone-functionalised carbons as new materials for electrochemical carbon dioxide capture. *J. Mater. Chem. A* **2023**, *11*, 16221–16232. [CrossRef]
58. Hartley, N.A.; Xu, Z.; Kress, T.; Forse, A.C. Correlating the structure of quinone-functionalized carbons with electrochemical CO_2 capture performance. *Mater. Today Energy* **2024**, *45*, 101689. [CrossRef]
59. Khan, M.A.; Xu, Z.; Muzammil, M.; Bird, S.; Munawar, M.; Salam, F.; Hartley, N.A.; Taylor, J.; Amin, K.; Ling, J.; et al. Electrochemical CO_2 Capture by a Quinone-based Covalent Organic Framework. *ChemRxiv* **2025**. [CrossRef]
60. Zhao, W.; Chen, T.; Wang, W.; Bi, S.; Jiang, M.; Zhang, K.Y.; Liu, S.; Huang, W.; Zhao, Q. Layer-by-Layer 2D Ultrathin Conductive $Cu_3(HHTP)_2$ Film for High-Performance Flexible Transparent Supercapacitors. *Adv. Mater. Interfaces* **2021**, *8*, 2100308. [CrossRef]
61. Choi, I.; Jung, Y.E.; Yoo, S.J.; Kim, J.Y.; Kim, H.-J.; Lee, C.Y.; Jang, J.H. Facile synthesis of M-MOF-74 (M=Co, Ni, Zn) and its application as an electrocatalyst for electrochemical CO_2 conversion and H_2 production. *J. Electrochem. Sci. Technol.* **2017**, *8*, 61–68. [CrossRef]
62. Chui, S.S.-Y.; Lo, S.M.-F.; Charmant, J.P.H.; Orpen, A.G.; Williams, I.D. A Chemically Functionalizable Nanoporous Material $[Cu_3(TMA)_2(H_2O)_3]_n$. *Science* **1999**, *283*, 1148–1150. [CrossRef]
63. DeCoste, J.B.; Peterson, G.W. Metal–organic frameworks for air purification of toxic chemicals. *Chem. Rev.* **2014**, *114*, 5695–5727. [CrossRef]
64. Hamon, L.; Serre, C.; Devic, T.; Loiseau, T.; Millange, F.; Férey, G.; De Weireld, G. Comparative study of hydrogen sulfide adsorption in the MIL-53(Al, Cr, Fe), MIL-47(V), MIL-100(Cr), and MIL-101(Cr) metal−organic frameworks at room temperature. *J. Am. Chem. Soc.* **2009**, *131*, 8775–8777. [CrossRef]
65. Manna, F.; Oggianu, M.; Auban-Senzier, P.; Novitchi, G.; Canadell, E.; Mercuri, M.L.; Avarvari, N. A highly conducting tetrathiafulvalene-tetracarboxylate based dysprosium(iii) 2D metal–organic framework with single molecule magnet behaviour. *Chem. Sci.* **2024**, *15*, 19247–19263. [CrossRef]
66. Ribeiro, C.; Tan, B.; Figueira, F.; Mendes, R.F.; Calbo, J.; Valente, G.; Escamilla, P.; Paz, F.A.A.; Rocha, J.; Dincă, M.; et al. Mixed Ionic and Electronic Conductivity in a Tetrathiafulvalene-Phosphonate Metal–Organic Framework. *J. Am. Chem. Soc.* **2025**, *147*, 63–68. [CrossRef]
67. Leong, Z.Y.; Yao, J.; Boon, N.; Eral, H.B.; Li, D.-S.; Hartkamp, R.; Yang, H.Y. Electrochemical Selective Removal of Oxyanions in a Ferrocene-Doped Metal–Organic Framework. *ACS Nano* **2024**, *18*, 29067–29077. [CrossRef]
68. Easley, A.D.; Li, C.-H.; Li, S.-G.; Nguyen, T.P.; Kuo, K.-H.M.; Wooley, K.L.; Tabor, D.P.; Lutkenhaus, J.L. Electron transport kinetics for viologen-containing polypeptides with varying side group linker spacing. *J. Mater. Chem. A* **2024**, *12*, 31871–31882. [CrossRef]
69. Hod, I.; Bury, W.; Gardner, D.M.; Deria, P.; Roznyatovskiy, V.; Wasielewski, M.R.; Farha, O.K.; Hupp, J.T. Bias-switchable permselectivity and redox catalytic activity of a ferrocene-functionalized, thin-film metal–organic framework compound. *J. Phys. Chem. Lett.* **2015**, *6*, 586–591. [CrossRef]
70. Sheberla, D.; Sun, L.; Blood-Forsythe, M.A.; Er, S.; Wade, C.R.; Brozek, C.K.; Aspuru-Guzik, A.; Dincă, M. High electrical conductivity in $Ni_3(2,3,6,7,10,11\text{-hexaiminotriphenylene})_2$, a semiconducting metal–organic graphene analogue. *J. Am. Chem. Soc.* **2014**, *136*, 8859–8862. [CrossRef] [PubMed]
71. Siu, T.C.; Su, T.A. Conductivity in Porous 2D Materials Made Crystal Clear. *ACS Central Sci.* **2020**, *6*, 11–13. [CrossRef]
72. Kirlikovali, K.O.; Goswami, S.; Mian, M.R.; Krzyaniak, M.D.; Wasielewski, M.R.; Hupp, J.T.; Li, P.; Farha, O.K. An Electrically Conductive Tetrathiafulvalene-Based Hydrogen-Bonded Organic Framework. *ACS Mater. Lett.* **2022**, *4*, 128–135. [CrossRef]
73. Shen, L.-J.; Luo, X.-F.; Xiao, X. Promoting the spatial arrangement of TTF to optimize the photocurrent response of MOFs. *J. Solid. State Chem.* **2024**, *339*, 124933. [CrossRef]
74. Karmakar, S.; Barman, S.; Rahimi, F.A.; Rambabu, D.; Nath, S.; Maji, T.K. Confining charge-transfer complex in a metal-organic framework for photocatalytic CO_2 reduction in water. *Nat. Commun.* **2023**, *14*, 4508. [CrossRef]
75. Wang, T.; Zhang, L.; Liu, J.; Li, X.-X.; Yuan, L.; Li, S.-L.; Lan, Y.-Q. A viologen-functionalized metal–organic framework for efficient CO_2 photoreduction reaction. *Chem. Commun.* **2022**, *58*, 7507–7510. [CrossRef]
76. Yang, X.; Yan, C.; Li, Z.; Li, X.; Yu, Q.; Sang, T.; Gai, Y.; Zhang, Q.; Xiong, K. Viologen-Based Cationic Metal–Organic Framework for Efficient $Cr_2O_7^{2-}$ Adsorption and Dye Separation. *Inorg. Chem.* **2021**, *60*, 5988–5995. [CrossRef]
77. Han, J.; Zhang, M.; Chen, G.; Zhang, Y.; Wei, Q.; Zhuo, Y.; Xie, G.; Yuan, R.; Chen, S. Ferrocene covalently confined in porous MOF as signal tag for highly sensitive electrochemical immunoassay of amyloid-β. *J. Mater. Chem. B* **2017**, *5*, 8330–8336. [CrossRef] [PubMed]

78. Doughty, T.; Zingl, A.; Wünschek, M.; Pichler, C.M.; Watkins, M.B.; Roy, S. Structural Reconstruction of a Cobalt- and Ferrocene-Based Metal–Organic Framework during the Electrochemical Oxygen Evolution Reaction. *ACS Appl. Mater. Interfaces* **2024**, *16*, 40814–40824. [CrossRef]
79. Straube, A.; Useini, L.; Hey-Hawkins, E. Multi-Ferrocene-Based Ligands: From Design to Applications. *Chem. Rev.* **2025**, *125*, 3007–3058. [CrossRef]
80. Fang, Z.; Deng, Z.; Wan, X.; Li, Z.; Ma, X.; Hussain, S.; Ye, Z.; Peng, X. Keggin-type polyoxometalates molecularly loaded in Zr-ferrocene metal organic framework nanosheets for solar-driven CO_2 cycloaddition. *Appl. Catal. B Environ.* **2021**, *296*, 120329. [CrossRef]
81. Castro-Castillo, C.; Espinoza-González, R.; Jofré-Ulloa, P.P.; Luis-Sunga, M.; Silva, N.; Suazo-Hernández, J.; Soler, M.; Isaacs, M.; García, G. Enhanced hydrogen generation through electrodeposited non-precious metal Zn/Cu-BTC metal-organic frameworks on indium tin oxide. *J. Power Sources* **2025**, *650*, 237480. [CrossRef]
82. Jayaramulu, K.; Mukherjee, S.; Morales, D.M.; Dubal, D.P.; Nanjundan, A.K.; Schneemann, A.; Masa, J.; Kment, S.; Schuhmann, W.; Otyepka, M.; et al. Graphene-Based Metal–Organic Framework Hybrids for Applications in Catalysis, Environmental, and Energy Technologies. *Chem. Rev.* **2022**, *122*, 17241–17338. [CrossRef]
83. Hwang, S.-M.; Choi, S.Y.; Youn, M.H.; Lee, W.; Park, K.T.; Gothandapani, K.; Grace, A.N.; Jeong, S.K. Investigation on electroreduction of CO_2 to formic acid using $Cu_3(BTC)_2$ metal–organic framework (Cu-MOF) and graphene oxide. *ACS Omega* **2020**, *5*, 23919–23930. [CrossRef]
84. Altintas, C.; Erucar, I.; Keskin, S. MOF/COF hybrids as next generation materials for energy and biomedical applications. *CrystEngComm* **2022**, *24*, 7360–7371. [CrossRef]
85. Demir, H.; Aksu, G.O.; Gulbalkan, H.C.; Keskin, S. MOF Membranes for CO_2 Capture: Past, Present and Future. *Carbon. Capture Sci. Technol.* **2022**, *2*, 100026. [CrossRef]
86. Seo, H. Molecular redox-active organic materials for electrochemical carbon capture. *MRS Commun.* **2023**, *13*, 994–1008. [CrossRef]
87. Lee, G.; Rasouli, A.S.; Lee, B.-H.; Zhang, J.; Won, D.H.; Xiao, Y.C.; Edwards, J.P.; Jung, E.D.; Sargent, E. CO_2 Electroreduction to Multicarbon Products from Carbonate Capture Liquid. *Joule* **2023**, *7*, 1277–1288. [CrossRef]
88. Wang, Y.-R.; Huang, Q.; He, C.-T.; Chen, Y.; Liu, J.; Shen, F.-C.; Lan, Y.-Q. Oriented electron transmission in polyoxometalate-metalloporphyrin organic framework for highly selective electroreduction of CO_2. *Nat. Commun.* **2018**, *9*, 4466. [CrossRef]
89. Jiao, L.; Seow, J.Y.R.; Skinner, W.S.; Wang, Z.; Jiang, H.-L. Metal-organic frameworks: Structures and functional applications. *Mater. Today* **2019**, *27*, 43–68. [CrossRef]
90. Souto, M.; Santiago-Portillo, A.; Palomino, M.; Vitórica-Yrezábal, I.J.; Vieira, B.J.C.; Waerenborgh, J.C.; Valencia, S.; Navalón, S.; Rey, F.; García, H.; et al. A highly stable and hierarchical tetrathiafulvalene-based metal–organic framework with improved performance as a solid catalyst. *Chem. Sci.* **2018**, *9*, 2413–2418. [CrossRef] [PubMed]
91. Wang, Z.; Zhou, Y.; Xia, C.; Guo, W.; You, B.; Xia, B.Y. Efficient Electroconversion of Carbon Dioxide to Formate by a Reconstructed Amino--Functionalized Indium–Organic Framework Electrocatalyst. *Angew. Chem. Int. Ed. Engl.* **2021**, *60*, 19107–19112. [CrossRef]
92. Huang, Z.; Yu, H.; Wang, L.; Liu, X.; Lin, T.; Haq, F.; Vatsadze, S.Z.; Lemenovskiy, D.A. Ferrocene-contained metal organic frameworks: From synthesis to applications. *Coord. Chem. Rev.* **2021**, *430*, 213737. [CrossRef]
93. Wenger, S.R.; Hall, L.A.; D'aLessandro, D.M. Mechanochemical Impregnation of a Redox-Active Guest into a Metal–Organic Framework for Electrochemical Capture of CO_2. *ACS Sustain. Chem. Eng.* **2023**, *11*, 8442–8449. [CrossRef]
94. Leong, C.F.; Faust, T.B.; Turner, P.; Usov, P.M.; Kepert, C.J.; Babarao, R.; Thornton, A.W.; D'ALessandro, D.M. Enhancing selective CO_2 adsorption via chemical reduction of a redox-active metal–organic framework. *Dalton Trans.* **2013**, *42*, 9831–9839. [CrossRef]
95. McDonald, T.M.; Mason, J.A.; Kong, X.; Bloch, E.D.; Gygi, D.; Dani, A.; Crocellà, V.; Giordanino, F.; Odoh, S.O.; Drisdell, W.S.; et al. Cooperative insertion of CO_2 in diamine-appended metal-organic frameworks. *Nature* **2015**, *519*, 303–308. [CrossRef]
96. Xie, L.S.; Skorupskii, G.; Dincă, M. Electrically Conductive Metal–Organic Frameworks. *Chem. Rev.* **2020**, *120*, 8536–8580. [CrossRef]
97. Micero, A.; Hashem, T.; Gliemann, H.; Léon, A. Hydrogen Separation Performance of UiO-66-NH_2 Membranes Grown via Liquid-Phase Epitaxy Layer-by-Layer Deposition and One-Pot Synthesis. *Membranes* **2021**, *11*, 735. [CrossRef]
98. Friebe, S.; Geppert, B.; Steinbach, F.; Caro, J. Metal–Organic Framework UiO-66 Layer: A Highly Oriented Membrane with Good Selectivity and Hydrogen Permeance. *ACS Appl. Mater. Interfaces* **2017**, *9*, 12878–12885. [CrossRef]
99. Tranchemontagne, D.J.; Mendoza-Cortes, J.; O'Keeffe, M.; Yaghi, O. Secondary building units, nets and bonding in the chemistry of metal–organic frameworks. *Chem. Soc. Rev.* **2009**, *38*, 1257–1283. [CrossRef] [PubMed]
100. Leonzio, G.; Hankin, A.; Shah, N. CO_2 electrochemical reduction: A state-of-the-art review with economic and environmental analyses. *Chem. Eng. Res. Des.* **2024**, *208*, 934–955. [CrossRef]

101. Yang, T.; Xiao, R.; Lu, F.; Ke, X.; Wang, M.; Wang, K. Conductivity enhancement mechanism and application of NiCo-MOF hollow sphere electrode materials in lithium-ion batteries. *New J. Chem.* **2025**, *49*, 6674–6683. [CrossRef]
102. Coudert, F.-X. Responsive metal–organic frameworks and framework materials: Under pressure, taking the heat, in the spotlight, with friends. *Chem. Mater.* **2015**, *27*, 1905–1916. [CrossRef]
103. Poloni, R.; Lee, K.; Berger, R.F.; Smit, B.; Neaton, J.B. Understanding trends in CO_2 adsorption in metal–organic frameworks with open-metal sites. *J. Phys. Chem. Lett.* **2014**, *5*, 861–865. [CrossRef]
104. Ma, X.; Liu, X.; Shi, S.; Gao, Z.; Yu, J.; Shan, B.; Mao, Y.; Zhu, B.; Shao, C.; Zuo, M.; et al. Two novel MOFs based on viologen ligand: Ammonia vapor sensing and photomagnetic properties. *J. Mol. Struct.* **2025**, *1340*, 142502. [CrossRef]
105. Li, L.; Shan, L.; Sheveleva, A.M.; He, M.; Ma, Y.; Zhou, Y.; Nikiel, M.; Lopez-Odriozola, L.; Natrajan, L.S.; McInnes, E.J.L.; et al. Control of evolution of porous copper-based metal–organic materials for electroreduction of CO_2 to multi-carbon products. *Mater. Adv.* **2023**, *4*, 1941–1948. [CrossRef]
106. Su, J.; Yuan, S.; Wang, H.-Y.; Huang, L.; Ge, J.-Y.; Joseph, E.; Qin, J.; Cagin, T.; Zuo, J.-L.; Zhou, H.-C. Redox-switchable breathing behavior in tetrathiafulvalene-based metal–organic frameworks. *Nat. Commun.* **2017**, *8*, 2008. [CrossRef] [PubMed]
107. Fu, T.; Wei, Y.-L.; Zhang, C.; Li, L.-K.; Liu, X.-F.; Li, H.-Y.; Zang, S.-Q. A viologen-based multifunctional Eu-MOF: Photo/electro-modulated chromism and luminescence. *Chem. Commun.* **2020**, *56*, 13093–13096. [CrossRef]
108. Wang, W.; Chen, D.; Li, F.; Xiao, X.; Xu, Q. Metal-organic-framework-based materials as platforms for energy applications. *Chem* **2024**, *10*, 86–133. [CrossRef]
109. Liu, L.; Zhang, J.; Cheng, X.; Xu, M.; Kang, X.; Wan, Q.; Han, B.; Wu, N.; Zheng, L.; Ma, C. Amorphous NH2-MIL-68 as an efficient electro- and photo-catalyst for CO_2 conversion reactions. *Nano Res.* **2023**, *16*, 181–188. [CrossRef]
110. Liu, K.-G.; Bigdeli, F.; Panjehpour, A.; Larimi, A.; Morsali, A.; Dhakshinamoorthy, A.; Garcia, H. Metal organic framework composites for reduction of CO_2. *Coord. Chem. Rev.* **2023**, *493*, 215257. [CrossRef]
111. Mamaghani, A.H.; Liu, J.; Zhang, Z.; Gao, R.; Wu, Y.; Li, H.; Feng, M.; Chen, Z. Promises of MOF-Based and MOF-Derived Materials for Electrocatalytic CO_2 Reduction. *Adv. Energy Mater.* **2024**, *14*, 2402278. [CrossRef]
112. Kornienko, N.; Zhao, Y.; Kley, C.S.; Zhu, C.; Kim, D.; Lin, S.; Chang, C.J.; Yaghi, O.M.; Yang, P. Metal–Organic Frameworks for Electrocatalytic Reduction of Carbon Dioxide. *J. Am. Chem. Soc.* **2015**, *137*, 14129–14135. [CrossRef] [PubMed]
113. Zheng, Y.-L.; Dai, M.-D.; Yang, X.-F.; Yin, H.-J.; Zhang, Y.-W. Copper(II)-Doped Two-Dimensional Titanium-Based Metal–Organic Frameworks toward Light-Driven CO_2Reduction to Value-Added Products. *Inorg. Chem.* **2022**, *61*, 13981–13991. [CrossRef] [PubMed]
114. Åhlén, M.; Cheung, O.; Xu, C. Low-concentration CO_2 capture using metal–organic frameworks—Current status and future perspectives. *Dalton Trans.* **2023**, *52*, 1841–1856. [CrossRef]
115. Fang, Z.-B.; Wang, L.; Li, Y.-H.; Zhu, Z.; Chen, W.; Jiang, H.-L. Boosting Interfacial Charge-Transfer Kinetics for Efficient Overall CO_2 Photoreduction via Rational Design of Coordination Spheres on Metal–Organic Frameworks. *J. Am. Chem. Soc.* **2020**, *142*, 12515–12523. [CrossRef]
116. Ingham, M.; Aziz, A.; Di Tommaso, D.; Crespo-Otero, R. Simulating excited states in metal organic frameworks: From light-absorption to photochemical CO_2 reduction. *Mater. Adv.* **2023**, *4*, 5388–5419. [CrossRef]

catalysts MDPI

Review

Navigating Alkaline Hydrogen Evolution Reaction Descriptors for Electrocatalyst Design

Samuel Akinlolu Ogunkunle [1], Fabien Mortier [2], Assil Bouzid [2], Jack Jon Hinsch [1], Lei Zhang [1], Zhenzhen Wu [1], Samuel Bernard [2], Yong Zhu [3] and Yun Wang [1,*]

1 Centre for Clean Environment and Energy, School of Environment and Science, Griffith University, Gold Coast Campus, Southport 4222, Australia; samuelakinlolu.ogunkunle@griffithuni.edu.au (S.A.O.); jack.hinsch@griffithuni.edu.au (J.J.H.); lei.zhang@griffith.edu.au (L.Z.); zhenzhen.wu@griffith.edu.au (Z.W.)
2 Institut de Recherche sur les Céramiques (IRCER), UMR CNRS 7315-Université de Limoges, 87068 Limoges, France; fabien.mortier@unilim.fr (F.M.); assil.bouzid@unilim.fr (A.B.); samuel.bernard@unilim.fr (S.B.)
3 School of Engineering and Built Environment, Griffith University, Gold Coast Campus, Southport 4222, Australia; y.zhu@griffith.edu.au
* Correspondence: yun.wang@griffith.edu.au

Abstract: The quest for efficient green hydrogen production through Alkaline Water Electrolysis (AWE) is a critical aspect of the clean energy transition. The hydrogen evolution reaction (HER) in alkaline media is central to this process, with the performance of electrocatalysts being a determining factor for overall efficiency. Theoretical studies using energy-based descriptors are essential for designing high-performance alkaline HER electrocatalysts. This review summarizes various descriptors, including water adsorption energy, water dissociation barrier, and Gibbs free energy changes of hydrogen and hydroxyl adsorption. Examples of how to apply these descriptors to identify the active site of materials and better design high-performance alkaline HER electrocatalysts are provided, highlighting the previously underappreciated role of hydroxyl adsorption-free energy changes. As research progresses, integrating these descriptors with experimental data will be paramount in advancing AWE technology for sustainable hydrogen production.

Keywords: green hydrogen; alkaline water electrolysis; descriptor; density functional theory; electrocatalyst design

1. Introduction

The shift towards hydrogen gas as an alternative energy source is a significant step in combating climate change. Green hydrogen via electrocatalytic water splitting is directly competitive with fossil fuels, which decreases overall greenhouse gas emissions [1]. The intricacies of electrocatalytic water splitting involve the hydrogen evolution reaction (HER) and the oxygen evolution reaction (OER) [2]. There are several types of water electrolyzers (WE): solid oxide water electrolyzer (SOWE), proton exchange membrane water electrolyzer (PEMWE), and alkaline electrolyzer (AWE) [3]. SOWE can be adapted for continuous operation in industrial areas under high temperatures. However, the technology has seen limited applications due to long start-up times, mechanical compatibility, and chemical instabilities [4,5]. PEMWE in acidic environments is the most efficient for green hydrogen production due to the abundance of the proton reactant in the solution [6]. However, the main challenge associated with PEMWE is the high cost of the electrocatalysts, which are based on platinum and platinum group metals (PGMs) to resist acid corrosion [7]. In contrast, alkaline HER in AWE offers a promising alternative allowing for the exploration of earth-abundant transition metal-based catalysts because their hydro-oxide counterparts are stable in this environment [8–10]. Shifting towards an alkaline environment will reduce reliance on expensive noble metals and reduce the hydrogen production cost. Green hydrogen via electrocatalytic water splitting is directly competitive with fossil fuels, which decreases overall greenhouse gas emissions.

The operational principle of AWEs involves the conduction of hydroxide anions through a liquid electrolyte between two electrodes [8]. The main challenges of the current AWE are its low current densities, the inability to operate with intermittent renewable energy sources, and slow kinetics [11]. These limitations are associated with early versions of atmospheric electrolyzers designed for continuous operation at industrial sites with stable and inexpensive electricity sources. Modern pressurized high-temperature AWEs have shown the capability to reach performance and adaptability on par with PEMWEs [12]. The adaptability to various energy sources, coupled with improved design and control systems, positions pressurized AWEs as a promising solution for sustainable and flexible hydrogen generation in the evolving energy technology landscape [12].

Moreover, the anion exchange membrane water electrolyzer (AEMWE) replaces the conventional diaphragm of AWE with an anion exchange membrane [13]. A membrane reduces the gas cross-over and the need for pressure difference operation between the anode and cathode. Additionally, the thickness of the anion exchange membrane is less than the diaphragm, which leads to a lower ohmic over-voltage. The AEMWE uses an alkaline solution with low concentration. Conventional electrolyzers use concentrated potassium hydroxide electrolytes, which increase corrosion. Therefore, AEMWE has several benefits, including no leakage, ease of installation, and control. However, the low energy conversion efficiency of AWEs due to the slow kinetics remains a bottleneck, requiring high-performance electrocatalyst design.

The pursuit of an effective catalyst for AWE is multifaceted, necessitating a blend of experimental and theoretical studies to unlock the full potential of electrocatalysts [8,11,14]. The challenges for the alkaline HER electrocatalyst design are twofold: the inherently lower reaction rates of catalysts and a limited understanding of the complex kinetics involved. The HER process varies significantly depending on the electrolyte medium, which can be either acidic or alkaline. In acidic media, the HER mechanism involves the electrochemical reduction of a proton (H^+), leading to the release of hydrogen gas (H_2) [15–17]. The HER efficiency is associated with the adsorption of hydrogen atom intermediate (H^*), which consists of the Volmer/Heyrovsky or Volmer/Tafel steps as delineated below [18]:

$$\text{Volmer reaction}: H^+ + e^- \rightarrow H^* \tag{1}$$

$$\text{Heyrovsky reaction}: H^* + H^+ + e^- \rightarrow H_2 \tag{2}$$

$$\text{Tafel reaction}: H^* + H^* \rightarrow H_2 \tag{3}$$

Conversely, alkaline HER operates through the electrochemical reduction of water molecules, resulting in the formation of hydrogen gas and hydroxide ions (OH^-) [19].

$$\text{Water adsorption}: H_2O + * \rightarrow H_2O^* \tag{4}$$

$$\text{Volmer reaction (I)}: H_2O^* + e^- \rightarrow H^* + OH^* + e^- \rightarrow H^* + OH^- \tag{5}$$

$$\text{Volmer reaction (II)}: H_2O^* + e^- \rightarrow H^* + OH^- \tag{6}$$

$$\text{Heyrovsky reaction}: H_2O + e^- + H^* \rightarrow H_2 + OH^- \tag{7}$$

$$\text{Tafel reaction}: H^* + H^* \rightarrow H_2 \tag{8}$$

The HER pathway in alkaline media involves complex Volmer-Heyrovsky or Volmer-Tafel steps, with the intermediate hydrogen formed after an initial water adsorption step (Equation (4)). Additionally, the adsorbed hydroxyl intermediate may form during the Volmer step if its binding energy with the active site is strong enough (Equation (5)). Otherwise, the solution-phase hydroxide (OH^-) can directly form after the water dissociation

(Equation (6)) during the Volmer reaction. The H_2 product can be produced through an adsorbed H atom intermediate and water in the Heyrovsky reaction (Equation (7)) or through the combination of two adsorbed H* atom intermediates in the Tafel reaction (Equation (8)). Figure 1 indicates the possible alkaline HER pathway. It shows that the performance of alkaline HER can be influenced by four main factors: (1) water adsorption strength on active sites, (2) water dissociation ability, (3) hydrogen binding energy, and (4) OH adsorption ability. To this end, the properties relevant to these four factors have been used to screen the alkaline HER electrocatalysts.

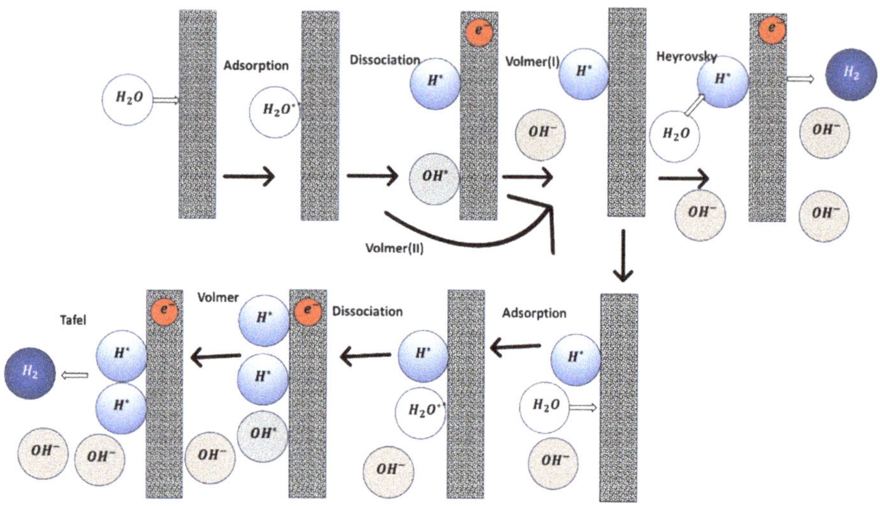

Figure 1. Possible hydrogen evolution reaction processes in alkaline media.

The step-by-step process that governs HER needs to be theoretically revealed by delving into the intricate details of reaction mechanisms through atomic-scale simulation [8]. Theoretical investigations can offer insights into the energetic aspects of water adsorption, water dissociation, intermediates adsorption, and hydrogen desorption [15,20–22]. This granular view allows researchers to identify and optimize the performance of active sites that influence both catalytic efficiency and stability. Such knowledge is instrumental in recognizing the rate-determining steps and potential energy barriers, which are critical for designing catalysts with optimal performance [15]. Furthermore, theoretical models serve as a blueprint for predicting the performance of different materials under various reaction conditions. Therefore, computational studies enable the fine-tuning of working conditions to enhance the overall efficiency of the process. Additionally, these studies contribute significantly to the green hydrogen field by offering a cost-effective strategy for material screening, circumventing the need for laborious and expensive experimental procedures [20,23]. Ultimately, the insights gained from theoretical research facilitate the development of innovative solutions for alkaline hydrogen production [8].

Recent advances have highlighted the importance of atomically precise electrocatalysts, which utilize theoretical descriptors to correlate the properties of catalysts with their alkaline HER performance [8,19,24]. Since different descriptors related to the different reaction steps shown in Figure 1 have been proposed and previously utilized [25–30], it is necessary to devise guidelines for selecting descriptors that can accurately predict the performance of HER electrocatalysts in alkaline conditions. Density functional theory (DFT) calculations are essential in computing the proposed energy descriptors related to water adsorption, water dissociation, intermediates adsorption [18,31–33], and product desorption. These descriptors are crucial to understanding alkaline HER electrocatalysts and electrocatalytic processes in the alkaline electrolyzer (AEL). Additionally, identifying important descriptors can improve computational accuracy while minimizing costs [23, 34]. While several reviews on the recent progress of alkaline HER electrocatalysis have been presented [8,11,35], a summary and discussion of the most appropriate descriptors for guiding the design of efficient AEL is still rare. To fill this gap, this review will provide a framework by which the desired theoretical descriptor can be selected via DFT computations to design electrocatalysts rather than a comprehensive overview of AWEs.

2. Theoretical Descriptors

As indicated by Equations (4)–(8) and Figure 1, the performance of alkaline HER electrocatalysts is a complex interplay of various factors. The water adsorption strength on active sites, the ability of the catalyst to dissociate water molecules, the hydrogen binding energy, and the OH adsorption processes are all critical in determining the efficiency of the HER process. These factors are intricately linked to the intrinsic properties of the electrocatalysts and their interaction with the operational reaction environment, e.g., electrolyte. Therefore, different theoretical energy-related descriptors have been employed as indicators to screen electrocatalysts for alkaline HER [25–30,36].

2.1. Water Adsorption and Dissociation

The first step of the electrocatalytic water splitting in an alkaline solution is water adsorption, followed by water dissociation to provide the H^+ cation for hydrogen production (Equations (4)–(6)). HER performance in alkaline or neutral media is primarily determined by the equilibrium between water dissociation (Volmer step) and the subsequent chemisorption of water-splitting intermediates (OH and H) on the surface of the electrocatalysts [37]. To this end, it is critical to strengthen water adsorption at the active sites and decrease the water dissociation activation energy during the Volmer step to increase the overall HER activities of these electrocatalysts. Consequently, the adsorption energy of water ($\Delta E_{H_2O^*}$) and its dissociation energy barriers (ΔE_a) have been used as descriptors to evaluate the performance of alkaline HER electrocatalysts. The $\Delta E_{H_2O^*}$ value can be calculated as shown below:

$$\Delta E_{H_2O*} = \frac{1}{n}\left(E_{H_2O*} - E_* - E_{H_2O}\right) \tag{9}$$

E_{H_2O*} and E_* are the energies of the surface with and without adsorbed water, respectively. E_{H_2O} and n represents the energy of an isolated water molecule and the number of adsorbed water molecules in each surface cell, respectively. According to Equation (9), a more negative value of ΔE_{H_2O*} corresponds to stronger water adsorption.

To calculate the water dissociation energy barrier on the active site, the nudged elastic band (NEB) is widely used [38,39]. The NEB method is highly effective for locating the maximum energy point between specified initial and final configurations of a given transition. The NEB approach for climbing images involves finding a transition path where the saddle point corresponds to the image with the highest energy. The image does not depict the spring forces acting on the band. The image attempts to optimize its energy along the band while minimizing it in all other directions. Once this image reaches convergence, it can be precisely located at the saddle point [40]. The climbing image NEB method enhances the possibility of finding the transition state (TS) for the reaction at the highest energy point. The ΔE_a value is the energy difference between the TS state and the surface with the adsorbed water.

2.2. Gibbs Free Energy Change of Hydrogen Atom Adsorption

Hydrogen atoms adsorbed at the active site are the HER intermediates. The adsorption and desorption of hydrogen atoms on the catalyst surface are commonly viewed as competing processes. During the HER process, the active catalytic site must have a strong affinity for a hydrogen atom. However, a weak adsorption between the active catalytic site and the hydrogen atom is required [41,42]. The high adsorption strength of hydrogen atoms can promote hydrogen atom intermediate formation but impede the formation of hydrogen molecules. Comparatively, weak adsorption can benefit the formation of H_2 but hinder the formation of hydrogen atom intermediates. A balance must be achieved where the hydrogen intermediate must form but not compete with/prevent the generation and release of the hydrogen molecule. As a result, the optimal adsorption strength of the hydrogen atom intermediate is essential for improving the HER. The adsorption strength of hydrogen atom intermediates on active sites in catalysts can be evaluated using the Gibbs free energy change of hydrogen atom adsorption (ΔG_{H^*}) [43].

The ΔG_{H^*} is obtained by using the computational hydrogen electrode (CHE) method:

$$\Delta G_{H^*} = \Delta E_{H^*} + \Delta ZPE_{H^*} - T\Delta S_{H^*} \tag{10}$$

where ΔE_{H^*} describes the binding energy, which is calculated as follows:

$$\Delta E_{H^*} = E_{H^*} - E_* - \frac{1}{2}E_{H_2} \tag{11}$$

Here, E_{H^*} is the total energy of the system with one adsorbed H atom and E_{H_2} represents the energy of an isolated H_2 molecule. ΔZPE_{H^*} is calculated by $\Delta ZPE_{H^*} = ZPE_{H^*} - \frac{1}{2}ZPE_{H_2}$, where the ZPE_{H^*} denotes the vibrational energy of the adsorbed H atom on the active site and the value of ZPE_{H_2} is 0.27 eV. Further, ΔS_{H^*} is the entropy change of H* adsorption, which is obtained by $1/2\ \Delta S_{H_2}$. The $T\Delta S_{H^*}$ value is equal to $1/2\ T\Delta S_{H_2}$, which is −0.20 eV at 300 K, obtained by Nørskov et al. [32]. The ideal ΔG_{H^*} value is 0 eV.

2.3. *Gibbs Free Energy Change of Hydroxyl Adsorption*

While hydrogen adsorption energy (ΔG_H) primarily determines the alkaline HER activity, the Gibbs free energy change of OH adsorption energy (ΔG_{OH^*}) is found to be equally important since it is responsible for the water dissociation and the amount of hydrogen available [44]. Therefore, the adsorption and desorption of both reaction intermediates (OH* and H*, see Equation (2)) dictate the overall alkaline HER efficiency [45].

The ΔG_{OH^*} is calculated by using the computational hydrogen electrode (CHE) method:

$$\Delta G_{OH^*} = \Delta E_{OH^*} + \Delta ZPE - T\Delta S \tag{12}$$

where ΔE_{OH^*} describes the binding energy of an OH group at the active site, which is calculated as follows:

$$\Delta E_{OH*} = \left(E_{OH*} - E_* - E_{H20} + \frac{1}{2}E_{H_2}\right) \tag{13}$$

where E_{OH^*} is the total energy of the system with one adsorbed OH intermediate, the $\Delta ZPE - T\Delta S$ value is 0.35 eV at 300 K, obtained by Nørskov et al. The ideal ΔG_{OH^*} value is −0.3 eV for the alkaline HER [46].

3. Applications of Descriptors

3.1. *Water Adsorption and Dissociation*

Figure 1 shows that water adsorption is the first step in generating hydrogen atom intermediates for alkaline HER. The H_2O dissociation becomes a rate-determining step that regulates the catalytic HER activity in alkaline media [47,48]. To this end, the adsorption energy of water and its dissociation energy barriers have been used as descriptors in some research to explain the performance of alkaline HER electrocatalysts.

For example, Chen et al. experimentally synthesized a Ni-doped Pt catalyst on carbon support at 500 °C (termed as NiPt-C-500), which has a low overpotential of 14.0 mV at 10 mA/cm^2 in the alkaline media [49], using the adsorption energy of water molecules and their dissociation energy barriers as descriptors through DFT computations at the PBE-GGA level to identify the active sites of the NiPt-C-500 catalyst. Figure 2a shows that H_2O can be adsorbed on the Ni sites with an adsorption energy of −0.48 eV. As a comparison, the water adsorption energy on Pt is +0.37 eV. The positive value suggests that the adsorption on Pt is energetically unfavorable. Moreover, a lower H_2O-dissociation barrier on Ni-doped Pt (111) of 0.17 eV was theoretically revealed compared to that on Pt(111), as illustrated in Figure 2b. This suggests a rapid dissociation of H_2O catalyzed by surface Ni atoms on the Pt surface in NiPt-C-500 electrocatalysts, which can increase the rate of H* generation for alkaline HER. Using the same descriptors, Yao et al. resorted to DFT computations to explain the high performance of Cr-doped Co_4N nanorods on carbon cloth (Cr-Co_4N) for alkaline HER [50]. Figure 2c shows that the calculated H_2O adsorption energy on the top of the Cr dopant within Cr-Co_4N is −0.63 eV, which is much lower than that on the surface Co of Co_4N (−0.18 eV). Therefore, the water adsorption on Cr dopants is substantially stronger than on undoped Co_4N. This is because the oxygen atom of water carries a negative Bader charge of −1.13 |e|. As a result, Cr atoms with a higher positive charge (0.72 |e|) than surface Co (0.21 |e|) can more strongly adsorb water molecules via higher electrostatic interactions. The enhanced adsorption strength of water can further facilitate the dissociation of the adsorbed water and lower the dissociation barrier, as illustrated in Figure 2d. It explains the experimental observation that introducing Cr dopants can boost the alkaline HER activity.

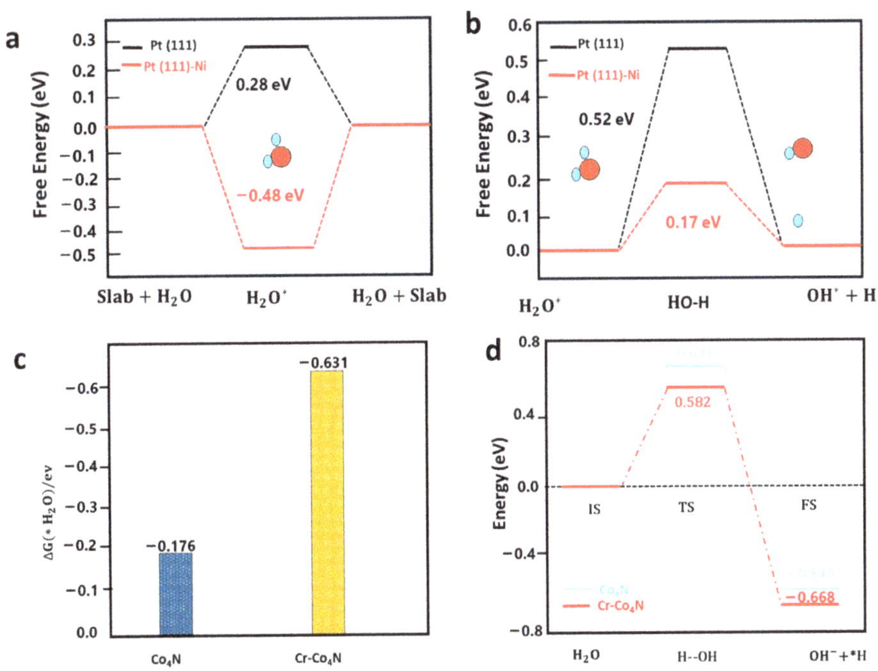

Figure 2. (**a**) DFT-calculated water adsorption energies (ΔE_{H_2O*}) and (**b**) water dissociation energy barriers (ΔE_a) on Pt (1 1 1) and Pt (1 1 1)-Ni surfaces. Reprinted with permission from Ref. [49]. Copyright 2023 Elsevier. (**c**) DFT-calculated water adsorption energies (ΔE_{H_2O*}) and (**d**) water dissociation energy barriers (ΔE_a) on Co_4N Cr-Co_4N. Reprinted with permission from Ref. [50]. Copyright 2019 Wiley.

More examples of studies using the $\Delta E_{H_2O^*}$ to identify the active sites of the alkaline HER electrocatalysts can be found in Table 1. Similarly, some examples of using the ΔE_a as the descriptor to explain the performance of alkaline HER electrocatalysts are listed in Table 2.

Table 1. Water adsorption energies as descriptors to identify the active site of the alkaline HER electrocatalyst.

Catalysts	DFT Method	Adsorption Site	$\Delta E_{H_2O^*}$(eV)	Overpotential@ Current Density of 10 mA cm^{-2}	Conclusion	Reference
Co-doped WO_2/Amorphous Co_xW	PBE	Co WO_2 (001) Co_xW	0.12 −0.66 −0.88	- 49 25.0	Amorphous CoxW is an active site for water adsorption.	ACS Appl. Mater. Interfaces (2019), 11, 38771 [51]
Cr-doped Ni(111)	-	Ni(111) Cr-doped Ni(111)	−0.21 −0.58	322.0 203.0	Co-dopants can strengthen water adsorption.	J. Am. Chem. Soc. (2020), 143, 1399 [52]
CuCoMo-doped Ni(111)	PBE	Ni-top Co-top Cu-top Mo-top	−0.23 −0.25 −0.15 −0.67	- - - -	Water is mainly adsorbed on Mo dopants.	Electroanal. Chem. (2019), 839, 224 [53]
N-doped Ni	PBE	Surface Ni Surface Ni next to N dopants	−0.30 −0.41	- 64.0	N dopants can strengthen the water adsorption on their neighboring Ni atoms.	J. Am. Chem. Soc. (2017), 139, 12283 [54]
Cu-Ru/$RuSe_2$	PBE	Cu-$RuSe_2$ Cu-Ru/$RuSe_2$	−0.40 −1.28	61.0 23.0	Cu-doped Ru/$RuSe_2$ displays a much stronger affinity to water.	Adv. Mater. (2023), 35, 2300980 [55]
Ru_{SA}@$Ti_3C_2O_2$	PBE	$Ti_3C_2O_2$ Ru_{SA}@$Ti_3C_2O_2$	−0.18 −1.47	- 40.3	Enhanced H_2O adsorption by single Ru atom (Ru_{SA}) on MXenes.	EcoMat (2023), 5, 12274 [56]

The overpotential values are reported based on the experimental results in the corresponding reference.

Table 2. The dissociation energy barrier of the adsorption water as descriptors to identify the active site of the alkaline HER electrocatalyst.

Catalyst	DFT Method	Active Site	ΔE_a(eV)	Overpotential (mV@ Current Density of 10 mA cm^{-2})	Conclusion	Reference
Co-doped WO_2 /Amorphous Co_xW	PBE	Co-WO_2 WO_2 (001) Co_xW	1.25 1.03 0.46	 49 25	Amorphous CoxW is an active site for water dissociation.	ACS Appl. Mater. & Interface. (2019), 11, 38771 [51]
W/WO_2	PBE-D3	W WO_2 W/WO_2	0.84 0.06 0.02	183 106 35	W/WO_2 interface is the active site that allows water dissociation.	Nat. Commun. (2023), 14, 5363 [57]
F-Ni_3S_4	PBE + U	Ni_3S_4 F-Ni_3S_4	1.05 0.55	112 29	F-Ni_3S_4 allows a better H_2O adsorption.	Adv. Funct. Mater. (2021), 31, 2008578 [58]
Co_2P/O-Co_2P	PBE	Co_2P O-Co_2P	0.88 0.51	247 160	Oxygen incorporation may induce a higher positive charge state Co, which could benefit water adsorption and weaken the O-H bond in adsorbed H_2O.	Adv. Mater. (2017), 29, 1606980 [59]
Cr-doped Ni(111)	-	Ni(111) Cr-doped Ni(111)	0.78 0.43	322 203	Co-dopant can facilitate water dissociation.	J. Am. Chem. Soc. (2020), 143, 1399 [52]
Bimetallic Nickel-Based Alloys	PBE	NiCu O-NiCu	1.02 0.65	85 23	Partial oxidization of CuNi alloy can facilitate water dissociation.	Angew. Chem. Int. Ed. (2022), 61, 202202518 [60]
N-modified Ni	PBE	Ni N-Ni	0.84 0.42	- 64	N dopant can greatly facilitate water dissociation on its neighbouring active site.	J. Am. Chem. Soc. (2017), 139, 12283 [54]
Cu-Ru/$RuSe_2$	PBE	Cu-$RuSe_2$ Cu-Ru/$RuSe_2$	0.31 0.10	61 23	Cu-dopant can promote water dissociation on Ru/$RuSe_2$.	Adv. Mater. (2023), 35, 2300980 [55]

Table 2. *Cont.*

Catalyst	DFT Method	Active Site	ΔE_a (eV)	Overpotential (mV@ Current Density of 10 mA cm^{-2})	Conclusion	Reference
NiP_2-FeP_2/Cu	PBE	NiP_2	0.52	37	The coupling between interface-rich NiP_2–FeP_2 and metallic Cu can synergistically accelerate water dissociation.	ACS Energy Lett. (2021), 6, 354 [61]
		FeP_2	0.49	-		
		NiP_2-FeP_2	0.40	-		
		NiP_2-FeP_2/Cu	0.16	23.6		
		F-Ni_3S_4	0.55	-		
High-entropy alloys (HEAs) $Pt_{18}Ni_{26}Fe_{15}Co_{14}Cu_{27}$	PBE	Fe	0.11	11	Adsorption of H_2O located on the Fe sites, which activates the dissociation of water molecules.	Nat. Commun. (2020), 11, 5437 [62]
2D transition-metal dichalcogenide	PBE	MoS_2	0.62	-	MoS_2 and WS_2 have similar activation energy.	J. Phys. Chem. C. (2022), 126, 5151 [63]
		$MoSe_2$	0.74	-		
		WS_2	0.60	-		
		WSe_2	0.84	-		

The overpotential values are reported based on the experimental results in the corresponding reference.

3.2. Hydrogen Atom Adsorption

The Gibbs free energy change of the intermediate H atom adsorption (ΔG_{H^*}) was proposed as a descriptor for HER by Nørskov and his co-workers. This descriptor has successfully identified the active site of acidic HER electrocatalysts and the relevant reaction mechanism [32]. A volcano plot is a valuable tool for understanding and optimizing catalyst performance in acidic HER. It is usually a plot of ΔG_{H*} against activities (such as overpotential, specific current density, and turnover frequency) [64,65]. The highest point on the volcano map is the ideal ΔG_{H*} value at which the catalytic activity is maximized. Catalysts located on the left side of the volcano plot exhibit excessive hydrogen binding (very strong), while those on the right side have insufficient hydrogen binding (very weak). According to the Sabatier principle, an optimum catalyst should exhibit adsorption energies that are neither too strong nor too weak. In the alkaline HER, the adsorption of H atoms at the active site is also important since it is related to the water dissociation, Volmer, Heyrovsky, and Tafel steps, as suggested by Equations (5)–(8) and Figure 1. To this end, the ΔG_{H^*} was also widely used in the analysis of the performance of the alkaline HER electrocatalysts [66].

For example, Chen et al. adopted a dual approach to simultaneously promote water dissociation and hydrogen desorption kinetics with Co-doped WO_2/amorphous Co_xW hybrid catalysts using DFT [51]. As shown in Figure 3a, the ΔG_{H^*} on amorphous Co_xW is -1.88 eV, indicating a strong hydrogen-binding interaction that prevents H_2 production and desorption (Figure 4a). The most catalytically active site on Co$-WO_2$(011) has a ΔG_{H^*} of -0.06 eV, closer to thermoneutral than Pt's -0.09 eV, indicating significant hydrogen adsorption capability. According to DFT calculations, the HER on Co-doped WO_2/amorphous CO_xW hybrid catalyst follows the Volmer-Tafel step, where water molecules are activated and cleaved to form H atoms on the surface of amorphous Co_xW. These H atoms then rapidly combine to form H_2 on the surface of Co-doped WO_2. He et al. incorporated a single Ru atom into Ni_5P_4 to achieve an effective electrocatalyst for alkaline HER [67]. The DFT calculations were conducted to comprehend the impact of single-atomic Ru incorporation into Ni_5P_4 on its catalytic and structural properties. The structures of Ni_5P_4 were determined, both with and without Ru incorporation, and they revealed several potential catalytic sites. Experimental analysis and the established model structural parameters were remarkably congruent. In the case of Ni_5P_4-Ru, the computed ΔG_{H^*} value at the Ru-doped sites was -0.30 eV, which was comparatively higher than the value of -0.42 eV observed at the site of P of pristine Ni_5P_4 (see Figure 3b). This finding indicated that the Ru dopant exhibited favorable energetics for the desorption of absorbed H atoms to form the H_2 gas.

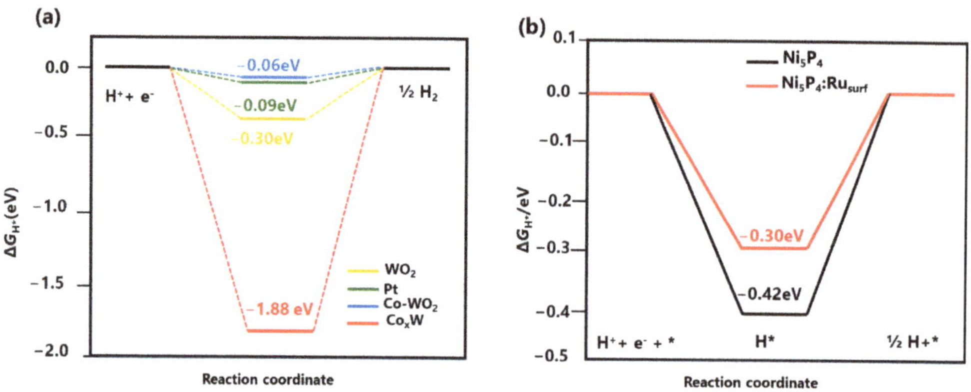

Figure 3. (**a**) DFT-calculated Gibbs free energy change of the intermediate H atom adsorption (ΔG_{H^*}) on WO_2, Co-WO_2, CoxW, and Pt. Reprinted with permission from Ref. [51]. Copyright 2019 American Chemical Society. (**b**) DFT-calculated Gibbs free energy change of the intermediate H atom adsorption (ΔG_{H^*}) on Ni_5P_4 and Ni_5P_4: Ru. Reprinted with permission from Ref. [67]. Copyright 2020 Wiley.

Table 3 lists more examples of using ΔG_{H^*} as the descriptor to identify the active site of the alkaline HER electrocatalysts and evaluate their performance.

Table 3. Gibbs free energy change of H intermediate adsorption as descriptors to identify the active site of the alkaline HER electrocatalyst.

Catalyst	DFT Method	Active Site	ΔG_{H^*}(eV)	Overpotentials (mV@ Current Density of −10 mA cm^{-2})	Conclusion	Reference
W/WO_2	DFT-D3	W	−0.51	183.0	W/WO_2 interface is the active site for HER.	Nat. Commun. (2023) 14, 5363 [57]
		WO_2	−1.24	106.0		
		W/WO_2 interface	−0.41	35.0		
Co-doped WO_2 /Amorphous Co_xW	PBE	Co-WO_2	−0.06	-	Co-WO_2 is the active site for H_2 formation.	ACS Appl. Mater. & Interface (2019), 11, 38771 [51]
		WO_2 (001)	−0.30	49.0		
		Co_xW	−1.88	-		
Cr-doped Ni(111)	-	Ni(111)	−0.23	322.0	Co dopant can reduce the hydrogen desorption rate.	J. Am. Chem. Soc. (2020), 143, 1399 [52]
		Cr-doped Ni(111)	−0.38	203.0		
CuCoMo-doped Ni(111)	PBE	Ni_3-fcc	−2.76 *	-	The preferential adsorption site is a Mo-coordinated fcc site.	Electroanal. Chem. (2019), 839, 224 [53]
		Cu_2Co-fcc	−2.51 *	-		
		Co_2Cu-fcc	−2.67 *	-		
		CuCoMo-fcc	−2.78 *	-		
Ni/NiCu/O-NiCu	PBE	Ni	−0.31	-	Cu can weaken the adsorption strength of hydrogen to benefit the H_2 production.	Angew. Chem. Int. Ed. (2022), 61, 202202518 [60]
		NiCu	−0.11	85.0		
		O-NiCu	−0.035	23.0		
Cu-Ru/$RuSe_2$	PBE	Cu-$RuSe_2$	0.50	61.0	Cu can weaken the adsorption strength of hydrogen to benefit the H_2 production.	Adv. Mater. (2023), 35, 2300980 [55]
		Cu-Ru/$RuSe_2$	0.19	23.0		

Table 3. *Cont.*

Catalyst	DFT Method	Active Site	ΔG_{H^*}(eV)	Overpotentials (mV@ Current Density of -10 mA cm^{-2})	Conclusion	Reference
NiP-FeP$_2$/Cu	PBE	NiP$_2$	−0.358	37.0	NiP-FeP$_2$/Cu interface effectively adsorbs generated H intermediate.	ACS Energy Lett. (2021), 6, 354. [61]
		FeP$_2$	−0.203	-		
		NiP-FeP$_2$	−0.043	-		
		NiP-FeP$_2$/Cu	−0.03	23.6		
O-Co$_2$P	PBE	Co$_2$P	0.70	247.0	Oxygen incorporation may induce a higher positive charge state Co to benefit water adsorption and weaken the O-H bond in adsorbed H$_2$O.	Adv. Mater. (2017), 29, 1606980 [59]
		O-Co$_2$P	0.19	160.0		
Ru$_{SA}$@Ti$_3$C$_2$O$_2$.	PBE	Ti$_3$C$_2$O$_2$	−0.33	-	Ru$_{SA}$ favours H$_2$ formation.	EcoMat (2023), 5, 12274 [56]
		Ru$_{SA}$@Ti$_3$C$_2$O$_2$	−0.07	40.3		
F-Ni$_3$S$_4$	PBE	Ni$_3$S$_4$	0.12	112.0	F-doping can promote water dissociation to increase the rate of Hads formation.	Adv. Funct. Mater. (2021), 31, 2008578 [58]
		F-Ni$_3$S$_4$	−0.034	29		
Metal-modified transition metal carbides	PW91	Pt/Mo$_2$C interface	0.01	-	The ΔG_{H^*} has a strong correlation with the alkaline HER performance.	ACS Catal. (2019), 9, 2415 [29]
		Pt/NbC	−0.28	-		
		Pt/TaC	−0.07	-		
		Pt/TiC	−0.03	-		
		Mo$_2$C(0001)	−0.75	-		
		NbC	−0.90	-		
		TaC	−0.99	-		
		TiC	−0.96	-		
		VC	−0.84	-		
		W$_2$C(0001)	−0.60	-		
		WC(0001)	−0.81	-		
Ni/Co-modified MoSe$_2$	PBE	MoSe$_2$	1.49	301	Partial substitution of Ni or Co atoms for Mo atoms can optimize the ΔG_{H^*} value.	Angew. Chem. (2020), 132, 15344 [27]
		Ni-MoSe$_2$	−0.43	98		
		Co-MoSe$_2$	−0.20	183		
FeCoNiCu$_{0.5}$	PBE	Ni	−0.08	436	ΔG_{H^*} value of −0.08 eV, close to that of the Pt catalyst.	J. Alloys Compd. (2024), 175356 [68]
		Fe	0.17	509		
		Co	−0.18	361		
		Cu	0.23	252		
		FeCoNiCu$_{0.5}$	−0.08	71		
2D transition-metal dichalcogenides	PBE	MoS$_2$	0.03	-	Except for WSe$_2$, the calculated ΔG_{H^*} of the rest is close to zero.	J. Phys. Chem. C. (2022), 126, 5151 [63]
		MoSe$_2$	0.09	-		
		WS$_2$	−0.03	-		
		WSe$_2$	0.24	-		

* The authors calculated the adsorption energy of the H atoms here. The overpotential values are reported based on the experimental results in the corresponding reference.

3.3. Hydrogen and Hydroxyl Adsorption

In recent studies, the volcano plot has been adopted to describe HER activities in alkaline media by plotting ΔG_{OH^*} against activities. In contrast to the volcano-shaped association observed with ΔG_{H^*}, Zhang et al. demonstrated that ΔG_{OH^*} exhibits a significantly weaker correlation with alkaline HER activity than the volcano-shaped relationship established with ΔG_{H^*} [29]. Au-modified TMCs exhibit OHBEs comparable to those of Pt and Pt-modified TMCs; however, the alkaline exchange current densities of Au/TMCs are two to three orders of magnitude lower than those of their Pt counterparts. Furthermore, carbides modified with Ag and Cu exhibit a broad spectrum of ΔG_{OH^*}, but there is no discernible HER trend within this spectrum. According to the research of Zhang et al., ΔG_{OH^*} is not an appropriate descriptor for alkaline HER on TMCs and metal-modified TMCs. It also implies that the adsorbed hydroxyl group does not directly participate in the rate-determining step of alkaline HER kinetics on these surfaces [29].

Markovic et al. were the first to study the impact of adsorbed OH on Pt-based electrocatalysts, emphasizing its poisoning effect on surface sites and its influence on the kinetics of the alkaline HER [69]. The hydrogen binding energy (HBE) and hydroxyl binding energy (OHBE) values for pure Ru and various Mo-Ru compounds were computed (Figure 4a) by Zhao et al. Through DFT, the optimal H and OH adsorption sites on distinct catalysts for alkaline HER were identified [70]. In the Mo-Ru-1 structure, one of the Ru atoms is doped with Mo, while Mo-Ru-2 and Mo-Ru-3 are doped with two and four atoms of Mo, respectively. Figure 4a displays the HBE (ΔG_{H^*}) values of Ru, Mo-Ru-1, Mo-Ru-2, and Mo-Ru-3, which are −0.64, −0.69, −0.67, and −0.76 eV, respectively. These results indicate that the adsorption energy of H is similar for all catalysts and nearly identical to experimental values [25,71,72]. The OHBE (ΔG_{OH^*}) values of Mo-Ru-1 (−0.53 eV), Mo-Ru-2 (−0.68 eV), and Mo-Ru-3 (−0.97 eV) exhibited a substantial increase after Mo-doping, in comparison to native Ru (−0.31 eV). Consequently, the OHBE progressively rose as the Mo atom content increased. Furthermore, the increased OHBE after Mo-doping can be ascribed to the augmented HOR/HER activity of Mo-Ru.

Zhang et al. designed high-efficiency alkaline HER electrocatalysts using a unique dual descriptor of optimal free energies (ΔG_{H^*} and ΔG_{OH^*}). Ni_3N surface reactivity was tailored using theory to balance the adsorption energies of hydrogen and hydroxyl species. Nickel-based materials are highly promising non-noble metal electrocatalysts for the hydrogen evolution reaction (HER) [73,74]. In this study, the metallic Ni_3N was considered a case study to illustrate the application of dual-descriptor-driven design. Ni_3N has been regarded as a suitable electrocatalyst for cleaving OH-H bonds in the Volmer phase. According to Figure 4b, the DFT calculations demonstrate that the Ni_3N (111) surface exhibits a facile hydroxyl adsorption free energy of 0.03 eV. This characteristic is advantageous for the energetics of water dissociation and subsequent hydroxyl desorption [75]. Such findings align with prior studies indicating that Ni_3N can function as a promoter of water dissociation [76]. Nevertheless, the DFT calculations indicate that the average G_{H^*} value observed on the surface of Ni_3N (111) is approximately −0.31 eV, which surpasses the ideal value of ΔG_{H^*} = 0 eV by a wide margin. The excessive reactivity of the surface results in unfavorable desorption of hydrogen and subsequent formation of H_2. This approach has successfully enhanced the hydrogen evolution reaction (HER) performance by incorporating various transition-metal dopants into Ni3N, including Mo-Ni_3N, W-Ni_3N, and V-Ni_3N [30]. The hydrogen and hydroxyl binding energy results on Mo-Ni_3N effectively balance the dual descriptors.

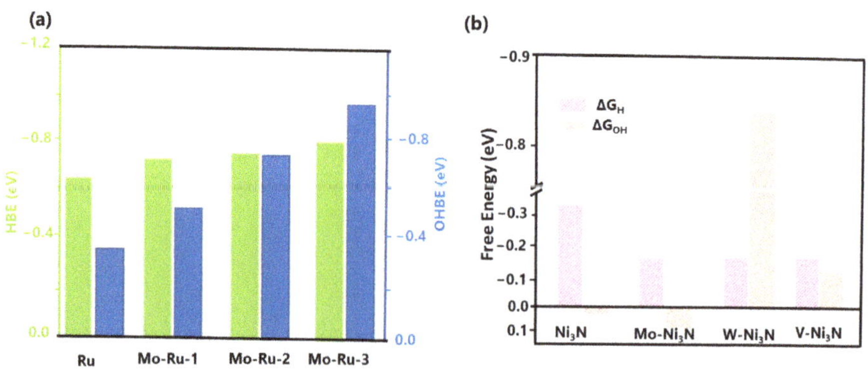

Figure 4. (**a**) DFT-calculated Gibbs free energy change of the intermediate H atom adsorption (ΔG_{H^*}, HBE) and the Gibbs free energy change of OH adsorption energy (ΔG_{OH^*}, OHBE) on Ru, Mo-Ru-1, Mo-Ru-2, Mo-Ru-3. Reprinted with permission from Ref. [70]. Copyright 2022 American Chemical Society. (**b**) DFT-calculated Gibbs free energy change of the intermediate H atom adsorption (ΔG_{H^*}) and the Gibbs free energy change of OH adsorption energy (ΔG_{OH^*}) on Ni_3N, Mo-Ni_3N, W-Ni_3N, and V-Ni_3N. Reprinted with permission from Ref. [30]. Copyright 2019 American Chemical Society.

More examples of using ΔG_{H^*} and ΔG_{OH^*} as the descriptor to identify the active site of the alkaline HER electrocatalysts and evaluate their performance are listed in Table 4.

Table 4. Gibbs free energy change of OH intermediate adsorption as descriptors to identify the active site of the alkaline HER electrocatalyst.

Catalyst	DFT Method	Active Site	ΔG_{OH^*}	Overpotentials (mV@ Current Density of -10 mA cm^{-2})	Conclusion	Reference
Cr-doped Ni(111)	-	Ni(111)	-3.23 *	322	Co-doping allows for a balance between facilitating the dissociation of water and preventing the poisoning effect.	J. Am. Chem. Soc., (2021), 143, 1399 [52]
		Cr-doped Ni(111)	-3.83 *	203		
Ru_{SA}@$Ti_3C_2O_2$/$Ti_3C_2O_2$	PBE	Ru_{SA}@$Ti_3C_2O_2$	-0.49	40.3	Ru improved adsorption ability of $Ti_3C_2O_2$ toward HO*.	EcoMat (2023), 5, 12274 [56]
		$Ti_3C_2O_2$	2.00	-		
Ni/Co-modified $MoSe_2$	PBE	Mo-$MoSe_2$	-1.74	301	Co-$MoSe_2$ is favorable for OH^- desorption in alkaline HER.	Angew. Chem. (2020), 132, 315344 [27]
		Co-$MoSe_2$	0.67	183		
		Ni-$MoSe_2$	1.22	98		
Metal-modified transition metal carbides	PW91	Pt	-2.39	-	The ΔG_{OH^*} does not show a strong correlation with the alkaline HER performance.	ACS Catal. (2019), 9, 2415 [29]
		Nb	-2.41	-		
		Ta	-2.51	-		
		Mo	-4.6	-		
		Nb	-4.85	-		
		Ta	-4.6	-		
		Ti	-4.85	-		
		V	-4.96	-		
		W	-4.91	-		
		W	-4.19	-		
		Ru_{SA}@ $Ti_3C_2O_2$	-0.49	-		
		Ni	0.17	-		
Ni-modified Pt (111)	PBE	Pt	1.05	28.9	The strong adsorption of OH on Ni indicates a fast H_2O dissociation.	J. Colloid Interface Sci. (2023) 650, 1715 [49]
		Ni-Pt	-0.51	14		

* The authors calculate the adsorption energy of the OH radical at the active site. The overpotential values are reported based on the experimental results in the corresponding reference.

4. Discussion

The selection of the right descriptor for screening alkaline HER electrocatalysts is a long-standing debate, especially using the ΔG_{OH^*} descriptor. For example, Zhang et al. demonstrated that ΔG_{OH^*} exhibits a significantly weaker correlation with alkaline HER activity than the volcano-shaped relationship established with ΔG_{H^*} (Figure 5a), which implies that the adsorbed hydroxyl group does not directly participate in the rate-determining step of alkaline HER kinetics on these surfaces [29]. Contrastingly, McCrum et al. found that the performance of alkaline HER electrocatalysts shows a volcano relationship between the natural logarithm of the experimentally measured rate of hydrogen evolution and the DFT-calculated ΔG_{OH^*} [19]. The reaction is bifunctional as it involves both ΔG_{H^*} and ΔG_{OH^*} on the too-strong OH binding side of the volcano. On the too-weak OH binding side of the volcano, ΔG_{H^*} is only apparently bifunctional; ΔG_{OH^*} is not a useful descriptor for evaluating the performance of catalysts. The different conclusions of these two studies may be ascribed to the scopes of ΔG_{OH^*} values. All the ΔG_{OH^*} values from Zhang et al. shown in Figure 5a are less than -1.0 eV, which belong to the too-strong OH binding side suggested in Figure 5b. As a result, no volcanic relationship can be identified in Figure 5a. Since all ΔG_{OH^*} values are far from the optimal value of -0.3 eV, the performance of these metal-modified transition metal carbides is more affected by the ΔG_{H^*} value, as suggested by Zhang et al. [29].

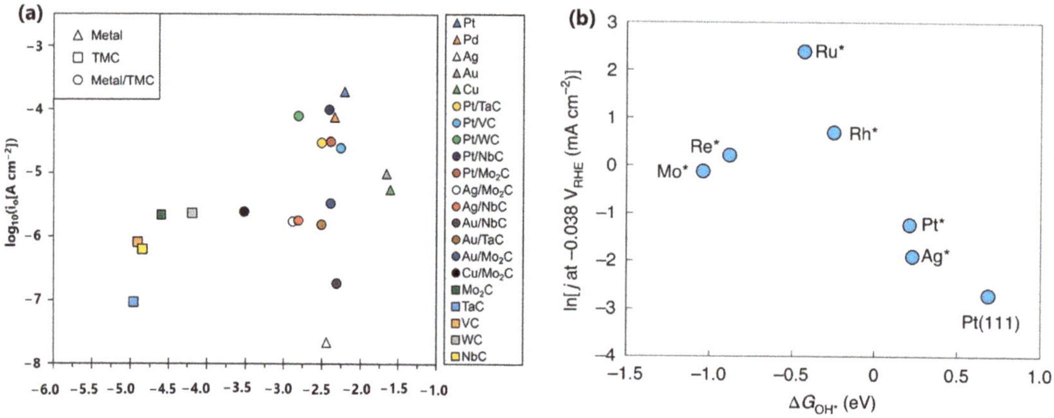

Figure 5. (**a**) The logarithm of the experimentally measured rate of hydrogen evolution of metal-modified transition metal carbides in 0.1 M KOH and the corresponding DFT-calculated ΔG_{OH^*}. Reprinted with permission from Ref. [29]. Copyright 2019 American Chemical Society. (**b**) The natural logarithm of the experimentally measured rate of hydrogen evolution of Pt (553), Pt(553) with Mo*, Re*, Ru*, Rh*, and Ag* adsorbed at the step, and Pt(111) in 0.1 M NaOH and the corresponding DFT-calculated ΔG_{OH^*}. Reprinted with permission from Ref. [19]. Copyright 2020 Springer Nature.

To better screen the alkaline HER electrocatalysts, McCrum et al. adopted a three-dimensional (3D) volcano plot depicting the rate of hydrogen evolution in relation to the ΔG_{OH^*} and ΔG_{H^*}, as illustrated in Figure 6a. Catalysts exhibiting minimal hydrogen and hydroxide binding energies yield the lowest rates, as indicated by the purple area on the upper right of Figure 6a. The catalysts show the highest performances when they bind hydrogen at an intermediate strength at 0 V_{RHE} (near 0 eV) and hydroxide strongly (near −0.3 eV) (yellow in the middle of Figure 6a). The too-strong H and OH binding leads the H_2 or OH* desorption to become the rate-determining step. Consequently, Figure 6a qualitatively represents trends in alkaline HER kinetics and provides design guidelines for efficient catalysts. For example, to improve the hydrogen evolution rate of the well-studied Pt(111) electrocatalysts in an alkaline solution, the ΔG_{H^*} must be increased by approximately 0.2 eV. In comparison, the ΔG_{OH^*} must be significantly reduced by approximately 0.9 eV through appropriate manipulation strategies such as doping, surface engineering, and defect engineering [19].

Following these guidelines, we employed both ΔG_{H^*} and ΔG_{OH^*} descriptors to investigate the catalytic HER performance of 1T′ transition metal dichalcogenides (TMDs such as $MoSe_2$, MoS_2, WSe_2, and WS_2) in an alkaline solution using DFT [77]. Our findings indicate that the pristine sulfides exhibited superior alkaline HER performance compared to their selenide counterparts. Nevertheless, the activities of all pristine 1T′ TMDs are insufficient to dissociate water (See Figure 6b). Defect engineering techniques were employed to improve the reactivity of TMD-based electrocatalysts. Our DFT results indicate that the reactivities of TMD materials can be enhanced by introducing single S/Se vacancy defects, as shown in Figure 6c. However, the rate-determining phase is the desorption of OH. The reactivities of active sites were further regulated to achieve optimal OH desorption by doping defective MoS_2 with late 3d transition metal (TM) atoms, particularly Cu, Ni, and Co, as illustrated in Figure 6d. Consequently, the TM-doped defective 1T′ MoS_2 can substantially improve the alkaline HER performance, which matches the recently reported experimental observations.

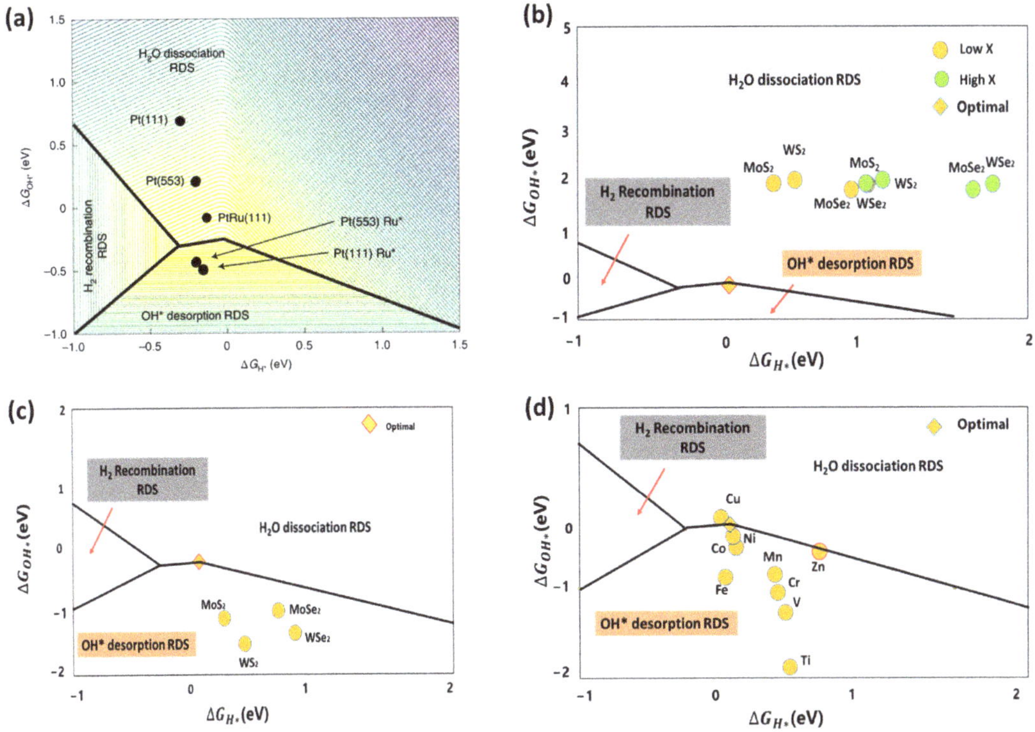

Figure 6. (**a**) Logarithm of the rate of hydrogen evolution (contours) as a function of DFT-calculated ΔG_{H^*} and ΔG_{OH^*} on Pt(111), Pt(553), Ru* adsorbed at the step of Pt(553), PtRu(111) alloy and Ru* clusters on Pt(111). Reprinted with permission from Ref. [19]. Copyright 2020 Springer Nature. (**b**) Plot of ΔG_{OH^*} against ΔG_{H^*} of pristine MX_2 TMDs (M = Mo, W and X = S, Se), (**c**) plot of ΔG_{OH^*} against ΔG_{H^*} of MX_2 TMDs with a single X vacancy; and (**d**) plot of ΔG_{OH^*} against ΔG_{H^*} of MX_2 TMDs with 3d TM dopant next to the single X vacancy. Reprinted with permission from Ref. [77]. Copyright 2024 IOPscience.

It is crucial to remember that each descriptor has limitations and challenges when studying alkaline HER in alkaline media. Descriptors such as water adsorption energy and water dissociation energy barriers serve as indicators of material reactivity. They are closely linked to the Volmer step, which encompasses the adsorption of water molecules and their subsequent dissociation into adsorbed hydrogen (H*) and hydroxide ions (OH^-). These steps are fundamental to the overall HER process, yet they do not provide a complete picture of the catalytic site's capabilities, especially concerning the Heyrovsky and Tafel steps. The Heyrovsky step involves the electrochemical desorption of H* to form hydrogen gas.

The ΔG_{H^*} has been the most prevalent descriptor for assessing HER performance at the Heyrovsky and Tafel steps. It offers a measure of the free energy change when H* is adsorbed on the catalyst surface, which is a critical factor in determining the rate of the HER. However, ΔG_{H^*} alone is insufficient for evaluating the Volmer step or the desorption efficiency of adsorbed OH, which are also essential for a complete understanding of the HER mechanism.

The Gibbs free energy of hydroxide adsorption (ΔG_{OH^*}) can be used to evaluate the desorption efficiency of adsorbed OH, providing insights into the potential for water dissociation, as suggested by the Bronsted-Evans-Polanyi (BEP) relationship [19]. This relationship posits a linear correlation between the activation energy of a reaction and the reaction enthalpy, allowing for the prediction of reaction barriers based on thermodynamic parameters. However, it does not offer information on the catalyst's performance during the Heyrovsky or Tafel steps. Therefore, relying solely on ΔG_{OH^*} would give an incomplete assessment of a catalyst's overall activity and efficiency.

Given these considerations, it is evident that no single descriptor can fully encapsulate the complexities of the HER process. Researchers must carefully select and combine multiple descriptors to gain a comprehensive understanding of the catalytic activity and to design more efficient catalysts. This approach allows for the evaluation of catalysts across all steps of the HER, ensuring a more accurate prediction of their performance in real-world applications.

It is also worth noting that different descriptors may be required to understand the performance of alkaline HER composite electrocatalysts. For example, Chen et al. designed a complicated composite electrocatalyst including a Co dopant on WO_2/amorphous Co_xW hybrid materials [51]. Using the water adsorption energy and dissociation energy barrier energy as the descriptor, they found that the amorphous CoxW is the active site. However, the analysis of the ΔG_{H^*} descriptor indicates that Co-doped WO_2 is the active site for H_2 formation. The DFT calculations then suggest that the individual component of this composite electrocatalyst has a different function. The synergy between them enables the high alkaline HER performance of this Co-Doped WO_2/amorphous Co_xW hybrid electrocatalyst.

5. Conclusions and Outlook

The hydrogen evolution reaction (HER) in alkaline media is a prominent method for large-scale hydrogen production from an electrolyzer. Alkaline Water Electrolysis (AWE) has demonstrated superior cost-effectiveness to acidic proton exchange membrane (PEM). One of the biggest challenges of AWE is ascribed to the relatively low energy conversion efficiency of their electrocatalysts. To address this issue, selecting the right computational tools for theoretical studies is essential in designing high-performance alkaline HER electrocatalysts. The descriptors used to evaluate the energy conversion efficiency of electrocatalysts include water adsorption energy, water dissociation barrier, Gibbs free energy change of hydrogen adsorption, and Gibbs free energy change of hydroxyl adsorption from DFT calculations. In the review, some of the latest examples have been used to illustrate the applications of different descriptors. It reveals that the hydroxyl (ΔG_{OH^*}) adsorption process is one of the important parameters often ignored in many previous theoretical studies. When the adsorption strength of hydroxyl is weak, it suggests a high water dissociation energy barrier due to the BEP relationship [19]. The water cannot effectively interact with the active site to provide an H atom intermediate. At the same time, the weak adsorption of OH also indicates that the dissolved hydroxide is via the Volmer (II) mechanism (Equation (6)), as shown in Figure 1. In this case, the ΔG_{H^*} becomes a more important descriptor to analyze and predict the efficiency of an active site. It explains the great matches between some experimental measurements and the trend of the ΔG_{H^*} descriptor in some combined studies.

The CHE model has evolved to include previously overlooked factors such as solvent and electrolyte effects, reflecting its adaptability [18]. Despite these advancements, a knowledge gap persists, particularly in the dynamic aspects of water splitting. A significant hurdle remains in developing a more accurate model of the electrode–electrolyte interface, which is crucial for pinpointing active sites in HER and connecting microscopic interactions with macroscopic observations. The electrified solid–liquid interface in alkaline HERs largely determines the charge transfer rate of electrochemical redox reactions [78]. Innovative approaches, such as engineering a localized acid-like environment within an alkaline medium, have significantly boosted HER performance significantly [79]. This is achieved by tailoring the local reaction conditions, which are critical for the process. Moreover, understanding the impact of variables like electrolyte concentration on HER allows for further refinement of reaction conditions [16]. The descriptors described in this review can only provide limited dynamic information about the electrochemical processes at the electrode–electrolyte interface [15,21,22]. This is because the atomistic models used in most of these studies make it difficult to comprehend the thermodynamic state and dynamic properties of interfacial processes [21]. The interaction between water molecules and the electrified surface must be investigated by examining scenarios in which water molecules are introduced to the catalyst surface in various configurations [80]. Insights into the hydrogen bonding network and structural characteristics of the water layer can be derived from the average dipole orientation of water molecules relative to the surface normal [80]. Furthermore, the complex relationship among adsorbed hydrogen, water molecules, and the electrochemical environment, which influences the behaviors of adsorption and the overall characteristics of the interface between solid and water, needs to be investigated [81]. To comprehend the activities of HER intermediates in alkaline media, it is critical to know the potential and pH-dependent adsorption energies of these intermediates [82]. To this end, it is clear that theoretical studies on alkaline HER are still an area ripe for further research.

The intersection of machine learning (ML) with density functional theory and multiscale modeling is a burgeoning field that holds great promise for advancing materials science and chemistry [83–86]. By leveraging ML's ability to analyze vast datasets and identify patterns, researchers can significantly reduce the computational resources required for DFT calculations. This synergy enables the prediction of complex chemical behaviors and the design of new materials with tailored properties. However, the success of the ML methods hinges on the quality and consistency of the data fed into the ML models [84]. As such, creating comprehensive and reliable databases is crucial for training algorithms that can accurately predict electrochemical behaviors and guide the development of efficient alkaline HER electrocatalysts. This integrated approach is set to transform the landscape of computational chemistry, offering a more streamlined and precise method for exploring the vast potential of chemical space. They can predict the performance of electrocatalysts, thereby informing the design of next-generation materials with enhanced catalytic properties. These concerted efforts in theoretical research are essential for the progression of clean energy technologies.

In sum, the exploration of electrocatalytic mechanisms in alkaline HER is a complex field that necessitates a comprehensive approach. Advancements beyond the DFT-calculated descriptors are critical for a deeper understanding. Operando simulations offer a dynamic perspective by considering actual working conditions, providing insights into the real-time structural and chemical changes during the reaction process. Meanwhile, ML-based force fields for classical molecular dynamics and Monte Carlo simulations represent a significant leap in mesoscale modeling, enabling simulations that capture the nuanced interactions within molecular systems. These ML models can bridge the gap between classical and quantum mechanical accuracy, offering a more detailed view of the catalytic processes. Lastly, ML-driven high-throughput screening is revolutionizing the way electrocatalysts are discovered and optimized. By analyzing vast datasets, ML algorithms can predict performance, stability, and efficiency, thereby accelerating the development of new materials for HER. Together, these methodologies form a multi-faceted approach that could significantly advance the field of electrocatalysis.

Author Contributions: Methodology, Formal analysis, Writing–original draft, S.A.O.; Writing—review & editing, F.M.; Writing—review & editing, Funding Request, A.B.; Writing—review & editing, J.J.H.; Writing—review & editing, L.Z.; Writing–review & editing, Z.W.; Writing—review & editing, S.B.; Writing—review & editing, Y.Z.; Project administration, Funding Request, Supervision, Conceptualization, Resources, Writing—review & editing, Y.W. All authors have read and agreed to the published version of the manuscript.

Funding: The authors acknowledge financial support from the Australian Research Council Discovery Project (Grant No. DP210103266, DP210104010). This research was conducted on the supercomputers in National Computational Infrastructure (NCI) in Canberra, Australia, which is supported by the Australian Commonwealth Government, and Pawsey Supercomputing Centre in Perth with the funding from the Australian government and the Government of Western Australia. Calculations were performed by using resources from Grand Equipement National de Calcul Intensif (GENCI, grant no. A0120913426). Computational resources provided by the computing facilities Mésocentre de Calcul Intensif Aquitain (MCIA) of the Université de Bordeaux and of the Université de Pau et des Pays de l'Adour. We acknowledge funding from CNRS Institute of Chemistry through the "International Emerging Actions 2022" mobility grant (2DH2 project) and from Agence national de recherche under the RECIFE ANR-DFG project (Grant Number ANR-21-CE08-0036-01).

Data Availability Statement: The data presented in this study are available on request from the corresponding author.

Conflicts of Interest: The authors declare no conflicts of interest.

References

1. de Kleijne, K.; Huijbregts, M.A.J.; Knobloch, F.; van Zelm, R.; Hilbers, J.P.; de Coninck, H.; Hanssen, S.V. Worldwide greenhouse gas emissions of green hydrogen production and transport. *Nat. Energy* **2024**, 1–14. [CrossRef]
2. Wei, C.; Xu, Z.J. *The Comprehensive Understanding of as an Evaluation Parameter for Electrochemical Water Splitting*; Wiley Online Library: Hoboken, NJ, USA, 2018; Volume 2, p. 1800168.
3. Horri, B.A.; Ozcan, H. Green hydrogen production by water electrolysis: Current status and challenges. *Curr. Opin. Green. Sust.* **2024**, *47*, 100932. [CrossRef]
4. Ghosh, S.; Basu, S. Solid Oxide Electrolysis Cell for Hydrogen Generation: General Perspective and Mechanism. In *Climate Action and Hydrogen Economy: Technologies Shaping the Energy Transition*; Springer: Berlin/Heidelberg, Germany, 2024; pp. 231–260.
5. Laguna-Bercero, M.A.; Wang, Y.; Zhou, X.-D.; Zhu, L. Fundamentals of Solid Oxide Electrolysis Cells (SOEC). In *High. Temperature Electrolysis*; Springer: Berlin/Heidelberg, Germany, 2023; pp. 5–34.
6. Tao, H.B.; Liu, H.; Lao, K.J.; Pan, Y.P.; Tao, Y.B.; Wen, L.R.; Zheng, N.F. The gap between academic research on proton exchange membrane water electrolysers and industrial demands. *Nat. Nanotechnol.* **2024**, *19*, 1074–1076. [CrossRef] [PubMed]
7. Yuan, S.; Zhao, C.F.; Li, H.Y.; Shen, S.Y.; Yan, X.H.; Zhang, J.L. Rational electrode design for low-cost proton exchange membrane water electrolyzers. *Cell Rep. Phys. Sci.* **2024**, *5*, 101880. [CrossRef]
8. Zheng, Y.; Jiao, Y.; Vasileff, A.; Qiao, S.Z. The Hydrogen Evolution Reaction in Alkaline Solution: From Theory, Single Crystal Models, to Practical Electrocatalysts. *Angew. Chem. Int. Edit.* **2018**, *57*, 7568–7579. [CrossRef]
9. Pitchai, C.; Vedanarayanan, M.; Chen, C.-M.; Gopalakrishnan, S.M. Enhanced electrochemical efficiency of the open porous sandrose structured electrocatalyst for sustainable hydrogen and oxygen evolution reactions. *Int. J. Hydrogen Energy* **2024**, *72*, 755–763. [CrossRef]
10. Pitchai, C.; Gopalakrishnan, S.M.; Chen, C.-M. Ultra-efficient Nitrogen-Doped Carbon Dots-Supported Nickel Sulfide as a Platinum-Free Electrocatalyst for Overall Water Splitting in Basic Medium. *Energy Fuels* **2024**, *38*, 2235–2247. [CrossRef]
11. Hu, C.L.; Zhang, L.; Gong, J.L. Recent progress made in the mechanism comprehension and design of electrocatalysts for alkaline water splitting. *Energy Environ. Sci.* **2019**, *12*, 2620–2645. [CrossRef]
12. Ehlers, J.C.; Feidenhans'l, A.A.; Therkildsen, K.T.; Larrazabal, G.O. Affordable Green Hydrogen from Alkaline Water Electrolysis: Key Research Needs from an Industrial Perspective. *ACS Energy Lett.* **2023**, *8*, 1502–1509. [CrossRef]
13. Du, N.Y.; Roy, C.; Peach, R.; Turnbull, M.; Thiele, S.; Bock, C. Anion-Exchange Membrane Water Electrolyzers. *Chem. Rev.* **2022**, *122*, 11830–11895. [CrossRef] [PubMed]
14. Li, J.; Jing, Z.; Bai, H.; Chen, Z.; Osman, A.I.; Farghali, M.; Rooney, D.W.; Yap, P.-S. Optimizing hydrogen production by alkaline water decomposition with transition metal-based electrocatalysts. *Environ. Chem. Lett.* **2023**, *21*, 2583–2617. [CrossRef]
15. Liu, J.; Wang, Y. Theoretical Identification and Understanding of Catalytic Active Sites for Water Splitting Reactions. *Catalysis* **2022**, *34*, 1–16. [CrossRef]
16. Wang, Y.; Liu, X.; Liu, J.X.; Al-Mamun, M.; Liew, A.W.C.; Yin, H.J.; Wen, W.; Zhong, Y.L.; Liu, P.R.; Zhao, H.J. Electrolyte Effect on Electrocatalytic Hydrogen Evolution Performance of One-Dimensional Cobalt-Dithiolene Metal-Organic Frameworks: A Theoretical Perspective. *ACS Appl. Energy Mater.* **2018**, *1*, 1688–1694. [CrossRef]
17. Liu, J.X.; Yin, H.J.; Liu, P.R.; Chen, S.; Yin, S.W.; Wang, W.L.; Zhao, H.J.; Wang, Y. Theoretical Understanding of Electrocatalytic Hydrogen Production Performance by Low-Dimensional Metal-Organic Frameworks on the Basis of Resonant Charge-Transfer Mechanisms. *J. Phys. Chem. Lett.* **2019**, *10*, 6955–6961. [CrossRef] [PubMed]
18. Skúlason, E.; Karlberg, G.S.; Rossmeisl, J.; Bligaard, T.; Greeley, J.; Jónsson, H.; Nørskov, J.K. Density functional theory calculations for the hydrogen evolution reaction in an electrochemical double layer on the Pt (111) electrode. *Phys. Chem. Chem. Phys.* **2007**, *9*, 3241–3250. [CrossRef]
19. McCrum, I.T.; Koper, M.T. The role of adsorbed hydroxide in hydrogen evolution reaction kinetics on modified platinum. *Nat. Energy* **2020**, *5*, 891–899. [CrossRef]
20. Tian, Y.; Zhang, S.; Wang, Y. Mathematical Modeling for Enhanced Electrochemical Properties. In *Metal-Air Batteries*; CRC Press: Boca Raton, FL, USA, 2023; pp. 45–63.
21. Wang, Y. *Multiscale Modeling of Electrochemical Reactions and Processes*; AIP Publishing LLC: Melville, NY, USA, 2021.
22. Hinsch, J.J.; Wang, Y. *DFT and Simulation of Solid-Liquid Interface Properties and Processes*; Elsevier: Amsterdam, The Netherlands, 2023.

23. Norskov, J.K.; Bligaard, T.; Rossmeisl, J.; Christensen, C.H. Towards the computational design of solid catalysts. *Nat. Chem.* **2009**, *1*, 37–46. [CrossRef]
24. Greeley, J.; Jaramillo, T.F.; Bonde, J.; Chorkendorff, I.; Nørskov, J.K. Computational high-throughput screening of electrocatalytic materials for hydrogen evolution. *Nat. Mater.* **2006**, *5*, 909–913. [CrossRef]
25. Zheng, J.; Nash, J.; Xu, B.; Yan, Y. Perspective—Towards establishing apparent hydrogen binding energy as the descriptor for hydrogen oxidation/evolution reactions. *J. Electrochem. Soc.* **2018**, *165*, H27. [CrossRef]
26. Fung, V.; Hu, G.; Wu, Z.; Jiang, D.-E. Descriptors for hydrogen evolution on single atom catalysts in nitrogen-doped graphene. *J. Phys. Chem. C* **2020**, *124*, 19571–19578. [CrossRef]
27. Mao, B.; Sun, P.; Jiang, Y.; Meng, T.; Guo, D.; Qin, J.; Cao, M. Identifying the transfer kinetics of adsorbed hydroxyl as a descriptor of alkaline hydrogen evolution reaction. *Angew. Chem.* **2020**, *132*, 15344–15349. [CrossRef]
28. Jia, Q.; Liang, W.; Bates, M.K.; Mani, P.; Lee, W.; Mukerjee, S. Activity descriptor identification for oxygen reduction on platinum-based bimetallic nanoparticles: In situ observation of the linear composition–strain–activity relationship. *ACS Nano* **2015**, *9*, 387–400. [CrossRef] [PubMed]
29. Zhang, Q.; Jiang, Z.; Tackett, B.M.; Denny, S.R.; Tian, B.; Chen, X.; Wang, B.; Chen, J.G. Trends and descriptors of metal-modified transition metal carbides for hydrogen evolution in alkaline electrolyte. *ACS Catal.* **2019**, *9*, 2415–2422. [CrossRef]
30. Zhang, B.; Wang, J.; Liu, J.; Zhang, L.; Wan, H.; Miao, L.; Jiang, J. Dual-descriptor tailoring: The hydroxyl adsorption energy-dependent hydrogen evolution kinetics of high-valance state doped Ni3N in alkaline media. *ACS Catal.* **2019**, *9*, 9332–9338. [CrossRef]
31. Skúlason, E.; Tripkovic, V.; Björketun, M.E.; Gudmundsdóttir, S.; Karlberg, G.; Rossmeisl, J.; Bligaard, T.; Jónsson, H.; Nørskov, J.K. Modeling the electrochemical hydrogen oxidation and evolution reactions on the basis of density functional theory calculations. *J. Phys. Chem. C* **2010**, *114*, 18182–18197. [CrossRef]
32. Nørskov, J.K.; Bligaard, T.; Logadottir, A.; Kitchin, J.; Chen, J.G.; Pandelov, S.; Stimming, U. Trends in the exchange current for hydrogen evolution. *J. Electrochem. Soc.* **2005**, *152*, J23. [CrossRef]
33. Ostergaard, F.C.; Bagger, A.; Rossmeisl, J. Predicting catalytic activity in hydrogen evolution reaction. *Curr. Opin. Electrochem.* **2022**, *35*, 101037. [CrossRef]
34. Norskov, J.K.; Abild-Pedersen, F.; Studt, F.; Bligaard, T. Density functional theory in surface chemistry and catalysis. *Proc. Natl. Acad. Sci. USA* **2011**, *108*, 937–943. [CrossRef]
35. Ďurovič, M.; Hnát, J.; Bouzek, K. Electrocatalysts for the hydrogen evolution reaction in alkaline and neutral media. A comparative review. *J. Power Sources* **2021**, *493*, 229708. [CrossRef]
36. Ledezma-Yanez, I.; Wallace, W.D.Z.; Sebastian-Pascual, P.; Climent, V.; Feliu, J.M.; Koper, M.T.M. Interfacial water reorganization as a pH-dependent descriptor of the hydrogen evolution rate on platinum electrodes. *Nat. Energy* **2017**, *2*, 1–7. [CrossRef]
37. Zhang, L.; Gao, X.; Zhu, Y.; Liu, A.; Dong, H.; Wu, D.; Han, Z.; Wang, W.; Fang, Y.; Zhang, J. Electrocatalytically inactive copper improves the water adsorption/dissociation on Ni 3 S 2 for accelerated alkaline and neutral hydrogen evolution. *Nanoscale* **2021**, *13*, 2456–2464. [CrossRef]
38. Guha, A.; Sahoo, M.; Alam, K.; Rao, D.K.; Sen, P.; Narayanan, T.N. Role of water structure in alkaline water electrolysis. *Iscience* **2022**, *25*, 104835. [CrossRef]
39. Zheng, X.; Yao, Y.; Ye, W.; Gao, P.; Liu, Y. Building up bimetallic active sites for electrocatalyzing hydrogen evolution reaction under acidic and alkaline conditions. *Chem. Eng. J.* **2021**, *413*, 128027. [CrossRef]
40. Henkelman, G.; Uberuaga, B.P.; Jónsson, H. A climbing image nudged elastic band method for finding saddle points and minimum energy paths. *J. Chem. Phys.* **2000**, *113*, 9901–9904. [CrossRef]
41. He, R.; Hua, J.; Zhang, A.; Wang, C.; Peng, J.; Chen, W.; Zeng, J. Molybdenum disulfide–black phosphorus hybrid nanosheets as a superior catalyst for electrochemical hydrogen evolution. *Nano Lett.* **2017**, *17*, 4311–4316. [CrossRef]
42. Lin, L.; Zhou, W.; Gao, R.; Yao, S.; Zhang, X.; Xu, W.; Zheng, S.; Jiang, Z.; Yu, Q.; Li, Y.-W. Low-temperature hydrogen production from water and methanol using Pt/α-MoC catalysts. *Nature* **2017**, *544*, 80–83. [CrossRef]
43. Cuenya, B.R.; Behafarid, F. Nanocatalysis: Size-and shape-dependent chemisorption and catalytic reactivity. *Surf. Sci. Rep.* **2015**, *70*, 135–187. [CrossRef]
44. Tian, X.; Zhao, P.; Sheng, W. Hydrogen evolution and oxidation: Mechanistic studies and material advances. *Adv. Mater.* **2019**, *31*, 1808066. [CrossRef]
45. Zhao, L.; Zhang, Y.; Zhao, Z.; Zhang, Q.-H.; Huang, L.-B.; Gu, L.; Lu, G.; Hu, J.-S.; Wan, L.-J. Steering elementary steps towards efficient alkaline hydrogen evolution via size-dependent Ni/NiO nanoscale heterosurfaces. *Natl. Sci. Rev.* **2020**, *7*, 27–36. [CrossRef]
46. Norskov, J.K.; Rossmeisl, J.; Logadottir, A.; Lindqvist, L.; Kitchin, J.R.; Bligaard, T.; Jonsson, H. Origin of the overpotential for oxygen reduction at a fuel-cell cathode. *J. Phys. Chem. B* **2004**, *108*, 17886–17892. [CrossRef]
47. Danilovic, N.; Subbaraman, R.; Strmcnik, D.; Chang, K.C.; Paulikas, A.; Stamenkovic, V.; Markovic, N.M. Enhancing the alkaline hydrogen evolution reaction activity through the bifunctionality of $Ni(OH)_2$/metal catalysts. *Angew. Chem.* **2012**, *124*, 12663–12666. [CrossRef]
48. Subbaraman, R.; Tripkovic, D.; Strmcnik, D.; Chang, K.-C.; Uchimura, M.; Paulikas, A.P.; Stamenkovic, V.; Markovic, N.M. Enhancing hydrogen evolution activity in water splitting by tailoring Li+-$Ni(OH)_2$-Pt interfaces. *Science* **2011**, *334*, 1256–1260. [CrossRef] [PubMed]

49. Chen, L.; Kang, L.; Cai, D.; Geng, S.; Liu, Y.; Chen, J.; Song, S.; Wang, Y. Ultrafine Pt-based catalyst decorated with oxygenophilic Ni-sites accelerating alkaline H2O dissociation for efficient hydrogen evolution. *J. Colloid Interface Sci.* **2023**, *650*, 1715–1724. [CrossRef]
50. Yao, N.; Li, P.; Zhou, Z.; Zhao, Y.; Cheng, G.; Chen, S.; Luo, W. Synergistically tuning water and hydrogen binding abilities over Co_4N by Cr doping for exceptional alkaline hydrogen evolution electrocatalysis. *Adv. Energy Mater.* **2019**, *9*, 1902449. [CrossRef]
51. Chen, J.; Jin, Q.; Li, Y.; Li, Y.; Cui, H.; Wang, C. Design superior alkaline hydrogen evolution electrocatalyst by engineering dual active sites for water dissociation and hydrogen desorption. *ACS Appl. Mater. Interfaces* **2019**, *11*, 38771–38778. [CrossRef]
52. Kim, J.; Jung, H.; Jung, S.-M.; Hwang, J.; Kim, D.Y.; Lee, N.; Kim, K.-S.; Kwon, H.; Kim, Y.-T.; Han, J.W. Tailoring binding abilities by incorporating oxophilic transition metals on 3D nanostructured Ni arrays for accelerated alkaline hydrogen evolution reaction. *J. Am. Chem. Soc.* **2020**, *143*, 1399–1408. [CrossRef]
53. Maslovara, S.; Aničijević, D.V.; Brković, S.; Georgijević, J.; Tasić, G.; Kaninski, M.M. Experimental and DFT study of CoCuMo ternary ionic activator for alkaline HER on Ni cathode. *J. Electroanal. Chem.* **2019**, *839*, 224–230. [CrossRef]
54. You, B.; Liu, X.; Hu, G.; Gul, S.; Yano, J.; Jiang, D.-E.; Sun, Y. Universal surface engineering of transition metals for superior electrocatalytic hydrogen evolution in neutral water. *J. Am. Chem. Soc.* **2017**, *139*, 12283–12290. [CrossRef]
55. Wang, K.; Zhou, J.; Sun, M.; Lin, F.; Huang, B.; Lv, F.; Zeng, L.; Zhang, Q.; Gu, L.; Luo, M. Cu-Doped Heterointerfaced Ru/RuSe2 Nanosheets with Optimized H and H2O Adsorption Boost Hydrogen Evolution Catalysis. *Adv. Mater.* **2023**, *35*, 2300980. [CrossRef]
56. Zou, Y.; Kazemi, S.A.; Shi, G.; Liu, J.; Yang, Y.; Bedford, N.M.; Fan, K.; Xu, Y.; Fu, H.; Dong, M. Ruthenium single-atom modulated Ti3C2Tx MXene for efficient alkaline electrocatalytic hydrogen production. *EcoMat* **2023**, *5*, e12274. [CrossRef]
57. Chen, Z.; Gong, W.; Wang, J.; Hou, S.; Yang, G.; Zhu, C.; Fan, X.; Li, Y.; Gao, R.; Cui, Y. Metallic W/WO2 solid-acid catalyst boosts hydrogen evolution reaction in alkaline electrolyte. *Nat. Commun.* **2023**, *14*, 5363. [CrossRef]
58. Wang, J.; Zhang, Z.; Song, H.; Zhang, B.; Liu, J.; Shai, X.; Miao, L. Water dissociation kinetic-oriented design of nickel sulfides via tailored dual sites for efficient alkaline hydrogen evolution. *Adv. Funct. Mater.* **2021**, *31*, 2008578. [CrossRef]
59. Xu, K.; Ding, H.; Zhang, M.; Chen, M.; Hao, Z.; Zhang, L.; Wu, C.; Xie, Y. Regulating water-reduction kinetics in cobalt phosphide for enhancing HER catalytic activity in alkaline solution. *Adv. Mater.* **2017**, *29*, 1606980. [CrossRef]
60. Wang, J.; Xin, S.; Xiao, Y.; Zhang, Z.; Li, Z.; Zhang, W.; Li, C.; Bao, R.; Peng, J.; Yi, J. Manipulating the Water Dissociation Electrocatalytic Sites of Bimetallic Nickel-Based Alloys for Highly Efficient Alkaline Hydrogen Evolution. *Angew. Chem. Int. Ed.* **2022**, *61*, e202202518. [CrossRef]
61. Kumar, A.; Bui, V.Q.; Lee, J.; Jadhav, A.R.; Hwang, Y.; Kim, M.G.; Kawazoe, Y.; Lee, H. Modulating interfacial charge density of NiP2–FeP2 via coupling with metallic Cu for accelerating alkaline hydrogen evolution. *ACS Energy Lett.* **2021**, *6*, 354–363. [CrossRef]
62. Li, H.; Han, Y.; Zhao, H.; Qi, W.; Zhang, D.; Yu, Y.; Cai, W.; Li, S.; Lai, J.; Huang, B. Fast site-to-site electron transfer of high-entropy alloy nanocatalyst driving redox electrocatalysis. *Nat. Commun.* **2020**, *11*, 5437. [CrossRef] [PubMed]
63. Wang, Z.; Tang, M.T.; Cao, A.; Chan, K.; Nørskov, J.K. Insights into the hydrogen evolution reaction on 2D transition-metal dichalcogenides. *J. Phys. Chem. C* **2022**, *126*, 5151–5158. [CrossRef]
64. Laursen, A.B.; Varela, A.S.; Dionigi, F.; Fanchiu, H.; Miller, C.; Trinhammer, O.L.; Rossmeisl, J.; Dahl, S. Electrochemical hydrogen evolution: Sabatier's principle and the volcano plot. *J. Chem. Educ.* **2012**, *89*, 1595–1599. [CrossRef]
65. Laursen, A.B.; Wexler, R.B.; Whitaker, M.J.; Izett, E.J.; Calvinho, K.U.; Hwang, S.; Rucker, R.; Wang, H.; Li, J.; Garfunkel, E. Climbing the volcano of electrocatalytic activity while avoiding catalyst corrosion: Ni3P, a hydrogen evolution electrocatalyst stable in both acid and alkali. *ACS Catal.* **2018**, *8*, 4408–4419. [CrossRef]
66. Durst, J.; Siebel, A.; Simon, C.; Hasché, F.; Herranz, J.; Gasteiger, H. New insights into the electrochemical hydrogen oxidation and evolution reaction mechanism. *Energy Environ. Sci.* **2014**, *7*, 2255–2260. [CrossRef]
67. He, Q.; Tian, D.; Jiang, H.; Cao, D.; Wei, S.; Liu, D.; Song, P.; Lin, Y.; Song, L. Achieving efficient alkaline hydrogen evolution reaction over a Ni5P4 catalyst incorporating single-atomic Ru sites. *Adv. Mater.* **2020**, *32*, 1906972. [CrossRef] [PubMed]
68. Guo, F.; Zhang, Y.; Shen, B.; Wan, L.; Hou, G.; Cao, H.; Zheng, G.; Zhao, Y.; Zhang, H. Multi-site synergistic hydrogen evolution reactions on porous homogeneous FeCoNiCu high-entropy alloys fabricated by solution combustion synthesis and hydrogen reduction. *J. Alloys Compd.* **2024**, *1002*, 175356. [CrossRef]
69. Markovic, N. *The Hydrogen Electrode Reaction and the Electrooxidation of CO and H_2/CO Mixtures on Well-Characterized Pt and Pt-Bimetallic Surfaces*; Lawrence Berkeley National Laboratory: Berkeley, CA, USA, 2001.
70. Zhao, Y.; Wu, D.; Luo, W. Correlating alkaline hydrogen electrocatalysis and hydroxide binding energies on Mo-modified Ru catalysts. *ACS Sustain. Chem. Eng.* **2022**, *10*, 1616–1623. [CrossRef]
71. Sheng, W.; Bivens, A.P.; Myint, M.; Zhuang, Z.; Forest, R.V.; Fang, Q.; Chen, J.G.; Yan, Y. Non-precious metal electrocatalysts with high activity for hydrogen oxidation reaction in alkaline electrolytes. *Energy Environ. Sci.* **2014**, *7*, 1719–1724. [CrossRef]
72. Sheng, W.; Myint, M.; Chen, J.G.; Yan, Y. Correlating the hydrogen evolution reaction activity in alkaline electrolytes with the hydrogen binding energy on monometallic surfaces. *Energy Environ. Sci.* **2013**, *6*, 1509–1512. [CrossRef]
73. Gong, M.; Wang, D.-Y.; Chen, C.-C.; Hwang, B.-J.; Dai, H. A mini review on nickel-based electrocatalysts for alkaline hydrogen evolution reaction. *Nano Res.* **2016**, *9*, 28–46. [CrossRef]
74. De, S.; Zhang, J.; Luque, R.; Yan, N. Ni-based bimetallic heterogeneous catalysts for energy and environmental applications. *Energy Environ. Sci.* **2016**, *9*, 3314–3347. [CrossRef]

75. Zhang, J.; Wang, T.; Liu, P.; Liu, S.; Dong, R.; Zhuang, X.; Chen, M.; Feng, X. Engineering water dissociation sites in MoS_2 nanosheets for accelerated electrocatalytic hydrogen production. *Energy Environ. Sci.* **2016**, *9*, 2789–2793. [CrossRef]
76. Wang, Y.; Chen, L.; Yu, X.; Wang, Y.; Zheng, G. Superb alkaline hydrogen evolution and simultaneous electricity generation by pt-decorated Ni3N nanosheets. *Adv. Energy Mater.* **2017**, *7*, 1601390. [CrossRef]
77. Ogunkunle, S.A.; Bouzid, A.; Hinsch, J.J.; Allen, O.J.; White, J.J.; Bernard, S.; Wu, Z.; Zhu, Y.; Wang, Y. Defect engineering of 1T′MX 2 (M = Mo, W and X = S, Se) transition metal dichalcogenide-based electrocatalyst for alkaline hydrogen evolution reaction. *J. Phys. Condens. Matter* **2024**, *36*, 145002. [CrossRef] [PubMed]
78. Wang, Y. *Numerical Simulation of Electrified Solid–Liquid Interfaces*; AIP Publishing LLC: Melville, NY, USA, 2021; pp. 3-1–3-18.
79. Tan, H.; Tang, B.; Lu, Y.; Ji, Q.Q.; Lv, L.Y.; Duan, H.L.; Li, N.; Wang, Y.; Feng, S.H.; Li, Z.; et al. Engineering a local acid-like environment in alkaline medium for efficient hydrogen evolution reaction. *Nat. Commun.* **2022**, *13*, 2024. [CrossRef]
80. Hinsch, J.J.; Bouzid, A.; Barker, J.C.; White, J.J.; Mortier, F.; Zhao, H.; Wang, Y. Revisiting the Electrified Pt (111)/Water Interfaces through an Affordable Double-Reference Ab Initio Approach. *J. Phys. Chem. C* **2023**, *127*, 19857–19866. [CrossRef]
81. Eslamibidgoli, M.J.; Eikerling, M.H. Approaching the self-consistency challenge of electrocatalysis with theory and computation. *Curr. Opin. Electrochem.* **2018**, *9*, 189–197. [CrossRef]
82. Duan, Z.; Henkelman, G. Theoretical resolution of the exceptional oxygen reduction activity of Au (100) in alkaline media. *ACS Catal.* **2019**, *9*, 5567–5573. [CrossRef]
83. Liu, T.; Zhao, X.; Liu, X.; Xiao, W.; Luo, Z.; Wang, W.; Zhang, Y.; Liu, J.-C. Understanding the hydrogen evolution reaction activity of doped single-atom catalysts on two-dimensional GaPS4 by DFT and machine learning. *J. Energy Chem.* **2023**, *81*, 93–100. [CrossRef]
84. Botu, V.; Batra, R.; Chapman, J.; Ramprasad, R. Machine Learning Force Fields: Construction, Validation, and Outlook. *J. Phys. Chem. C* **2016**, *121*, 511–522. [CrossRef]
85. Raccuglia, P.; Elbert, K.C.; Adler, P.D.F.; Falk, C.; Wenny, M.B.; Mollo, A.; Zeller, M.; Friedler, S.A.; Schrier, J.; Norquist, A.J. Machine-learning-assisted materials discovery using failed experiments. *Nature* **2016**, *533*, 73–76. [CrossRef]
86. Ooka, H.; Wintzer, M.E.; Nakamura, R. Non-Zero Binding Enhances Kinetics of Catalysis: Machine Learning Analysis on the Experimental Hydrogen Binding Energy of Platinum. *ACS Catal.* **2021**, *11*, 6298–6303. [CrossRef]

catalysts

MDPI

Review

Design and Synthesis of Self-Supported Water-Splitting Transition Metal-Based Electrocatalysts via Electrospinning

Sai Che *,†, Yu Jia † and Yongfeng Li *

State Key Laboratory of Heavy Oil Processing, China University of Petroleum, Beijing 102249, China; jywsrycup@163.com
* Correspondence: sche@cup.edu.cn (S.C.); yfli@cup.edu.cn (Y.L.)
† These authors contributed equally to this work.

Abstract: Recent advances in transition metal-based electrocatalysts have significantly enhanced the hydrogen evolution reaction (HER) and oxygen evolution reaction (OER) in water electrocatalysis. Self-supported electrodes, where active sites are directly integrated with substrates, offer superior kinetics and stability compared to traditional powder-based electrocatalysts. The electrospinning technique is particularly effective for fabricating self-supported electrocatalysts with high surface areas, porosity, and uniform distribution of active sites, leading to improved catalytic performance. Despite extensive research on self-supported electrocatalysts, a comprehensive review focusing on those developed via electrospinning remains scarce. This review provides a detailed overview of the electrospinning process, the fundamental principles of water electrocatalysis, and recent progress in the development of transition metal-based electrocatalysts fabricated through this approach.

Keywords: electrospinning; transition metal catalysts; self-supported electrodes; electrocatalysis; water splitting

1. Introduction

The global shift toward sustainable energy solutions has highlighted the importance of hydrogen as a clean and efficient energy carrier [1–4]. Hydrogen, with its abundant feedstock [4], high energy density [5], and environmental benefits [6], holds significant potential for addressing the growing demand for renewable energy [7]. Among the various methods for hydrogen production, water electrolysis [8] stands out as a sustainable approach using water as a feedstock to generate hydrogen and oxygen [9,10]. This process involves two key reactions: the cathodic hydrogen evolution reaction (HER) and the anodic oxygen evolution reaction (OER). However, both reactions are highly energy-intensive, requiring operating voltages of 1.8–2.0 V, which is significantly higher than the theoretical 1.23 V [3]. To enhance the efficiency of water electrolysis, highly active electrocatalysts are required to improve the sluggish reaction kinetics and reduce overpotential [11]. While noble metal-based catalysts demonstrate outstanding performance, their high cost and limited availability restrict widespread use [12]. As a result, there is a growing need to develop low-cost, highly efficient catalysts for water electrolysis.

The current research on water-splitting electrocatalysts has focused largely on transition metal compounds, including sulfides [13], selenides [14,15], phosphides [16,17], oxides, nitrides [8,18], carbides [19], etc. However, the performance of these noble metal-free catalysts in terms of activity and durability remains below expectations. The effectiveness of electrocatalysts depends on several key factors: the intrinsic properties of active sites [20], the number of active sites [21], the mass transfer efficiency [22], and the electron transfer ability [23]. Various strategies have been proposed to address these factors. Among them, the design of nanostructures appears as one of the most widely explored approaches [24]. However, nanostructured electrocatalysts have some limitations for practical applications. For instance, nanoparticles are typically cast on a substrate using polymer binders, which increase electrical resistance, block active sites, and hamper mass transport. Additionally, the low conductivity of many electrocatalysts necessitates the use of conductive additives, which not only add complexity to the preparation process but also reduce the loading of active components. These drawbacks increase production costs and can result in poor mechanical stability due to the potential detachment of catalyst particles.

In light of these challenges, self-supported electrodes, where catalytically active phases are grown directly on conductive substrates, offer a promising alternative [25]. These electrodes offer several key advantages over conventional powder-based catalysts: (1) simplification of the electrode preparation process [26], (2) increased loading of active sites [27], (3) greater exposure of active sites by eliminating the involvement of binders [28], (4) improved charge transfer through uniform distribution of the catalyst on the substrate, (5) the ability to engineer hydrophilic or hydrophobic surfaces by adjusting the morphology and microstructure [29], and (6) enhanced stability through the immobilization of catalysts on the substrate, leading to superior electrocatalytic performance.

The methods for preparing self-supported electrocatalytic materials generally fall into two approaches: top-down and bottom-up. The top-down strategy involves dispersing active substances in solvents through mechanical or chemical processes, followed by deposition on the substrate to form thin membrane materials [30–32]. In contrast, bottom-up methods focus on building membrane materials through techniques like chemical vapor deposition, filtration, or electrostatic spinning [33]. Among these, electrostatic spinning has gained attention as a versatile and efficient technique to construct adjustable chemical components and well-defined micro/nanostructures [34].

Despite significant advances in the development of self-supported electrocatalysts, there is still a lack of comprehensive reviews focusing on those created through electrospinning. Therefore, we exclusively focus on electrospun self-supported electrocatalysts, a specialized domain that has hitherto received scant comprehensive examination. In this paper, we initiate our exploration by delving into the foundational principles of electrospinning, followed by furnishing a comprehensive elucidation of the principles governing electrolytic water splitting. Subsequently, we undertake a meticulous review of the recent advancements in the structural design and electrocatalytic properties of transition metal-based self-supported electrocatalysts fabricated via electrospinning. Finally, we delineate prospective trajectories for enhancing the efficiency and performance of these materials, proffering incisive insights for the design and development of next-generation electrocatalysts (Figure 1). This thorough investigation endeavors to bridge extant knowledge gaps and galvanize further innovation within the realm of sustainable energy electrocatalysis.

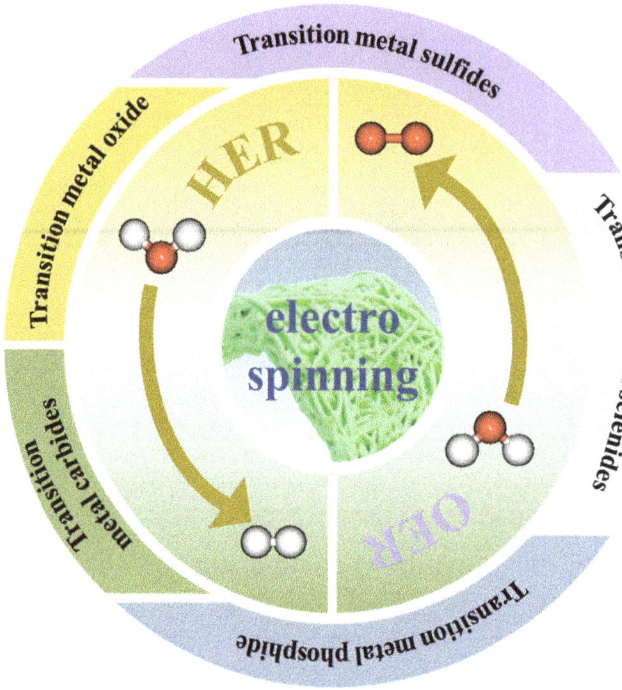

Figure 1. Schematic diagram showing the classification of transition metal-based water-splitting electrocatalysts with the employment of electrospinning technique.

2. Basics of Electrostatic Spinning

Electrospinning is derived from the electrospray technique, where high-voltage electrostatic forces are applied to propel liquid jets [35]. The equipment used for electrospinning consists of three main components: a high-voltage source, a nozzle, and a collector. The high-voltage source, which can be either Direct Current or Alternating Current, generates tens of kilovolts between the spinneret and the collector [36]. This creates a strong electrostatic field that counteracts the surface tension of the polymer solution at the nozzle, forming the Taylor cone [37]. When the electrostatic force surpasses the threshold, a jet with charge is propelled from the tip of the Taylor cone [38]. As the solution jet travels, the solvent evaporates, leaving behind charged polymer fibers that are stretched and thinned due to the movement of the jet. These fibers accumulate on a grounded collector as randomly oriented microfibers or nanofiber webs [39,40]. The final morphology of the fibers is influenced by several factors, including the properties of the polymer solution (such as molecular weight, viscosity, surface tension, and conductivity) and the processing parameters (such as voltage, distance between the nozzle and collector, and flow rate) [41,42].

With the increasing demand for one-dimensional nanomaterials, electrospinning has emerged as one of the fastest-growing and most versatile methods for nanofiber production. A wide range of nanofibers can be synthesized through this technique, including carbon-based fibers, metal oxides, polymers, and hybrid materials [43].

A variety of polymers are readily soluble in the corresponding organic solvents and can be spun into micro-nanofibers via electrospinning. Polymer precursors commonly involved in this process include the following: 1. natural components (i.e., hyaluronic acid (HA), chitosan, alginate, gelatin, lignin, collagen, cellulose, and chitin [44]); and 2.

synthetic polymers [45–49] (i.e., polyacrylonitrile (PAN), polymethylmethacrylate (PMMA), polyethylene oxide (PEO), polyvinyl alcohol (PVA), polystyrene (PS), polyvinylidene fluoride (PVDF), polyvinylpyrrolidone (PVP), poly(lactic acid) (PLA), polyvinyl chloride (PVC), and polycaprolactone (PCL)). In addition, inorganic nanofibers, including various metallic compounds, carbon, and their composites, are often produced by electrospinning with subsequent post-processing techniques such as air annealing or oxygen-free carbonization at elevated temperatures. These processes enable the creation of high-performance nanofibers with enhanced structural integrity. Nanofibers can serve not only as functional electrocatalysts but also as support matrices for other active substances [50]. In many cases, the performance of these materials can be further enhanced by combining electrospinning with various post-processing techniques, such as carbonization, sulfuration, selenization, phosphorization, self-assembly, chemical vapor deposition, chemical etching, and solvothermal growth [51]. These combined methods are crucial for creating high-performance composites with optimized electrocatalytic properties.

In summary, electrospinning is a versatile method for producing one-dimensional nanofibers from various materials, offering precise control over fiber morphology for applications like electrocatalysis. The technique allows the creation of high-performance fibers, especially when combined with post-processing methods such as carbonization. However, it requires the careful optimization of parameters like solution properties and voltage, which can be complex and time-consuming. Challenges include achieving uniform fiber morphology and reproducibility, particularly with hybrid materials. Despite these issues, ongoing advancements in process control and material design promise scalable production of high-performance nanofibers, with potential for innovations in energy storage and conversion technologies.

3. Principle of Electrocatalytic Water Splitting and Evaluation Criteria

3.1. HER Mechanism

HER is the cathodic half-reaction of water electrocatalysis. It is a multi-step electrochemical process that occurs at the electrode/electrolyte interface following two main reaction pathways. The first step is the Volmer reaction, which is then followed by either the Tafel or Heyrovsky process [52]. The mechanism of hydrogen production differs depending on the pH of the medium (Figure 2). In acidic media, hydrogen is primarily generated by the reduction of protons (H^+), while in alkaline media, hydrogen is produced through the reduction of water (H_2O) [53]. In addition, the HER process in neutral solutions closely resembles the mechanism in alkaline media. At each step, an electrocatalyst is essential to lower the energy barrier, accelerating the reaction and enhancing the efficiency of hydrogen generation [54].

(1) Acidic media

In acidic media, solvated protons serve as the main reactants in HER. The process consists of three fundamental steps [55,56].

Volmer

$$H^+ + e^- + * = H_{ad} \quad (1)$$

Tafel

$$H_{ad} + H_{ad} = H_2 + 2* \quad (2)$$

Heyrovsky

$$H_{ad} + H^+ + e^- = H_2 + * \quad (3)$$

The first step of the reaction is the Volmer reaction (Equation (1)), an electrochemical adsorption process where hydrated protons (H^+) combine with electrons at the surface of active sites, resulting in the adsorption of hydrogen atoms (H_{ad}). Next, depending on the concentration of these adsorbed atoms, hydrogen is either released through the Tafel pathway (Equation (2)) or the Heyrovsky pathway (Equation (3)) [57]. If the coverage of H_{ad} is high, neighboring hydrogen atoms directly couple to form hydrogen gas (H_2) in a process known as chemical desorption (Tafel reaction). Conversely, when the coverage of H_{ad} is low, the adsorbed hydrogen atoms combine with additional H^+ and e^-, producing H_2 through an electrochemical desorption process (Heyrovsky reaction) [57].

(2) Alkaline media

Compared to acidic media, the HER process in alkaline solutions is kinetically slower. This is due to the additional energy required to break the covalent bonds in H_2O for adsorption, a process more challenging than proton adsorption in acidic environments [54]. In alkaline media, the HER follows two key steps [3,54]:

Volmer

$$H_2O + e^- + * = H_{ad} + OH^- \tag{4}$$

Heyrovsky

$$H_{ad} + H_2O + e^- = H_2 + OH^- + * \tag{5}$$

In the first Volmer reaction, water molecules undergo adsorption and dissociation at the surface of the active sites (Equation (4)), resulting in the formation of H_{ad} and hydroxide ions (OH^-). This is considered the rate-determining step (RDS) in alkaline HER. In the Heyrovsky step, H_{ad} reacts with water, gaining an electron to produce H_2 and additional OH^- (Equation (5)). While the overall HER process in alkaline solutions is similar to that in acidic media, the need to break H_2O bonds leads to slower kinetics, which makes the reaction more difficult. Despite this, alkaline conditions are less corrosive, which can enhance the long-term stability of the electrocatalysts compared to their performance in acidic environments [58].

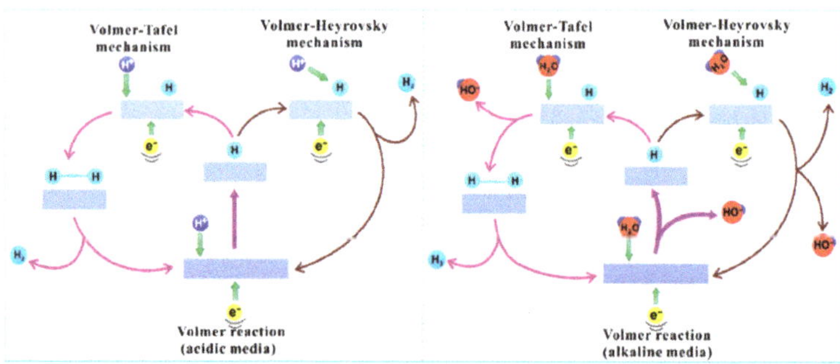

Figure 2. Mechanism of hydrogen evolution on the surface of an electrode in acidic (**left**) and alkaline (**right**) solutions [59]. Reproduced with permission. Copyright 2019 American Chemical Society.

3.2. OER Mechanism

Compared to HER at the cathode, anodic OER is more complex with multiple intermediates involving a four-electron transfer process [60]. The slower reaction kinetics and intricate mechanisms make OER more challenging than HER [61]. Similarly to the HER process, the OER mechanism depends heavily on the surface structure of the electrocatalyst and the pH of the electrolyte. Throughout the OER process, several intermediates can form, such as absorbed hydroxyl groups ($M - OH$), adsorbed oxygen ($M - O$), and adsorbed perovskite species ($M - OOH$). Oxygen (O_2) is eventually produced either by the direct combination of two $M - OH$ intermediates or through the decomposition of $M - OOH$ in acidic or alkaline media [62]. The OER mechanism in acidic and alkaline media is described as follows [3,55]:

(1) Acidic media:

$$H_2O + M = M - OH + H^+ + e^- \tag{6}$$

$$M - OH = M - O + H^+ + e^- \tag{7}$$

$$M - O + H_2O = M - OOH + H^+ + e^- \tag{8}$$

$$M - OOH = M + O_2 + H^+ + e^- \tag{9}$$

(2) Alkaline media:

$$M + OH^- = M - OH + e^- \tag{10}$$

$$M - OH + OH^- = M - O + H_2O + e^- \tag{11}$$

$$M - OOH^- = M - OOH + e^- \tag{12}$$

$$M - OOH + OH^- = M + O_2 + H_2O + e^- \tag{13}$$

Due to the involvement of multiple intermediates, it can be difficult to pinpoint the rate-determining step for the OER and fully describe its kinetics. Understanding the interactions between the electrocatalyst surface and these intermediates is crucial for optimizing the overall OER performance [53,63]. In this context, density functional theory (DFT) calculations are often used to analyze the adsorption energies of intermediates and to model the reaction mechanism of OER electrocatalysts. To date, plenty of work has been performed in modeling the thermodynamics of electrochemical reactions [64]. Among electrocatalysts, Ru/Ir-based oxides have proven to be the most effective in acidic media. However, most non-precious metal catalysts are not stable in acidic environments, particularly at high current densities, due to their susceptibility to dissolution. As a result, transition metal-based OER electrocatalysts are more favorable for use in alkaline electrolytes, where they exhibit better stability and performance [65].

To sum up, HER and OER are essential components of electrocatalytic water splitting, with each reaction mechanism varying based on the pH of the electrolyte. In acidic media, HER proceeds via the Volmer, Tafel, and Heyrovsky steps, primarily involving proton reduction, while in alkaline media, water reduction predominates, leading to slower kinetics due to the added energy required to break water molecules. The OER mechanism, involving a more complex four-electron transfer process, also depends on the pH and electrocatalyst surface structure, with multiple intermediates such as M-OH, M-O, and M-OOH formed during the process. OER in acidic media is typically dominated by Ru/Ir-based catalysts, which exhibit superior activity but suffer from instability under high current densities. In contrast, transition metal-based catalysts perform better in alkaline media, offering improved stability but requiring further optimization to enhance performance. Understanding the interactions between catalysts and reaction intermediates, often analyzed through DFT, is key to advancing the efficiency and stability of both HER and OER electrocatalysts for scalable water-splitting applications.

4. Synthesis and Morphologies of Water-Splitting Electrocatalysts

To date, various techniques have been employed to develop efficient water-splitting electrocatalysts, including electrospinning, hydrothermal, and sol–gel procedures. Electrospinning empowers the fabrication of materials with distinct one-dimensional nanofiber architectures. Through the precise modulation of process parameters, it facilitates the controlled growth of crystals, resulting in highly crystalline and well-ordered structures. Concomitantly, it endows the materials with favorable porosity characteristics and extensive surface areas, both of which are cardinal for augmenting catalytic efficiency. In the case of hydrothermal methods, by precisely manipulating the temperature and pressure within a confined reaction environment, materials with specific crystalline phases and defect structures can be synthesized, thereby exerting a deterministic influence on their physicochemical properties. The sol–gel process, meanwhile, offers the advantage of fine-grained control over chemical composition and homogeneity, ensuring the uniformity and integrity of the final product, all of which cumulatively contribute to the overall performance.

With respect to morphology, diverse nanostructures assume indispensable roles. Nanofibers, owing to their elongated and continuous morphology, provide an optimized conduit for charge transfer, effectively minimizing electrical resistance and expediting electron mobility. Nanoparticles, characterized by their diminutive size and concomitantly high surface–volume ratio, expose a copious number of active sites, thereby catalyzing efficient catalytic reactions. Nanorods, with their anisotropic geometric configuration, can be strategically oriented to enhance the interaction with reactants, facilitating enhanced mass transport phenomena. Collectively, the judiciously designed synthesis methods and meticulously optimized morphologies operate in concert, synergistically bolstering the electrochemical performance of water-splitting electrocatalysts, thereby laying a solid foundation for the pursuit of more efficient and sustainable energy conversion pathways.

5. Water-Splitting Electrocatalysts via Electrostatic Spinning

Noble metal catalysts remain the most effective for water electrolysis, but their high cost and limited availability hinder widespread application. As a result, research in recent years has focused on developing catalysts made from inexpensive and abundant elements, particularly for renewable energy technologies [66]. Transition metals emerge as intriguing catalysts due to their unique electronic configurations. They can either provide d-orbital lone-pair electrons, acting as nucleophiles, or utilize empty d-orbitals as electrophilic sites during chemical reactions. The diverse chemisorption bonds formed on the d-orbitals of these transition metals enable molecular activation, reducing the activation energy of complex reactions [67]. Furthermore, different transition metal catalysts can alter the free energy of adsorption within the same reaction system, thereby influencing the reaction rate.

Among all the construction strategies, electrostatic spinning proves to be effective in generating high-performance self-supported transition metal electrocatalysts for water splitting. 1. Compared to other electrocatalytic reactions, water splitting involves two half-reactions, HER and OER, each with distinct kinetic barriers and mechanistic requirements. These dual demands necessitate bifunctional catalysts with complementary properties. The hierarchical porous structures and tunable compositions offered by electrospinning are particularly advantageous in addressing these dual challenges. 2. The interconnected porous networks reduce diffusion resistance for reactants and intermediates, thus enhancing mass and charge transport, which is particularly crucial for the OER. 3. Self-supported architectures minimize the detachment of active materials, a common issue in the harsh oxidative environment of water splitting, achieving significantly improved long-term stability.

The water splitting performances of electrospun transition metal electrocatalysts are summarized in Tables 1 and 2 as well as Figure 3.

Table 1. HER performance of reported electrospun transition metal-based electrocatalysts.

Catalysts	Preparation Methodology	Morphology	Electrolyte	Overpotential at 10 mA cm^{-2} (mV)	Tafel Slope (mV dec^{-1})	References
CNF/Co_3S_4/MoS_2	Electrospinning and high-temperature carbonization followed by hydrothermal method	Nanofiber sandwich structure of "substrate/accelerator/catalyst"	1 mol KOH	80	99.2	[68]
NiCoP	Electrospinning with subsequent heat treatment of the precursor prepared	NiCoP polycrystalline fibers	1 mol KOH	141	66	[69]
HEMP CNFs	Electrospinning and in situ phosphating	Three-dimensional networks of self-supporting HEMP nanofibers	1 mol KOH	26	50.9	[70]
Fe_7Se_8/CNT/C	Electrospinning combined with in situ selenide process.	Uniformly distributed Fe_7Se_8 nanoparticles on fibrous network	1 mol KOH	138	114	[71]
CM/Mo_xC@NC	Electrospinning method followed by specified calcination	Nanofiber interlinked network	0.5 mol H_2SO_4	149	86	[72]
Sn/Mo_2C-CNFs	Electrospinning and carbonization process	Sn/Mo_2C nanoparticles encapsulated in carbon nanofibers of a 2D structure	1 mol KOH	144	49.8	[73]
CoN/MoC/NMCNFs	Electrospinning and carbonization process	1D N-doped multi-channel carbon nanofibers embedded with MoC/CoN nanoparticles	1 mol KOH	92.5	107.9	[74]
Mott–Schottky Ni/CeO_2@N-CNFs	Electrospinning method followed by specified calcination	Ni/CeO_2 hetero-nanoparticles immobilized onto N-doped carbon nanofibers	1 mol KOH	100	85.7	[75]
Ni-MNCNF	Combination of coaxial electrospinning and pyrolysis	Nitrogen-doped carbon nanofibers with multi-channels loaded with nanoscale Ni nanoparticles	1 mol KOH	65	130.3	[76]
4%-SPCNF	Extraction of regenerated silk fibroin, followed by the electrospinning and carbonization	Bundled nanofibers	1 mol KOH	401.3	138.43	[77]
Fe_3C-Mo_2C/NC-0.15	Electrospinning and carbonization process	Nanoparticles within a porous carbon nanofiber matrix	0.5 mol H_2SO_4	116	43	[78]
NiCoP/CNF900	Electrospinning and carbonization process	NiCoP encapsulated with the carbon nanofibers	1 mol KOH	130	83	[79]
Co/CoP@NC	Electrospinning followed by carbonization and phosphorization	Co/CoP nanoparticles encapsulated in carbon nanofibers	1 mol KOH	180	60.3	[80]
Mo_xC-NPC-1.0-800	Electrospinning and carbonization process	Porous carbon fibers with wrapped Mo_2C/MoC nanoparticles	1 mol KOH	71	114.1	[81]

Table 1. *Cont.*

Catalysts	Preparation Methodology	Morphology	Electrolyte	Overpotential at 10 mA cm^{-2} (mV)	Tafel Slope (mV dec^{-1})	References
C/Ni/NiP_{red}	Electrospinning followed by carbonization and H_2 reduction	Incorporation of nanoparticles in the carbon matrix	0.5 mol H_2SO_4	262	45.4	[82]
MoS_2/CoS_2-2@HCNFs	Electrospinning and high-temperature carbonization followed by hydrothermal method	Hollow nanofibers with dispersed nanosheets	1 mol KOH	124	88.7	[83]
CoS-2@CNF	Extraction of regenerated silk fibroin, followed by the electrospinning, carbonization, and hydrothermal method	Homogeneous dispersion of small CoS nanoparticles onto the rough surface of carbon nanofibers	1 mol KOH	168	265.01	[84]
CoMoS@CNF	Electrospinning and high-temperature carbonization followed by hydrothermal method	Carbon nanofibers with nanosheet	1 mol KOH	105.2	152.8	[85]
Fe-CoP/PCNF	Electrospinning followed by carbonization and phosphorization process	Porous nanofibers with dispersed nanoparticles	0.5 mol H_2SO_4	151	53.9	[86]
FeNiP/NPC800	Electrospinning and carbonization process	Nanofibers with rough surfaces and dispersed nanoparticles	0.5 mol H_2SO_4	277	95	[87]

Table 2. OER performance of reported electrospun transition metal-based electrocatalysts.

Catalysts	Preparation Methodology	Morphology	Electrolyte	Overpotential at 10 mA cm^{-2} (mV)	Tafel Slope (mV dec^{-1})	References
Fe(III)/NiS	Electrospinning technique followed by alkaline hydrothermal method	One-dimensional hollow structure with vertical nanosheets on the surface	1 mol KOH	264	77.6	[88]
Co-C/Ni_1Fe_1S	Electrospinning–carbonization–electrodeposition strategy	Hierarchical heterostructure with Co-C nanofibers core and amorphous NiFeS nanosheets sheath	1 mol KOH	233	53.1	[89]
NiCoP	Electrospinning with subsequent heat treatment of the precursor prepared	NiCoP polycrystalline fibers	1 mol KOH	540	70	[69]
Cobalt phosphate fibrous network	Simple precipitation method and electrospinning technique	Hydrated 1D cobalt phosphate fibrous network	1 mol KOH	245	229	[90]
Se-$CoSe_2$/CNFs	Electrospinning combined with in situ growth of ZIF-67 strategy, followed by carbonization and a facile selenization process	One-dimensional CNFs with ZIF-67 as a template	0.1 mol KOH	325	60.89	[91]

Table 2. *Cont.*

Catalysts	Preparation Methodology	Morphology	Electrolyte	Overpotential at 10 mA cm^{-2} (mV)	Tafel Slope (mV dec^{-1})	References
LFNOSe-III NFs	Electrospinning and subsequent heating treatment	Porous perovskite nanofibers	1 mol KOH	287	87	[92]
Co-$ZIF_{1.5}$/$10CNF_2$	Electrospinning of CNF and Co-ZIF and calcination as a bifunctional catalyst	Co nanoparticles uniformly dispersed and encapsulated inside the CNFs	0.1 mol KOH	390	122	[93]
Mott–Schottky Ni/CeO_2@N-CNFs	Electrospinning method followed by specified calcination	Ni/CeO_2 hetero-nanoparticles immobilized onto N-doped carbon nanofibers	1 mol KOH	230	54.2	[75]
S/LMO-E	Electrospinning and sulfurization treatment	Nanorods and tiny nanoparticles	1 mol KOH	314	47	[94]
SNCF-NF-Co_3O_4@N_2	Electrospinning followed by co-precipitation method and sintering under N_2	Heterostructure with oxygen vacancies	0.1 mol KOH	343	71.3	[95]
CoNC/NCNTs@CNF	Electrospinning method followed by pyrolysis process	3D callistemon-like structure with 1D nanofiber matrix	1 mol KOH	390	82	[96]
NiFe@NCNFs	Electrospinning process and subsequent heat treatment	One-dimensional mesoporous structures with metal components encapsulated in the N-doped carbon nanofibers	1 mol KOH	294	52	[97]
CoP/NC	Electrospinning followed by MOF fabrication	Ultra-small CoP nanoparticles incorporated into nitrogen-doped carbon nanofibers derived from ZIF-67	1 mol KOH	290	62	[98]

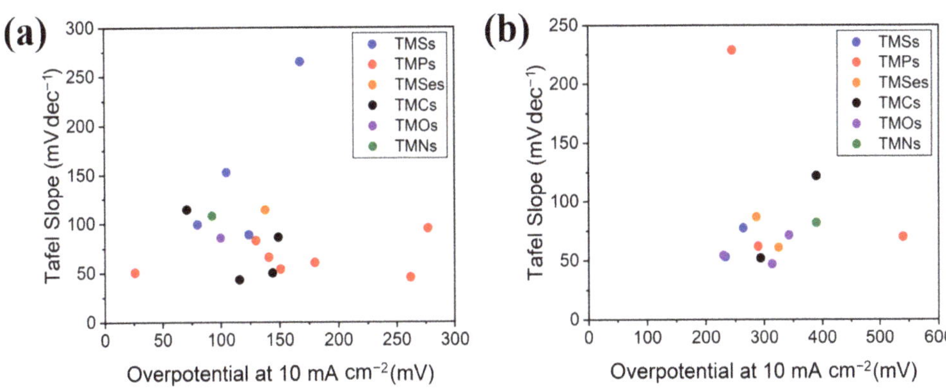

Figure 3. Summary of the water splitting performance of the reported electrospun transition metal-based electrocatalysts with the relationship of overpotential at 10 mA cm^{-2} and Tafel slope. (**a**) HER performance; (**b**) OER performance.

5.1. Transition Metal Sulfides (TMSs)

TMSs have garnered significant attention due to their exceptional electrical conductivity, thermal and mechanical stability, and redox capabilities. These properties make them highly attractive for applications in catalysis, energy conversion, storage, and even optoelectronic devices [99–101]. When combined with electrospun nanofibers, these metal sulfides can form innovative nanostructures, leading to hybrid materials with unique interfaces such as core–shell, Janus, and hierarchical structures. These hybrid materials exhibit enhanced electron transfer and abundant active sites, essential for catalysis and other energy-related applications [102]. The strong interaction between the different components in these materials often leads to electronic modulation and the reconstruction of active sites, improving the overall catalytic performance.

Electrospun nanofibers not only provide structural stability but also serve as an excellent support matrix for metal sulfides, often leading to synergistic effects that enhance the long-term stability of these hybrid catalysts. For example, Zhong et al. synthesized [88] novel Fe(III)/NiS electrocatalysts via electrospinning combined with an interfacial adsorption method (Figure 4a–c). The resulting tubular morphology and highly active Ni(Fe)OOH during oxygen evolution led to a low overpotential of 264 mV at 10 mA·cm^{-2} (η_{10}), outperforming conventional RuO_2 and other metal sulfide-based electrocatalysts. Li et al. developed [89] a catalyst (Co-C/NiFeS nanofibers) composed of carbon nanofiber-supported amorphous NiFeS nanosheets embedded with cobalt nanoparticles using a combination of electrospinning, carbonization, and electrodeposition. The optimized catalyst demonstrated impressive OER performance in 1 mol·L^{-1} KOH, with a low η_{10} of 233 mV and a Tafel slope of 53.1 mV·dec^{-1}, along with excellent HER activity.

Among TMSs, molybdenum disulfide (MoS_2) has attracted considerable interest due to its layered structure and favorable electron arrangement [103]. The Gibbs free energies of hydrogen adsorption (ΔG_{H^*}) at the Mo and S atomic edges of MoS_2 are near zero, making these sites highly active for HER catalysis. However, MoS_2 suffers from low current-carrying density and poor electrical conductivity, which limit its catalytic performance [104]. To address these limitations, researchers have focused on synthesizing MoS_2 nanomaterials with exposed edges on high-conductivity carbon nanofibers (CNFs) to improve HER efficiency. Guan et al. demonstrated [68] this by growing MoS_2 in situ on nitrogen-doped nanofibers with Co_3S_4 nanoparticles via electrospinning, carbonization, and hydrothermal processes (Figure 4d). The resulting CNF/Co_3S_4/MoS_2 sandwich structure allowed molybdenum disulfide to act as the primary HER catalyst, while Co_3S_4 nanoparticles accelerated electron transfer from the substrate to the catalyst, promoting water molecule decomposition. The CNF/Co_3S_4/MoS_2 composites exhibited a η_{10} of only 80 mV and a Tafel slope of 99.2 mV·dec^{-1} (Figure 4e,f). Due to the in situ growth, Co_3S_4 particles remained stably embedded in the CNF, facilitating prolonged and efficient HER activity (Figure 4h,i).

As a conclusion, TMSs are renowned for their exceptional electrical conductivity, stability, and redox properties, making them attractive for catalysis and energy conversion applications. When combined with electrospun nanofibers, TMSs form hybrid nanostructures with enhanced catalytic performance due to improved electron transfer and abundant active sites. These hybrid materials benefit from synergistic effects that enhance stability and overall efficiency, addressing some of the limitations of TMSs such as poor conductivity and low current-carrying capacity while improving long-term catalytic performance.

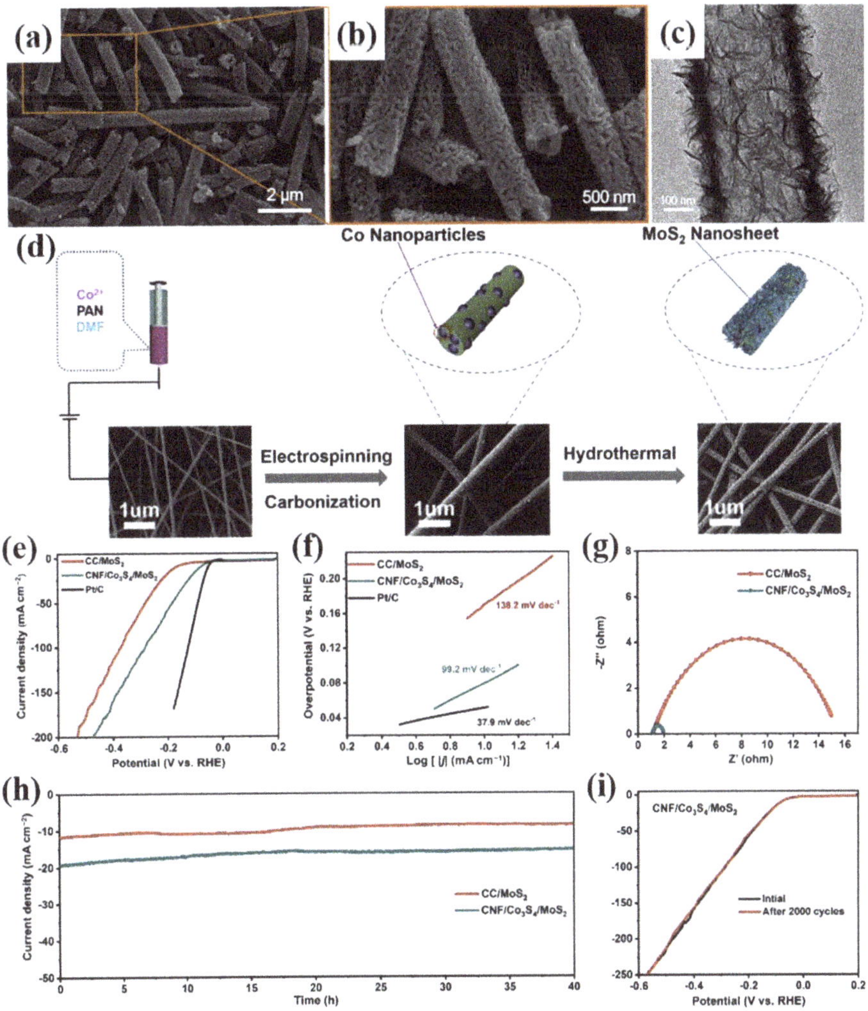

Figure 4. (**a**,**b**) SEM and (**c**) TEM images of Fe(III)/NiS-18 NTs. Reproduced with permission. Copyright 2023, Elsevier. (**d**) The schematic illustration of the CNF/Co_3S_4/MoS_2 synthesis. For CNF/Co_3S_4/MoS_2, CC/MoS_2, and Pt/C, the LSV (**e**), the corresponding Tafel slope curves (**f**), and the EIS (**g**) curves in 1 M KOH. (**h**) The time–current density curves of CNF/Co_3S_4/MoS_2 and CC/MoS_2 under a constant voltage. (**i**) Comparison of the LSV curves before and after 2000 cycles of CV. Reproduced with permission. Copyright 2022, Elsevier.

5.2. Transition Metal Phosphides (TMPs)

TMPs have garnered considerable attention due to their unique HER and OER catalytic properties [105]. The negatively charged phosphorus (P) sites in TMPs are known to weaken the metal-hydride (M-H) binding strength, facilitating hydrogen desorption and enhancing HER activity [106,107]. Simultaneously, the positively charged metal sites in TMPs can form metal-oxygen/hydroxyl radicals, which serve as active species during OER, significantly improving the electrocatalytic efficiency [108].

A key strategy to further enhance the performance of TMPs is their combination with conductive carbon-based materials such as carbon nanotubes, carbon nanofibers, and graphene. These conductive matrices provide an excellent platform to boost electrical conductivity and facilitate charge/mass transport. Electrospun nanofibers, with their high specific surface area and rapid charge transfer capabilities, are especially well suited for this purpose [109,110]. For instance, Yang et al. synthesized [69] NiCoP electrocatalysts using a needle-free electrospinning technique. These electrocatalysts exhibited impressive HER activity, with low overpotentials of 141 mV in alkaline media and 146 mV in acidic environments at a current density of 10 $mA \cdot cm^{-2}$. The corresponding Tafel slopes were 53.0 and 97.8 $mV \cdot dec^{-1}$, respectively, indicating its fast reaction kinetics. In addition, the bimetallic phosphides demonstrated excellent OER performance in 1 M KOH and sustained durability during long-term stability tests. The bimetallic phosphides also show excellent oxygen precipitation reactivity in the alkaline medium 1 M KOH and satisfactory durability in long-term stability tests. The innovative fibrous structure of these electrocatalysts plays a crucial role in modulating their electronic properties, enhancing the availability of electrochemically active sites, and improving their overall catalytic performance in water-splitting reactions. Furthermore, the fibrous architecture improves the conductivity, utilization, and mechanical stability of the electrodes while also promoting better permeability and access to reactive sites.

Sankar et al. demonstrated [90] another approach by mixing synthesized cobalt phosphate powder with polymers and solvents to form fibers through electrospinning (Figure 5a). They successfully produced a network of hydrated, one-dimensional cobalt phosphate fibers (Figure 5b–d) which exhibited promising electrocatalytic activity. This material achieved a relatively low η_{10} of 245 mV in 1 M KOH (Figure 5e), along with a small Tafel slope of 229 $mV \cdot dec^{-1}$. In addition, the wrapping of TMPs by nanofibers not only enhanced the electrocatalytic performance but also protected the TMPs from corrosion in alkaline environments, significantly improving the stability of the composite catalyst (Figure 5f). Wang et al. took a more advanced approach by preparing [70] three-dimensional (3D) networks of self-supported high-entropy metal phosphide (HEMP) nanofibers via electrospinning and in situ phosphorylation. These nanofibers showed an outstanding HER performance with a η_{10} as low as 26 mV and a Tafel slope of 50.9 $mV \cdot dec^{-1}$. This exceptional catalytic activity was attributed to the synergistic effects of multiple metals and the formation of metal-P bonds. The introduction of electronegative P atoms modulated the electronic structure of the catalyst, inducing lattice strain and lowering the free energy of hydrogen adsorption. This enhanced the overall electrocatalytic performance, making the HEMP nanofibers highly efficient for hydrogen evolution.

To sum up, TMPs have gained attention for their unique catalytic properties in both HER and OER. The phosphorus sites in TMPs help weaken metal-hydride bonds, promoting efficient hydrogen desorption, while metal sites facilitate oxygen species formation, enhancing OER efficiency. Combining TMPs with conductive carbon-based materials, such as electrospun nanofibers, improves their electrical conductivity, charge transport, and catalytic performance. The fibrous structures further enhance the active site availability, stability, and overall efficiency of TMP-based electrocatalysts for water-splitting applications.

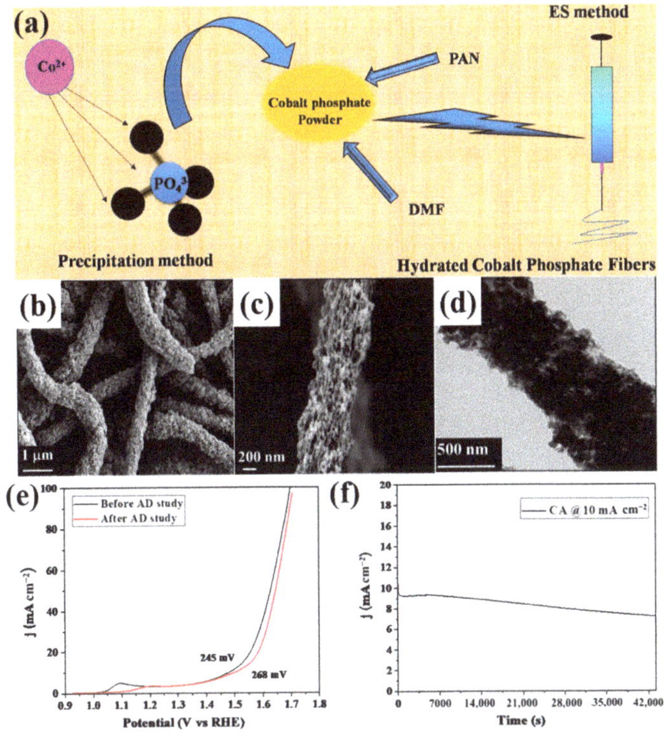

Figure 5. (**a**) Schematic illustration of the synthetic procedure for making hydrated cobalt phosphate fibers. (**b**,**c**) The FE-SEM images and (**d**) HR-TEM images of the hydrated cobalt phosphate fibers. (**e**) OER LSV curve and (**f**) stability study of the cobalt phosphate fibrous network. Reproduced with permission. Copyright 2021, Elsevier.

5.3. Transition Metal Selenides (TMSes)

In the structure of TMSes, the Se-H bonds in TMSes are weaker than those found in TMSs and TMPs, resulting in improved hydrogen dissociation efficiency and excellent HER activity [111]. In the context of OER, TMSes facilitate more efficient molecular oxygen transport and accelerate OER kinetics. This is largely due to the cumulative 3d-2p repulsion between negatively charged selenium sites and the transition metals, which efficiently release oxygen from the active sites, contributing to superior OER performance. These characteristics make TMSes highly valuable in the field of overall water-splitting electrocatalysis [112].

The integration of electrospinning techniques with TMSes further enhances their catalytic potential. Electrospun nanofibers help increase the electrochemically active surface area, providing more active sites for catalysis, reducing contact resistance, and facilitating faster charge transfer [113,114]. For example, Cui et al. synthesized [91] selenium-doped $CoSe_2$ embedded in CNFs (Se-$CoSe_2$/CNFs) using electrospinning coupled with the ZIF-67 in situ growth strategy, followed by carbonization and a selenization process. The selenium doping not only modulated the electronic structure of $CoSe_2$ but also improved the reaction kinetics and optimized the adsorption of intermediates. The one-dimensional (1D) CNFs provided enhanced conductivity and structural stability, while the ZIF-67 template prevented the agglomeration of the materials. As a result, the Se-$CoSe_2$/CNF catalysts demonstrated excellent OER performance, requiring only 325 mV overpotential at 10 mA·cm^{-2}.

Li et al. prepared [92] porous $LaFe_{1-x}Ni_xO_3$ (LFNO) chalcogenide nanofibers through electrospinning and subsequent heat treatment with selenium doping (Figure 6a–d). The selenium-doped LFNO nanofibers required a low η_{10} of just 287 mV with a Tafel slope of 87 mV and excellent stability in alkaline medium (Figure 6e–g). The high specific surface area provided by the porous nanofibers improved the mass-specific activity of the catalysts. Additionally, selenium doping enhanced the intrinsic conductivity of $LaFe_{0.25}Ni_{0.75}O_3$, improving charge transfer kinetics and mass transfer efficiency (Figure 6h–m).

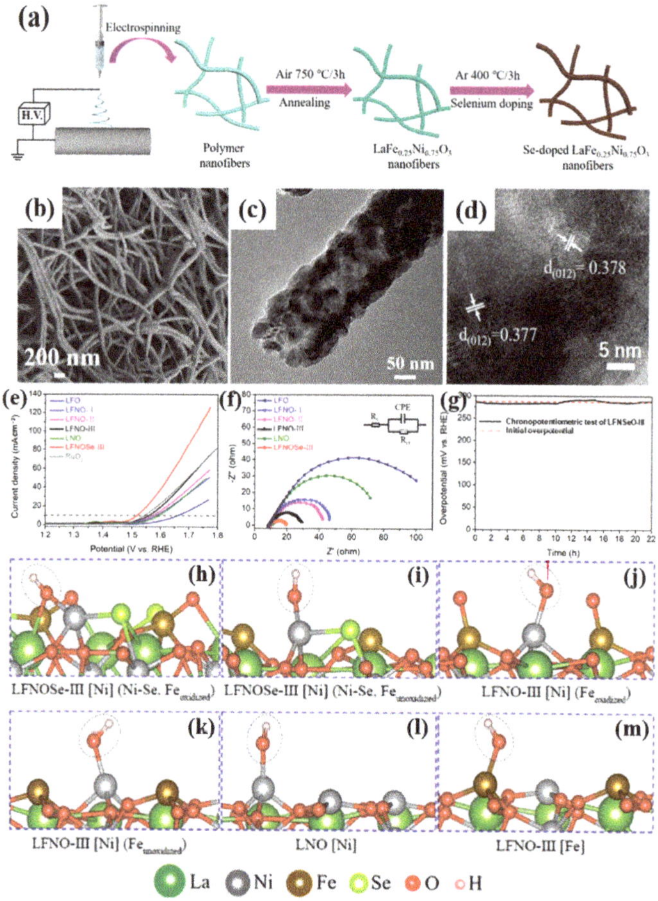

Figure 6. (**a**) Schematic illustration of the processing of Se-doped $LaFe_{0.25}Ni_{0.75}O_3$ nanofibers. (**b**) The SEM image, (**c**) TEM image, and (**d**) HRTEM image of LFNOSe-III. (**e**) The polarization curves and (**f**) Nyquist plots of LFO, LFNO-I, LFNO-II, LFNO-III, LNO, and LFNOSe-III. (**g**) The chronopotentiometric measurements of LFNOSe-III at 10 mA cm^{-2}. (**h–m**) Corresponding adsorption geometries of *O and/or *OH intermediates on catalysts. Reproduced with permission. Copyright 2020, American Chemical Society.

As a brief summary, the integration of electrospinning with TMSes significantly enhances their electrocatalytic properties. By increasing surface area and facilitating faster charge transfer, electrospun nanofibers improve both hydrogen evolution and oxygen evolution reactions. Selenium doping optimizes reaction kinetics while enhancing structural stability and conductivity for better performance in water splitting.

5.4. Transition Metal Carbides (TMCs)

The growing interest in various TMCs, such as Mo_2C, W_2C, and Fe_3C, has led to extensive research into their water-splitting capabilities [115] due to their metal-like electronic structures, excellent hydrogen adsorption properties, high stability, and low cost [116–118]. The carbon-based framework in TMCs can effectively adjust the distribution of metal ions, resulting in a more uniform distribution of active sites [93,117,119]. For instance, He et al. synthesized [93] a cowpea-like Co-ZIF/CNF bifunctional catalyst using a straightforward electrospinning and calcination method.

Among different categories of TMCs, molybdenum carbide stands out as one of the highest-performing electrocatalysts, attributed to its d-band electronic structure, which closely resembles that of platinum [72]. However, the carbonization of Mo-based compounds at high temperatures often results in the aggregation and overgrowth of MoC [120]. To mitigate this issue, electrostatic spinning technology has been employed to produce metal-embedded carbon nanofibers. The limited mobility of metal particles on the carbon surface helps maintain a high dispersion of nanoparticles, effectively suppressing aggregation [121]. Zhang et al. developed [73] a bimetallic carbon-based electrocatalyst consisting of Sn/Mo_2C composite nanoparticles encapsulated in carbon nanofibers. This unique structure prevents the binding and aggregation of β-Mo_2C with the metal nanoparticles, ensuring the structural stability of the electrocatalyst and reducing the energy barrier during the water-splitting process. The catalyst demonstrated impressive HER performance in alkaline media, requiring a low η_{10} of 144 mV overpotential and exhibiting a Tafel slope of 49.8 mV·dec $^{-1}$ over 60 h of stable operation. Wang et al. further advanced this field by preparing [74] one-dimensional nitrogen-doped multichannel carbon nanofibers encapsulating MoC/CoN nanoparticles (Figure 7a–c). This design helped avoid nanoparticle aggregation while increasing the specific surface area of the catalysts, resulting in the better accessibility of active site (Figure 7d–g). The resulting CoN/MoC/NMCNFs exhibited enhanced HER performance in both alkaline and acidic electrolytes, achieving a η_{10} of only 92.5 mV and 199.2 mV, respectively.

To sum up, TMCs, such as Mo_2C and W_2C, are gaining attention for their exceptional stability, hydrogen adsorption, and low-cost properties, making them ideal for water-splitting applications. Electrospinning aids in producing metal-embedded carbon nanofibers, which prevent nanoparticle aggregation, enhancing the catalyst's performance and structural stability.

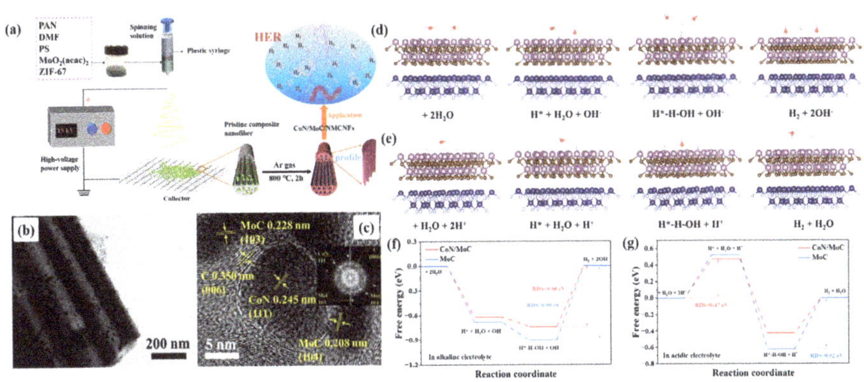

Figure 7. (**a**) Synthetic process of CoN/MoC/NMCNFs electrocatalysts by electrospinning. (**b**,**c**) HR-TEM images of CoN/MoC/NMCNFs. (**d**,**e**) Schematic configurations and (**f**,**g**) free energy profiles of (**d**,**f**) alkaline HER and (**e**,**g**) acidic HER through the Volmer–Heyrovsky–Rideal pathway on the CoN/MoC surface. Reproduced with permission. Copyright 2023, Elsevier.

5.5. Transition Metal Oxides (TMOs)

TMOs are widely utilized in electrocatalysis due to their availability, lower cost compared to noble metals, and their ability to provide d orbitals for oxygen binding [122–125]. For example, Li et al. developed [75] a Mott–Schottky-type electrocatalyst by immobilizing Ni/CeO_2 heterogeneous nanoparticles on nitrogen-doped carbon nanofibers (N-CNFs). This catalyst design significantly enhanced the charge transfer rate, optimized the chemisorption energy of reaction intermediates, and accelerated the reaction kinetics. As a result, the Ni/CeO_2@N-CNFs exhibited excellent catalytic activity for both HER and OER in alkaline solutions, achieving overpotentials of 100 mV for HER and 230 mV for OER at 10 $mA{\cdot}cm^{-2}$.

Among the various TMO catalysts, perovskite-type oxides, such as monocalcocite (ABO_3) [126], dicalcocite ($A_2B_2O_6$) [127], Ruddlesden–Popper chalcocite (A_2BO_4) [128], and pyrochlore-type oxides ($A_2B_2O_7$) [129], have gained recognition as excellent electrocatalysts. These materials are notable for their low-cost transition metals, robust skeletal structures, intrinsic oxygen vacancies, versatile electronic configurations, and combinatorial flexibility [130,131]. Wang et al. explored this further by selecting [94] appropriate doping elements based on ionic radius, electronegativity, and oxygen vacancy formation energy. They introduced oxygen vacancies into high-entropy perovskite oxides (HEPOs) using a two-step strategy via electrospinning. This approach optimized the number of orbital-filling electrons, significantly increased the active surface area, boosted oxygen vacancy content, and improved conductivity. In addition, the pronounced pH dependence and tetramethylammonium cation inhibition clearly identify the involvement of lattice oxygen-mediated mechanism. The obtained catalyst exhibited a highly efficient OER catalyst with a η_{10} of 314 mV at 10 $mA{\cdot}cm^{-2}$ and a Tafel slope of 47 $mV{\cdot}dec^{-1}$. This work highlights the potential of high-entropy catalysts for efficient water splitting, providing a promising path for the production of renewable hydrogen. Yuan et al. also contributed to this area by preparing [95] the Co-based perovskite of $SrNb_{0.1}Co_{0.7}Fe_{0.2}O_{3-\delta}$ (SNCF) nanofibers through electrospinning (Figure 8a), significantly increasing the specific surface area and providing more active sites. Co_3O_4 nanoparticles were then uniformly anchored onto the SNCF nanofibers using a co-precipitation method (Figure 8b–d), followed by sintering under a nitrogen atmosphere. The resulting SNCF-NF-Co_3O_4@N_2 heterostructures contained numerous oxygen vacancies, which contributed to the excellent OER catalytic activity and long-term stability (Figure 8e,f), with a low η_{10} of 340 mV.

As a conclusion, TMOs are widely used in electrocatalysis due to their low cost, availability, and ability to provide oxygen-binding d orbitals. In particular, perovskite-type oxides stand out for their structural stability, oxygen vacancies, and versatile electronic configurations. Electrospinning techniques enhance their performance by increasing surface area and optimizing active sites.

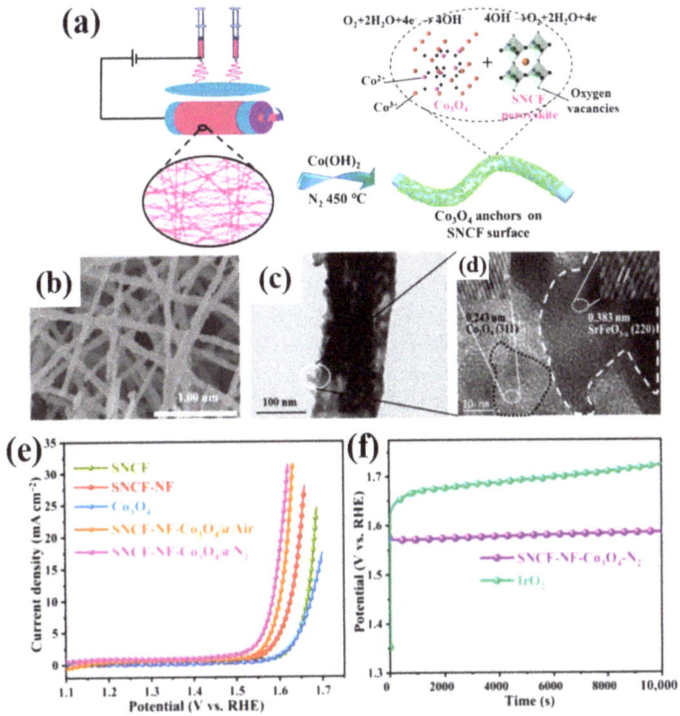

Figure 8. (**a**) Schematic illustration of the preparation of the SNCF-NF-Co_3O_4@N_2 sample. (**b**) The SEM, (**c**) TEM, and (**d**) HETEM images of SNCF-NF-Co_3O_4@N_2. (**e**) The OER LSV curves, and (**f**) OER stability test of SNCF-NF-Co_3O_4@N_2. Reproduced with permission. Copyright 2024, Elsevier.

5.6. Transition Metal Nitrides (TMNs)

TMNs are intriguing compounds formed by the incorporation of nitrogen into the transition metal lattice. This insertion significantly alters the material's properties, blending the characteristics typical of covalent compounds, ionic crystals, and transition metals [132,133]. Nanostructured TMNs are typically synthesized via the temperature-programmed nitridation of oxidized precursors, a process known as a "local structural reaction" [134,135]. During nitridation, nitrogen atoms expand the metal lattice, weaken interatomic forces, and redistribute d-band electrons, resulting in a high-density state close to the Fermi level [136], resulting in enhanced electrocatalytic performance [137]. However, TMNs usually suffer from high overpotentials, particle agglomeration, and suboptimal catalytic performance. Addressing these limitations has been the focus of recent research, with strategies aimed at modifying the composition, structure, morphology, and surface properties of TMNs to enhance their catalytic activity. For example, Yao et al. designed and synthesized [96] 1D CNFs using electrospinning, forming a unique callistemon-like nanostructure. This structure featured the in situ formation of N-doped carbon nanotubes (NCNTs) that integrated into a 3D hybrid nanostructure. The presence of NCNTs enabled high mass transport and an increased electrochemically active region, while the Co-N-doped graphitic carbon at the tips of the structure exposed catalytic sites more effectively to the electrolyte. This configuration also facilitated fast electron and ion transport, contributing to enhanced catalytic performance. The resulting electrocatalyst exhibited a η_{10} of 390 mV for OER, with a Tafel slope of 82 mV·dec^{-1}. For HER, the catalyst achieved a η_{10} of 190 mV with a Tafel slope of 117 mV·dec^{-1}. Wei et al. constructed [97] porous nitrogen-doped carbon nanofibers (NCNFs) encapsulated with nickel–iron (NiFe) alloy nanoparticles (NiFe@NCNFs) as OER electrocatalysts (Figure 9a). The combination of a 1D mesoporous structure and the N-doped carbon nanofibers, along with the optimized binary metal composition (Figure 9b–d), resulted in an optimal synergistic effect (Figure 9e,f). This led to enhanced OER performance, with a low overpotential of 294 mV at a current density of 10 mA·cm^{-2}. The catalyst also demonstrated excellent durability in alkaline solution, maintaining performance for over 10 h. Moreover, the two-electrode water electrolyzer with a NiFe@NCNF anode and Pt/C cathode showed extraordinary overall water splitting performance, with an initial cell voltage of only 1.531 V at 10 mA cm^{-2}, as well as long-term stability up to 20 h (Figure 9g,h). Feng et al. utilized [98] electrospinning combined with metal–organic framework (MOF) preparation to create hybridized CoP/N-doped carbon (NC) nanofiber composites, designed as bifunctional oxygen electrocatalysts. The CoP/NC catalysts exhibited good electrical conductivity thanks to the exposure of more active sites. As a result, excellent OER performance was achieved in alkaline solution with a η_{10} of 290 mV and a Tafel slope of 62 mV·dec^{-1}.

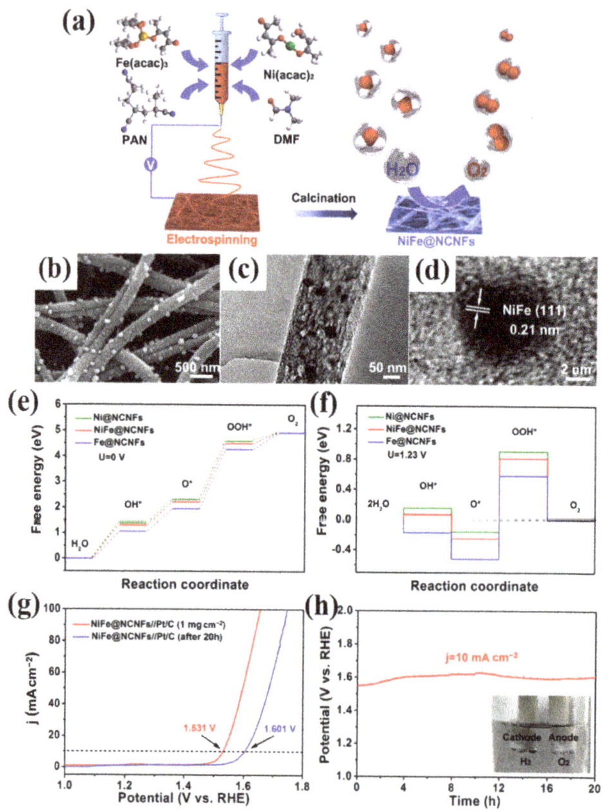

Figure 9. (**a**) Schematic illustration of the synthesis process of the NiFe@NCNFs. (**b**) The SEM and (**c**,**d**) TEM images of the NiFe@NCNFs. Free energy profiles (**e**) at the zero potential (U = 0 V) and (**f**) equilibrium potential (U = 1.23 V). (**g**) The overall water splitting LSV curves of the NiFe@NCNFs in a two-electrode device. (**h**) The overall water splitting chronopotentiometric curve of the NiFe@NCNFs at 10 mA cm^{-2} in 1.0 M KOH. Reproduced with permission. Copyright 2020 American Chemical Society.

To sum up, TMNs are created by incorporating nitrogen into transition metal lattices, enhancing electrocatalytic performance. However, challenges like high overpotentials and particle agglomeration remain. Recent advancements, such as using electrospun nanostructures with nitrogen doping or metal alloy integration, address these limitations, improving catalytic activity and stability for OER and HER.

6. Conclusions

In summary, this review has furnished a comprehensive and meticulously detailed exposition of the recent extraordinary advancements in electrospun electrocatalysts for water splitting. During the past five years, a heterogeneous assemblage of transition metal sulfides, phosphides, carbides, selenides, nitrides, oxides, and affiliated compounds has emerged as pre-eminent contenders in the realm of contemporary research. Electrospinning, as a highly innovative and propitious technique within the purview of electrocatalysis, has unequivocally manifested its distinctive capabilities in endowing materials with one-dimensional architectures. These architectures not only augment electron and mass transport kinetics but also engender expansive surface areas and an exceptional aptitude for the fine-grained manipulation of chemical compositions and active sites. Through an exhaustive exploration of the fundamental principles underpinning electrospinning, a profound dissection of the characteristic traits of transition metal-based compounds, and a lucid elucidation of the merits attendant to their integration, a resilient knowledge foundation has been established, poised to galvanize further investigative pursuits.

Nevertheless, significant impediments endure that necessitate immediate redress. Firstly, from a mechanical vantage point, electrospun materials presently exhibit a relatively diminished mechanical strength when juxtaposed with other self-supporting electrodes. This inadequacy severely circumscribes their direct and unimpeded incorporation into practical energy conversion apparatuses. Consequently, future research endeavors must be steered towards devising novel stratagems and materials engineering paradigms to fortify their mechanical integrity. This could potentially encompass the exploration of composite materials, nanostructured reinforcements, or avant-garde fabrication procedures capable of conferring augmented tensile and compressive strength.

Secondly, environmental variables, with humidity being a particularly prominent factor, pose a formidable barricade to the consistent quality and reproducibility of electrospun fibers. Even within exquisitely controlled laboratory environs furnished with state-of-the-art humidity regulation apparatus, fluctuations in humidity can instigate disruptions in the spinning process, precipitating erratic fiber morphologies and, ineluctably, elevated energy consumption. To surmount this obstacle, it is of paramount urgency to integrate real-time environmental monitoring systems with automated feedback loops. Such systems would facilitate the dynamic modulation of humidity levels during electrospinning, thereby ensuring a stable and regulated milieu. Additionally, the development of closed-loop electrospinning chambers equipped with advanced dehumidifiers or humidifiers is indispensable. These chambers would not only stabilize environmental conditions but also vouch for the safe production of electrospun fibers with uniform morphology and quality. Moreover, we advocate the exploration and development of temperature-controlled spinning heads and adaptive nozzle designs. These innovative constituents would possess the capacity to modulate solution flow rates in real time in response to environmental perturbations, potentially augmented by machine learning algorithms. These algorithms could prognosticate and optimize process parameters predicated on continuous environmental inputs and historical data, thereby effecting a revolutionary transformation of the electrospinning process.

Furthermore, while electrospinning has demonstrated its efficacy in fabricating one-dimensional nanostructures, our current capabilities in effecting precise control over surface morphology, defect generation, and grain boundary manipulation remain nascent. Future work must pivot on the design and implementation of advanced synthesis methodologies and cutting-edge surface/interface engineering tactics. This could potentially entail the utilization of templating techniques, molecular self-assembly, or precision etching to engineer nanostructures with hitherto unprecedented complexity and control. By attaining such precise control, we can unlock novel tiers of electrocatalytic performance.

Finally, from an ecological and economic perspective, the preponderant reliance on organic solvents in electrospinning represents a significant Achilles' heel. These solvents not only inflate production costs but also engender grave toxicity and safety hazards, both during the manufacturing process and in potential end-of-life disposal scenarios. Therefore, the urgent development of environmentally innocuous, green solvents that can satisfy the exacting process requirements is not merely a desirable aspiration but an ineluctable exigency. Such solvents would underpin the scalability and long-term viability of electrospinning for electrocatalyst fabrication, paving the way for a more sustainable future.

Overall, by methodically addressing these multifaceted challenges, we can unlock the heretofore unexploited full potential of electrospun electrocatalysts in water-splitting and energy conversion applications, heralding a new era of clean energy technology.

Author Contributions: S.C.: writing—review and editing and conceptualization. Y.J.: writing—original draft, formal analysis, and investigation. Y.L.: Writing—review and editing and supervision. S.C. and Y.J. contributed equally to this work. All the authors discussed the results and commented on the manuscript. All authors have read and agreed to the published version of the manuscript.

Funding: This work was funded by the National Natural Science Foundation of China, grant numbers No. 22108301, 22238012, and 22178384.

Data Availability Statement: All the relevant data used in this study have been provided in the form of figures and tables in the published article, and all the data provided in the present manuscript are available to whom it may concern.

Conflicts of Interest: The authors declare no conflicts of interest.

References

1. Zhang, Y.; Baik, J.M.; Park, H. Spontaneous Heterophase Atomic Structure Engineering of NiMo(S)/NiMoP for Enhanced Overall Electrochemical Water Splitting in Neutral Media. *Appl. Catal. B Environ. Energy* **2024**, *359*, 124469. [CrossRef]
2. Gao, X.; Chen, Y.; Wang, Y.; Zhao, L.; Zhao, X.; Du, J.; Wu, H.; Chen, A. Next-Generation Green Hydrogen: Progress and Perspective from Electricity, Catalyst to Electrolyte in Electrocatalytic Water Splitting. *Nano-Micro Lett.* **2024**, *16*, 237. [CrossRef]
3. Li, S.; Li, E.; An, X.; Hao, X.; Jiang, Z.; Guan, G. Transition Metal-Based Catalysts for Electrochemical Water Splitting at High Current Density: Current Status and Perspectives. *Nanoscale* **2021**, *13*, 12788–12817. [CrossRef]
4. Zeng, Y.; Zhao, M.; Huang, Z.; Zhu, W.; Zheng, J.; Jiang, Q.; Wang, Z.; Liang, H. Surface Reconstruction of Water Splitting Electrocatalysts. *Adv. Energy Mater.* **2022**, *12*, 2201713. [CrossRef]
5. Wang, C.; Humayun, M.; Debecker, D.P.; Wu, Y. Electrocatalytic Water Oxidation with Layered Double Hydroxides Confining Single Atoms. *Coord. Chem. Rev.* **2023**, *478*, 214973. [CrossRef]
6. Wu, Z.; Zhao, Y.; Xiao, W.; Fu, Y.; Jia, B.; Ma, T.; Wang, L. Metallic-Bonded Pt–Co for Atomically Dispersed Pt in the Co_4N Matrix as an Efficient Electrocatalyst for Hydrogen Generation. *ACS Nano* **2022**, *16*, 18038–18047. [CrossRef] [PubMed]
7. You, M.; Du, X.; Hou, X.; Wang, Z.; Zhou, Y.; Ji, H.; Zhang, L.; Zhang, Z.; Yi, S.; Chen, D. In-Situ Growth of Ruthenium-Based Nanostructure on Carbon Cloth for Superior Electrocatalytic Activity Towards Her and Oer. *Appl. Catal. B Environ.* **2022**, *317*, 121729. [CrossRef]
8. Zhang, P.; Liu, Y.; Liang, T.; Ang, E.H.; Zhang, X.; Ma, F.; Dai, Z. Nitrogen-Doped Carbon Wrapped Co-Mo_2C Dual Mott–Schottky Nanosheets with Large Porosity for Efficient Water Electrolysis. *Appl. Catal. B Environ.* **2021**, *284*, 119738. [CrossRef]

9. Wang, H.-F.; Chen, L.; Pang, H.; Kaskel, S.; Xu, Q. Mof-Derived Electrocatalysts for Oxygen Reduction, Oxygen Evolution and Hydrogen Evolution Reactions. *Chem. Soc. Rev.* **2020**, *49*, 1414–1448. [CrossRef]
10. Wang, C.; Zhang, Q.; Yan, B.; You, B.; Zheng, J.; Feng, L.; Zhang, C.; Jiang, S.; Chen, W.; He, S. Facet Engineering of Advanced Electrocatalysts toward Hydrogen/Oxygen Evolution Reactions. *Nano-Micro Lett.* **2023**, *15*, 52. [CrossRef] [PubMed]
11. Yu, Z.-Y.; Duan, Y.; Feng, X.-Y.; Yu, X.; Gao, M.-R.; Yu, S.-H. Clean and Affordable Hydrogen Fuel from Alkaline Water Splitting: Past, Recent Progress, and Future Prospects. *Adv. Mater.* **2021**, *33*, 2007100. [CrossRef] [PubMed]
12. Kawashima, K.; Márquez, R.A.; Smith, L.A.; Vaidyula, R.R.; Carrasco-Jaim, O.A.; Wang, Z.; Son, Y.J.; Cao, C.L.; Mullins, C.B. A Review of Transition Metal Boride, Carbide, Pnictide, and Chalcogenide Water Oxidation Electrocatalysts. *Chem. Rev.* **2023**, *123*, 12795–13208. [CrossRef] [PubMed]
13. Nguyen, D.C.; Luyen Doan, T.L.; Prabhakaran, S.; Tran, D.T.; Kim, D.H.; Lee, J.H.; Kim, N.H. Hierarchical Co and Nb Dual-Doped MoS_2 Nanosheets Shelled Micro-TiO_2 Hollow Spheres as Effective Multifunctional Electrocatalysts for HER, OER, and ORR. *Nano Energy* **2021**, *82*, 105750. [CrossRef]
14. Chang, Y.; Zhai, P.; Hou, J.; Zhao, J.; Gao, J. Excellent Her and Oer Catalyzing Performance of Se-Vacancies in Defects-Engineered $PtSe_2$: From Simulation to Experiment. *Adv. Energy Mater.* **2022**, *12*, 2102359. [CrossRef]
15. Xue, H.; Yang, T.; Zhang, Z.; Zhang, Y.; Geng, Z.; He, Y. Stimulate the Hidden Catalysis Potential and Exposure of Nickel Site in NiSe@CNTs Result in Ultra-High HER/OER Activity and Stability. *Appl. Catal. B Environ.* **2023**, *330*, 122641. [CrossRef]
16. Zhu, E.; Shi, C.; Yu, J.; Jin, H.; Zhou, L.; Yang, X.; Xu, M. Simultaneous Regulation of Thermodynamic and Kinetic Behavior on FeN_3P_1 Single-Atom Configuration by Fe_2P for Efficient Bifunctional ORRr/OER. *Appl. Catal. B Environ. Energy* **2024**, *347*, 123796. [CrossRef]
17. Cao, L.-M.; Zhang, J.; Ding, L.-W.; Du, Z.-Y.; He, C.-T. Metal-Organic Frameworks Derived Transition Metal Phosphides for Electrocatalytic Water Splitting. *J. Energy Chem.* **2022**, *68*, 494–520. [CrossRef]
18. Zhang, R.; Liu, W.; Zhang, F.-M.; Yang, Z.-D.; Zhang, G.; Zeng, X.C. COF-C_4N Nanosheets with Uniformly Anchored Single Metal Sites for Electrocatalytic OER: From Theoretical Screening to Target Synthesis. *Appl. Catal. B Environ.* **2023**, *325*, 122366. [CrossRef]
19. Yu, Y.; Zhou, J.; Sun, Z. Novel 2d Transition-Metal Carbides: Ultrahigh Performance Electrocatalysts for Overall Water Splitting and Oxygen Reduction. *Adv. Funct. Mater.* **2020**, *30*, 2000570. [CrossRef]
20. Lao, M.; Li, P.; Jiang, Y.; Pan, H.; Dou, S.X.; Sun, W. From Fundamentals and Theories to Heterostructured Electrocatalyst Design: An in-Depth Understanding of Alkaline Hydrogen Evolution Reaction. *Nano Energy* **2022**, *98*, 107231. [CrossRef]
21. Liang, W.; Zhou, M.; Lin, X.; Xu, J.; Dong, P.; Le, Z.; Yang, M.; Chen, J.; Xie, F.; Wang, N.; et al. Nickel-Doped Tungsten Oxide Promotes Stable and Efficient Hydrogen Evolution in Seawater. *Appl. Catal. B Environ.* **2023**, *325*, 122397. [CrossRef]
22. Guo, T.; Fei, H.; Liu, R.; Liu, F.; Wang, D.; Wu, Z. Constructing Molybdenum Vacancy Defect for Mop with Optimized P-Band Center Towards High-Efficiency Hydrogen Evolution. *Appl. Catal. B Environ.* **2024**, *343*, 123480. [CrossRef]
23. Cai, C.; Liu, K.; Zhang, L.; Li, F.; Tan, Y.; Li, P.; Wang, Y.; Wang, M.; Feng, Z.; Motta Meira, D.; et al. Atomically Local Electric Field Induced Interface Water Reorientation for Alkaline Hydrogen Evolution Reaction. *Angew. Chem. Int. Ed.* **2023**, *62*, e202300873. [CrossRef] [PubMed]
24. Nie, N.; Zhang, D.; Wang, Z.; Yu, W.; Ge, S.; Xiong, J.; Gu, Y.; Yang, B.; Lai, J.; Wang, L. Stable PtNb-Nb_2O_5 Heterostructure Clusters @CC for High-Current-Density Neutral Seawater Hydrogen Evolution. *Appl. Catal. B Environ.* **2022**, *318*, 121808. [CrossRef]
25. Sun, H.; Yao, B.; Han, Y.; Yang, L.; Zhao, Y.; Wang, S.; Zhong, C.; Chen, J.; Li, C.-P.; Du, M. Multi-Interface Engineering of Self-Supported Nickel/Yttrium Oxide Electrode Enables Kinetically Accelerated and Ultra-Stable Alkaline Hydrogen Evolution at Industrial-Level Current Density. *Adv. Energy Mater.* **2024**, *14*, 2303563. [CrossRef]
26. Wang, J.; Zang, N.; Xuan, C.; Jia, B.; Jin, W.; Ma, T. Self-Supporting Electrodes for Gas-Involved Key Energy Reactions. *Adv. Funct. Mater.* **2021**, *31*, 2104620. [CrossRef]
27. Shi, H.; Sun, X.-Y.; Liu, Y.; Zeng, S.-P.; Zhang, Q.-H.; Gu, L.; Wang, T.-H.; Han, G.-F.; Wen, Z.; Fang, Q.-R.; et al. Multicomponent Intermetallic Nanoparticles on Hierarchical Metal Network as Versatile Electrocatalysts for Highly Efficient Water Splitting. *Adv. Funct. Mater.* **2023**, *33*, 2214412. [CrossRef]
28. Zhou, X.; Mo, Y.; Yu, F.; Liao, L.; Yong, X.; Zhang, F.; Li, D.; Zhou, Q.; Sheng, T.; Zhou, H. Engineering Active Iron Sites on Nanoporous Bimetal Phosphide/Nitride Heterostructure Array Enabling Robust Overall Water Splitting. *Adv. Funct. Mater.* **2023**, *33*, 2209465. [CrossRef]
29. Zhou, Q.; Xu, C.; Hou, J.; Ma, W.; Jian, T.; Yan, S.; Liu, H. Duplex Interpenetrating-Phase FeNiZn and $FeNi_3$ Heterostructure with Low-Gibbs Free Energy Interface Coupling for Highly Efficient Overall Water Splitting. *Nano-Micro Lett.* **2023**, *15*, 95. [CrossRef]

30. Zhang, L.; Hu, Z.; Huang, J.; Chen, Z.; Li, X.; Feng, Z.; Yang, H.; Huang, S.; Luo, R. Experimental and DFT Studies of Flower-Like Ni-Doped Mo_2C on Carbon Fiber Paper: A Highly Efficient and Robust HER Electrocatalyst Modulated by $Ni(NO_3)_2$ Concentration. *J. Adv. Ceram.* **2022**, *11*, 1294–1306. [CrossRef]
31. Moreno-García, P.; Gálvez-Vázquez, M.d.J.; Prenzel, T.; Winter, J.; Gálvez-Vázquez, L.; Broekmann, P.; Waldvogel, S.R. Self-Standing Metal Foam Catalysts for Cathodic Electro-Organic Synthesis. *Adv. Mater.* **2024**, *36*, 2307461. [CrossRef] [PubMed]
32. Liu, W.; Wang, X.; Qu, J.; Liu, X.; Zhang, Z.; Guo, Y.; Yin, H.; Wang, D. Tuning Ni Dopant Concentration to Enable Co-Deposited Superhydrophilic Self-Standing Mo_2C Electrode for High-Efficient Hydrogen Evolution Reaction. *Appl. Catal. B Environ. Energy* **2022**, *307*, 121201. [CrossRef]
33. An, W.Y.; Kim, S.; Lee, W.; Choi, S.; Choi, S.R.; Yoo, S.; Han, J.W.; Li, O.L.; Park, J.-Y. Electrically and Morphologically Tailored Misfit-Layered Structure $Gd_{0.3}Ca_{2.7}Co_{3.82}Cu_{0.18}O_9$ Nanofibers as Efficient Oxygen Catalysts for Zinc-Air Batteries. *Appl. Catal. B Environ. Energy* **2024**, *358*, 124354. [CrossRef]
34. Zhong, M.; Xu, M.; Ren, S.; Li, W.; Wang, C.; Gao, M.; Lu, X. Modulating the Electronic Structure of $Ni(OH)_2$ by Coupling with Low-Content Pt for Boosting the Urea Oxidation Reaction Enables Significantly Promoted Energy-Saving Hydrogen Production. *Energy Environ. Sci.* **2024**, *17*, 1984–1996. [CrossRef]
35. Wang, J.; Wang, Z.; Ni, J.; Li, L. Electrospinning for Flexible Sodium-Ion Batteries. *Energy Storage Mater.* **2022**, *45*, 704–719. [CrossRef]
36. Ahmadijokani, F.; Molavi, H.; Bahi, A.; Fernández, R.; Alaee, P.; Wu, S.; Wuttke, S.; Ko, F.; Arjmand, M. Metal-Organic Frameworks and Electrospinning: A Happy Marriage for Wastewater Treatment. *Adv. Funct. Mater.* **2022**, *32*, 2207723. [CrossRef]
37. Li, C.; Qiu, M.; Li, R.; Li, X.; Wang, M.; He, J.; Lin, G.; Xiao, L.; Qian, Q.; Chen, Q.; et al. Electrospinning Engineering Enables High-Performance Sodium-Ion Batteries. *Adv. Fiber Mater.* **2022**, *4*, 43–65. [CrossRef]
38. Wang, J.; Chen, X.; Wang, J.; Cui, X.; Wang, Z.; Zhang, G.; Lyu, W.; Shkunov, M.; Silva, S.R.P.; Liao, Y.; et al. Electrospinning Engineering of Gas Electrodes for High-Performance Lithium–Gas Batteries. *Carbon Energy* **2024**, *6*, e572. [CrossRef]
39. Zhu, J.; Yan, C.; Li, G.; Cheng, H.; Li, Y.; Liu, T.; Mao, Q.; Cho, H.; Gao, Q.; Gao, C.; et al. Recent Developments of Electrospun Nanofibers for Electrochemical Energy Storage and Conversion. *Energy Storage Mater.* **2024**, *65*, 103111. [CrossRef]
40. Zong, D.; Zhang, X.; Yin, X.; Wang, F.; Yu, J.; Zhang, S.; Ding, B. Electrospun Fibrous Sponges: Principle, Fabrication, and Applications. *Adv. Fiber Mater.* **2022**, *4*, 1434–1462. [CrossRef]
41. Li, S.; Guo, H.; He, S.; Yang, H.; Liu, K.; Duan, G.; Jiang, S. Advanced Electrospun Nanofibers as Bifunctional Electrocatalysts for Flexible Metal-Air (O_2) Batteries: Opportunities and Challenges. *Mater. Des.* **2022**, *214*, 110406. [CrossRef]
42. Zhang, J.; Yu, M.; Tao, S. Advanced Electrospinning Nanomaterials: From Spinning Fabrication Techniques to Electrochemical Applications. *Nano Res.* **2024**, *17*, 7077–7116. [CrossRef]
43. Lei, W.; Li, H.; Tang, Y.; Shao, H. Progress and Perspectives on Electrospinning Techniques for Solid-State Lithium Batteries. *Carbon Energy* **2022**, *4*, 539–575. [CrossRef]
44. Angel, N.; Li, S.; Yan, F.; Kong, L. Recent Advances in Electrospinning of Nanofibers from Bio-Based Carbohydrate Polymers and Their Applications. *Trends Food Sci. Technol.* **2022**, *120*, 308–324. [CrossRef]
45. Fu, Y.; Liu, J.; Zou, J.; Xu, S.; Wei, Y.; Zhang, W.; Li, D. Electrospun PAN/$BaTiO_3$/Mxene Nanofibrous Membrane with Significantly Improved Piezoelectric Property for Self-Powered Wearable Sensor. *Chem. Eng. J.* **2024**, *489*, 151495. [CrossRef]
46. Ziyadi, H.; Baghali, M.; Bagherianfar, M.; Mehrali, F.; Faridi-Majidi, R. An Investigation of Factors Affecting the Electrospinning of Poly (Vinyl Alcohol)/Kefiran Composite Nanofibers. *Adv. Compos. Hybrid Mater.* **2021**, *4*, 768–779. [CrossRef] [PubMed]
47. Hu, X.; Xu, Y.; Wang, J.; Ma, J.; Wang, L.; Jiang, W. In Situ Fabrication of Superfine Perovskite Composite Nanofibers with Ultrahigh Stability by One-Step Electrospinning toward White Light-Emitting Diode. *Adv. Fiber Mater.* **2023**, *5*, 183–197. [CrossRef]
48. Bhatta, T.; Maharjan, P.; Cho, H.; Park, C.; Yoon, S.H.; Sharma, S.; Salauddin, M.; Rahman, M.T.; Rana, S.M.S.; Park, J.Y. High-Performance Triboelectric Nanogenerator Based on Mxene Functionalized Polyvinylidene Fluoride Composite Nanofibers. *Nano Energy* **2021**, *81*, 105670. [CrossRef]
49. Yin, J.; Xu, L.; Ahmed, A. Batch Preparation and Characterization of Electrospun Porous Polylactic Acid-Based Nanofiber Membranes for Antibacterial Wound Dressing. *Adv. Fiber Mater.* **2022**, *4*, 832–844. [CrossRef]
50. Liu, X.; Zhang, Y.; Guo, X.; Pang, H. Electrospun Metal–Organic Framework Nanofiber Membranes for Energy Storage and Environmental Protection. *Adv. Fiber Mater.* **2022**, *4*, 1463–1485. [CrossRef]
51. Shankar, E.G.; Vasant, P.M.; Yu, J.S. Binder-Free Electrospun Nickel Cerium Selenide Nanofiber Electrodes Based on Voltage-Stimulated Diameter Refinement for Solar-Charged Quasi-Solid-State Wearable Supercapacitors. *Adv. Fiber Mater.* **2024**, *6*, 106–119. [CrossRef]
52. Wang, Y.Z.; Yang, M.; Ding, Y.-M.; Li, N.-W.; Yu, L. Recent Advances in Complex Hollow Electrocatalysts for Water Splitting. *Adv. Funct. Mater.* **2022**, *32*, 2108681. [CrossRef]

53. Gao, G.; Chen, X.; Han, L.; Zhu, G.; Jia, J.; Cabot, A.; Sun, Z. Advances in Mofs and Their Derivatives for Non-Noble Metal Electrocatalysts in Water Splitting. *Coord. Chem. Rev.* **2024**, *503*, 215639. [CrossRef]
54. Li, J.; Zhang, J.; Shen, J.; Wu, H.; Chen, H.; Yuan, C.; Wu, N.; Liu, G.; Guo, D.; Liu, X. Self-Supported Electrocatalysts for the Hydrogen Evolution Reaction. *Mater. Chem. Front.* **2023**, *7*, 567–606. [CrossRef]
55. Li, W.; Wang, C.; Lu, X. Integrated Transition Metal and Compounds with Carbon Nanomaterials for Electrochemical Water Splitting. *J. Mater. Chem. A* **2021**, *9*, 3786–3827. [CrossRef]
56. Wang, S.; Lu, A.; Zhong, C.J. Hydrogen Production from Water Electrolysis: Role of Catalysts. *Nano Converg* **2021**, *8*, 4. [CrossRef]
57. Sun, H.; Yan, Z.; Liu, F.; Xu, W.; Cheng, F.; Chen, J. Self-Supported Transition-Metal-Based Electrocatalysts for Hydrogen and Oxygen Evolution. *Adv. Mater.* **2020**, *32*, e1806326. [CrossRef] [PubMed]
58. Yin, H.; Rong, F.; Xie, Y. A Review of Typical Transition Metal Phosphides Electrocatalysts for Hydrogen Evolution Reaction. *Int. J. Hydrogen Energy* **2023**, *52*, 350–375. [CrossRef]
59. Zhu, J.; Hu, L.; Zhao, P.; Lee, L.Y.S.; Wong, K.-Y. Recent Advances in Electrocatalytic Hydrogen Evolution Using Nanoparticles. *Chem. Rev.* **2020**, *120*, 851–918. [CrossRef]
60. Seh, Z.W.; Kibsgaard, J.; Dickens, C.F.; Chorkendorff, I.; Norskov, J.K.; Jaramillo, T.F. Combining Theory and Experiment in Electrocatalysis: Insights into Materials Design. *Science* **2017**, *355*, eaad4998. [CrossRef] [PubMed]
61. Li, H.; Lin, Y.; Duan, J.; Wen, Q.; Liu, Y.; Zhai, T. Stability of Electrocatalytic OER: From Principle to Application. *Chem. Soc. Rev.* **2024**, *53*, 10709–10740. [CrossRef]
62. Sahoo, D.P.; Das, K.K.; Mansingh, S.; Sultana, S.; Parida, K. Recent Progress in First Row Transition Metal Layered Double Hydroxide (LDH) Based Electrocatalysts Towards Water Splitting: A Review with Insights on Synthesis. *Coord. Chem. Rev.* **2022**, *469*, 214666. [CrossRef]
63. Liang, Z.; Shen, D.; Wang, L.; Fu, H. Recent Progress of Cobalt-Based Electrocatalysts for Water Splitting: Electron Modulation, Surface Reconstitution, and Applications. *Nano Res.* **2024**, *17*, 2234–2269. [CrossRef]
64. Lv, X.-W.; Tian, W.-W.; Yuan, Z.-Y. Recent Advances in High-Efficiency Electrocatalytic Water Splitting Systems. *Electrochem. Energy Rev.* **2023**, *6*, 23. [CrossRef]
65. Fan, C.; Zang, Z.; Zhang, X. Non-Metal Doping Regulation in Transition Metal and Their Compounds for Electrocatalytic Water Splitting. *Int. J. Hydrogen Energy* **2024**, *56*, 1273–1283. [CrossRef]
66. Wang, C.; Guo, W.; Chen, T.; Lu, W.; Song, Z.; Yan, C.; Feng, Y.; Gao, F.; Zhang, X.; Rao, Y.; et al. Advanced Noble-Metal/Transition-Metal/Metal-Free Electrocatalysts for Hydrogen Evolution Reaction in Water-Electrolysis for Hydrogen Production. *Coord. Chem. Rev.* **2024**, *514*, 215899. [CrossRef]
67. Zhou, B.; Gao, R.; Zou, J.-J.; Yang, H. Surface Design Strategy of Catalysts for Water Electrolysis. *Small* **2022**, *18*, 2202336. [CrossRef]
68. Guan, J.; Chen, W.; Fang, Y.; Wang, L.; Fu, Y.; Guo, B.; Zhang, M. Electrospinning Derivative Fabrication of Sandwich-Structured CNF/Co_3S_4/MoS_2 as Self-Supported Electrodes to Accelerate Electron Transport in HER. *Int. J. Hydrogen Energy* **2022**, *47*, 14930–14941. [CrossRef]
69. Bera, C.; Streckova, M.; Orinakova, R.; Guboova, A.; Bystron, T.; Girman, V.; Kromka, F.; Podobova, M.; Bouzek, K. NiCoP Fibers as Novel Catalysts for Hydrogen Evolution in Alkali and Acidic Environment. *Int. J. Hydrogen Energy* **2024**, *60*, 118–132. [CrossRef]
70. Wang, P.; Zheng, J.; Li, X.-H.; Cui, W.-B.; Liu, J.-H.; Wan, Y.; Zhang, J.; Yamauchi, Y.; Wang, Z.-L.; Niu, M.; et al. Carbon Nanofiber Catalysts Containing High-Entropy Metal Phosphides with Low-Content Ru for Highly Efficient Hydrogen Evolution Reaction. *Rare Met.* **2025**, *44*, 324–335. [CrossRef]
71. Kong, X.; Wang, L.; Xi, Z.; Liu, Y.; Zhou, Y.; Wan, Z.; Chen, X.; Li, S.; Rong, L. Fe7se8 Nanoparticles Encapsulated by CNT Reinforced Fibrous Network with Advanced Sodium Ion Storage and Hydrogen Evolution Reaction. *Electrochim. Acta* **2023**, *439*, 141660. [CrossRef]
72. Chen, J.; Zhang, H.; Yu, J.; Guan, D.; She, S.; Zhou, W.; Shao, Z. Self-Catalyzed Formation of Strongly Interconnected Multiphase Molybdenum-Based Composites for Efficient Hydrogen Evolution. *Carbon Energy* **2022**, *4*, 77–87. [CrossRef]
73. Zhang, L.; Wei, K.; Ma, J.; Wang, J.; Liu, Z.; Xing, R.; Jiao, T. Coupled Sn/Mo_2C Nanoparticles Wrapped in Carbon Nanofibers by Electrospinning as High-Performance Electrocatalyst for Hydrogen Evolution Reaction. *Appl. Surf. Sci.* **2021**, *566*, 150754. [CrossRef]
74. Wang, L.; He, W.; Yin, D.; Zhang, H.; Liu, D.; Yang, Y.; Yu, W.; Dong, X. Con/Moc Embedded in Nitrogen-Doped Multi-Channel Carbon Nanofibers as an Efficient Acidic and Alkaline Hydrogen Evolution Reaction Electrocatalysts. *Renew. Sustain. Energy Rev.* **2023**, *181*, 113354. [CrossRef]

75. Li, T.; Yin, J.; Sun, D.; Zhang, M.; Pang, H.; Xu, L.; Zhang, Y.; Yang, J.; Tang, Y.; Xue, J. Manipulation of Mott−Schottky Ni/CeO_2 Heterojunctions into N-Doped Carbon Nanofibers for High-Efficiency Electrochemical Water Splitting. *Small* **2022**, *18*, 2106592. [CrossRef] [PubMed]
76. Liao, Y.; Kang, T.; Liu, F.; Zhou, S.; Liu, X.; Shen, S.; Tang, Z. Ni Nanoparticles Embedded in Multi-Channel Carbon Nanofibers: Self-Supporting Electrodes for Bifunctional Catalysis of Hydrogen and Oxygen Evolution Reactions. *J. Alloys Compd.* **2024**, *998*, 174894. [CrossRef]
77. He, H.; Zhang, Y.; Zhang, W.; Li, Y.; Zhu, X.; Wang, P.; Hu, D. Porous Carbon Nanofibers Derived from Silk Fibroin through Electrospinning as N-Doped Metal-Free Catalysts for Hydrogen Evolution Reaction in Acidic and Alkaline Solutions. *ACS Appl. Mater. Interfaces* **2021**, *14*, 834–849. [CrossRef] [PubMed]
78. Lin, H.; Zhang, W.; Shi, Z.; Che, M.; Yu, X.; Tang, Y.; Gao, Q. Electrospinning Hetero-Nanofibers of Fe_3C-Mo_2C/Nitrogen-Doped-Carbon as Efficient Electrocatalysts for Hydrogen Evolution. *ChemSusChem* **2017**, *10*, 2597–2604. [CrossRef] [PubMed]
79. Surendran, S.; Shanmugapriya, S.; Sivanantham, A.; Shanmugam, S.; Kalai Selvan, R. Electrospun Carbon Nanofibers Encapsulated with NiCoP: A Multifunctional Electrode for Supercapattery and Oxygen Reduction, Oxygen Evolution, and Hydrogen Evolution Reactions. *Adv. Energy Mater.* **2018**, *8*, 1800555. [CrossRef]
80. Li, Y.; Li, H.; Cao, K.; Jin, T.; Wang, X.; Sun, H.; Ning, J.; Wang, Y.; Jiao, L. Electrospun Three Dimensional Co/CoP@Nitrogen-Doped Carbon Nanofibers Network for Efficient Hydrogen Evolution. *Energy Storage Mater.* **2018**, *12*, 44–53. [CrossRef]
81. Zhang, Y.; Kong, D.; Bo, L.; Shi, W.; Guan, X.; Wang, Y.; Lei, Z.; Tong, J. Electrospinning Preparation of N, P Dual-Doped Molybdenum Carbide/Porous Carbon Fibers with Highly Improved Electrocatalytic Activity for Hydrogen Evolution Reaction. *ACS Appl. Energy Mater.* **2021**, *4*, 13051–13060. [CrossRef]
82. Streckova, M.; Mudra, E.; Orinakova, R.; Markusova-Buckova, L.; Sebek, M.; Kovalcikova, A.; Sopcak, T.; Girman, V.; Dankova, Z.; Micusik, M.; et al. Nickel and Nickel Phosphide Nanoparticles Embedded in Electrospun Carbon Fibers as Favourable Electrocatalysts for Hydrogen Evolution. *Chem. Eng. J.* **2016**, *303*, 167–181. [CrossRef]
83. Li, L.; Yu, M.; Gao, C.; Yan, X.; Zhang, N.; Bao, J.; Zhang, X. MoS_2/CoS_2 with Heterostructure In-Situ Grown on the Surface of Electrospinning Carbon Nanotube for Efficient Hydrogen Evolution Reaction. *J. Alloys Compd.* **2024**, *1003*, 175530. [CrossRef]
84. Xiao, W.; Li, Y.; Zhang, Y.; Wang, P.; Qi, N. Silk Derived Electrospun Carbon Nanofibers Loaded with CoS Nanoparticles as Efficient Hydrogen Evolution Reaction Catalysts in Acidic and Alkaline Media. *Fuel* **2024**, *369*, 131797. [CrossRef]
85. Yu, S.; Zou, Y.; Wang, Q.; Xu, J.; Xiang, C.; Xu, F.; Sun, L.; Yang, F. Self-Supported Co–Mo Sulfide in Electrospun Carbon Nanofibers as Electrocatalysts for Hydrogen Evolution Reaction in Alkaline Medium. *J. Alloys Compd.* **2022**, *911*, 165094. [CrossRef]
86. Chen, X.; Yu, B.; Dong, Y.; Zhu, X.; Zhang, W.; Ramakrishna, S.; Liu, Z. Electrospun Porous Carbon Nanofibers Decorated with Iron-Doped Cobalt Phosphide Nanoparticles for Hydrogen Evolution. *J. Alloys Compd.* **2022**, *918*, 165733. [CrossRef]
87. Gong, Y.; Xu, L.-H.; Li, J.; Shan, D. Confinement of Transition Metal Phosphides in N, P-Doped Electrospun Carbon Fibers for Enhanced Electrocatalytic Hydrogen Evolution. *J. Alloys Compd.* **2021**, *875*, 159934. [CrossRef]
88. Zhong, M.; Li, W.; Chen, J.; Ren, S.; Qi, R.; Wang, C.; Lu, X. Surface Reconstruction of Fe(III)/NiS Nanotubes for Generating High-Performance Oxygen-Evolution Catalyst. *Sep. Purif. Technol.* **2023**, *310*, 123164. [CrossRef]
89. Li, W.; Zhong, M.; Chen, X.; Ren, S.; Yan, S.; Wang, C.; Lu, X. Hierarchical Amorphous Bimetallic Sulfide Nanosheets Supported on Co-C Nanofibers to Synergistically Boost Water Electrolysis. *Sci. China Mater.* **2023**, *66*, 2235–2245. [CrossRef]
90. Sankar, S.S.; Rathishkumar, A.; Geetha, K.; Kundu, S. Electrospinning as a Tool in Fabricating Hydrated Porous Cobalt Phosphate Fibrous Network as High Rate OER Electrocatalysts in Alkaline and Neutral Media. *Int. J. Hydrogen Energy* **2021**, *46*, 10366–10376. [CrossRef]
91. Cui, W.; Xu, S.; Bai, J.; Li, C.; Sun, X. Zif-67-Derived Se-Doped CoSe2 Grown on Carbon Nanofibers as Oxygen Electrocatalysts for Rechargeable Zn–Air Batteries. *New J. Chem.* **2024**, *48*, 4310–4319. [CrossRef]
92. Li, Z.; Xue, K.-H.; Wang, J.; Li, J.-G.; Ao, X.; Sun, H.; Song, X.; Lei, W.; Cao, Y.; Wang, C. Cation and Anion Co-Doped Perovskite Nanofibers for Highly Efficient Electrocatalytic Oxygen Evolution. *ACS Appl. Mater. Interfaces* **2020**, *12*, 41259–41268. [CrossRef] [PubMed]
93. He, H.; Lei, Y.; Liu, S.; Thummavichai, K.; Zhu, Y.; Wang, N. Tunable Active-Sites of Co– Nanoparticles Encapsulated in Carbon Nanofiber as High Performance Bifunctional OER/ORR Electrocatalyst. *J. Colloid Interface Sci.* **2023**, *630*, 140–149. [CrossRef]
94. Wang, Z.; Han, S.; Zhang, Y.; Wang, X.; Bai, Q.; Wang, Y. Constructing Oxygen Vacancies by Selective Anion Doping in High Entropy Perovskite Oxide for Water Splitting. *Renew. Energy* **2024**, *233*, 121180. [CrossRef]
95. Yuan, B.; Zou, J.; Jian, J.; Wang, J.; Xu, H.; Zhang, X.; Zhang, C.; Miao, H.; Yuan, J. Promoting the Bifunctional Oxygen Catalytic Activity of the Perovskite and Co_3O_4 Heterostructure by Introducing Oxygen Vacancies. *Int. J. Hydrogen Energy* **2024**, *80*, 801–809. [CrossRef]

96. Yao, X.; Li, J.; Zhu, Y.; Li, L.; Zhang, W. Designed Synthesis of Three-Dimensional Callistemon-Like Networks Structural Multifunctional Electrocatalyst: Graphitic-Carbon-Encapsulated Co Nanoparticles/N-Doped Carbon Nanotubes@Carbon Nanofibers for Zn-Air Batteries Application. *Compos. Part B Eng.* **2020**, *193*, 108058. [CrossRef]
97. Wei, P.; Sun, X.; Liang, Q.; Li, X.; He, Z.; Hu, X.; Zhang, J.; Wang, M.; Li, Q.; Yang, H.; et al. Enhanced Oxygen Evolution Reaction Activity by Encapsulating Nife Alloy Nanoparticles in Nitrogen-Doped Carbon Nanofibers. *ACS Appl. Mater. Interfaces* **2020**, *12*, 31503–31513. [CrossRef] [PubMed]
98. Feng, L.; Ding, R.; Chen, Y.; Wang, J.; Xu, L. Zeolitic Imidazolate Framework-67 Derived Ultra-Small CoP Particles Incorporated into N-Doped Carbon Nanofiber as Efficient Bifunctional Catalysts for Oxygen Reaction. *J. Power Sources* **2020**, *452*, 227837. [CrossRef]
99. Fan, K.; Zong, L.; Liu, J.; Chuang, C.-H.; Dong, M.; Zou, Y.; Xu, Y.; Fu, H.Q.; Zhang, L.; Wang, L.; et al. In Situ Reconstruction to Surface Sulfide Adsorbed Metal Scaffold for Enhanced Electrocatalytic Hydrogen Evolution Activity. *Adv. Energy Mater.* **2024**, *14*, 2400052. [CrossRef]
100. Bao, W.; Liu, J.; Ai, T.; Han, J.; Hou, J.; Li, W.; Wei, X.; Zou, X.; Deng, Z.; Zhang, J. Unveiling the Role of Surface Self-Reconstruction of Metal Chalcogenides on Electrocatalytic Oxygen Evolution Reaction. *Adv. Funct. Mater.* **2024**, *34*, 2408364. [CrossRef]
101. Yi, L.; Nie, K.; Li, B.; Zhang, Y.; Hu, C.; Hao, X.; Wang, Z.; Qu, X.; Liu, Z.; Huang, W. Tailoring Copper Single-Atoms-Stabilized Metastable Transition-Metal-Dichalcogenides for Sustainable Hydrogen Production. *Angew. Chem. Int. Ed.* **2025**, *64*, e202414701. [CrossRef] [PubMed]
102. Zhu, Z.; Hao, J.; Zhu, H.; Sun, S.; Duan, F.; Lu, S.; Du, M. In Situ Fabrication of Electrospun Carbon Nanofibers–Binary Metal Sulfides as Freestanding Electrode for Electrocatalytic Water Splitting. *Adv. Fiber Mater.* **2021**, *3*, 117–127. [CrossRef]
103. Zhou, Q.; Wang, Z.; Yuan, H.; Wang, J.; Hu, H. Rapid Hydrogen Adsorption-Desorption at Sulfur Sites Via an Interstitial Carbon Strategy for Efficient Her on MoS_2. *Appl. Catal. B Environ.* **2023**, *332*, 122750. [CrossRef]
104. Chen, B.; Hu, P.; Yang, F.; Hua, X.; Yang, F.F.; Zhu, F.; Sun, R.; Hao, K.; Wang, K.; Yin, Z. In Situ Porousized MoS_2 Nano Islands Enhance HER/OER Bifunctional Electrocatalysis. *Small* **2023**, *19*, 2207177. [CrossRef]
105. Li, S.; Xia, T.; Dou, Y.; Xie, Y.; Wang, J.; Zhao, H.; Huo, L. Phosphatizing Engineering of Perovskite Oxide Nanofibers for Hydrogen Evolution Reaction to Achieve Extraordinary Electrocatalytic Performance. *Adv. Funct. Mater.* **2022**, *32*, 2112164. [CrossRef]
106. Ding, M.; Wei, Z.; Zhao, W.; Lu, Q.; Lu, C.; Zhou, M.; Liu, D.; Yang, H. Bimetallic $Ni_{2-x}Co_xP$ Carbon Nanofibers Network: Solid–Solution Alloy Nano-Architecture as Efficient Electrocatalyst for Water Splitting. *Green Chem.* **2024**, *26*, 7789–7798. [CrossRef]
107. Huang, T.; Xu, G.; Ding, H.; Liu, X.; Zhang, L. NiFeP Nanosheet Arrays Supported on Nickel Nitrogen Codoped Carbon Nanofiber as an Efficient Bifunctional Catalyst for Overall Water Splitting. *Sep. Purif. Technol.* **2025**, *354*, 129094. [CrossRef]
108. Liu, D.; Xu, G.; Yang, H.; Wang, H.; Xia, B.Y. Rational Design of Transition Metal Phosphide-Based Electrocatalysts for Hydrogen Evolution. *Adv. Funct. Mater.* **2023**, *33*, 2208358. [CrossRef]
109. Liu, X.; Xu, G.; Ding, H.; Zhang, L.; Huang, T. NiCoP Nanoparticles Embedded in Coal-Based Carbon Nanofibers as Self-Supporting Bifunctional Electrocatalyst toward Water Splitting. *Int. J. Hydrogen Energy* **2023**, *48*, 35064–35074. [CrossRef]
110. Lei, Y.; Wang, Q.; Peng, S.; Ramakrishna, S.; Zhang, D.; Zhou, K. Electrospun Inorganic Nanofibers for Oxygen Electrocatalysis: Design, Fabrication, and Progress. *Adv. Energy Mater.* **2020**, *10*, 1902115. [CrossRef]
111. Karingula, S.; Kummari, S.; Kotagiri, Y.G.; Bhookya, T.; Gobi, K.V. Development of 1d-Metalloid-Induced Highly Porous Carbon Nanofiber Conjugated with PEDOT Polymer through Concurrent Selenization of Zif-67 for Energy Storage and Green H_2 Production. *Small* **2024**, *20*, 2400812. [CrossRef] [PubMed]
112. Jiang, Y.; Gao, S.; Liu, X.; Wang, Y.; Zhou, S.; Liu, Q.; Abdukayum, A.; Hu, G. Recent Achievements in Selenium-Based Transition Metal Electrocatalysts for pH-Universal Water Splitting. *Nano Res.* **2024**, *17*, 5763–5785. [CrossRef]
113. Wang, Z.; Liu, S.; Duan, W.; Xing, Y.; Hu, Y.; Ma, Y. Transition Metal Selenides as Catalysts for Electrochemical Water Splitting. *Int. J. Hydrogen Energy* **2024**, *60*, 1414–1432. [CrossRef]
114. Peng, X.; Yan, Y.; Jin, X.; Huang, C.; Jin, W.; Gao, B.; Chu, P.K. Recent Advance and Prospectives of Electrocatalysts Based on Transition Metal Selenides for Efficient Water Splitting. *Nano Energy* **2020**, *78*, 105234. [CrossRef]
115. Gong, T.; Zhang, J.; Liu, Y.; Hou, L.; Deng, J.; Yuan, C. Construction of Hetero-Phase Mo_2C-CoO@N-CNFs Film as a Self-Supported Bi-Functional Catalyst Towards Overall Water Splitting. *Chem. Eng. J.* **2023**, *451*, 139025. [CrossRef]
116. Huang, T.; Xu, G.; Ding, H.; Zhang, L.; Wei, B.; Liu, X. Ultrafine Cobalt Molybdenum Phosphide Nanoparticles Embedded in Crosslinked Nitrogen-Doped Carbon Nanofiber as Efficient Bifunctional Catalyst for Overall Water Splitting. *J. Colloid Interface Sci.* **2022**, *625*, 956–964. [CrossRef] [PubMed]
117. Wang, X.; Zhang, X.; Fu, G.; Tang, Y. Recent Progress of Electrospun Porous Carbon-Based Nanofibers for Oxygen Electrocatalysis. *Mater. Today Energy* **2021**, *22*, 100850. [CrossRef]

118. Wang, J.; Kong, H.; Zhang, J.; Hao, Y.; Shao, Z.; Ciucci, F. Carbon-Based Electrocatalysts for Sustainable Energy Applications. *Prog. Mater. Sci.* **2021**, *116*, 100717. [CrossRef]
119. Zhong, R.; Zhi, C.; Wu, Y.; Liang, Z.; Tabassum, H.; Zhang, H.; Qiu, T.; Gao, S.; Shi, J.; Zou, R. Atomic Fe-N_4 Sites on Electrospun Hierarchical Porous Carbon Nanofibers as an Efficient Electrocatalyst for Oxygen Reduction Reaction. *Chin. Chem. Lett.* **2020**, *31*, 1588–1592. [CrossRef]
120. Lin, Q.; Shang, C.; Chen, Z.; Wang, X.; Zhou, G. Boron-Doped Molybdenum Carbide as a pH-Independent Electrocatalyst for the Hydrogen Evolution Reaction. *Int. J. Hydrogen Energy* **2020**, *45*, 30659–30665. [CrossRef]
121. Zhang, H.; Wang, L.; He, W.; Liu, D.; Shao, H.; Yu, W.; Yin, D.; Dong, X. Mo_2C Regulated by Cobalt Components Doping in N-Doped Hollow Carbon Nanofibers as an Efficient Electrocatalyst for Hydrogen Evolution Reaction. *Int. J. Hydrogen Energy* **2022**, *47*, 29337–29347. [CrossRef]
122. Han, N.; Zhang, W.; Guo, W.; Pan, H.; Jiang, B.; Xing, L.; Tian, H.; Wang, G.; Zhang, X.; Fransaer, J. Designing Oxide Catalysts for Oxygen Electrocatalysis: Insights from Mechanism to Application. *Nano-Micro Lett.* **2023**, *15*, 185. [CrossRef]
123. Han, J.; Guan, J. Multicomponent Transition Metal Oxides and (oxy)Hydroxides for Oxygen Evolution. *Nano Res.* **2023**, *16*, 1913–1966. [CrossRef]
124. Zhang, Y.; Fu, Q.; Song, B.; Xu, P. Regulation Strategy of Transition Metal Oxide-Based Electrocatalysts for Enhanced Oxygen Evolution Reaction. *Acc. Mater. Res.* **2022**, *3*, 1088–1100. [CrossRef]
125. Wu, Y.; Guo, C.; Yao, R.; Zhang, K.; Li, J.; Liu, G. Modulating Carrier Oxygen Vacancies to Enhance Strong Oxide-Support Interaction in IrO_2/Nb_2O_{5-X} Catalysts for Promoting Acidic Oxygen Evolution Reaction. *Adv. Funct. Mater.* **2024**, *34*, 2410193. [CrossRef]
126. Zhao, J.-W.; Shi, Z.-X.; Li, C.-F.; Ren, Q.; Li, G.-R. Regulation of Perovskite Surface Stability on the Electrocatalysis of Oxygen Evolution Reaction. *ACS Mater. Lett.* **2021**, *3*, 721–737. [CrossRef]
127. McGuire, S.C.; Wesley, W.; Sasaki, K.; Tong, X.; Wong, S.S. Yttrium-Based Double Perovskite Nanorods for Electrocatalysis. *ACS Appl. Mater. Interfaces* **2022**, *14*, 30914–30926. [CrossRef]
128. Xu, X.; Pan, Y.; Zhong, Y.; Ran, R.; Shao, Z. Ruddlesden–Popper Perovskites in Electrocatalysis. *Mater. Horiz.* **2020**, *7*, 2519–2565. [CrossRef]
129. Tian, L.; Zhang, Z.; Liu, S.; Li, G.; Gao, X. High-Entropy Perovskite Oxide Nanofibers as Efficient Bidirectional Electrocatalyst of Liquid-Solid Conversion Processes in Lithium-Sulfur Batteries. *Nano Energy* **2023**, *106*, 108037. [CrossRef]
130. Fu, X.; Lü, S.; Meng, X.; Sun, C.; Wei, M.; Jiang, H.; Gong, W. High-Entropy Cobalt-Free Perovskite as a High-Performing Nanofiber Cathode for Solid Oxide Fuel Cells. *J. Mater. Chem. A* **2024**, *12*, 27452–27463. [CrossRef]
131. Fu, L.; Zhou, J.; Zhou, L.; Yang, J.; Liu, Z.; Wu, K.; Zhao, H.; Wang, J.; Wu, K. Facile Fabrication of Exsolved Nanoparticle-Decorated Hollow Ferrite Fibers as Active Electrocatalyst for Oxygen Evolution Reaction. *Chem. Eng. J.* **2021**, *418*, 129422. [CrossRef]
132. Yang, H.-M.; Weng, C.-C.; Wang, H.-Y.; Yuan, Z.-Y. Transition Metal Nitride-Based Materials as Efficient Electrocatalysts: Design Strategies and Prospective Applications. *Coord. Chem. Rev.* **2023**, *496*, 215410. [CrossRef]
133. Tian, D.; Denny, S.R.; Li, K.; Wang, H.; Kattel, S.; Chen, J.G. Density Functional Theory Studies of Transition Metal Carbides and Nitrides as Electrocatalysts. *Chem. Soc. Rev.* **2021**, *50*, 12338–12376. [CrossRef] [PubMed]
134. Wei, P.; Zhuge, X.; Li, Q.; Sun, X.; Liu, W.; Liang, K.; Han, J.; Ren, Y.; Huang, Y. Interface Engineering and Nanoconfinement Strategies to Synergistically Enhance Hydrogen Evolution in Acidic and Basic Media. *J. Colloid Interface Sci.* **2024**, *662*, 814–821. [CrossRef]
135. Li, R.; Li, Y.; Yang, P.; Ren, P.; Wang, D.; Lu, X.; Xu, R.; Li, Y.; Xue, J.; Zhang, J.; et al. Synergistic Interface Engineering and Structural Optimization of Non-Noble Metal Telluride-Nitride Electrocatalysts for Sustainably Overall Seawater Electrolysis. *Appl. Catal. B Environ.* **2022**, *318*, 121834. [CrossRef]
136. Theerthagiri, J.; Lee, S.J.; Murthy, A.P.; Madhavan, J.; Choi, M.Y. Fundamental Aspects and Recent Advances in Transition Metal Nitrides as Electrocatalysts for Hydrogen Evolution Reaction: A Review. *Curr. Opin. Solid State Mater. Sci.* **2020**, *24*, 100805. [CrossRef]
137. Lin, L.; Piao, S.; Choi, Y.; Lyu, L.; Hong, H.; Kim, D.; Lee, J.; Zhang, W.; Piao, Y. Nanostructured Transition Metal Nitrides as Emerging Electrocatalysts for Water Electrolysis: Status and Challenges. *EnergyChem* **2022**, *4*, 100072. [CrossRef]

catalysts

MDPI

Review

Substrate Engineering of Single Atom Catalysts Enabled Next-Generation Electrocatalysis to Power a More Sustainable Future

Saira Ajmal [1,2], Junfeng Huang [1,*], Jianwen Guo [1], Mohammad Tabish [3], Muhammad Asim Mushtaq [3], Mohammed Mujahid Alam [4,5] and Ghulam Yasin [3,6,*]

1 School of Mechanical Engineering, Dongguan University of Technology, Dongguan 523808, China
2 Department of Precision Machinery and Precision Instrumentation, School of Engineering Science, University of Science and Technology of China, Hefei 230029, China
3 State Key Laboratory of Chemical Resource Engineering, Beijing University of Chemical Technology, Beijing 100029, China
4 Department of Chemistry, College of Science, King Khalid University, Abha 61413, Saudi Arabia
5 Research Center for Advanced Materials Science (RCAMS), King Khalid University, Abha 61413, Saudi Arabia
6 School of Environmental Science and Engineering, Tianjin University, Tianjin 300350, China
* Correspondence: 20221460@dgut.edu.cn (J.H.); yasin.bzu@hotmail.com or yasin@mail.buct.edu.cn (G.Y.)

Abstract: Single-atom catalysts (SACs) are presently recognized as cutting-edge heterogeneous catalysts for electrochemical applications because of their nearly 100% utilization of active metal atoms and having well-defined active sites. In this regard, SACs are considered renowned electrocatalysts for electrocatalytic O_2 reduction reaction (ORR), O_2 evolution reaction (OER), H_2 evolution reaction (HER), water splitting, CO_2 reduction reaction (CO_2RR), N_2 reduction reaction (NRR), and NO_3 reduction reaction (NO_3RR). Extensive research has been carried out to strategically design and produce affordable, efficient, and durable SACs for electrocatalysis. Meanwhile, persistent efforts have been conducted to acquire insights into the structural and electronic properties of SACs when stabilized on an adequate matrix for electrocatalytic reactions. We present a thorough and evaluative review that begins with a comprehensive analysis of the various substrates, such as carbon substrate, metal oxide substrate, alloy-based substrate, transition metal dichalcogenides (TMD)-based substrate, MXenes substrate, and MOF substrate, along with their metal-support interaction (MSI), stabilization, and coordination environment (CE), highlighting the notable contribution of support, which influences their electrocatalytic performance. We discuss a variety of synthetic methods, including bottom-up strategies like impregnation, pyrolysis, ion exchange, atomic layer deposition (ALD), and electrochemical deposition, as well as top-down strategies like host-guest, atom trapping, ball milling, chemical vapor deposition (CVD), and abrasion. We also discuss how diverse regulatory strategies, including morphology and vacancy engineering, heteroatom doping, facet engineering, and crystallinity management, affect various electrocatalytic reactions in these supports. Lastly, the pivotal obstacles and opportunities in using SACs for electrocatalytic processes, along with fundamental principles for developing fascinating SACs with outstanding reactivity, selectivity, and stability, have been highlighted.

Keywords: single-atom catalysts; support engineering; regulation strategies; coordination environment; next-generation electrocatalysis

1. Introduction

In the last decade, the use of fossil fuels has expanded to provide primary energy needs. However, fossil fuels such as coal and natural gas were generating major environmental difficulties around the world, including greenhouse gas and carbon dioxide (CO_2) emissions [1]. The quest for sustainable energy fosters the growth of innovative catalytic materials for different electrochemical reactions such as the O_2 reduction reaction (ORR), O_2 evolution reaction (OER), H_2 evolution reaction (HER), water splitting, CO_2 reduction reaction (CO_2RR), N_2 reduction reaction (NRR), and NO_3 reduction reaction (NO_3RR). Table 1 illustrates the key electrocatalytic reactions, such as ORR, OER, HER, CO_2RR, NRR, and NO_3RR, and their intermediates for both acid and alkaline electrolytes. These reactions are crucial chemical transformation reaction processes that maintain eco-friendly energy conversions, in which the electrocatalytic activity, selectivity, and stability of electrocatalysts play an integral role in optimizing energy efficiencies and system performance [2]. Thus, the development of efficient and cost-effective electrocatalysts is critical and is emerging as a lucrative research area.

Single-atom catalysts (SACs) have recently sparked widespread attention, owing primarily to their distinct structural attributes, which encompass maximized metal usage, optimized reactive active sites, and promising electrocatalytic capabilities [3]. SACs constitute an innovative category of catalyst that involves strong interactions among different atomic species and nearby single atoms (SAs) on a substrate. SACs are disseminated over a variety of support substrates/materials, including carbon, metal oxides, alloys, transition metal dichalcogenides (TMD), MXenes, and metal–organic framework (MOF)-based supports [4]. Accordingly, SA-support interactions can have a remarkable impact on the electrocatalytic activity and durability of dispersed SACs [5]. Comprehending the strong metal-support interactions (SMSI) and the influence of the supports on their characteristics is critical for modifying SACs' catalytic performance, stability, and selectivity. Because of their major effects on electrocatalytic performance, support and SMSI have been thoroughly studied [6]. Furthermore, it has been proven that the supports in SACs can efficiently affect the electrical characteristics to improve the transfer of electrons during electrochemical reactions, allocating them with optimum adsorption and/or desorption energy of their intermediate species [7]. Furthermore, different support materials can stabilize different atomic species to avoid aggregation, significantly improving the electrochemical robustness of SACs [8]. Additionally, extensive surface imperfections in support can significantly alter the local CE and its electronic structure, contributing to the presence of numerous coordinatively unsaturated sites and vacancies [9].

However, there are no limits to the quest for improved electrocatalytic performance. Dual and triple atom catalysts (DACs and TACs) are a newer type of electrocatalysis that can overcome the limitations of SACs. They have characteristics in common with SACs, including the ability to maximize metal dispersion, circumvent the conventional scaling relations that restrict extended surfaces, provide adjacent sites that are required for specific reaction mechanisms, and increase the tunability of the electronic structure and binding energies [10].

This review is distinguished by its thorough and multifaceted discussion of SACs in electrocatalytic applications. It contrasts with traditional reviews that concentrate solely on the electrocatalytic activity of SACs or their synthesis methods. Instead, it provides a comprehensive overview of the interaction between substrates and SACs, highlighting the significance of MSI, stabilization mechanisms, and CEs that influence electrocatalytic performance. The review distinctly merges insights from both bottom-up and top-down synthetic methodologies. Furthermore, it elucidates the study of diverse substrates such as carbon-based materials, metal oxides, alloys, TMDs, MXenes, and MOF-based substrates and emphasizes how the unique features of these substrates and their interactions with SACs affect reactivity and stability. Moreover, it concurrently explores the influence of different regulatory methods such as morphology engineering, vacancy engineering, heteroatom doping, facet engineering, and crystallinity control on these substrates in different electrocatalytic reactions, including ORR, OER, HER, CO_2RR, NRR, and NO_3RR, offering a framework for the development of advanced SACs with remarkable performance. The review ends by summarizing key insights and presenting future directions to address challenges and broaden the use of SACs in real-world applications. In contrast to prior reviews, which frequently focus on specific regulation strategies for SACs or individual electrocatalytic reactions, this review distinguishes itself by systematically linking substrate selection and structural engineering strategies to electrocatalytic performance [11]. While other studies discuss SACs without thoroughly investigating the influence of substrates or addressing the synergistic effects of regulatory techniques across different substrates and electrocatalytic processes, this work presents a thorough approach [3]. It combines substrate-specific analysis, recent synthetic techniques, and regulatory approaches to provide a comprehensive explanation of electrocatalytic reactions. Furthermore, it addresses the role of diverse substrates in overall water-splitting reactions, emphasizing the effectiveness of various performance-enhancing techniques. Unlike others that just focus on the synthesis techniques of SACs and their use for specific electrochemical reactions [4], it analyzes the performance of DACs and DSACs in different electrocatalytic reactions, challenges, and prospects of each substrate in different performance regulatory methods, filling a significant void in the literature. This thorough discussion not only emphasizes the key features of SAC design but also gives insights into overcoming recurrent challenges in the substrates, distinguishing this review from others in the sector.

Table 1. Overview of different electrocatalytic reactions and their intermediates in acidic and alkaline electrolytes.

Reaction	Electrolyte	Half-Reaction (Equation)	Reaction Pathways	Ref.
ORR	Acidic Alkaline	$O_2 + 4H^+ + 4e^- \rightarrow 2H_2O$ $O_2 + 2H_2O + 4e^- \rightarrow 4OH^-$	Direct 4-electron pathway or 2-electron pathway (H_2O_2 intermediate) Same as acidic, but OH^- formation is dominant	[12]
OER	Acidic Alkaline	$2H_2O \rightarrow O_2 + 4H^+ + 4e^-$ $4OH^- \rightarrow O_2 + 2H_2O + 4e^-$	Oxygen evolution via water oxidation Oxygen evolution via hydroxide oxidation	[13]
HER	Acidic Alkaline	$2H^+ + 2e^- \rightarrow H_2$ $2H_2O + 2e^- \rightarrow H_2 + 2OH^-$	Proton reduction Water reduction	[14]
CO_2RR	Acidic Alkaline	$CO_2 + 2H^+ + 2e^- \rightarrow CO + H_2O$ $CO_2 + H_2O + 2e^- \rightarrow CO + 2OH^-$	Multiple pathways: CO, CH_4, HCOOH, etc., based on the catalyst Similar to hydroxide products	[15]
NRR	Acidic Alkaline	$N_2 + 6H^+ + 6e^- \rightarrow 2NH_3$ $N_2 + 6H_2O + 6e^- \rightarrow 2NH_3 + 6OH^-$	Ammonia production via stepwise hydrogenation A similar pathway with water-splitting	[16]
NO_3RR	Acidic Alkaline	$NO_3^- + 10H^+ + 8e^- \rightarrow NH_4^+ + 3H_2O$ $NO_3^- + 6H_2O + 8e^- \rightarrow NH_4^+ + 10OH^-$	Stepwise reduction to NH_4^+ or N_2 Similar reduction with hydroxide products	[17]

2. Scope of This Review

As a revolutionary class of materials with great potential for a broad range of electrocatalytic reactions, SACs have undoubtedly emerged as a new frontier in the area of catalysis. Numerous SACs have been effectively created, and they all have unique performance traits that tackle important issues in environmental and energy conversion applications. The majority of reviews tend to concentrate mostly on the synthesis, characterization, and application aspects of SACs, even if this topic is developing quickly and there are an increasing number of fascinating review articles published one after the other. Nevertheless, there is still a lack of comprehensive research on how various support materials affect SACs' electrocatalytic efficiency.

We aim to fill this gap in this review by providing a thorough analysis of how diverse supports, including carbon-based materials (graphene, graphitic carbon nitride), metal-based oxides (FeO_x, γ-Al_2O_3, and Fe $(OH)_x$), alloy-based materials, TMDs, MXenes (carbides and carbonitrides), and MOFs, affect the performance of SACs in various electrocatalytic reactions (Figure 1). In addition to giving SAs stability, these support materials actively work to improve the electrocatalysts' electroactivity, selectivity, and overall efficiency. Employing mechanisms such as electronic interactions, the creation of new alloy phases, the enhancement of charge transfer, the generation of new interface sites, the optimization of atomic use, and the support of nanoscale geometric topologies could modify the local catalytic environment of SAC supports.

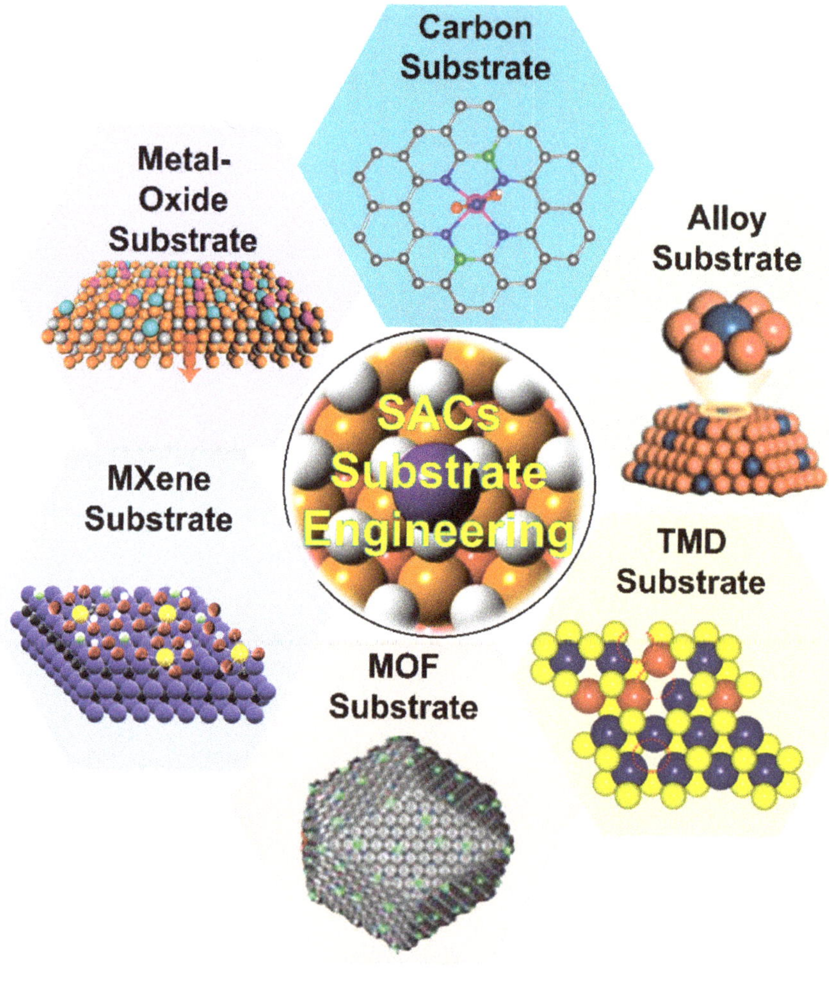

Figure 1. Engineering the various substrates used in SACs for next-generation electrocatalysis. Figure parts taken with permission from [18–24].

To further improve the performance of SACs, we also explore the many approaches that may be used on these supports, such as heteroatom doping, vacancy formation, facet and morphology engineering, and crystallinity control. These tactics affect the supports' electrical characteristics, improve their capacity to stabilize and attach individual atoms, maximize the local CE surrounding the active sites, and in certain situations, even serve as active catalytic sites. This mini-review clarifies the critical role that support engineering plays in pushing the limits of SAC performance in electrocatalysis by methodically examining the synergistic link between SACs and their supports.

3. Synthesis of SACs

The synthesis procedures for developing SACs are critical but difficult, as they need the dispersion of single metal sites onto solid supports while preventing metal aggregation. Various methods for fabricating SACs have been constructed, which are extensively classified into two main strategies: bottom-up (impregnation, pyrolysis, ion-exchange method, atomic layer deposition (ALD), and electrochemical deposition) and top-down (host-guest strategy, atom trapping, ball milling method, chemical vapor deposition (CVD), and abrasion method).

3.1. Bottom-Up Strategies

The bottom-up strategy is widely used for developing SACs. It entails the adsorption of metal precursors onto substrates, followed by various reduction procedures to confine isolated metal atoms on the defect sites of the support. Metal atoms are atomically scattered in the precursors, making it easier to generate an SA structure in the final SAC. The most popular bottom-up technique for producing carbon-based SACs is to carbonize predesigned metal precursors at high temperatures, such as MOFs, metal-porphyrin compounds, and tiny organic molecules. The impregnation method (Figure 2a) is a wet chemistry technique for immobilizing highly distributed active species on supports. In this approach, the metal precursor is combined with the supports, and metal atoms or ions are gradually impregnated onto the supports' surfaces or the entire framework [25]. The pyrolysis process (Figure 2b) is commonly used to create atomically distributed materials. These materials can be generated by decomposing various precursors at high temperatures in a gas atmosphere (like Ar and NH_3), resulting in atomically dispersed catalysts from a variety of precursors, including metal complexes, polymers, MOFs, and so on [26]. Furthermore, to create SACs, the ion exchange method (Figure 2c) substitutes single metal atoms for metal ions that are bound within a host material. This is usually accomplished through a controlled exchange process where the precursor material is loaded with metal ions, which are then exchanged with SAs by treating the material with an appropriate agent [27]. Furthermore, a strong method for creating uniformly dispersed SA thick films that can be used to create noble metal-based SACs is the ALD method (Figure 2d). The metal precursors undergo vaporization, deposition, and confinement on the solid substrate during the synthesis process [28]. Recently, the strong electrostatic adsorption (SEA) method has developed as a viable way for attaching metal precursors onto metal oxide supports to synthesize SACs. By altering the pH at the support's point of zero charge (PZC), the surface becomes positively or negatively charged, allowing for electrostatic adsorption of oppositely charged metal precursors. SEA stabilizes single metal atoms by strong electrostatic interactions, reducing aggregation during synthesis phases such as mixing, drying, and reduction. This approach assures monolayer coverage of metal ions on the support, which promotes the production of highly scattered SA sites. SEA has been used successfully to synthesize SACs such as Pt and Pd on substrates such as silica, titania, alumina, and carbon [29]. Last but not least is electrochemical deposition (Figure 2e), an electrochemical process that involves reducing the cations in the cathode and dissolving and re-depositing the reactive metal anodes to load onto the cathode surface [30].

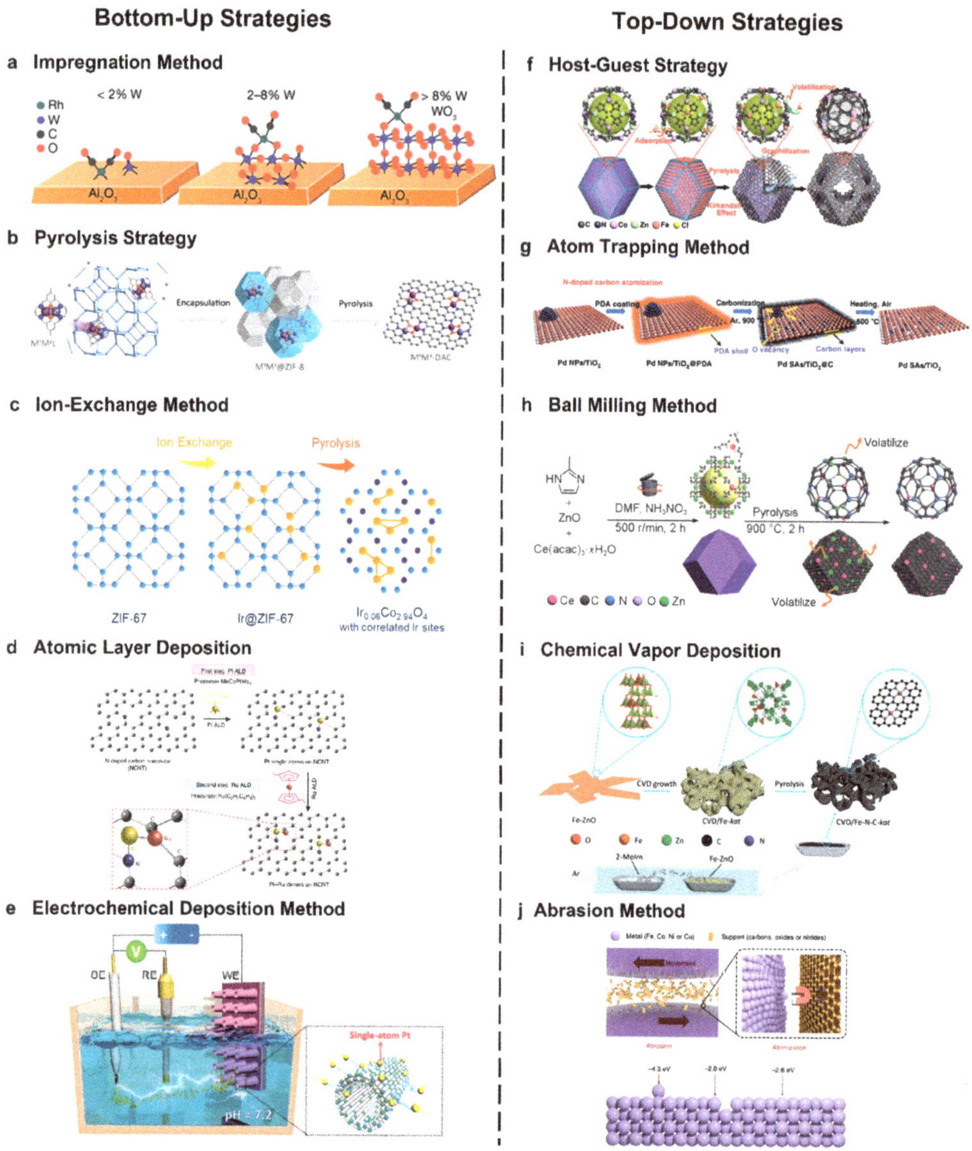

Figure 2. Bottom-up strategies. (**a**) Impregnation method. Reprinted from Ref. [25] with copyright permission from Nature. (**b**) Pyrolysis strategy. Reprinted from Ref. [26] with copyright permission from The American Chemical Society. (**c**) Ion-exchange strategy. Modified from Ref. [27] with copyright permission from The American Chemical Society. (**d**) Atomic layer deposition. Reprinted from Ref. [28] with copyright permission from Nature. (**e**) Electrochemical deposition method. Modified from Ref. [30] with copyright permission from John Willey and Sons. Top-down strategies. (**f**) Host-guest strategy. Modified from Ref. [31] with copyright permission from The American Chemical Society. (**g**) Atom trapping method. Reprinted from Ref. [32] with copyright permission from Nature. (**h**) Ball milling method. Modified from Ref. [33] with copyright permission from Springer. (**i**) Chemical vapor deposition. Modified from Ref. [34] with copyright permission from John Willey and Sons. (**j**) Abrasion method. Reprinted from Ref. [35] with copyright permission from Nature.

3.2. Top-Down Strategies

The top-down approach is completely different since it uses metal nanoparticles or even larger-scale metal elements as the precursors and uses various techniques to transform them into isolated locations on the support. Top-down approaches simplify and lower the cost of the entire production process since they do not require atomically distributed metal atoms in precursors, unlike bottom-up solutions. Furthermore, when bulk metals are reduced to single atoms in the structure, many flaws and vacancies may be produced. Metal precursors are encapsulated within a porous host material using the host-guest strategy method (Figure 2f), which uses the spatial confinement and unique interaction between the guest and host to regulate the stabilization and dispersion of the metal atoms at the atomic level [31]. Furthermore, the many vacancies in reducible supports are usually exploited by the atom trapping technique (Figure 2g), which was initially created to manufacture platinum group metals (PGMs) SACs. PGMs move, release volatile metal oxides, and are then confined by the defective support after annealing [32]. Additionally, the synthesis of SACs can be performed via ball milling (Figure 2h), a scalable, environmentally benign, and reasonably priced mechanochemical process that uses mechanical energy to break and rebuild chemical bonds without the need for extra solvents, templates, or additives [33]. Vaporized precursors are introduced into a CVD reactor (Figure 2i), where they attach to a substrate that is kept at a high temperature. After that, a solid coating is formed on the substrate by the adsorbed molecules either thermally decomposing or reacting with other gases or vapors [34]. The acid leaching method is an essential step in the synthesis of SACs, as it removes bulk or aggregated metal particles while leaving isolated metal atoms on the support material. During this process, the catalyst is treated with acids such as hydrochloric acid (HCl) or nitric acid (HNO_3). These acids selectively dissolve bigger metal particles due to their increased reactivity, leaving single metal atoms tightly bonded to the support via powerful coordination bonds. This approach improves not just the dispersion of active sites but also the catalyst's surface area and accessibility. Furthermore, acid leaching can eliminate contaminants and improve the structural integrity of the support material, which is critical to achieving improved electrocatalytic performance [36]. Last but not least, the abrasion technique (Figure 2j) uses metal balls as a source of the active metal as well as a means of directly atomizing bulk metals onto various supports without the need for a solvent or the production of waste or byproducts [35].

There are specific challenges for both top-down and bottom-up approaches to SAC synthesis. It can be difficult to achieve uniform atomic dispersion with large metal loading without compromising stability in bottom-up techniques such as impregnation, pyrolysis, ion exchange, ALD, and electrochemical deposition. These methods frequently include expensive, intricate synthesis and less control over the location of atoms. Due to their high energy requirements, top-down methods such as host-guest strategy, atom trapping, ball milling, CVD, and abrasion have scalability problems that restrict large-scale manufacturing. Both approaches struggle to maintain controlled dispersion and stability, which can lead to aggregation. Optimizing precursors and synthesis conditions for effective atom use and reduced energy costs is one of the upcoming advancements in bottom-up techniques. Energy-efficient electrochemical deposition may expand its use, while improvements in ion exchange and ALD substrates may allow for precise atomic control. Aggregation may be decreased for top-down approaches by including new host matrices and improving atom trapping. Ball milling and CVD modifications that are scalable and energy-efficient may improve the sustainability of large-scale SAC synthesis. For the development of SACs with improved stability and catalytic performance for energy applications, these developments are crucial.

4. Substrate's Engineering of SACs

In SACs, the substrate is essential because it not only offers a strong and conductive foundation to anchor active metal sites but also develops a specific environment for a variety of electrocatalytic processes, including water splitting, ORR, OER, HER, CO_2RR, NRR, and NO_3RR. Materials like carbon supports, which are renowned for their superior conductivity, chemical stability in both acidic and alkaline conditions, and mechanical robustness, provide the perfect foundation for SAs with careful substrate design. Furthermore, metal oxides can effectively bind isolated atoms due to their abundance of defect sites and surface hydroxyl (-OH) groups. They can also improve catalytic performance by utilizing characteristics like surface acidity, basicity, and redox activity. Furthermore, in single-atom alloys (SAAs), distinct atomic and electronic structures that promote effective electrocatalysis are produced by strong metallic contacts between isolated metal atoms and support. The large surface areas, high conductivity, and chemical stability of TMDs render them optimal substrates for atomically dispersed metals, particularly when combined with isolated metal sites or introduced vacancies that raise the local electronic density for enhanced activity. Interestingly, monolayer TMD substrates optimize available surface area and provide a distinct structure for in-depth analysis of electrochemical phenomena. Furthermore, MXenes and microporous materials such as MOFs serve as dependable supports by taking advantage of the coordinative environment surrounding SACs to produce active centers with special characteristics that either resemble homogeneous catalysts or deviate from the behavior of bulk metals, as in the case of zero-valent metal centers. In the meantime, SAC substrates' performance in electrocatalytic reactions was further improved by implementing several techniques, including morphology, vacancy engineering, heteroatom doping, facet engineering, and crystallinity management. High-performance electrocatalyst design is made possible by this customized substrate engineering in SACs. The different advantages and disadvantages of SACs on different supports by using some performance regulatory approaches are illustrated in Table 2.

4.1. Carbonaceous Substrate

Carbon-based materials can generally be used as supports in SAC-based electrocatalysis catalysts. The robust covalent bonding between carbon atoms creates the distinctive chemical and physical properties of carbon nanostructures that confer their unique physical and chemical characteristics. The uniform dispersed single metal atoms on the carbon matrix can modify the electrical configurations and geometric shapes of reactive catalyst sites. The impact of varying core atom coordination on a particular electrochemical reaction must be considered while designing a catalyst. The reaction's trajectory can even be altered by varying electron transfer quantities. We will discuss how various techniques, like morphology, vacancy engineering, and heteroatom doping in various electrocatalytic activities, can be used to tune the carbon matrix of SACs.

Table 2. Advantages and disadvantages of SACs of different approaches based on diverse supports.

Approach	Advantages and Disadvantages	Supports					
		Carbon Support	**Metal Oxide Support**	**Alloy Support**	**Transition Metal Dichalcogenides (TMDs) Support**	**MXenes Support**	**Metal–Organic Framework (MOF) Support**
Morphology Engineering	Advantages	High surface area, tunable porosity, lightweight, enhanced mass transport	Diverse morphologies, robustness, wide bandgap tunability	Tunable surface, improved electron transport, alloy synergy effects	Layered structures, abundant active edges, high catalytic activity	High conductivity, layered structure, enhanced active site exposure	Tailorable structure, ultrahigh porosity, and versatility in design
	Disadvantages	Limited chemical functionality; weak interaction with metal atoms	Brittle nature, low conductivity	Complexity in precise morphology tuning, potential phase segregation	Instability under certain conditions, higher synthesis cost	Susceptibility to oxidation, challenging defect creation	Structural fragility in liquid-phase reactions, scalability issues
Vacancy Engineering	Advantages	Improves catalytic activity and conductivity; lightweight	Enhances active site density and binding energies	Provides electronic tunability; synergetic effects enhance catalytic performance	Boosts catalytic sites at edges and vacancies; enables fine-tuning of electronic states	Promotes metallic conductivity and excellent adsorption properties	Facilitates enhanced adsorption properties and active site creation
	Disadvantages	Limited vacancy stability under reaction conditions	Difficulty in maintaining vacancy stability over time	Formation of vacancies can weaken overall structural integrity	Complex fabrication methods, moderate thermal stability	High sensitivity to oxidative and acidic conditions	Hard to control vacancies precisely
Heteroatom Doping	Advantages	Enhances electronic properties, provides functional groups for improved binding	Tailors catalytic activity via bandgap engineering	Strengthens catalytic performance by introducing new alloy phases and interactions	Modifies electronic structure to improve catalytic selectivity and activity	Facilitates synergistic effects with intrinsic conductivity	Allows diverse doping combinations, enhancing versatility
	Disadvantages	Limited thermal stability; uneven dopant dispersion	High temperature is often required for dopant diffusion	Dopant clustering may reduce uniformity and effectiveness	Reduced mechanical stability due to strain effects	Possible degradation of intrinsic conductivity upon heavy doping	May alter intrinsic stability, reducing long-term durability
Facet Engineering	Advantages	Enhances site-specific catalytic activity; tunable electronic interactions	Enables site-selective reactions and improved stability	Offers higher selectivity and activity by controlling atomic arrangements	Improves edge-site activity with facet exposure	Specific facet control for electronic property optimization	Facilitates selective catalysis with exposed active facets
	Disadvantages	Facet exposure control can be challenging in porous carbon	Limited options for controlling facets in certain oxides	Surface diffusion may compromise facet selectivity	Facet engineering may reduce stability in some catalytic environments	Limited reproducibility in precise facet exposure	Achieving stable facet configurations can be difficult
Crystallinity Control	Advantages	Enhances electronic conductivity and stability	Improves charge transport and catalytic stability	Reduces grain boundaries, improving structural integrity and activity	Increases intrinsic activity by reducing defects while maintaining conductivity	Enables fine-tuning of electronic and catalytic properties	Allows structural and catalytic property optimization
	Disadvantages	Difficult to achieve uniform crystallinity in disordered carbons	Requires high temperatures, challenging scalability	Balancing crystallinity with alloy composition can be challenging	Limited options for achieving precise control	Susceptible to oxidation and degradation	Trade-off between crystallinity and porosity

4.1.1. Morphology Engineering

Through the optimization of surface area, porosity, and accessibility of active sites, morphological engineering encompasses several dimensions, such as three-dimensional (3D), two-dimensional (2D), and one-dimensional (1D) carbon supports for SACs that are integral in improving electrocatalytic performance. Three-dimensional-based catalysts encompass porous carbon scaffolds, foams, and networks that can be customized for particular electrocatalytic purposes. These materials facilitate efficient mass movement and support a substantial quantity of active sites. These dimension structures offer superior mechanical qualities and extensive accessible surface areas, essential for improving reaction rates, particularly in different electrocatalytic processes due to the porous architecture facilitating efficient electron transport [5]. For example, Gong et al. [37] developed a catalyst denoted as PdZn-N-CMK-3-D with a dual catalytic site and achieved high faradic efficiency (FE) of up to 97.14% and durability in reducing CO_2 to carbon monoxide (CO) due to the PdZn DAC's synergistic effect in between the thermodynamic pathway, which reduces the activation energy of *COOH intermediate species while concurrently generating CO. Furthermore, Wei Yu and his team [38] synthesized a 3D PtSAC@CC supported on carbon cloth (CC) for HER and showed 83.0 at η = 50 mV. In addition, a 3D-based catalyst known as Fe-N-C/N-OMC, which was developed by Han and his team [39], is embedded in an N-doped ordered mesoporous carbon (OMC) framework exhibiting elevated mass activity and half-wave potential towards ORR, because high Fe and N loadings create accessible active sites with high density, graphitic N dopants, which increase the intrinsic activity of FeN_4 sites, and the OMC structure facilitates electron and mass transport in 3D interconnected pores (Figure 3a,a1). Additionally, robust electrocatalytic performance for N_2RR has been reported for an isolated SA involving an Fe atom site supported on the N-doped-C-framework (ISAS-Fe/NC), and the electrocatalyst achieved an excellent NH_3 yield rate of 62.9 $\pm$ 2.7 μg h^{-1} $mgcat^{-1}$ and long-term stability lasting beyond 24 h and high FE of 18.6 $\pm$ 0.8% [40]. Furthermore, Li and his team [41] created a 3D-based electrocatalyst, Co@CNB-Nx, which stabilized on N-doped carbon nanoboxes (CNB) promoting high efficiency as a bifunctional electrocatalyst for overall water splitting (OWS), driving a current density of 10 mA cm^{-2} at 1.59 V owing to the uniformly dispersed Co atoms, which could be effectively and stably riveted on the CNB because of the dependable metal-nonmetal bonds between the Co-SA and the carbon substrate. Similarly, a CoSAs@NC-920 stabilized by N-doped carbon (NC) was synthesized by Lv and colleagues [42] and because of its Co-N_3C_1 SA configuration, large specific surface area, abundant N-active species, and 3D interconnected structures, it demonstrated exceptional ORR/OER performance (Figure 3b,b1). Preserving the structural integrity of 3D materials under severe reaction conditions and achieving a homogeneous distribution of active sites might be difficult. To solve this problem, 2D carbon-based morphologies such as graphene, graphene oxide, and layered carbon structures demonstrate substantial surface areas and improved electron conductivity inside the plane of the structure and show promising performance in electrocatalysis, in contrast to 3D carbon supports for SACs. Two-dimensional catalysts significantly enhance charge distribution and promote efficient charge transfer. Their elevated surface-to-volume ratio provides a substantial density of available active sites, enhancing catalytic activity in different electrocatalytic reactions [5]. For example, CuNi/rGO showed excellent electrocatalytic performance in ORR and OER, which was synthesized by Musa and colleagues [43] and achieved a lower wave potential and overpotential because of the stronger interactions of Cu and Ni with rGO. The primary active sites for ORR/OER in FeNi-NC are Fe-N_4 and Ni-N_4, and their coexistence enhances bifunctional electrocatalytic activity. Additionally, FeNi-NC supported on N-doped flowering hollow carbon spheres demonstrated good OER and ORR activity [44]. In another study, 2D-Pt SAC demonstrated increased HER activity supported by the N-doped carbon support [45] (Figure 3c,c1). Furthermore, it was demonstrated that Ni-SA-based porous carbon nanosheets had N/S co-coordination for CO_2 reduction, achieving over 95% CO selectivity rate and −7.8 mA cm^{-2} current density. These results were ascribed to the nanosheets' sheet-like shape, long-distance conductivity, and the abundant Ni-N_3S SA sites with high activity [46] (Figure 3d,d1). The computational study was also carried out in 2D SAC supported on a carbon matrix in different electrocatalytic reactions and proved enhanced performance [47–49]. However, the tendency of 2D materials to agglomerate or stack without appropriate surface modifications may restrict their effective surface area and electrocatalytic performance. In contrast to 2D, 1D structures, including nanowires, nanotubes, and nanorods, possess a high surface area and exhibit significant structural anisotropy. These morphologies offer a high concentration of active sites along their length and can enhance electron transport along the structural axis. This renders them suitable for electrocatalytic processes necessitating great charge mobility for electrochemical reactions [50]. For instance, Co, Zn SAs@Co-CNTs were synthesized by Huang et al. [51] and stabilized on Co-embedded carbon nanotubes (CNTs). Owing to the synergetic interactions between Co and Zn DASs, they exhibit superior ORR ($E_{1/2}$ = 0.92 V) and HER (10 = 89 mV; ~10 = 142 mV) activities with long-term operational stability. They also lower the adsorption–desorption energy barrier of H* in the HER and accelerate the formation of OOH* and the dissociation of OH* in the ORR. Furthermore, Wu and colleagues [52] synthesized bimetallic Fe-Ru SAC dispersed on N-doped carbon nanorod spheres and exhibited a maximized NH_3 yield of 43.9 μg h^{-1} mg^{-1} with an optimized FE of 29.3% for the NRR owing to robust synergistic interactions between the electronic structure of Fe sites, which were effectively influenced by Ru atoms, resulting in stronger N_2 adsorption and a shift in the d-band center, thus improving NRR performance (Figure 3e,e1). Moreover, Liu and his team [53] created a CuO-based catalyst with Pd SAs dispersed on its interlayer unsaturated bonds (Pd-CuO), and the NO_3 to NH_3 conversion could be improved and promoted by the added shear stress and the dynamic effects of support formation to the Pd SAs situated on dislocations, reaching a maximum NH_3 production and optimized FE (4.2 mol $gcat^{-1}$ h^{-1} and 90%) (Figure 3f,f1). Furthermore, Chen et al. [54] synthesized a catalyst named Fe-N_x-S supported by CNT with atomically dispersed Fe species, and the presence of S dopants and abundant Fe-N_x species modified the electronic structure of the Fe-SAs, resulting in remarkable OER activity with a low overpotential of 326 mV at 10 mA cm^{-2} and stability.

The impact of various dimensions, 1D, 2D, and 3D, on electrocatalytic performance in carbonaceous supports is substantial since the catalyst's geometry and structure can affect charge transfer, active site accessibility, and reaction kinetics. For example, 1D structures dominate in rapid charge transfer along their length, whereas 2D-based catalysts offer an expanded surface area for active sites and enhanced electron conductivity in the plane, which is beneficial for reactions necessitating numerous active sites [55]. Further, 2D dimensions typically exhibit superior electrical properties, whereas 3D morphologies excel in enhancing mass transfer, reaction kinetics, and long-term stability. Three-dimensional structures are more appropriate for processes requiring intricate mass movement. Although 1D materials facilitate superior charge transfer, 3D structures often exhibit enhanced catalytic effectiveness owing to improved mass transport and increased surface area. The stability and practical scalability of 3D structures render them advantageous for extensive applications [56].

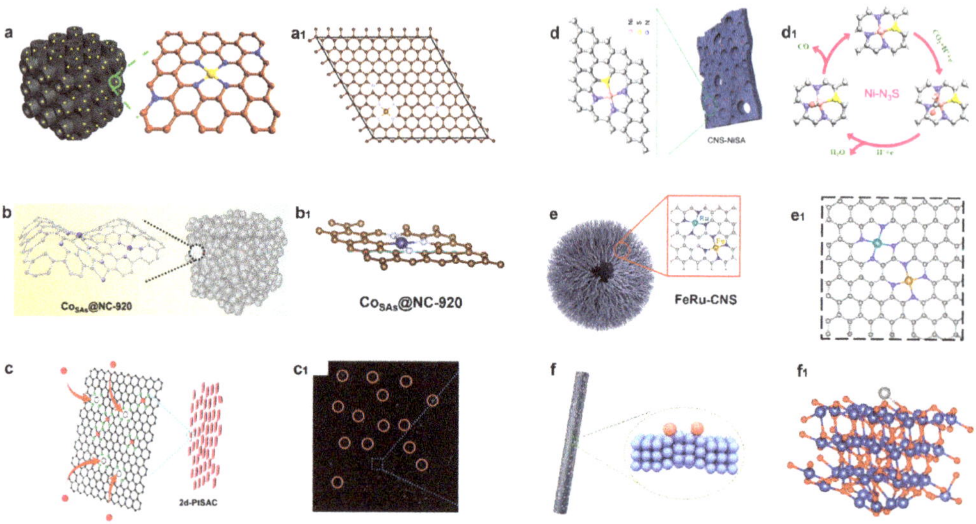

Figure 3. (**a**) Structure of Fe-N-C/N-OMC; (**a1**) configuration for Fe-N-C/N-OMC. Modified from Ref. [39] with copyright permission from Elsevier. (**b**) Structure of CoSAs@NC-920; (**b1**) structure of CoSAs@NC-920. Modified from Ref. [42] with copyright permission from Elsevier. (**c**) Structure of 2D-Pt SAC; (**c1**) HAADF-STEM for 2D-Pt SAC. Modified from Ref. [45] with copyright permission from John Willey and Sons. (**d**) Structure of CNS-NiSA; (**d1**) CO_2 reduction on Ni-N3S sites. Modified from Ref. [46] with copyright permission from Elsevier. (**e**) Structure of FeRu-CNS; (**e1**) atomic structure of FeRu-CNS. Reprinted from Ref. [52] with copyright permission from The Royal Society of Chemistry. (**f**) Structure of Pd-CuO; (**f1**) model of CuO-swa*-Pd. Modified from Ref. [53] with copyright permission from John Willey and Sons.

4.1.2. Vacancy Engineering

Vacancies are regarded as a type of crystal defect that can modify the electron distribution, lattice structure, and elemental composition. As a result, voids are typically complicated and require more careful design control. There are three types of vacancies: cationic, anionic, and mixed vacancies, depending on the charge polarity of the atoms in the structure. One efficient way to maximize the local environment for electrocatalysis is to introduce vacancies in the SA carbon matrix. For instance, to improve the ORR activity and stability, Tian et al. [57] showed that the generation of carbon vacancy (Cv) on Fe-N_4 atomic sites prevents the metal center from dissolving and may lower the adsorption energy barrier of OH*. Because of the good performance in ORR achieved by utilizing a C_{vac}, Guo and his colleagues [58] also utilized a C_{vac} in Fe-N_4-VC SAC next to the metal center in a DFT study and showed that elevated symmetry of the Fe-N_4 atomic site tuned the d-band center, hampered the adsorption of advantageous intermediates, and resultantly increased the ORR electrocatalytic performance. Moreover, Fe-N-C catalysts with pyridinic N vacancies (FeNC-VN) were synthesized by Lyu and associates [59] (Figure 4a). Adjacent VN sites are predicted by first-principles calculations to greatly improve ORR kinetics and stability by modestly weakening *OH binding energies and effectively enhancing the covalency of Fe-Nx moieties (Figure 4a1). Additionally, a catalyst with numerous catalytic reactive sites, containing the fast transfer of charge and with enhanced performance in OER was synthesized and denoted as δ-MnO2/CoNP@CoSAs-NCNTs/CC which was supported on a 3D-based structure exhibiting O_{vac} requires only 165 mV overpotential to reach a current density of 10 mA cm^{-2} [60]. Moreover, a trifunctional catalyst for OER/ORR/ZAB was synthesized by Yi and his group [61] named $ZnIn_2S_4$/PC, and due to the combined action of N and F atoms, doped Sb-SAs, and sulfur vacancies (Svac), it reduces the energy potential barrier of intermediate products, consequently achieving remarkable performance (Figure 4b,b1). Furthermore, the divacancy of graphene defects (DG) successfully trapped Ni atoms to generate a Ni-C_4 site, which enhanced HER activity [62]. In another study, a Pt-C_3 configuration was developed by Yao and colleagues [63], which contains C_{vac} in the DG matrix and traps a large number of Pt-SAs, lowers the adsorbed H^+ reduction, has a stronger ability to capture electrons, and accelerates the desorption of H_2, simultaneously improving HER activity. Additionally, an effective dual-site HER catalyst with a core–shell flatform structure was synthesized using an O_{vac}-containing $LaCeO_x$ material that combined with N-doped graphene along with the dispersion of Ru SAs denoted as $LaCeO_x$ @NGr/Ru_1 (Figure 4c). The H_2 was produced by using $LaCeO_x$ @NGr/Ru_1, which causes the adsorption and dissociation of water on the surface of O_{vac}-rich $LaCeO_x$, along with the diffusion of produced H* species towards Ce^{3+}-N-Ru_1 bridges, and H_2 evolute on the Ru_1 sites (Figure 4c1) [64]. Moreover, the DFT calculation for the SACs supported on an Ovs-enriched graphene matrix showed improved performance in the CO_2 reduction to CO product [65]. Similarly, Rong et al. [66] developed a Ni-N_3-Vac SAC with three N atoms demonstrating exceptionally high CO_2 reduction to CO, exhibiting FE above 90% due to the presence of Ni-N_3 vacancy. In addition, PSB-CuN_3 with local planar symmetry was synthesized and showed exceptionally high electrocatalytic activity and stability for CO2RR to formic acid (HCOO) because of the local D4h symmetry, which was broken to produce planar-like CuN_3 and CuN_2 moieties with lower local C_2v symmetry, resulting in charge redistribution and improved performance [67] (Figure 4d,d1). Additionally, TMs used as SA materials with abundant vacancies anchor on carbon substrates that provide localized d-orbitals [68,69] that effectively transfer electrons between TM and N_2, weakening the N_2 triple bond and enhancing their performance in NRR. In contrast, Ik Kwon [70] reported that the integrated kinetic evaluations with electronic and thermodynamic analyses on various SAC candidates, using boron carbon nitride (BCN) as a substrate, exhibited a high turnover frequency (TOF) ($3.1 \times 10^{-6}\ s^{-1}$) among various vacancy sites, with Cr anchored to BCN exhibiting the least amount of TM aggregation over time (Figure 4e,e1). Further, TM-SACs dispersed on diverse N-doped graphene (NDG) systems with different combinations of pyridinic, graphitic N-doped, and pyrrolic structures with single and double vacancies were reported, and among them, Ru- and Ir-SACs/PD_2cGn showed high NO_3 reduction and stability [71]. Superior NO_3RR performance was also achieved at low overpotentials by the atomically dispersed Fe- SAs coordinated with P and N stabilized on hollow carbon nanocages, and modulating the P enhanced the Fe center electron density, which promotes the transfer of protons and adsorption of NO_3 ions along with their related intermediates species on the surface of the electrocatalyst [72] (Figure 4f,f1). Chen and his team [73] synthesized a Cu-N-C catalyst featuring Cu-N_3 coordination that attains a remarkable CO FE of 98% and exhibits exceptional stability (FE remains above 90% for 20 h) due to the Cu-N_3 coordination that may be stabilized by an extended carbon plane with six nitrogen vacancies, whereas three empty N sites are spontaneously protonated during the CO_2 reduction reaction. Consequently, the hydrogen bonds established between the adsorbed *COOH and neighboring protons markedly diminish the energy barrier for *COOH production. Further, Guo [74] and colleagues developed Ni-SAC@NC that exhibits superior performance for CO_2RR with a maximum FECO of 95% due to the defect-Ni-N_3 configuration serving as the highly active site in Ni-SAC@NC for CO_2 reduction reactions. Moreover, Liu and co-workers [75] synthesized a highly durable and efficient Fe-N-C catalyst and showed good performance in ORR by transforming defect-laden pyrrolic

N-coordinated FeN_4 sites into extremely stable pyridinic N-coordinated FeN_4 sites. Further, by employing first-principles DFT calculations, Mavrikakis and researchers [76] constructed a pH-dependent microkinetic model to assess the comparative efficiency of SA-TM-based catalysts integrated within fourfold N-substituted double carbon vacancies in graphene for the OER, and reaction routes that involve intermediates co-adsorbed on the metal site are favored across all TMs, and these routes result in improvements in catalytic activity.

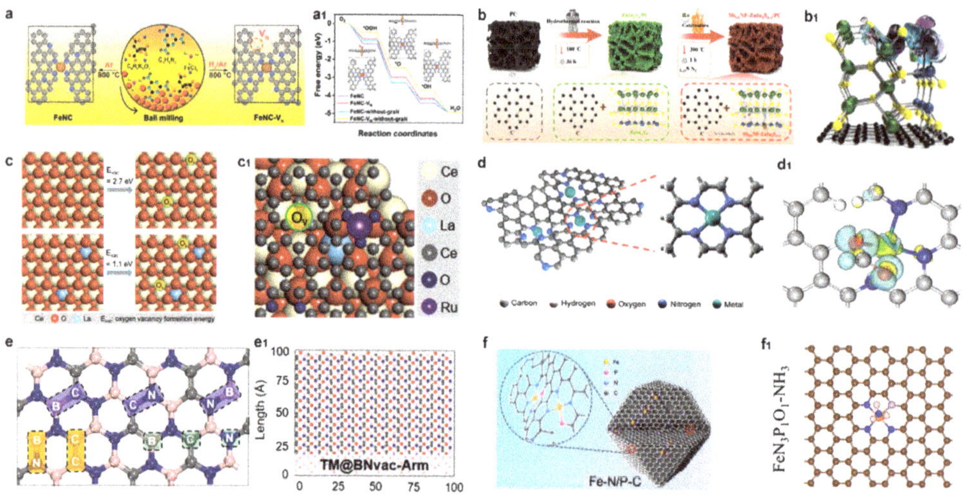

Figure 4. (**a**) Synthesis of FeNC-V_N and FeNC; (**a1**) free energy for the ORR catalyzed by different catalysts. Reproduced from Ref. [59] with copyright permission from The American Chemical Society. (**b**) Synthesis of Sb_{SA}NF-$ZnIn_2S_{4-x}$/PC; (**b1**) density difference of Sb_{SA}NF-$ZnIn_2S_{4-x}$/PC. Reprinted from Ref. [61] with copyright permission from Elsevier. (**c**) O_v formation in CeO_2 and LaCeOx; (**c1**) structure for the LaCeOx @NGr/Ru_1. Modified from Ref. [64] with copyright permission from Elsevier. (**d**) Structure of PSB-CuN_3; (**d1**) differential charge density and projected density of states for the Cu center. Adapted from Ref. [67] with copyright permission from Nature. (**e**) BCN substrate features different vacancy sites; (**e1**) lattice structures of TM-C_2BN@BNvac-Arm. Reprinted from Ref. [70] with copyright permission from Elsevier. (**f**) Structure of Fe-N/P-C; (**f1**) model of $FeN_3P_1O_1$-NH_3. Adapted from Ref. [72] with copyright permission from John Willey and Sons.

4.1.3. Heteroatom Doping

Although carbon-based materials showed good support for SACs, SACs find it challenging to adhere to these materials because of the weak metal-support interaction. Heteroatom doping is the most widely used technique to modify the CE of SACs and is a straightforward and efficient way to accomplish successful anchoring of SACs. It offers metal atom coordination sites that can greatly increase the resilience of SACs. Additionally, the CE of SACs can be controlled by judiciously regulating the atomic size and heteroatom electronegativity. By regulating the concentrations of N, S, P, and B in carbon-based SACs, their catalytic performance can be markedly enhanced. N doping, the most extensively researched heteroatom, increases the electron density at active sites, hence promoting the adsorption of essential reactants like oxygen or N. It possesses high electronegativity, facilitating electron localization and generating highly polarized environments that stabilize SAs and offer several active sites owing to diverse nitrogen configurations (e.g., pyridinic-N, graphitic-N, and pyrrolic-N). Its capacity to stabilize SAs through coordination bonds renders it essential in electrochemical reactions. For example, Zhang et al. [77] produced Co-N/C-based catalysts, and due to the occurrence of N coordinated site, it reduced the binding energy of the d-band center of the Co-

N_4 site, thereby reducing the binding energy between the intermediates and Co-N_4 catalytic site consequentially increased O_2 reduction catalytic activity. Outstanding electrocatalytic performance on SA electrocatalysts by using N dopants on the carbon matrix for OER has been reported, such as N-doped-holy-graphene matrix (M-NHGFs) [78], Co/Ru SAs-N-C [79], and N-Co-Mo-GF/CNT [80]. For an effective electrocatalytic H_2 evolution, Xue's group [81] binds Mo-SAs atomic sites with co-coordinated N, C, and O atoms stabilized on N-doped carbon fibers (NCF) denoted as Mo@NMCNFs, and these fibers demonstrated exceptional stability and great HER activity with an overpotential of 66 mV. Furthermore, Kuang et al. [82] dispersed PtSACs on N-doped materials to synthesize mesoporous-hollow carbon spheres named Pt_1/NMHCS that demonstrated remarkable stability and attained an overpotential of 40 mV at 10 mA cm^{-2} for the HER. Musa and colleagues [83] synthesized atomically dispersed Fe-N_4 embedded on CNTs denoted as FeSA@CNTs (Figure 5a). Utilizing the protective properties of the CNTs structure through anchoring Fe-N_4 and facilitating electrolyte access to the active areas, the synthesized catalyst demonstrated superior ORR catalytic performance and electrochemical stability due to the Fe-N_4, which serves as the primary active site for catalysis and contains square planar in shape (D4h), and the interaction between the frontier orbitals of Fe in Fe-N_4 (dz^2) and the LUMO of O_2 (π-orbital), thereby promoting the catalysts' effective catalytic activity (Figure 5a1,a2). In contrast to N, doping with S induces distinct electrical and geometric alterations owing to its greater atomic radius and softer Lewis base characteristics. These characteristics improve selectivity and activity in different electrocatalytic reactions as the S possesses lower electronegativity than nitrogen, creating electron-rich regions that facilitate charge transfer. S atoms can establish robust coordination with metals, resulting in a synergistic effect that enhances catalytic activity. The doping process frequently leads to structures abundant in defects owing to the greater atomic scale. For instance, for the OER/ORR application, theoretical research was conducted on the electrocatalytic mechanism of S-doping in TM-N-C SACs, and the electrons flow from S sites to the substrate, and the presence of S dopant enhanced the geometry configuration, stability, and charge redistribution in TM-N-C SACs (Figure 5b–b2) [84]. Additionally, a dual single atom (DSA) denoted as Fe, Co/DSA-NSC was doped with N and S dispersed on porous carbon nanosheets, and these unsymmetrically organized N and S coordinated Fe/Co atomic sites boosted the transfer of charge, lowered the adsorption energy of intermediates, and enhanced the oxygenated reaction kinetics [85]. Some studies have shown outstanding performance in diverse electrochemical reactions by using S and N as the dopants on different carbon substrates in SACs, DACs, and DSACs [86–88]. P doping is recognized for facilitating significant electronic redistribution, which frequently enhances the binding energy modulation of intermediates, especially in reactions such as HER and ORR. Its capacity to affect charge transfer kinetics is notable among dopants. P imparts electron-rich properties akin to S, although it exhibits more robust metal-phosphorus interactions. P doping augments surface basicity, hence facilitating intermediate adsorption and catalytic reaction mechanisms. An increased atomic radius induces lattice aberrations, enhancing the accessibility of catalytic sites. Additionally, the role of P doping in Fe and Co bimetallic SACs, known as P/FeCo-NC HER-catalysts, is reported by Tao and his team [89]. The Fe-P-Co site was demonstrated as the most effective HER activity by modifying the electrical interactions and orbital hybridization in the P bridge between the Fe and Co sites, increasing the H_2 adsorption energy. P and Co co-regulate the electrical characteristics of the Fe sites, resulting in an ideal H_2 binding energy (Figure 5c–c2). To investigate the impact of distinct CE of Ni-SACs on water-splitting catalytic activity, Wang et al. [90] prepared a variety of Ni-SACs doped with different heteroatoms into porous CNFs, and among these, the Ni-SAC with the Ni-N_3-P structure (Ni-N, P/CNFs) demonstrated the best OWS performance. Furthermore, P doping was also investigated by Zhou and his team [91] by using P/Fe-N-C in OER. In addition, the N, S, and P, B doping endows a distinctive ability to generate p-type semiconducting properties, rendering it exceptionally efficient in processes where charge transfer regulation is essential. B is an electron-deficient element that creates Lewis acidic sites, hence improving the adsorption of reactants and enhancing the dispersion and stabilization of SAs. The capacity to take electrons alters the electronic configuration of metal centers, enhancing catalytic efficiency. For example, by adding single TM atoms and different NM atoms onto GY substrates, Long Li and his group [92] developed different SACs for NRR, and, in particular, Os-B-GY demonstrated exceptionally effective NRR activity with a low overpotential of 0.13 V within the distal pathway (Figure 5d–d2). The co-doping of TMs and B can modify the d-orbital electronic structure of transition metals, promoting moderately positive charge concentrations via a localized boron coordination framework. Moreover, the robust d-2π* interaction between TM and N_2 promotes electronic back donation, thereby activating N_2 and preparing it for NRR. Some studies have promising performance by using B and N as dopants in different electrocatalytic reactions. Furthermore, it is reported that rationally controlling the CE of SACs can successfully enhance the FE of NRR. For example, coordination with a B dopant in Fe-B_2N_2 can significantly enhance the Fe metal center, exhibiting a low limiting potential of −0.65 V, indicating its high NRR catalytic activity [93]. Additionally, Xin and colleagues [94] synthesized Mn-SACs by anchoring the Mn-O_3N_1 moieties on porous carbon, and the Mo-B_2N_2 configuration lowered the energy needed for N≡N bond-breaking and boosted the adsorption of N_2 at the active site, resulting in enhanced N_2 reduction to NH_3 with a yield rate of 66.41 μg h^{-1} $mgcat^{-1}$. Moreover, Ye and co-workers [95] synthesized B, N co-doped Mo-SAC denoted as Mo/BCN for electrocatalytic N_2 reduction. SACs have recently been shown to have favorable action on NO_3RR, and there is still a need for improved catalysts with increased stability and activity. To prepare Fe-BCN catalysts, Lu and his team [96] added B to the NC matrix, and consequently B improved the electronic structure around Fe SAs and Fe-N, lowering the reaction barrier and increasing NO_3RR activity, demonstrating excellent NO_3RR performance (NH_3 yield of 2.17 mg h^{-1} cm^{-2}, FE of 97.48%). Furthermore, by mounting a Ni single site to BNG supports, Ajmal and colleagues [19] developed a boron-spanning method to modify the coordination structure of Ni SACs and developed a rational NiN_4B_2 structure and allowed for the incorporation of active sites and electric fields into a catalyst. It is found that the electric field distribution surrounding the B-atom next to N-atoms coordinated with the Ni-atom is responsible for the charge distribution over the Ni single site in NiSA@BNG. The adsorption of NO_3^-on the NiSA@BNG catalyst is facilitated by the existence of locally polarized B-N bonds, which are found to be essential in producing electric fields and require only −0.42 eV energy for NiSA@BNG. Some non-metal (NM) dopant SACs have recently been produced to support the high electrocatalytic activity and high selectivity of CO_2RR. For instance, a computational screening method was developed by Lu et al. [97] in which different NM dopants adjusted the CE of various SAs, and among these TM@N_4, Zn@N_4-Cl_1(II) with 0.15 V, Ni@N_3-B_1 with 0.43 V, Fe@N_4-F_1(I) with 0.75 V, and Cu@N_4-O_1(III) with 0.49 V, exhibit the lowest overpotentials for CO_2 reduction to CO. Additionally, the axial Cl atom, co-coordinated with Fe and N-atoms, represented as FeN4Cl/NC, showed a partial current of 9.78 mA cm^{-2} and a FE of 90.5% for CO due to the Cl atom, which caused a negative shift in the d-band center, resulting in a weakened binding strength between the intermediates and the Fe atom [98]. Furthermore, Ma et al. [99] conducted DFT calculations and examined the different axial ligands that affected the CO_2 performances, and among them, the M-N_4 configuration showed that the axial ligands could alter the metal center's electrical and magnetic characteristics, changing the activity and selectivity for CO_2 to CO. Furthermore, an F-doped Ni-SA denoted as Ni-SAs@FNC with high surface Ni-N_4 loading and exhibiting an ultrathin nanosheet structure was synthesized by Guo Han and colleagues [100]. Owing to the high Ni atom utilization, unique morphology, modulated Ni-N_4 configuration, and superior F-tuning effect, it allowed it to effectively boost the electroreduction of CO_2 to CO, as evidenced by its high FECO of approximately 95%. Furthermore, a multi-heteroatom doping of N, P, and Cl was investigated by Sui and co-workers [101], denoted as Ag_1-h-NPClSC anchoring on a hollow carbon matrix, and the synergetic effect between different heteroatoms and Ag SAs dramatically optimized the ORR catalytic activity and durability.

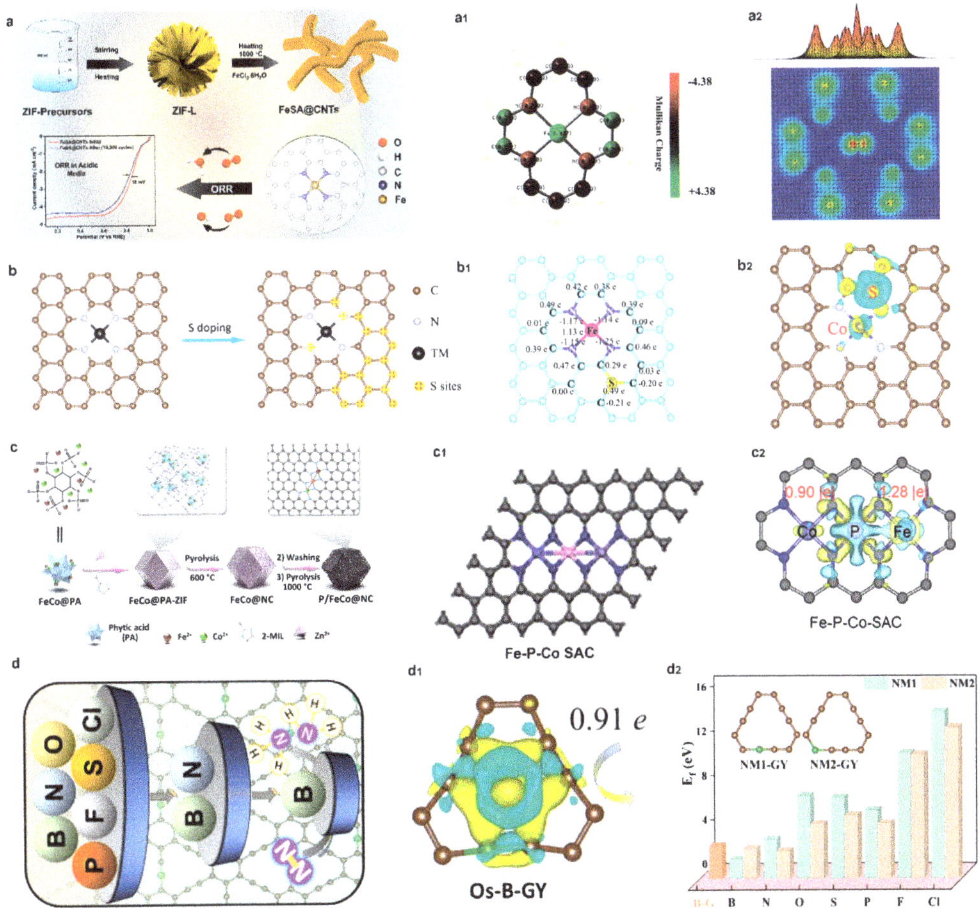

Figure 5. (**a**) Synthesis of FeSA@CNTs; (**a1**) charge profile for FeN_4 active site; (**a2**) Fourier image for FeN_4 active site. Reproduced from Ref. [83] with copyright permission from Elsevier. (**b**) S-undoped and doped TM-N-C SAC structure; (**b1**) Bader charge distribution of S-doped Fe-N-C; (**b2**) distribution of charge density S-doped SACs. Reprinted from Ref. [84] with copyright permission from The Royal Society of Chemistry. (**c**) Structure of P/FeCo-NC; (**c1**) top view of Fe-P-Co-SAC, (**c2**) charge density difference (CDD) in Fe-P-Co-SAC. Modified from Ref. [89] with copyright permission from John Willey and Sons. (**d**) Diagram of NM-GY-SACs; (**d1**) distribution of charge density in Os-B-GY; (**d2**) formation energies of NM-GY and B-G. Reprinted from Ref. [92] with copyright permission from Elsevier.

In conclusion, the introduction of non-metal heteroatoms like N, S, P, and B markedly affects the efficacy of SACs in diverse electrocatalytic reactions by altering their electronic structure and catalytic characteristics. N doping generally increases the electron density at the metal center, facilitating charge transfer and enhancing activity in electrochemical processes. S and P with lower electronegativity than nitrogen generate electron-rich regions that stabilize intermediates in processes. Moreover, S can provide dual functions by improving both catalytic activity and hydrophobicity. B, by inducing electron shortage, can modulate the adsorption energy of intermediates, so enhancing selectivity and efficiency. The heteroatoms together affect the local coordination environment, active site accessibility, and reaction kinetics, resulting in improved catalytic activity, selectivity, and stability in various electrocatalytic applications.

Carbon-supported SACs exhibit various morphologies, vacancy types, and doping configurations, showing considerable potential for catalytic applications owing to their high specific surface areas, swift mass transfer rates, and strong structural stability. Nonetheless, other issues remain that must be addressed. A significant problem is that the interior spaces of 1D geometries are frequently underutilized, resulting in diminished atomic utilization efficiency due to inadequate exposure to active sites, which can negatively impact catalytic performance. Moreover, 2D carbon substrates often undergo restacking due to robust π-π interactions during thermal treatments and catalytic reactions, requiring the development of effective ways to alleviate this phenomenon. In 3D porous carbon frameworks, numerous individual metal atoms become encapsulated within the carbon layers during high-temperature pyrolysis, complicating their complete exposure to catalytic reactions. Moreover, in the realm of morphological engineering, attaining exact control over the nanoscale configurations and forms of carbon substrates is essential for enhancing atomic dispersion and accessibility, which directly affects catalytic performance. Likewise, vacancy engineering is limited by the stability of active sites under diverse electrochemical circumstances, potentially leading to structural failure or deterioration, thereby compromising long-term performance. Furthermore, heteroatom doping encounters similar difficulties; the integration of dopants like N, S, and P requires meticulous adjustment to improve electron dispersion while maintaining conductivity. Despite the exploration of several methodologies for heteroatom doping, the predominant methods for synthesizing carbon-based SACs depend on high-temperature pyrolysis. Future initiatives should concentrate on advancing synthesis processes to provide stable, uniformly dispersed vacancies and customized doping profiles, perhaps resulting in SACs with enhanced durability and reactivity. Examining the interplay among these three engineering strategies may improve the accessibility of catalytic sites and refine electronic properties, potentially advancing the scalability and applicability of SACs in various electrochemical reactions for energy conversion applications.

4.2. Metal-Oxide-Based Substrate

Engineered metal oxides have gained attention recently as a possible option for creating high-performance HER electrocatalysts. The distinct benefits of earth-abundant metal oxides include great chemical and electrochemical stability, strong metal-support interaction (MSI), flexible tunability, structural diversity, and ease of synthesis. They are also inexpensive. However, some techniques like morphological tuning, phase, creating oxygen vacancies, doping (metal and non-metal), and controlling crystallinity are frequently used to improve the properties of pristine metal oxides, which are inert to electrocatalytic reactions because of their inadequate electrical conductivity, restricted number of catalytic active sites, and weak H_2 adsorption capacity. It is notably recognized that the electrocatalytic efficiency of a catalyst is highly associated with the electronic properties of the active sites, suggesting that the electrocatalytic activity of SACs can be competently tuned by maximizing the metal-metal (M-M) robust interaction. Following this guideline, diverse efforts were focused on the development of advanced transition metal compounds (TMCs)-based SACs by judiciously modulating the M-M and M-TMCs interactions. Significant advancements in the synthesis of advanced SACs have been achieved during the past decade due to rigorous efforts.

4.2.1. Morphology Engineering

A key tactic for improving electrocatalytic performance in SAC-based metal oxides is morphology engineering, which maximizes surface area, electrical characteristics, and active site accessibility. The density of exposed SAs can be greatly increased by modifying morphology at nanoscale dimensions, such as via hollow structures, nanowires, or nanosheets, which facilitates effective reactant adsorption and electron transport. These morphological modifications frequently improve mass transport efficiency, speed up reaction kinetics, and lower diffusion resistance in electrocatalytic reactions. Catalytic activity and selectivity in electrocatalytic reactions depend on precise control over defect sites and electronic structures, which can be achieved by tailoring morphologies. For instance, Zhou's group [102] recently synthesized 3D Pt SACs by dispersing Pt SAs on NiO/Ni heterostructure, which was developed on Ag nanowires and demonstrated exceptional performance and 30 h durability, exhibiting a lower ΔGH^* of 0.07 eV for Pt SAs and dissociating energy barrier for water attributed to the presence of dual active reactive sites in the Ni and O-vacancies-rich NiO sites for HER. Additionally, a 3D-based cerium hydroxide-doped copper catalyst was synthesized by Sargent and his team [103] and yielded a 43% FE. Mechanistic studies revealed that H_{ad} on Cu prefers the ethanol pathway over the ethylene pathway by targeting the Cu-C link of the *HCCOH intermediate. Additionally, using polystyrene nanospheres known as SA-FeN_4O_1-C, Xu and associates [104] synthesized a 3D-frame structure with organized pore channels, and the catalyst's higher overpotential in ORR and OER is attributed to the special CE of the SA active site and 3D framework. Furthermore, Zheng et al. [105] deposited Ru-SA onto an Ov-riched SnO_2 matrix and synthesized a 2D-based Ru/SnO_{2-x} catalyst (Figure 6a). Ru SAs near Ov and the SnO_{2-x} substrate undergo a minimal transfer of electrons from Ru to O atoms, leading to electron loss of Ru atoms (Figure 6a1), and the catalyst produced a remarkable 32.4 $\mu g\ h^{-1}\ mgcat^{-1}$ of NH_3, along with outstanding electrochemical stability and a FE of 16.2%.

4.2.2. Vacancy Engineering

A great way to control the CE, electrical characteristics, and catalytic activity of SACs is by defect engineering. Surface imperfections, for instance, might influence the density of charge close to the Fermi level and encourage molecular orbital hybridization, leading to more unsaturated coordination sites and suspended bonds, even at very low concentrations. C, O, N, S, cationic vacancy, and anionic vacancy defects are among the defect types employed to control the CE of SACs. To utilize uniformly dispersed Pt SAs on titania nanosheets ($Pt_1O_1/Ti_{1-x}O_2$), Lu and colleagues [106] used Tivac-enriched nanosheets, and these negatively charged Tivac readily trapped positively charged Pt^{+2} ions, and an ultrahigh activity and exceptional stability (300,000 cycles) were promoted by the generation of the Pt-O dual atomic site by intrinsic charge transfer. Moreover, Wu and his team [107] synthesized a catalyst denoted as Ni/M-CeO_2 SACs in which the OV concentration was controlled by a doping series of different metal cations over CeO_2 surfaces. Among them, the Ni/CeO_2 SACs presented high initial activity in CO_2 reduction. Yuan's team [108] synthesized Ru SA onto Ov-rich cobalt oxides (Ru/Co_3O_{4-x}), and the Ov causes electrons to delocalize around the Co atoms. This delocalization is further improved after loading Ru atoms, because of the coexistence of Ovs and Ru atoms, and achieves a low potential (280 mV @10 mA cm^{-2}) and long-term durability for OER (Figure 6b,b1). Additionally, by controlling the electronic MSI, Yao and colleagues [109] described an intrinsic oxide anchoring technique to secure ligand-free, uniform Ru SAs on the amorphous layered surface of monolithic Ti support (Figure 6c). The strong electronic MSI was demonstrated by a clear charge rearrangement across the interface between the Ru SA and TiO_x/Ti, which generated a charge transfer from Ru SA to TiO_x/Ti and achieved a good performance in NO_3RR (Figure 6c1). In addition, Wang et al. [110] synthesized an electrocatalyst with Cu-Sn dual atomic active sites utilized by double Ovacs on CeO_{2-x} and these Ovacs are crucial in controlling the electrocatalytic performance and stability, highlighting the significance of controlling the electronic MSI between Cu-Sn diatoms and CeO_{2-x} and achieving a 90.0% FE for CO_2 conversion to formate.

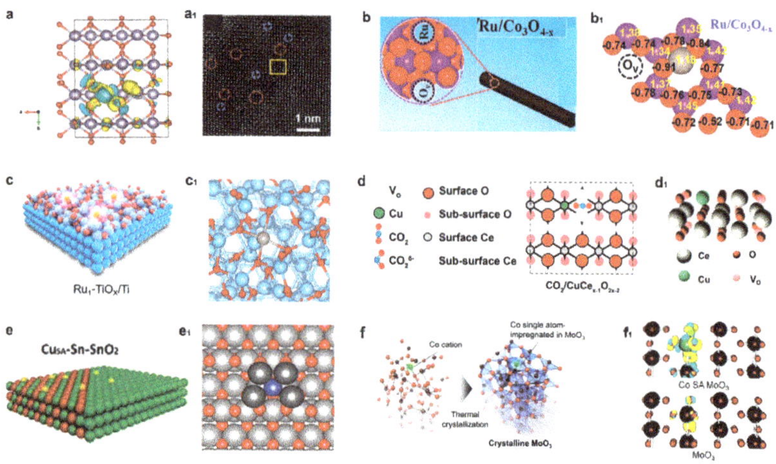

Figure 6. (**a**) Charge density evaluation of Ru/SnO_{2-x}; (**a1**) AC-STEM image of Ru/SnO_{2-x}. Reprinted from Ref. [105] with copyright permission from Elsevier. (**b**) Structure of Ru/Co_3O_{4-x}; (**b1**) Bader charge numbers of Ru/Co_3O_{4-x}. Modified from Ref. [108] with copyright permission from The American Chemical Society. (**c**) Structure of Ru_1-TiO_x/Ti; (**c1**) difference in charge density of Ru_1-TiO_x/Ti. Adapted from Ref. [109] with copyright permission from The American Chemical Society. (**d**,**d1**) Model of Cu-CeO_2. Reprinted from Ref. with copyright permission from The American Chemical Society. (**e**) Structure of CuSA-Sn-SnO_2; (**e1**) DFT model of CuSA-Sn-SnO_2. Modified from Ref. [111] with copyright permission from Elsevier. (**f**) Structure of Co SA MoO_3; (**f1**) charge density difference in Co SA MoO_3 and MoO_3. Modified from Ref. [112] with copyright permission from Elsevier.

4.2.3. Heteroatom Doping

Through altering the electrical environment surrounding the active metal sites, heteroatom doping in SACs has become a calculated strategy to improve their electrocatalytic performance. The electronic structure of the catalyst can be adjusted to increase its conductivity, stability, and reactivity in electrocatalytic reactions by adding NM atoms, such as S, P, N, or any other metal or NM, into the metal oxide support material. For example, the Ir-Co_3O_4 exhibited superior electrocatalytic OER performance and stability exhibited by the 3d-TM-based oxides doped with single noble metal atoms reported by Wang and his team [113]. Additionally, due to the surface hydroxylation and doping of Mo and W SAs, Yuan Lu and colleagues [114] created a highly efficient bifunctional water electrolysis-based catalyst, and due to the synergistic interactions between Mo and W SAs with the co-doped surface, hydroxylated NiFe oxide denoted as FN-MoWact maintains its stability after 100 h of operation and achieves ultralow overpotentials for OER, HER, and water splitting (1.504 and 1.729 V to deliver current densities of 10 and 500 mA cm^{-2}). In addition, Yang and his group [115] showed that Cu-doped, single-atomic Cu-substituted CeO_2 and multiple Ovac-bound CeO_2 may maximize CO_2 electrocatalytic reduction to CH_4 and produce a remarkably efficient catalytic center for CO_2 activation and adsorption with a FE as high as 58%, indicating tremendous potential for designing more highly efficient electrocatalyst active centers to increase activity and selectivity (Figure 6d). Further, Cu-SACs supported on Sn-SnO_2, for instance, were synthesized together with the Cu doping denoted as CuSA-Sn-SnO_2. Cu-SAs modulated the electronic structure, which effectively influences the adsorption and stabilization of *CO intermediates and resultantly enhanced the performance of CO_2 (Figure 6e,e1) [111].

4.2.4. Facet Engineering

Facet engineering in metal-oxide-based SACs is critical for tuning catalytic activity because it exposes specific crystallographic planes that optimize the interaction of single metal atoms with the support. Different facets of metal oxides have distinct electronic structures, surface energies, and atomic arrangements, which influence the CE and electronic properties of the anchored single metal sites. For instance, Zhang et al. [116] reported the CEs of Pt SAs distributed on anatase TiO_2 supports with controlled shapes, specifically focusing on the (001) and (101) crystal facets. Among them, Pt SAs located on the (001) surface show reduced reactivity for CO oxidation, primarily due to restricted accessibility to gas-phase species. Further, Hu and his team [117] demonstrate that the crystal facets of CeO_2 play a crucial role in determining the form of Pd species. On the CeO_2(100) surface, Pd is present as isolated single atoms, whereas on CeO_2 (111), Pd tends to form clusters. This variation influences the catalytic behavior in N-alkylation reactions, demonstrating that as the catalytic component size reduces to the single-atom level, the impact of crystal facets becomes more significant and can potentially dominate the catalytic performance. Moreover, Chen and colleagues [118] reported that the electronic structure of Pt SAs (Pt_1) is affected by the CeO_2 facets. Pt_1 on CeO_2 (110) exhibits a positive charge, whereas, on CeO_2 (100), it tends toward a metallic state. These variations arise from the differences in Pt-O-Ce microstructures and electron transfer behaviors. The positively charged Pt1 shows weak CO adsorption, which favors oxidation, while metallic Pt_1 enhances H_2 dissociation, aiding CO oxidation and lowering the reaction barrier. Lastly, Li and co-workers [119] reported the SA-Pt catalysts supported on the CeO_2 surface with facets such as (111), (110), and (100) by using DFT calculations, and on all three ceria surfaces, it is thermodynamically advantageous to replace a lattice Ce ion with a single Pt atom. Whereas the Pt atom produces a planar structure with a lowered oxidation state of +2 on the (110) surface due to the spontaneous creation of surface peroxide (O_2^{2-}) species, it adopts a + 4 oxidation state on the (111) and (100) surfaces after replacing Ce^{4+} and resultantly enhanced performance.

4.2.5. Crystallinity Control

Because it directly affects the electrical structure, stability, and accessibility of the active site, crystallinity management in SACs is essential for improving electrocatalytic activity. A well-defined atomic arrangement is usually ensured by the high crystallinity in SACs, which results in consistent coordination environments around the single metal atoms. For the best catalytic activity, precise electronic tuning is necessary, and this homogeneity aids in this process. Furthermore, a highly crystalline structure can offer strong support, improving SA stability in challenging electrochemical environments and reducing aggregation. But it is also critical to strike the desired balance between defect engineering and crystallinity; flaws like vacancies or grain boundaries can provide special electronic states that improve adsorption and charge transfer. Therefore, improving SACs for effective and long-lasting electrocatalytic applications requires deliberate flaw engineering in addition to regulated crystallinity. For instance, Xu and colleagues [120] developed Pt SAs on an amorphous MoO_x substrate and, due to the robust Pt-O bonding contact, obtained a good performance in HER with an overpotential of 19 mV at 10 mA cm^{-2}. Furthermore, to synthesize a Co-SA catalyst, Hyuk Moon and associates [112] used an amorphous-to-crystalline phase transition in the MoO_3 matrix (Figure 6f). The difference in charge densities of Co SA MoO_3 shows that it has improved hydrogen covalent bonding characteristics, and these H^+ adsorption results provide more evidence for Co SA MoO_3's strong HER activity (Figure 6f1).

SACs anchored on metal oxides exhibit considerable potential yet encounter numerous obstacles. A problem exists in morphological engineering, which, although capable of augmenting active surface area and modifying electrical characteristics, faces difficulties in attaining uniform and accurate shapes across diverse scales. Another critical strategy is vacancy engineering, which is vital for enhancing charge transfer and generating several active sites. Controlling the distribution and ensuring the stability of vacancies under reaction conditions presents significant challenges. Further, heteroatom doping presents challenges, as variations in the incorporation and activation of dopants can adversely impact catalytic efficiency and longevity. Moreover, achieving a balance between crystallinity and the introduction of defects and dopants is a challenging endeavor; an excess of defects may undermine structural integrity, whilst insufficient crystallinity might result in diminished conductivity. Future endeavors should prioritize the synergistic implementation of these engineering methodologies, concentrating on synthesis techniques that provide meticulous control over morphology, vacancy distribution, and dopant incorporation. The progression of in situ characterization techniques may yield real-time insights into structural and functional alterations, hence aiding the creation of SAC-supported metal oxides with improved activity and stability for electrocatalytic applications.

4.3. *Alloy-Based Substrate*

A single-atom alloy (SAA) is an innovative advancement in SACs. It amalgamates the advantages of conventional alloy catalysts with SACs, wherein minimal quantities of reactive metal centers are uniformly scattered inside inert metal matrices. In 2012, Sykes and his team [121] effectively dispersed isolated Pd atoms on a pristine Cu (111) active surface and introduced the term SAA. Subsequently, other SAAs have been documented and extensively employed in the electrocatalytic domain due to their distinctive physicochemical characteristics. This distinctive electronic configuration presents a possibility to disrupt the linear scaling correlations between the activation and reaction energy barrier. Simultaneously, SAA catalysts produce an intrinsic synergistic effect as dopant host metal atoms, and SAs on the reaction surface function as dual-catalytic centers for the activation of various supports. By separating the dissociation and catalytic active sites, the SAA catalyst presents the intriguing potential to circumvent linear scaling connections between the reaction energy barrier and the stability of intermediates species, which constrain the catalytic reactivity and selectivity of numerous catalysts. Additionally, we will examine the SAA structure by employing several techniques, including morphological manipulation, vacancy engineering, heteroatom doping, facet engineering, and crystallinity regulation in diverse electrocatalytic activities.

4.3.1. Morphology Engineering

Through the strong electronic MSI, modifying the SAA's shape and nanostructure may alter the microenvironment and, consequently, its intrinsic activity. Due to their distinct advantages in electrocatalytic applications, SAA comes in a variety of morphologies, including 1D, which includes nanofibers, nanotubes, nanorods, and nanowires; 2D, which are nanosheets; and 3D, which is a hierarchical porous structure. For instance, a 1D SAA comprising Pt-Co dual reactive sites enclosed in N-doped graphite-CNTs, synthesized by Cheng et al. [122], exhibited exceptional ORR performance, attaining an ORR mass activity of 0.81 A mg^{-1}Pt. Furthermore, the direct conversion of NO_3 to NH_3 was accomplished by utilizing a 2D-based SA Ni-alloyed Cu electrocatalyst, exhibiting a high FE of approximately 100% and a yield rate of 326.7 μmol h^{-1} cm^{-2}. This performance is attributed to the robust interaction between Ni and the essential NOOH* reaction intermediates, which facilitated a reduction in reactions' limiting potential, consequently enhancing the preference for NH_3 synthesis [123]. Further, Li and his team [124] synthesized a $PdFe_1$ SAA metallene, initially developed as a potent and stable electrocatalyst for NRR, achieving an NH_3 yield of 111.9 μg $h^{-1}$$mg^{-1}$, a FE of 37.8%, and sustained stability over 100 h of electrolysis. This performance is attributed to the Pd co-coordinated Fe SAs serving as active sites, facilitating optimized N_2 activation through N_2-to-Fe σ-donation, alongside excellent thermodynamic stability, which collectively contributes to the excellent performance and durability of $PdFe_1$ in NRR applications. Mao and colleagues [125] similarly synthesized Co-substituted Ru nanostructures, which serve as extremely efficient Pt-free catalysts for HER, necessitating a low overpotential of 13 mV to attain a current density of 10 mA cm^{-2}. Furthermore, a CO_2-reduction-induced surface restructuring technique was accomplished through the innovative design of a nanoporous AgCu SAA catalyst, which demonstrates a self-reinforcing selectivity for CO_2 reduction, coinciding with the transfer of Cu to the reaction surface, and exhibits exceptional performance, with a CO FE of 97.5% [126] (Figure 7a,a1). Lastly, Gao and colleagues [127] synthesized a novel catalytic motif denoted as core/shell diluted SAA and ordered SAA. The Pauli repulsion from the fully occupied metal d-states resulted in a weaker *N binding, while the *NO_3 binding was boosted by the d-states hybridization in the ordered CuAu alloy layers. The tuning of the adsorption of important NO_3RR intermediates was not significantly impacted by the surface strain on Cu/CuAu-ordered SAA. Consequently, the outstanding ammonia selectivity and yield from NO_3RR on these Cu/CuAu-ordered SAA catalysts mark a significant advancement in the creation of core/shell, ordered intermetallic, and SAA catalysts.

4.3.2. Vacancy Engineering

The incorporation of vacancy engineering in SAAs is an effective approach for optimizing the electrocatalytic performance of these materials by generating active sites that improve catalytic activity, stability, and selectivity. The introduction of various vacancies, usually by the elimination of specific atoms in the alloy matrix, modifies the electronic structure and CE surrounding the individual atom sites. This structural alteration can reduce the activation energy of many electrocatalytic reactions, including O_2 and H_2 evolution, as well as CO_2 reduction, by enhancing the adsorption energies of reaction intermediates. For instance, Yao et al. [128] proposed the incorporation of Ru SAs into a Pt-rich environment ($PtCu_x/Pt_{skin}$) to mitigate the oxidation of Ru. This process increased the d-band center of Pt-atom vacancy, attributed to electron transfer between Pt and Ru. Consequently, the Ru_1-Pt_3 Cu configuration exhibited a lowered 90 mV overpotential to achieve a current density of 10 mA cm^{-2}. Moreover, Zhao and his team [129] reported a catalyst with a hierarchical pore structure denoted as Cu_1Au SAA as an effective electrocatalyst for CO_2 reduction and demonstrate exceptional CO_2 reduction performance, achieving nearly 100% CO FE, attributed to its hierarchically structured porous morphology with abundant vacancies and 3D catalytic reactive active sites. Furthermore, Zhang and colleagues [130] synthesized Au_1Cu (111) SAA with surface Covac (VCu-Au_1Cu SAAs), demonstrating an exceptional NH_3 with a FE of 98.7% and a yield of 555 μg h^{-1} cm^{-2} (Figure 7b). The local electronic structure of VCu-Au_1Cu

SAAs was synergistically influenced by individual gold sites and Covac, hence facilitating the generation of *H that can be efficiently employed by NO^{3-} and its intermediates (Figure 7b1).

4.3.3. Heteroatom Doping

The doping of different heteroatoms N, S, or B in SAAs incorporates foreign atoms into the alloy matrix, facilitating improved regulation of the electronic structure between the dopant and adjacent atoms, active sites, and surface characteristics for electrocatalytic processes, hence increasing electron density at specific places and optimizing binding energies for required reaction intermediates. For example, Matthew and colleagues [131] conducted a DFT-based investigation that clarifies the thermodynamics of the ORR on Ag- and Au-based SAAs. They examine SAAs composed of Ni, Rh, Pd, Co, Pt, and Co incorporated as SAs into the (111) surfaces, revealing that the PdAu SAA demonstrates a marginally reduced theoretical overpotential and improved selectivity for 4-electron ORR activity. Furthermore, a study by Cheng's group [132] was conducted on the competitive pathways for CO_2 reduction by using SAAs which reveals that doping of Au and Ag with different TMs facilitates tandem catalysis, and among them, the Au or Ag dispersed on the substrate reduced CO_2 to CO*, which is subsequently reduced by H_2 at the dopant metal to yield C1 hydrocarbons. Furthermore, Zheng and his team [133] introduced a SA Pb-alloyed Cu catalyst denoted as Pb_1Cu that can selectively reduce CO_2 into HCOOH with a high FE of around 96% wherein the Pb_1Cu electrocatalyst facilitates the conversion process on the modulated Cu sites. Thus, alloying Cu with isolated heteroatoms facilitates the C protonation of CO_2 to HCOO* on activated Cu sites, leading to exclusive electrochemical conversion of CO_2 to HCOOH with very high activity. Moreover, Wu et al. [128] synthesized a series of single Ru-doped $PtCu_x/Pt_{skin}$ SAAs and among them the Ru_1-Pt_3Cu catalyst exhibiting the optimal OER performance, characterized by an overpotential of 220 mV at 10 mA cm^{-2}. Recently, the advancement of bifunctional electrocatalysts that operate effectively for both HER and OER across a broad pH spectrum has garnered significant attention. Liang's group [134] recently synthesized a Ru-doped Co SAA catalyst encased in an N-doped carbon sheath, utilized for OWS. This catalyst exhibited exceptional bifunctional electrocatalytic activity achieving a minimal cell voltage of 1.55 V at 10 mA cm^{-2}, attributed to the doped Ru SAs that effectively reduced the adsorption of H* and other oxygenated reaction intermediates species. Recent studies have indicated a markedly improved reactivity and selectivity of SACs for N_2 reduction. Chen's team [135] developed a series of SAA model systems M/Au(111) and assessed their NRR performance using DFT, revealing that the incorporation of TM SAs on Au(111) substantially enhances the N_2 binding affinity and reduces the activation energy barrier of the initial hydrogenation step. Furthermore, Huang et al. [136] developed a series of $NiCoRu_x$/SP and, in particular, $NiCoRu_{0.2}$/SP exhibited optimized activity attaining a lowered overpotential of 59 mV at 10 mA cm^{-2} due to the incorporation of Ru atoms into the NiCo support would create a robust interaction at the Ru-Ni/Co interface thereby enhancing the HER activity (Figure 7c,c1).

4.3.4. Facet Engineering

Facet engineering is considered a highly effective strategy, as the crystal facet constitutes the most fundamental component of a SAC-based alloy. The surface atomic arrangement and configuration directly influence the reactant's and intermediates adsorption and activation energy, which fundamentally dictate catalytic activity. Facet engineering has demonstrated its efficacy in improving the performance of electrocatalysts for various electrocatalytic reactions. For example, Du and his team [139] discovered that a Cu(100) surface doped with Pd-SA exhibited enhanced catalytic activity and selectivity for NO_3RR. Moreover, Feng and his team [140] demonstrated that the Fe- and Ru-decorated Cu (211) step edges were the most efficient in facilitating the CO_2 reduction process to CH_4. In addition, Zheng's group [141] reported the development of Pt/Pd SAA catalysts, successfully dispersing Pt atoms onto Pd NPs with (100) and (111) facet surfaces, resulting in significantly enhanced HER and ORR activity. Moreover, Yao and his team [128] synthesized a series of Pt-Cu alloys along with the dispersion of Ru SAs and a controlled amount of the step, kink, and corner defects exposed on the facet's active surface demonstrated that these processes augmented the surface defects which further function as anchoring reactive sites for dispersed Ru_1, thereby enhancing OER performance (Figure 7d,d1). Furthermore, Gao and colleagues [137] synthesized a series of copper-based SAA, incorporating doped TMs and exposing surfaces with various Miller indices, utilizing DFT calculations and among them W_1-Cu(111)) with −0.42V, Re_1-Cu(111) with −0.38 V, Os_1-Cu(211) with −0.49 V, Os_1-Cu(110) with −0.48V, and Re_1-Cu(100) with −0.35V exhibiting low limiting potentials for the N_2 reduction reaction (Figure 7e).

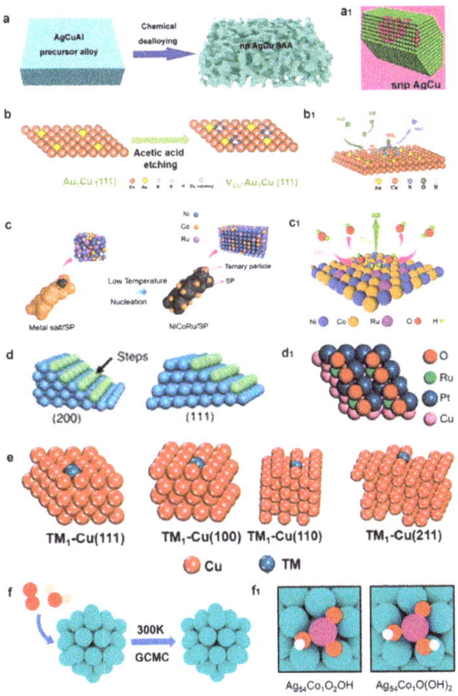

Figure 7. (**a**) Fabrication of np AgCu SAA, (**a1**) Structure of np AgCu SAA. Modified from Ref. [126] with copyright permission from Elsevier. (**b**) Synthesis of the VCu-Au_1Cu SAAs; (**b1**) mechanism on VCu-Au_1Cu SAAs for NO_3RR. Modified from Ref. [130] with copyright permission from Elsevier. (**c**) Synthesis of NiCoRu/SP; (**c1**) mechanism of water splitting on $NiCoRu_{0.2}$/SP. Modified from Ref. [136] with copyright permission from Elsevier. (**d**) Exposed steps on diverse facets for various catalysts; (**d1**) structure of $PtCu_3$ (111). Modified from Ref. [128] with copyright permission from Nature. (**e**) Models of Cu-based SAAs with four different Miller indices. Modified from Ref. [137] with copyright permission from Elsevier. (**f**) Grand canonical Monte Carlo (GCMC) simulation on the $Ag_{54}Co_1$ cluster; (**f1**) $Co_1O_x(OH)_y$ complexes configuration. Modified from Ref. [138] with copyright permission from The American Chemical Society.

4.3.5. Crystallinity Control

Regulating crystallinity in SAAs is essential for enhancing electrocatalytic activity, as it affects active site distribution, electrical characteristics, and stability. Regulating the crystalline phase and grain boundaries allows for the precise adjustment of the atomic dispersion of single atoms within the alloy matrix, which directly influences the local electronic environment of catalytic sites and, consequently, the reaction kinetics. For example, Pu et al. [138] conducted DFT calculations demonstrating that the $Ag_{54}M_1O_x(OH)_y$ complex serves as the crucial catalytic active site for facilitating the ORR in Ag-based SAA cluster electrocatalysts. The GCMC simulation demonstrates that the atomic O* species have a limited capacity to adsorb onto the $Ag_{54}Co_1$ cluster owing to the endothermic Gibbs free adsorption energy (Figure 7f). The suitable orbital hybridization between dopant metal and OH in the MO_x(OH)y complexes suggests that the $Ag_{54}Co_1$, $Ag_{54}Pd_1$, and $Ag_{54}Au_1$ clusters are the most effective Ag SAA catalysts for the ORR (Figure 7f1).

Besides the enhanced catalytic performance of alloy-based substrates in electrocatalysis, they still encounter numerous obstacles such as: (A) Understanding how the morphology of alloy-based SACs varies throughout prolonged catalytic operations is crucial to addressing the problem of catalyst deactivation. (B) It is difficult to predict how vacancies would affect the overall catalytic performance of alloy-based SACs due to the lack of clarity surrounding their function and interactions with other catalytic sites. (C) When heteroatoms are added, the metal lattice may experience strain and deformation, affecting the structural stability of alloy-based SACs and reducing the catalyst's activity. (D) Heteroatoms may cause strain and distortions in the metal lattice, which could reduce the durability and risk the structural stability of alloy-based SACs. (E) Surface reconstruction may cause alloy-based SACs' crystallinity to deteriorate across several catalytic cycles, potentially leading to a decrease in active sites and catalytic activity. Future considerations may improve the efficiency of SAA-based substrates by (A) better control over the morphology could help alloy-based SACs continue to maintain over time, preventing deactivation and making them more appropriate for widespread use in the energy application. (B) The development of new catalysts could be accelerated by using machine learning techniques to optimize vacancy engineering and determine the best vacancy kinds and densities for different catalytic reactions. (C) To improve catalytic performance, multi-element heteroatom doping in alloy-based SACs can optimize electrical and geometric features. (D) Multi-element heteroatom doping in alloy-based SACs can tune the geometric and electrical properties for increased catalytic performance. (E) The logical design of optimal compositions and structures can be facilitated by using advanced computational techniques to predict the crystallinity and stability of alloy SACs, offering important insights before experimental synthesis.

4.4. TMD-Based Substrate

TMDs are an essential category of 2D materials, which exhibit a high surface area, robust interaction with light fields, and elevated charge carrier mobility, rendering them optimal support for the incorporation of SACs. TMD materials generally manifest as MX_2, with M denoting the core TM atom and X signifying chalcogen atoms (S, Se, or Te). However, the saturated coordination atoms on the TMD substrate fundamentally restrict the accessible active sites for electrocatalytic processes. Consequently, the technique of utilizing TMD-supported TM atoms to fabricate SACs demonstrates considerable potential [142]. In recent years, various sophisticated TMD-loaded SA electrocatalysts have been developed, employing diverse techniques as outlined below.

4.4.1. Morphology Engineering

The intrinsic anisotropic behavior of TMDs in SACs renders morphological factors crucial in optimizing their catalytic performance. To fully exploit the potential of active edge sites and optimize the available surface area, various TMD-based SAC topologies have been developed, including nanosheets and hollow nanoboxes. For instance, Ling and co-workers [143] investigated the potential of the MoS_2 monolayer in the 1T′ phase for accommodating single TM atoms and assessed the ORR catalytic activity of these dispersed TM atoms through DFT calculations. Among the studied configurations, Pd@1T′ and Cu@1T′ exhibited the most favorable ORR performance, with overpotentials of 0.32 V and 0.41 V. Further, Liu and co-workers [144] predict that a single 3D TM co-doped with O-atoms in MoS_2 (3d-TMO_6@MoS_2) undergoes a transformation from a hexa-coordinated to a tetra-coordinated structure as oxidative intermediates approach, and among these, TM = Co, Mn, and Fe atoms exhibit superior catalytic activity for the OER. Additionally, Pattengale's group [145] utilized 1T-MoS_2 as a substrate to incorporate Ni SAs into its basal plane, substituting Mo atoms (Ni@1T-MoS_2) and the resulting catalyst demonstrates significantly improved HER performance. The advancement of MoS_2 edge-supported SACs is essential for water splitting. Xu and his team [146] reported various single TM SACs embedded on MoS_2 edges as bifunctional electrocatalysts for OWS through DFT calculations, and among them, Pt-SAs containing T1-vacancy exhibits the lowest theoretical overpotential for H_2/O_2 (−0.10/0.46 V). Furthermore, Ren and colleagues [147] examined a range of SAs immobilized on MoS_2 nanosheets for CO_2 conversion to methane through DFT calculations, with Ni@MoS_2 with −0.45 V, Fe@MoS_2 with −0.39 V, and Co@MoS_2 with −0.24 V exhibiting limiting potentials for methane production. Ultimately, it has been determined that SA supported on several chalcogenides (WS_2, WSe_2, and $MoSe_2$) exhibits exceptional activity for CO_2RR. Utilizing first-principles high-throughput computations, Yang et al. [148] examined the catalytic activity of several 2D TM-SAs supported on a MoS_2 monolayer denoted as MoS_2 (SA@MoS_2-B) and the optimized configuration of Mo@MoS_2-M for NRR (Figure 8a). The dispersed TM-SAs on the sulfur-sulfur bridge site of MoS_2 (SA@MoS_2-B) facilitate charge transfer from the Mo-SA to N_2, which is crucial for both activating the nitrogen triple bond by populating the π antibonding orbital and stabilizing the adsorbed N_2 through Coulombic interactions (Figure 8a,a1), and among these configurations, Mo@MoS_2-M exhibits good stability by using the distal pathway for catalyzing N_2 reduction, with a calculated overpotential of 0.28 V. Further, Li and his team [149] developed a 2D Fe-MoS_2 catalyst for the NO_3RR, demonstrating exceptional performance with a maximum FE of 98% for the conversion of NO_3 to NH_3 (Figure 8b,b1). The energy levels of the α-spin and β-spin d atomic orbitals of Fe-MoS_2 significantly overlap with the $2\pi^*$ orbitals of NO, resulting in the splitting and rearrangement of orbitals that generates new d-π^* antibonding and bonding orbitals (Figure 8b2). Besides the MoS_2, WS_2-based SA was also reported for NO_3 reduction reaction by using DFT calculations [150].

4.4.2. Vacancy Engineering

Generating defects, such as vacancies, is an effective method for altering the electrical structure of SACs anchored on TMDs. Different concerted efforts have been undertaken to develop regulated techniques for the introduction of atomic vacancies in TMDs with SACs, thereby facilitating various electrocatalytic activities. For instance, Wang et al. [151] synthesized SA Ru-doped MoS_2 catalysts with a high 2H phase content denoted as Ru@2H-MoS_2 and Ru0.10@2H-MoS_2 exhibits the lowest HER overpotentials of 137, 51, and 168 mV at 10 mA cm^{-2} due to the incorporation of Ru which creates additional reactive active sites and vacancies, thereby enhancing the HER performance. Additionally, various single TM atoms immobilized on MoS_2 edges have been evaluated as bifunctional electrocatalysts for OWS by DFT calculations [146] (Figure 8c,c1). The diffusion barrier of Pt on T1-vacancy is lower at 1.82 eV (Figure 8c2) and exhibited the lowest theoretical overpotential for the H_2/O_2 at −0.10/0.46 V for OWS, attributed to the adjusted band center of the single metal atom. Furthermore, Zhang and colleagues [152] utilized spin-polarized DFT-D_3 calculations to systematically examine the reaction mechanism of electrocatalytic CO_2RR on M@WTe_2 with a Te vacancy, finding that Ni@WTe_2 exhibits remarkable selectivity for HCOOH with an ultra-low overpotential of −0.11 V. Moreover, Jiang's team [153] indicated that a non-contacting TM atom adsorbed on the Svac of the MoS_2 basal plane functions as a spin-active probe by causing significant spin polarization on the exposed Mo atoms. The electronic spin moment of the Mo atom can be adjusted across a wide range by altering the type of TM atom and its adsorption site. Consequently, the TM's spin alignment properties significantly enhance the adsorption and polarization of N_2, facilitating the reduction of N_2 to NH_3. Furthermore, Yang and his team [154] adjust the phase of substrate MoS_2, adorned with a B-SA referred to as B@2H-MoS_2 and B@1T-MoS_2, by introducing Svac. Among these, B@1T-MoS_2 exhibits superior NO_3RR activity and remarkable stability, characterized by a metastable foundation, as validated by migration energy barrier and ab initio molecular dynamics (AIMD) simulations.

4.4.3. Heteroatom Doping

A key strategy for improving the catalytic reactivity of SACs supported by TMDs is the careful regulation of their physicochemical properties via heteroatom doping in their inert basal planes. The incorporation of metallic or non-metallic heteroatoms into the lattice structure can enhance H_2 adsorption affinity by modifying the electronic band structures of catalytically inactive basal planes. Striking a precise balance is crucial, since excessive doping may compromise lattice stability, consequently lowering catalytic reactivity and shortening the overall longevity of TMD-supported SACs. Therefore, the heteroatom doping technique must be well-optimized for the designated target materials. For example, Zhang and his team [155] synthesized an SA Ru catalyst supported on MoS_2, referred to as SA-Ru-MoS_2, which demonstrates a low overpotential of 76 mV at 10 mA cm^{-2}, attributed to the doping of Ru-SA which induces a phase transition in MoS_2 and the formation of Svac thereby markedly boosting the performance of inert 2D MoS_2 for HER. Moreover, Zhu and colleagues [156] investigated the potential of utilizing SA-dispersed Cu or Co on a 2H-MoS_2 monolayer denoted as Co@2H-MoS_2 and Cu@2H-MoS_2 for the OWS and these catalysts modulated the electronic band structure and charge density distribution of 2H-MoS_2 and the dispersion of Co or Cu SAs can activate inert in-plane S-atoms thereby enhancing the catalytic activity for both H_2 and O_2 reactions. Additionally, Hao et al. [157] developed an innovative heterostructure of MoS_2-Mo_2C nanosheets with minimal Ru loading, encapsulated atop TiN nanorod arrays supported by a carbon composite matrix (Figure 8d,d1). The ΔGH* value of Ru-MoS_2-Mo_2C was nearly 0, indicating its potential to attain an optimum ΔGH* value for superior HER performance (Figure 8d2) due to a synergistic effect that enhanced reaction kinetics for both the H_2 and O_2 evolution reaction as well as for OWS. Furthermore, the electrocatalytic NRR was performed utilizing a series of TM-doped Haeckelite-MoS_2 nanosheets (TM@HL-MoS_2) through DFT calculations and among these, Sc@HL-MoS_2 emerges as the most effective NRR catalyst, as it facilitates the activation of adsorbed N_2 molecules, particularly through the enzymatic pathway of the side-on configuration, which is distinguished by the lowest free energy (merely 0.001 eV) and an exceptionally low initial potential (−0.28 V) [158]. Magnetic field-enhanced heterogeneous SA spin catalysts increase metal utilization and reaction efficiency. The requirement for a high density of scattered active sites with both long-range ferromagnetic ordering and short-range quantum spin interactions makes their design difficult. In that case, Sun and his team [159] synthesized several SA spin catalysts with extensively adjustable substitutional magnetic atoms (M_1) in a MoS_2 host. Among all the M_1/MoS2 species, global room-temperature ferromagnetism is the result of the distorted tetragonal shape that Ni_1/MoS_2 adopts and this structure causes ferromagnetic coupling to neighboring Ni_1 sites and nearby S atoms consequently facilitates spin-selective charge transfer, resulting in triplet O_2. Additionally, the OER magnetocurrent is enhanced by approximately 2880% over Ni_1/MoS_2 in both saltwater and pure water splitting cells by a modest magnetic field of around 0.5 T due to a field-induced spin alignment and spin density optimization over S active sites resulting from field-regulated S(p)-Ni(d) hybridization which is responsible for a significant magnetic-field-enhanced OER performance over Ni_1/MoS_2.

4.4.4. Facet Engineering

Facet engineering is an effective approach for regulating the proportion of targeted crystal planes on the surfaces of SACs. Surface energy and atomic coordination can be modified by selectively exposing particular crystal facets, resulting in enhanced reactivity and selectivity in diverse electrochemical reactions. This method enables the incorporation of SACs onto selectively chosen surfaces of TMDs, enhancing electron transport and charge separation. Furthermore, facet engineering can enhance the stability of single atoms, inhibit their aggregation, and preserve elevated catalytic performance. For example, Qi and his team [160] developed a SA array catalyst, SA Co-D 1T MoS_2, for HER utilizing the top-down assembly/leaching methodology (Figure 8e,e1). The adsorption of H_2 at the upright position of Co on MoS_2 (111) is less favorable with the preference for the tilted top site linked to the ensemble effect at the Co-MoS_2 (111) interface. The energy acquired by transitioning from the upright to the tilted position is linked to the electrostatic attraction between the negatively charged proton and the positively charged S adjacent to Co (Figure 8e2), and the elevated HER activity of this SAC primarily results from an ensemble effect, facilitated by the synergy between the cobalt adatom and the sulfur of the MoS_2 (111) support, achieved through the modulation of the hydrogen-binding mode at the interface.

4.4.5. Crystallinity Control

Regulating crystallinity in TMD with SACs is essential for enhancing their catalytic efficiency in diverse electrochemical processes. Optimizing temperature and reaction time parameters might improve the uniformity and stability of the SACs inside the TMD matrix, hence promoting optimal atomic dispersion and interaction between the active sites and the matrix. Pan et al. [161] indicated that Pt SAs were incorporated into the 2H-MoS_2 lattice without disrupting its crystalline structure. Theoretical studies indicate a universal process of "onsite electrostatic polarization," wherein electrostatic forces substantially polarize distributions of charge at individual atomic sites and modify the reaction kinetics of rate-determining steps, resulting in enhanced reaction performance. The field-induced on-site polarization provides a distinctive approach for modeling catalytic activities in natural enzyme systems using dynamic, quantitative, precise external electric fields (Figure 8f,f1).

SACs-based TMDs encounter numerous obstacles in electrocatalysis such as (a) the reproducibility of TMD-based SACs with defined morphologies poses a considerable difficulty since minor alterations in experimental conditions can lead to huge inconsistencies in the catalyst's performance. (b) The production of TMD-based SACs with regulated morphology necessitates intricate processes that could limit their scalability by requiring costly precursors or producing low yields. (c) developing vacancies in SAC-based TMDs to enhance catalytic activity by introducing supplementary active sites is difficult. (d) It is difficult to predict how heteroatom doping would impact the catalytic activity and selectivity of SACs since the precise electrical changes caused by doping in TMDs are yet unclear. (e) Further, it is still very difficult to distribute heteroatoms uniformly within the TMD framework. (f) It is still difficult to achieve selective exposure of particular facets during the synthesis of TMD-based SACs. (g) Defects in TMD can increase catalytic activity, but if left uncontrolled, they can weaken SACs' crystallinity, which lowers their catalytic activity. The efficiency of SACs based on TMDs in electrocatalysis can be improved by; (A) the incorporation of throughput automation and high-throughput screening techniques may facilitate the swift optimization of synthesis methods, resulting in the scalable production of TMD-based SACs with accurate morphological control. (B) by regulating vacancy formation via etching or chemical processes, the type and density of vacancies can be refined in TMD-based SACs. (C) by improving their electrical characteristics and increasing catalytic efficiency, dual doping with distinct heteroatoms may improve the performance of TMD-based SACs, especially for complex processes. (D) The type, concentration, and distribution of heteroatoms at the atomic scale can be precisely controlled by developing high-throughput screening techniques. (E) The catalytic activity of TMD-based SACs can be increased synergistically by combining facet engineering with other techniques like doping or vacancy engineering, which will boost performance in a variety of processes. (F) This can be lessened by combining TMDs with additional materials, including metal oxide or carbon-based supports, which can help maintain the crystalline structure, reduce undesired phase shifts, and improve the stability and efficiency of SACs.

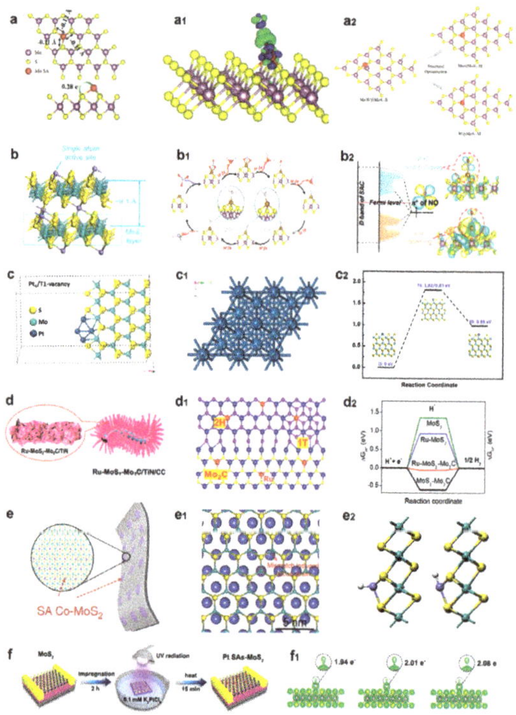

Figure 8. (**a**) Structure of Mo@MoS_2-M, (**a1**) Charge redistribution on Mo@MoS_2-M, (**a2**) Structural optimization of the Mo and W SA in Mo | W@MoS_2-B. Modified from Ref. [148] with copyright permission from Elsevier. (**b**) Structure of Fe-MoS_2; (**b1**) mechanism for the NO_3RR on M-MoS_2; (**b2**) interaction scheme in-between M-MoS2 and NO. Modified from Ref. [149] with copyright permission from John Willey and Sons. (**c**) Structure of Pt_4/1T-vacancy; (**c1**) top view of Pt (111) slab; (**c2**) diffusion barrier of Pt on T1-vacancy. Modified from Ref. [146] with copyright permission from The Royal Society of Chemistry. (**d**) Synthesis of Ru-MoS_2-Mo2C; (**d1**) Ru-MoS_2-Mo_2C structure; (**d2**) ΔGH* of different catalysts for HER. Reprinted from Ref. [157] with copyright permission from Elsevier. (**e**) Structure of SA Co-D 1T MoS_2; (**e1**) illustration of Co and 2H MoS_2; (**e2**) H_2 adsorption modes on the SA Co-MoS_2. Modified from Ref. [160] with copyright permission from Nature. (**f**) Synthesis of Pt SAs-MoS_2; (**f1**) charge analysis of Pt SAs-MoS_2. Modified from Ref. [161] with copyright permission from Nature.

4.5. MXene-Based Substrate

Two-dimensional TM carbides, nitrides, and carbonitrides characterized by the generic formula $M_{n+1}X_nT_X$ (where n = 1–3; M denotes a TM; X represents C and/or N; T signifies surface functional groups such as oxygen, hydroxyl, or fluorine) have proven to be advanced support for electrocatalytic reactions due to their remarkable electrical, mechanical, catalytic, and thermal features, possessing a substantial surface area, elevated active site density, exceptional durability, and hydrophilic characteristics; hence, these distinctive traits render them suitable as supports for SACs [162]. We elaborate below on several ways employed to boost their performance in distinct electrocatalytic activities.

4.5.1. Morphology Engineering

Morphology engineering is crucial for improving the performance of MXene-based SACs in electrocatalysis by adjusting their surface characteristics, structure, and electrical interactions. MXenes, characterized by their distinctive 2D layered architecture and elevated conductivity, provide an outstanding substrate for the immobilization of single atoms, hence optimizing active site accessibility. Furthermore, controlling morphology can enhance the accessibility of reactants to active sites and promote charge transfer, leading to reduced overpotentials and improved electrocatalytic performance. Kuznetsov's group [163] synthesized Co-substituted MXene catalysts designated as Mo_2CT_x: Co where the substitution of Mo with Co significantly alters the electronic configuration and affects the adsorption-free energies of H_2 on adjacent O_2 atoms, thereby enhancing the suitability of these active sites for HER electrocatalysis. Similarly, Kuznetsov et al. [164] prepared a multilayered Mo_2CT_x: Fe by substituting Mo sites in Mo_2CT_x, exhibiting good catalytic activity and selectivity in the reduction of O_2 to hydrogen peroxide (Figure 9a–a2). Furthermore, Huang and co-workers [165] employed innovative 2D Mo_2CO_2 and Ti_2CO_2 MXene monolayers as effective electrocatalysts for N_2 reduction reactions utilizing first-principles computational methods. The incorporation of Mo and Ru onto Mo_2CO_2 and Ti_2CO_2 resulted in improved electrochemical activity for N_2RR and due to the elevated conductivity of Mo_2CO_2, Ru@Mo_2CO_2 exhibited a reduced negative potential. Moreover, Zhao and his team [166] reported an SA-Cu immobilized MXene for the electrocatalytic reduction of CO_2 to methanol, achieving a high FE of 59.1% and demonstrating commendable stability due to the SA Cu's unsaturated electronic coordination sites ($Cu\delta+$, $0 < \delta < 2$), which facilitates a low activation energy for the RDS with the conversion of HCOOH* to the absorbed CHO* reaction intermediate species), thereby enabling efficient electrocatalytic CO_2 reduction to methanol (Figure 9b–b2).

4.5.2. Vacancy Engineering

Vacancy engineering is essential for improving the performance of MXenes-based SACs in electrocatalysis. The electrical structure and surface properties of MXenes can be considerably modified by the strategic introduction of vacancies, including C, O, or TM vacancies. These vacancies establish distinct coordination sites that secure individual metal atoms, inhibiting their aggregation and improving their catalytic stability. For instance, Zhang's group [167] synthesized double TM-SA dispersed on MXene denoted as $Mo_2TiC_2T_x$, characterized by abundant Mo vacancies and exposed basal planes in the outer layers, and achieved an H_2 evolution with low overpotentials (30mV @10 mA cm^{-2}, and 77 mV @100 mA cm^{-2}) owing to the interaction between surface functional groups and protons of $Mo_2TiC_2T_x$ (Figure 9c–c2). Moreover, Chen and colleagues [168] prepared a Mxene monolayer containing Ti vacancy resulting in remarkable stability and exceptional CO2RR activity. Further, Wang and his team [169] examined the NRR performance of TMn embedded in defective $V_3C_2O_2$ through DFT calculations, and these defects modify the d-band center and generate electron-deficient sites, enhancing N_2 activation, and among the series, $W_3@V_3C_2O_2$ with −0.31 V, $Mo_2@V_3C_2O_2$ with −0.31, $Mo_3@V_3C_2O_2$ with −0.33 V, and $W_2@V_3C_2O_2$ with −0.34 V exhibit commendable activity and selectivity, with the lowest limiting potentials (Figure 9d–d2). Additionally, the NO_3RR mechanism of SACs is methodically investigated by employing DFT to model single TM atoms on MXene with Ov, and among them, Cu/Ov-MXene with −0.34V and Ag/Ov-MXene with −0.24V demonstrate remarkable efficiency for NO_3RR to NH_3 [170].

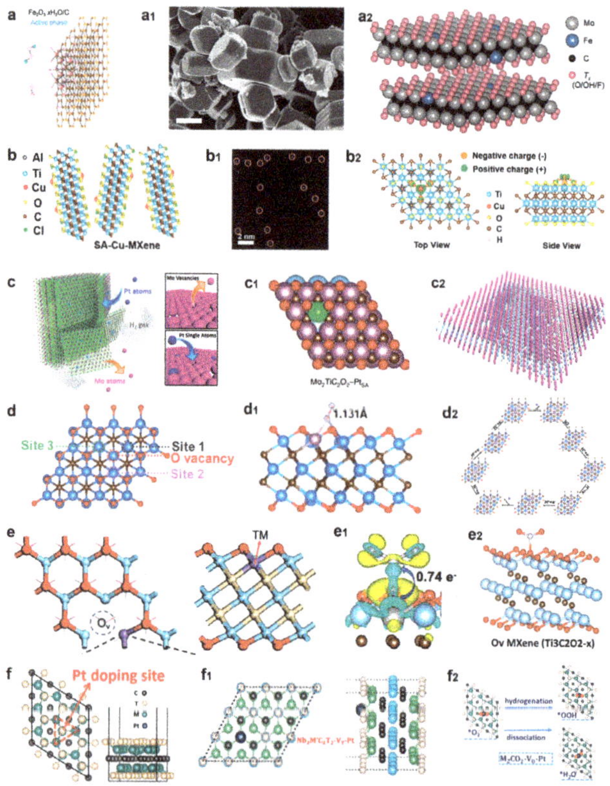

Figure 9. (**a**) Structure of $Fe_2O_3.xH_2O/C$, (**a1**) SEM of Mo_2CT_x:Fe structure; (**a2**) Mo_2CT_x:Fe structure. Reprinted from Ref. [164] with copyright permission from The American Chemical Society. (**b**) Structure of SA-Cu-MXene; (**b1**) AC HAADF-STEM image of SA-Cu-MXene; (**b2**) difference in charge density in SA-Cu-MXene. Modified from Ref. [166] with copyright permission from The American Chemical Society. (**c**) Mechanism of MXene with immobilized Pt SAs; (**c1**) top view $Mo_2TiC_2O_2$-PtSA; (**c2**) structure of the $Mo_2TiC_2T_x$-VMo. Reprinted from Ref. [167] with copyright permission from Nature. (**d**) Top view of $V_3C_2O_2$ monolayer with O vacancy; (**d1**) side views of optimized adsorption configurations on TMn@$V_3C_2O_2$; (**d2**) NRR mechanism on TMn@$V_3C_2O_2$. Reprinted from Ref. [169] with copyright permission from Elsevier. (**e**) Atomic structure of TM/Ov-MXene; (**e1**) difference in charge density on Ag/Ov-MXene; (**e2**) structure of Ag/Ov-MXene. Modified from Ref. [170] with copyright permission from John Willey and Sons. (**f**) Structure of M_2CT_2-VT-Pt; (**f1**) structure of $Nb_2Ta_2C_3T_2$-VT-Pt; (**f2**) reaction paths and energy barriers of O_2 hydrogeneration or dissociation on M_2CT_2-VT-Pt. Reprinted from Ref. [171] with copyright permission from The American Chemical Society.

4.5.3. Heteroatom Doping

Doping heteroatoms in MXenes-based substrates with SACs is a promising strategy to improve the catalytic performance and selectivity of these materials. Recently, several atoms, including N, S, P, Cu, and Fe, have been included in SACs-based MXenes to enhance their performance in diverse electrocatalytic reactions by adjusting their electronic structure, altering the elemental composition, and optimizing surface chemistry. Additionally, in the MXenes lattice and CE, the SACs can stabilize individual metal atoms through robust covalent connections, serving as active sites for electrocatalytic processes. For instance, Gao et al. [170] documented the O vacancy TM-doped MXene for NO_3RR utilizing DFT, revealing that Cu/Ov-MXene with −0.34V and Ag/Ov-MXene with −0.24V demonstrate remarkably low limiting potentials for NO3RR along with the adsorption energy of NO_3^- on Cu/Ov-MXene and Ag/Ov-MXene that exceeds that of the proton, hence promoting NO_3RR (Figure 9e–e2). Further, to demonstrate the significance of N and S doping, Ramalingam's group [172] synthesized RuSA-$Ti_3C_2T_X$ and RuSA-N-$Ti_3C_2T_X$ incorporating N and S into the MXene matrix. The supported MXene enhances the electrical interaction between Ru SAs and the MXene substrate, optimizing ΔGH* to practically zero. Nonetheless, investigations into MXene materials remain nascent relative to other 2D supports. Kan and his team [171] described the construction of Pt SACs utilizing standard M_2CT_2 MXenes to investigate their catalytic characteristics for the ORR and the OER. The Pt atom was identified as the active site in the process. The catalytic capabilities predominantly depend on the submetal (M) of the MXenes, with Nb_2CF_2-VF-Pt and Cr_2CF_2-VF-Pt identified as effective bifunctional catalysts for ORR/OER. Both are predicated on the F terminator. All O-terminated systems demonstrated inadequate ORR or OER performance owing to the reduced charge density surrounding the Pt atom and the C layer (Figure 9f–f2).

SACs utilizing MXenes by incorporating different strategies face some challenges that encompass (1) the correlation between the morphology of MXenes as a support and their efficiency as SACs-based catalysts is complex. (2) Another significant issue in optimizing morphology for MXene-based SACs is maintaining the structural integrity of MXenes under harsh catalytic conditions. (3) Further, incorporating vacancies into the MXene structure containing SACs may trigger phase changes, thereby reducing the catalyst's efficiency. (4) Nonetheless, not all vacancies exert identical effects differences in the nature and magnitude of vacancies can result in varying influences on the electronic structure and reactivity of them. Examining the impact of vacancy dimensions, morphology, and positioning on catalytic efficiency continues to be a significant challenge. (5) The influence of heteroatoms on the catalytic characteristics of MXenes-based SACs is not well understood. (6) Furthermore, attaining a high concentration of heteroatoms in MXenes's SACs structure without reducing the catalyst's stability or causing dopant aggregation is still a challenge. The performance of SACs based on MXenes in electrocatalysis can be enhanced by: 1. enhanced comprehension of the structure-function relationship can improve the creation of MXenes with specialized morphologies tailored for distinct catalytic processes by increasing the quantity of MXene layers, introducing surface imperfections, or altering edge configurations. 2. Their morphology can be optimized by regulating the number of layers, and thickness of MXenes, which can augment surface area, improve electron conductivity, and elevate the accessibility of active sites. 3. By meticulously regulating the concentration and types of vacancies, one can mitigate the probability of unfavorable phase transitions. 4. Moreover, by meticulously controlling vacancy dimensions and configurations, SACs-based MXenes can be tailored to establish certain electronic environments that improve selectivity. 5. By utilizing characterization techniques, such as DFT can provide significant insights into the impact of heteroatoms on the catalytic properties of SACs based MXenes. 6. Enhanced doping methodologies, including the utilization of low-temperature treatments or the application of coordination chemistry for superior heteroatom binding, may provide elevated doping concentrations while preserving the structural integrity of MXenes utilizing SACs.

4.6. MOF-Based Substrate

MOFs are advanced, porous, crystalline-based materials, consisting of diverse metal atoms and organic/inorganic ligands. The multi-component and well-defined structure allows for precise regulation of electrochemical characteristics through fundamental principles of molecular chemistry. In the past few years, due to several beneficial attributes, such as diverse functional groups, structured pores, unsaturated metal active sites, and plentiful surface-active sites, MOF, and MOF-derived materials have been investigated and utilized as novel substrates for SACs in various electrochemical reactions [3].

4.6.1. Morphology Engineering

The tailoring of morphology in SACs that anchor MOF-based substrates is crucial for improving catalytic efficacy and stability in diverse electrocatalytic processes. The structural dimensions of SACs, including nanofibers, nanowires, nanosheets, and nano boxes, have been optimized to synergistically improve catalytic performance. This optimization facilitates fast contact efficiency between the catalyst and reactants, minimizes mass transport distances, and ensures continuous electron transport pathways, thereby maximizing active sites and accessibility. For instance, a series of Fe_x-PCN-222, have been prepared by using a novel mixed-ligand strategy for ORR as reported by Jiao et al. [173] (Figure 10a). The hierarchical porous structure of FeSA-N-C, characterized by high-density Fe-N_4 active sites and orientated open mesoporous channels enhance elevated mass transfer, would significantly expedite the ORR process (Figure 10a1,a2). Another investigation demonstrates that enveloping a trinuclear $Fe^{III}2Fe^{II}$ complex within the channels of distinctive MOF results in the formation of iron nanoclusters stabilized on a carbon layer, synthesizing a DAC referred to as Fe^{II} by Zn^{II}/Co^{II}, which displays effective OER activity [174]. Moreover, Yuan and colleagues [175] synthesized a rhombic dodecahedral catalyst, termed Co-SAs/N-C, using high-temperature pyrolysis, which has exceptional catalytic performance owing to the robust interaction between Co-SAs/N-C and Ru clusters. Furthermore, Jiao and his team [176] constructed a 3D SA Fe_1-Ni_1-N-C catalyst with N-bridged Fe and Ni SAPs for CO_2RR by direct pyrolysis of MOFs (Figure 10b,b1). The Fe-N-N-Ni structure exhibits the most uniform CE and interatomic distance and owing to the synergistic effect of adjacent Fe and Ni SA pairs in Fe_1-Ni_1-N-C demonstrates markedly enhanced performance for the CO_2 reduction (Figure 10b2,b3). In addition, Tao and his team [177] documented that a 3D single Ru site dispersed on N-doped porous carbon significantly enhanced the electroreduction of aqueous N_2 to NH_3, yielding an NH_3 formation rate of 3.665 mg NH_3 h^{-1} mg^{-1} and attained an NH_3 FE of approximately 21% at a low overpotential of 0.17 V. Similarly, Zhang and his team [178] developed 2D Fe SACs featuring distinctive FeN_2O_2 coordination via direct pyrolysis of MOFs demonstrating a remarkable FE of approximately 92% and an elevated NH_3 yield rate of 46 mg h^{-1} mg cat^{-1}, attributed to the presence of O_2-atoms in FeN_2O_2 that modulate the d-band center of Fe, thereby influencing the adsorption activation energies of the NO_3RR intermediates.

4.6.2. Vacancy Engineering

The advancement of SACs with a precisely defined microenvironment has demonstrated advantages from the deliberate creation of surface imperfections on supports. These surface defects function as "traps" for the accumulation of metal atoms, aiding in this endeavor. The modification of SA site microenvironments predominantly occurs at defects, C_{vac}, S_{vac}, O_{vac}, or metal cations depending on the type of supports utilized. Cationic vacancies have been extensively employed to fabricate SACs with unique local environments based on MOFs. For instance, Shankar and his team [179] described a straightforward MOF-assisted electrodeposition technique for synthesizing Fe SA catalysts on C-Ni heterostructure nanosheets containing Ovs, which exhibit effective electrocatalytic activity for OER and HER, with low overpotentials of approximately 246 and 164 mV at 10 mA cm^{-2} attributed to the synergistic properties of Fe SA sites on the C-Ni matrix, high-valent Fe^{4+} centers, characterized by a substantial volume of active sites, electrophilicity, rapid transfer of charge, and a minimal energy barrier for the rate-limiting *OOH formation, thereby enhancing the water-splitting reaction. Further, Ni- SA with varying N coordination numbers (designated as Ni-N_x-C) dispersed on faulty carbon supports generated from MOFs with Zn vacancies. The Ni-N_3-C catalyst, characterized by a reduced nitrogen coordination number around nickel atoms, exhibits enhanced catalytic efficacy for CO_2 reduction, and a lower free energy change of 0.66 eV for the production of COOH* in Ni-N_3-C when utilized as an electrocatalyst [180] (Figure 10c–c3).

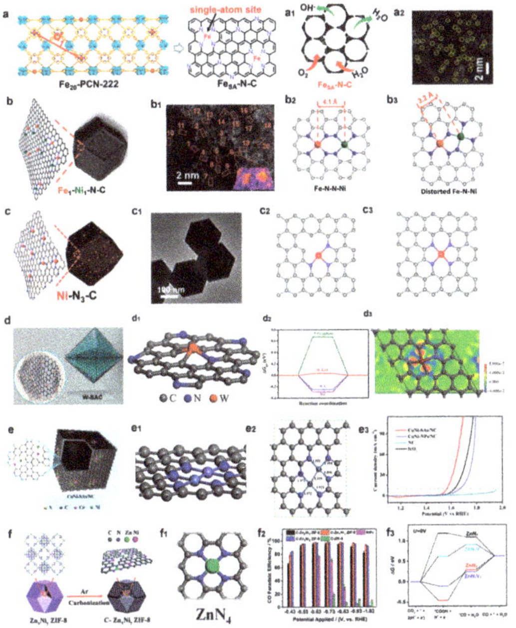

Figure 10. (**a**) Structure of FeSA-N-C, (**a1**) Mechanism of ORR in FeSA-N-C, (**a2**) AC HAADF-STEM image of FeSA-N-C. Reprinted from Ref. [173] with copyright permission from John Willey and Sons. (**b**) Structure of Fe_1-Ni_1-N-C, (**b1**) AC HAADF-STEM for Fe_1-Ni_1-N-C, (**b2**) Model of Fe_1-Ni_1-N-C, (**b3**) Model of distorted Fe-Ni-N. Modified from Ref. [176] with copyright permission from The American Chemical Society. (**c**) Structure of Ni-N_3-C, (**c1**) AC HAADF-STEM of Ni-N_3-C, (**c2**) Geometric configurations of Ni-N_3-C, (**c3**) Geometric configurations of Ni-N_4-C. Adapted from Ref. [180] with copyright permission from John Willey and Sons. (**d**) Structure of W-SAC, (**d1**) Model of the W-SAC, (**d2**) Gibbs free energy (ΔGH*) of H2 adsorption on the different catalysts, (**d3**) Differential charge density for W-SAC. Modified from Ref. [181] with copyright permission from John Willey and Sons. (**e**) Structure of CoNi-SAs/NC, (**e1,e2**) Diatomic Ni-Co-N_6 models, (**e3**) LSV curves of different catalysts. Modified from Ref. [182] with copyright permission from John Willey and Sons. (**f**) Structure of C-$Zn_x$$Ni_y$ ZIF-8, (**f1**) Configuration of ZnN_4, (**f2**) Electrocatalytic performances of the CO_2RR over the catalysts, (**f3**) Energy diagram of C-$Zn_x$$Ni_y$ ZIF-8. Modified from Ref. [183] with copyright permission from The Royal Society of Chemistry.

4.6.3. Heteroatom Doping

Incorporating heteroatoms such as S, O, and P into coordinated atoms for a singular metal center anchored on a MOF substrate can serve as an effective technique for engineering the CE. Controlling the quantity and types of these heteroatoms in the coordinating layer is crucial for improving the activity and selectivity of SACs. This approach has been extensively utilized in numerous electrocatalytic processes. For example, Sun and coworkers [184] illustrated the fabrication of Pt SACs using an ALD technique, and the metal F-NC substrate is expected to form a strong covalent bond with Pt SAs due to its high density of N doping sites, resulting in enhanced support contact that improves the electrocatalytic performance and stability of Pt SACs for the ORR. Moreover, Chen and his team [181] synthesized a W-SAC by using W as the dopant for effective electrochemical HER, demonstrating a low overpotential of 85 mV at a current density of 10 mA cm^{-2} (Figure 10d,d1). The optimal active site of W-SAC is the coordinated C atom adjacent to the N atom, exhibiting a remarkably low ΔGH^* value of merely 0.033 eV. The difference in charge density before and after W SA doping indicated that the stabilization of H is attributable to the augmented electron density on the C atoms coordinated to W, which can significantly enhance the HER activity. (Figure 10d2,d3).

Furthermore, Han and colleagues [182] deposit dopamine on Co-Ni dual atomic active sites immobilized on N-doped hollow carbon denoted as the CoNi-SAs/NC hybrid, which exhibits exceptional electrocatalytic performance for bifunctional O_2 reactions attributed to the uniformly dispersed SAs with an elevated reactivity and attributed to the dual atomic Co-Ni sites, which lowered the activation energy barrier and enhanced reaction rates, along with the pre-adsorption of OH* on the catalyst surface, significantly optimizing electrocatalytic activity (Figure 10e–e3). Additionally, N co-coordinated Co or Fe atoms, isolated inside carbon matrices (M-N-C), featuring both bulk and edge-hosted M-N_4 coordination, were synthesized, and among the different configurations, Fe is inherently active in M-N_4 for the CO_2 reduction to CO, achieving a CO FE of 93% [185]. Additionally, an Fe porphyrinic MOF was decorated on the SA Fe immobilized on N-doped carbon catalysts (Fe_1-N-C). Leveraging the atomically dispersed SA Fe sites, hierarchical pore structure, and excellent conductivity, Fe_1-N-C exhibits an FE of 4.51% and an NH_3 yield rate of 1.56×10^{-11} mol cm^{-2} s^{-1} [186].

4.6.4. Crystallinity Control

Regulating crystallinity in MOFs that include SACs is essential for enhancing their efficacy in diverse catalytic applications. The crystallinity of MOFs profoundly impacts their structural integrity, porosity, and accessibility of active sites, hence influencing the overall catalytic activity and selectivity of the included SACs. Strategies for attaining regulated crystallinity involve adjusting synthesis parameters, including temperature, solvent selection, and precursor concentration, alongside post-synthetic changes. For example, Bao et al. [183] synthesized coordinatively unsaturated Ni-N sites embedded in porous carbon through pyrolysis, demonstrating that the CO current density increases with overpotential, achieving 71.5 ± 2.9 mA cm^{-2} while sustaining a high CO FE of 92.0–98.0%. DFT calculations indicate that the CO_2 reduction reaction occurs readily over coordinatively unsaturated Ni-N sites, thereby overcoming the limitations between FE and current density in the CO_2 reduction reaction (Figure 10f–f3).

In addition, the promising performance of MOF-based SACs in electrocatalysis, they still encounter numerous obstacles that need to be addressed. 1. Achieving consistent distribution of metal sites within the MOF is challenging, frequently resulting in performance differences. 2. Precise control over morphology is another considerable challenge for them. 3. The introduction of vacancies, whether at metallic sites or linkers, might undermine the crystallinity and stability of the material, potentially resulting in structural alterations under reaction conditions. 4. Moreover, achieving a uniform distribution of vacancies and their efficient utilization for SA anchoring poses challenges, as the misplacement of vacancies may lead to inactive or inadequately distributed metal sites, hence diminishing the catalyst's efficiency. 5. Enhancing the interaction between dopants and metal sites is essential to avert undesirable side effects or instability. 6. Another problem is ensuring the long-term stability of these doped sites under catalytic conditions, as dopants may migrate or leach during reaction cycles, leading to a decline in catalytic activity. 7. Attaining high crystallinity in MOFs may restrict the accessibility of active sites, hence impeding overall efficiency. Prospective developments in SACs-based MOFs for electrocatalysis appear: i. Improved precision in morphology control would facilitate the development of hybrid structures, wherein MOFs with precisely designed shapes and sizes might be combined with conductive materials or other catalytic supports. ii. Improvements in synthesis methods, such as selective etching or thermal processing, may facilitate the accurate formation of vacancies, resulting in predictable structural effects and maintaining crystallinity and stability. iii. Improvements in doping methods, including post-synthesis alteration and in situ doping, will provide enhanced regulation of dopant distribution and concentration. iv. The integration of highly crystalline MOFs with amorphous or semi-crystalline phases or incorporating specific, adjustable dopants in high-crystalline MOFs to establish localized sites with improved accessibility to active sites can enhance the crystallinity of MOF-based SACs.

5. Conclusions

This paper thoroughly details the progress in the domain of SACs, highlighting the essential function of substrate engineering in improving electrocatalytic efficiency across many reactions, including ORR, OER, HER, water splitting, CO_2RR, NRR, and NO_3RR. This review elucidates how substrate engineering is fundamental for attaining high catalytic efficiency and stability in SACs, emphasizing substrate properties like electronic structure, surface morphology, and chemical composition, which profoundly affect the interaction between SAs and their support matrix, thereby determining their catalytic behavior. We have entailed a comprehensive analysis of various substrates, including carbon, metal oxide, alloy-based, TMD-based, MXenes, and MOF substrates, each providing distinct properties that enhance the performance and functionality of SACs.

Additionally, we have discussed synthesis strategies, classified into bottom-up and top-down methodologies. Bottom-up strategies encompass techniques such as the impregnation method, pyrolysis, the ion exchange method, ALD, and electrochemical deposition. Top-down strategies encompass the host-guest strategy, atom trapping method, ball milling method, CVD, and abrasion technique, all aimed at developing SACs. Both approaches are evaluated for their capacity to attain uniform dispersion and robust anchoring of SAs on substrates, hence ensuring long-term stability and elevated activity. Furthermore, we have reviewed the engineering of these substrates through various strategies, including morphology and vacancy engineering, heteroatom doping, facet engineering, and crystallinity control in multiple electrocatalytic reactions. These approaches are utilized to improve interactions with SAs, prevent aggregation, and sustain atomic dispersion under reaction conditions while underscoring their critical role in optimizing SAC performance. Finally, we have outlined several problems and future perspectives in the advancement of SACs, connecting fundamental understanding with practical applications.

6. Challenges

Substantial progress has been achieved in the catalytic efficiency of SACs in recent years by the application of efficient synthesis techniques. Significant obstacles persist in attaining a more accurate and logical design of CE for SACs, potentially obstructing further advancement. These challenges are as follows:

1. It is still difficult to study the many reactions of intermediate species adsorbed on the surfaces of catalytic reaction sites, even with the introduction of several enhanced characterization techniques to look at the influence of matrix on electrocatalytic properties.

2. The intricacies of precisely regulating the support's electrical structure, surface morphology, and interactions with metal atoms make it extremely difficult to tune the properties of supports for SACs (Figure 11).

3. One major issue with electrocatalytic processes is the leaching of metal atoms from the substrate, which leads to the loss of active sites and decreased catalyst stability because of insufficient bonds between the metal and the support (Figure 11).

4. The support material, which acts as a ligand and interacts directly with the metal atoms in atomically dispersed metal catalysis, has a major impact on the electrocatalytic efficiency. Because different supports provide different local CEs, this dependency adds complexity and causes variations in catalytic behavior even when the same metal is used. To achieve improved catalytic performance, it is essential but challenging to comprehend and control these local interactions, such as nearby redox-active species, binding components, and coordination patterns.

5. The efficient dispersion and utilization of metal centers in SACs are limited by the absence of suitable active locations on supports for anchoring SA. Since insufficient binding sites prevent the effective distribution of metal atoms on the support, this restriction may lead to subpar catalytic activity and performance (Figure 11).

6. The compatibility between the substrate and the reactive site is still unclear despite a great deal of research and the usage of a variety of materials for SACs.

7. A limited range of supports, including metal oxides, MXenes, carbon-based, and TMDs (particularly MoS_2)-based supports, is necessary for the stabilization of the majority of SACs (Figure 11).

8. Weak metal-support interactions in SACs might result in instability, which can cause metal atoms to migrate or aggregate during catalytic processes, decreasing catalytic efficiency and impeding SACs' long-term vitality (Figure 11).

9. In real-world applications, anchoring two or more sites on various supports to simultaneously increase electrical and catalytic activity is a significant challenge.

10. To enable higher SAC loading, a variety of synthetic methods have been developed to create uniform and adaptable single metal sites with built-in dopants and external ligands acting as anchoring points. However, little is known about how these changes impact the electrical and geometrical structures and how they therefore affect catalytic performance and selectivity.

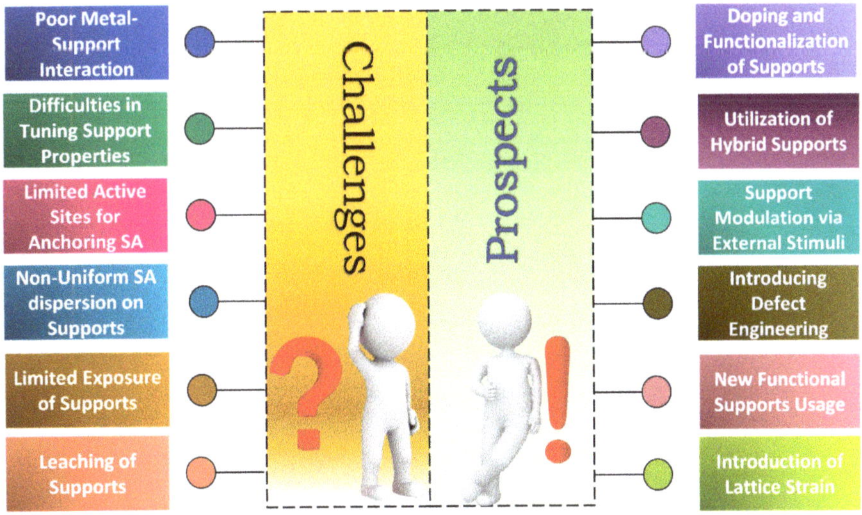

Figure 11. Challenges and future prospects of SACs in different supports.

7. Future Perspectives

In addition to the challenges, we have identified the following key prospects for the development of this rapidly growing field:

1. More convergent probe approaches are needed to capture transitory conditions on SACs. A greater emphasis should be placed on creating in situ characterization methods and monitoring the dynamic structure of SAs during the operational process.

2. Hybrid support materials, such as TMDs, alloys, and MXenes, can be precisely modified to optimize their properties (Figure 11).

3. Increasing MSI through functionalization, lattice strain, or dopants can improve bonding and reduce metal leaching on various supports (Figure 11).

4. Advanced computer modeling and in situ characterization approaches can provide valuable insights into the local CEs of atomically dispersed metals. These strategies can help to optimize support parameters, ensuring consistent catalytic performance across different supports.

5. Changing the supports of SACs using external variables such as temperature, pressure, light, or electric fields can improve catalytic efficiency, selectivity, and stability in energy applications (Figure 11).

6. To fully profit from monatomic catalysts, it is important to understand the compatibility between support and atomic sites and choose appropriate supports for metal atom deposition.

7. New functional support materials with complex surface topologies, including hierarchical and porous designs, are urgently needed (Figure 11).

8. Increasing MSI through surface modifications, chemical interactions, or dopants can increase metal atom stability and dispersion in SACs (Figure 11).

9. Novel support modification strategies, such as enhanced roughness, optimized morphology, and tailored surface chemistry, are of great interest. These modifications can create highly active reactive sites, allowing for stable and accurate attachment of dual or multimetal centers.

10. Engineering approaches, such as adding flaws or specialized anchoring locations, can improve SAC efficiency, durability, and catalytic efficacy (Figure 11).

11. To develop effective catalysts, real-time investigation of reaction intermediates, catalytic active sites, and fundamental mechanisms, as well as in operando or situ characterization techniques, are crucial.

Author Contributions: S.A., conceptualization, preparation, writing—original draft; J.H. and J.G., writing—review and editing, supervision; M.T. and M.A.M., formal analysis, writing—review and editing; M.M.A., review and editing; G.Y., review — editing, supervision, funding acquisition, project administration. All authors have read and agreed to the published version of the manuscript.

Funding: This work is sponsored by the Research Fund for International Scientists, National Natural Science Foundation of China.

Data Availability Statement: All data used in this work is available in the article.

Acknowledgments: G.Y. is grateful to the kind invitation from the editorial office for this contribution. This work was supported by the Research Fund for International Scientists (RFIS-Grant numbers: 52150410410) National Natural Science Foundation of China. M.M. Alam extends appreciation to the Deanship of Scientific Research and Graduate Studies at King Khalid University for funding this research work through a large research project under grant number RGP2/139/1446.

Conflicts of Interest: The authors declare no conflicts of interest.

References

1. Li, J.; Chen, C.; Xu, L.; Zhang, Y.; Wei, W.; Zhao, E.; Wu, Y.; Chen, C. Challenges and Perspectives of Single-Atom-Based Catalysts for Electrochemical Reactions. *JACS Au* **2023**, *3*, 736–755. [CrossRef] [PubMed]
2. Wang, Y.; Su, H.; He, Y.; Li, L.; Zhu, S.; Shen, H.; Xie, P.; Fu, X.; Zhou, G.; Feng, C.; et al. Advanced Electrocatalysts with Single-Metal-Atom Active Sites. *Chem. Rev.* **2020**, *120*, 12217–12314. [CrossRef]
3. Xu, H.; Zhao, Y.; Wang, Q.; He, G.; Chen, H. Supports promote single-atom catalysts toward advanced electrocatalysis. *Coord. Chem. Rev.* **2022**, *451*, 214261. [CrossRef]
4. Matthews, T.; Mashola, T.A.; Adegoke, K.A.; Mugadza, K.; Fakude, C.T.; Adegoke, O.R.; Adekunle, A.S.; Ndungu, P.; Maxakato, N.W. Electrocatalytic activity on single atoms catalysts: Synthesis strategies, characterization, classification, and energy conversion applications. *Coord. Chem. Rev.* **2022**, *467*, 214600. [CrossRef]
5. Gawande, M.B.; Fornasiero, P.; Zbořil, R. Carbon-Based Single-Atom Catalysts for Advanced Applications. *ACS Catal.* **2020**, *10*, 2231–2259. [CrossRef]
6. Ling, C.; Niu, X.; Li, Q.; Du, A.; Wang, J. Metal-Free Single Atom Catalyst for N_2 Fixation Driven by Visible Light. *J. Am. Chem. Soc.* **2018**, *140*, 14161–14168. [CrossRef]
7. Jiang, K.; Luo, M.; Peng, M.; Yu, Y.; Lu, Y.-R.; Chan, T.-S.; Liu, P.; de Groot, F.M.F.; Tan, Y. Dynamic active-site generation of atomic iridium stabilized on nanoporous metal phosphides for water oxidation. *Nat. Commun.* **2020**, *11*, 2701. [CrossRef]
8. Liang, L.; Jin, H.; Zhou, H.; Liu, B.; Hu, C.; Chen, D.; Wang, Z.; Hu, Z.; Zhao, Y.; Li, H.-W.; et al. Cobalt single atom site isolated Pt nanoparticles for efficient ORR and HER in acid media. *Nano Energy* **2021**, *88*, 106221. [CrossRef]
9. Liu, D.; Barbar, A.; Najam, T.; Javed, M.S.; Shen, J.; Tsiakaras, P.; Cai, X. Single noble metal atoms doped 2D materials for catalysis. *Appl. Catal. B Environ.* **2021**, *297*, 120389. [CrossRef]
10. Roth-Zawadzki, A.M.; Nielsen, A.J.; Tankard, R.E.; Kibsgaard, J. Dual and Triple Atom Electrocatalysts for Energy Conversion (CO2RR, NRR, ORR, OER, and HER): Synthesis, Characterization, and Activity Evaluation. *ACS Catal.* **2024**, *14*, 1121–1145. [CrossRef]
11. Zhang, W.; Zhao, Y.; Huang, W.; Huang, T.; Wu, B. Coordination environment manipulation of single atom catalysts: Regulation strategies, characterization techniques and applications. *Coord. Chem. Rev.* **2024**, *515*, 215952. [CrossRef]
12. Monini, V.; Bonechi, M.; Bazzicalupi, C.; Bianchi, A.; Gentilesca, P.; Giurlani, W.; Innocenti, M.; Meoli, A.; Romano, G.M.; Savastano, M. Oxygen reduction reaction (ORR) in alkaline solution catalysed by an atomically precise catalyst based on a Pd(ii) complex supported on multi-walled carbon nanotubes (MWCNTs). Electrochemical and structural considerations. *Dalton Trans.* **2024**, *53*, 2487–2500. [CrossRef]
13. Hao, Y.; Cao, X.; Lei, C.; Chen, Z.; Yang, X.; Gong, M. Chemical oxygen species on electrocatalytic materials during oxygen evolution reaction. *Mater. Today Catal.* **2023**, *2*, 100012. [CrossRef]

14. Jayabal, S.; Saranya, G.; Wu, J.; Liu, Y.; Geng, D.; Meng, X. Understanding the high-electrocatalytic performance of two-dimensional MoS2 nanosheets and their composite materials. *J. Chem. A* **2017**, *5*, 24540–24563. [CrossRef]
15. Nitopi, S.; Bertheussen, E.; Scott, S.B.; Liu, X.; Engstfeld, A.K.; Horch, S.; Seger, B.; Stephens, I.E.L.; Chan, K.; Hahn, C.; et al. Progress and Perspectives of Electrochemical CO_2 Reduction on Copper in Aqueous Electrolyte. *Chem. Rev.* **2019**, *119*, 7610–7672. [CrossRef]
16. Wen, J.; Zuo, L.; Sun, H.; Wu, X.; Huang, T.; Liu, Z.; Wang, J.; Liu, L.; Wu, Y.; Liu, X.; et al. Nanomaterials for the electrochemical nitrogen reduction reaction under ambient conditions. *Nanoscale Adv.* **2021**, *3*, 5525–5541. [CrossRef]
17. Zhang, R.; Li, C.; Cui, H.; Wang, Y.; Zhang, S.; Li, P.; Hou, Y.; Guo, Y.; Liang, G.; Huang, Z.; et al. Electrochemical nitrate reduction in acid enables high-efficiency ammonia synthesis and high-voltage pollutes-based fuel cells. *Nat. Commun.* **2023**, *14*, 8036. [CrossRef]
18. Zhang, Z.; Feng, C.; Wang, D.; Zhou, S.; Wang, R.; Hu, S.; Li, H.; Zuo, M.; Kong, Y.; Bao, J.; et al. Selectively anchoring single atoms on specific sites of supports for improved oxygen evolution. *Nat. Commun.* **2022**, *13*, 2473. [CrossRef]
19. Ajmal, S.; Kumar, A.; Mushtaq, M.A.; Tabish, M.; Zhao, Y.; Zhang, W.; Khan, A.S.; Saad, A.; Yasin, G.; Zhao, W. Uniting Synergistic Effect of Single-Ni Site and Electric Field of B- Bridged-N for Boosted Electrocatalytic Nitrate Reduction to Ammonia. *Small* **2024**, *20*, 2310082. [CrossRef]
20. Wang, C.; Wang, K.; Feng, Y.; Li, C.; Zhou, X.; Gan, L.; Feng, Y.; Zhou, H.; Zhang, B.; Qu, X.; et al. Co and Pt Dual-Single-Atoms with Oxygen-Coordinated Co–O–Pt Dimer Sites for Ultrahigh Photocatalytic Hydrogen Evolution Efficiency. *Adv. Mater.* **2021**, *33*, 2003327. [CrossRef]
21. Zhang, X.; Cui, G.; Feng, H.; Chen, L.; Wang, H.; Wang, B.; Zhang, X.; Zheng, L.; Hong, S.; Wei, M. Platinum–copper single atom alloy catalysts with high performance towards glycerol hydrogenolysis. *Nat. Commun.* **2019**, *10*, 5812. [CrossRef]
22. Chen, Y.-H.; Qi, M.-Y.; Li, Y.-H.; Tang, Z.-R.; Wang, T.; Gong, J.; Xu, Y.-J. Activating two-dimensional Ti3C2Tx-MXene with single-atom cobalt for efficient CO_2 photoreduction. *Cell Rep. Phys. Sci.* **2021**, *2*, 100371. [CrossRef]
23. Jiang, K.; Luo, M.; Liu, Z.; Peng, M.; Chen, D.; Lu, Y.-R.; Chan, T.-S.; de Groot, F.M.F.; Tan, Y. Rational strain engineering of single-atom ruthenium on nanoporous MoS2 for highly efficient hydrogen evolution. *Nat. Commun.* **2021**, *12*, 1687. [CrossRef]
24. Zhao, C.; Dai, X.; Yao, T.; Chen, W.; Wang, X.; Wang, J.; Yang, J.; Wei, S.; Wu, Y.; Li, Y. Ionic Exchange of Metal–Organic Frameworks to Access Single Nickel Sites for Efficient Electroreduction of CO_2. *J. Am. Chem. Soc.* **2017**, *139*, 8078–8081. [CrossRef]
25. Ro, I.; Qi, J.; Lee, S.; Xu, M.; Yan, X.; Xie, Z.; Zakem, G.; Morales, A.; Chen, J.G.; Pan, X.; et al. Bifunctional hydroformylation on heterogeneous Rh-WOx pair site catalysts. *Nature* **2022**, *609*, 287–292. [CrossRef] [PubMed]
26. Zhang, Y.-X.; Zhang, S.; Huang, H.; Liu, X.; Li, B.; Lee, Y.; Wang, X.; Bai, Y.; Sun, M.; Wu, Y.; et al. General Synthesis of a Diatomic Catalyst Library via a Macrocyclic Precursor-Mediated Approach. *J. Am. Chem. Soc.* **2023**, *145*, 4819–4827. [CrossRef]
27. Shan, J.; Ye, C.; Chen, S.; Sun, T.; Jiao, Y.; Liu, L.; Zhu, C.; Song, L.; Han, Y.; Jaroniec, M.; et al. Short-Range Ordered Iridium Single Atoms Integrated into Cobalt Oxide Spinel Structure for Highly Efficient Electrocatalytic Water Oxidation. *J. Am. Chem. Soc.* **2021**, *143*, 5201–5211. [CrossRef]
28. Zhang, L.; Si, R.; Liu, H.; Chen, N.; Wang, Q.; Adair, K.; Wang, Z.; Chen, J.; Song, Z.; Li, J.; et al. Atomic layer deposited Pt-Ru dual-metal dimers and identifying their active sites for hydrogen evolution reaction. *Nat. Commun.* **2019**, *10*, 4936. [CrossRef]
29. DeRita, L.; Dai, S.; Lopez-Zepeda, K.; Pham, N.; Graham, G.W.; Pan, X.; Christopher, P. Catalyst Architecture for Stable Single Atom Dispersion Enables Site-Specific Spectroscopic and Reactivity Measurements of CO Adsorbed to Pt Atoms, Oxidized Pt Clusters, and Metallic Pt Clusters on TiO2. *J. Am. Chem. Soc.* **2017**, *139*, 14150–14165. [CrossRef]
30. Zhang, L.; Han, L.; Liu, H.; Liu, X.; Luo, J. Potential-Cycling Synthesis of Single Platinum Atoms for Efficient Hydrogen Evolution in Neutral Media. *Angew. Chem. Int. Ed.* **2017**, *56*, 13694–13698. [CrossRef]
31. Wang, J.; Huang, Z.; Liu, W.; Chang, C.; Tang, H.; Li, Z.; Chen, W.; Jia, C.; Yao, T.; Wei, S.; et al. Design of N-Coordinated Dual-Metal Sites: A Stable and Active Pt-Free Catalyst for Acidic Oxygen Reduction Reaction. *J. Am. Chem. Soc.* **2017**, *139*, 17281–17284. [CrossRef]
32. Zhou, H.; Zhao, Y.; Xu, J.; Sun, H.; Li, Z.; Liu, W.; Yuan, T.; Liu, W.; Wang, X.; Cheong, W.-C.; et al. Recover the activity of sintered supported catalysts by nitrogen-doped carbon atomization. *Nat. Commun.* **2020**, *11*, 335. [CrossRef]
33. Zhang, X.; Zhong, Y.; Chen, H.; Cheng, Y.; Sun, Q.; Zhang, H.; He, Q.; Zhang, Y.; Guo, G.; He, X.; et al. Synthesis of Nitrogen-doped Carbon Supported Cerium Single Atom Catalyst by Ball Milling for Selective Oxidation of Ethylbenzene. *Chem. Res. Chin. Univ.* **2022**, *38*, 1258–1262. [CrossRef]
34. Liu, S.; Wang, M.; Yang, X.; Shi, Q.; Qiao, Z.; Lucero, M.; Ma, Q.; More, K.L.; Cullen, D.A.; Feng, Z.; et al. Chemical Vapor Deposition for Atomically Dispersed and Nitrogen Coordinated Single Metal Site Catalysts. *Angew. Chem. Int. Ed.* **2020**, *59*, 21698–21705. [CrossRef]
35. Han, G.-F.; Li, F.; Rykov, A.I.; Im, Y.-K.; Yu, S.-Y.; Jeon, J.-P.; Kim, S.-J.; Zhou, W.; Ge, R.; Ao, Z.; et al. Abrading bulk metal into single atoms. *Nat. Nanotechnol.* **2022**, *17*, 403–407. [CrossRef]

36. Hai, X.; Zhao, X.; Guo, N.; Yao, C.; Chen, C.; Liu, W.; Du, Y.; Yan, H.; Li, J.; Chen, Z.; et al. Engineering Local and Global Structures of Single Co Atoms for a Superior Oxygen Reduction Reaction. *ACS Catal.* **2020**, *10*, 5862–5870. [CrossRef]
37. Gong, Y.; Yang, X.; Zhang, Y.; Li, J.; Xu, L.; Li, S.; Fu, H.; An, X. Constructing asymmetric-electron-density Pd-Zn dual-atoms on N-doped defective mesoporous carbon for electrochemical reduction of CO_2. *Chem. Eng. J.* **2024**, *500*, 156603. [CrossRef]
38. Yu, P.-W.; Elmas, S.; Roman, T.; Pan, X.; Yin, Y.; Gibson, C.T.; Andersson, G.G.; Andersson, M.R. Highly active platinum single-atom catalyst grafted onto 3D carbon cloth support for the electrocatalytic hydrogen evolution reaction. *Appl. Surf. Sci.* **2022**, *595*, 153480. [CrossRef]
39. Han, J.; Bao, H.; Wang, J.-Q.; Zheng, L.; Sun, S.; Wang, Z.L.; Sun, C. 3D N-doped ordered mesoporous carbon supported single-atom Fe-N-C catalysts with superior performance for oxygen reduction reaction and zinc-air battery. *Appl. Catal. B Environ.* **2021**, *280*, 119411. [CrossRef]
40. Lü, F.; Zhao, S.; Guo, R.; He, J.; Peng, X.; Bao, H.; Fu, J.; Han, L.; Qi, G.; Luo, J.; et al. Nitrogen-coordinated single Fe sites for efficient electrocatalytic N_2 fixation in neutral media. *Nano Energy* **2019**, *61*, 420–427. [CrossRef]
41. Li, T.; Ren, S.; Zhang, C.; Qiao, L.; Wu, J.; He, P.; Lin, J.; Liu, Y.; Fu, Z.; Zhu, Q.; et al. Cobalt single atom anchored on N-doped carbon nanoboxes as typical single-atom catalysts (SACs) for boosting the overall water splitting. *Chem. Eng. J.* **2023**, *458*, 141435. [CrossRef]
42. Lv, C.; Li, B.; Ren, Y.; Zhang, G.; Lu, Z.; Li, L.; Zhang, X.; Yang, X.; Yu, X. A "MOF-plus-MOF" strategy to synthesize Co-N_3C1 single-atom catalyst for rechargeable Zn-air battery. *Chem. Eng. J.* **2024**, *495*, 153670. [CrossRef]
43. Musa, S.; Pirzada, B.M.; Talib, S.H.; Anjum, D.H.; Haija, M.A.; Mohamed, S.; Qurashi, A. Growth of copper-nickel (Cu-Ni) dual atom catalysts over graphene variants as active anodes for clean oxygen generation: Integrative experimental and computational validation. *Nano Energy* **2024**, *125*, 109479. [CrossRef]
44. Liu, C.; Wu, S.; Tian, S.; Yang, J.; Li, J.; Guan, Q.; Yin, F.; Xiang, X.; Wang, Y.; Meng, X.; et al. Structurally optimized rosette-like microspheres carbon with Fe-Ni single atom sites for bifunctional oxygen electrocatalysis in Zinc-Air batteries. *Chem. Eng. J.* **2024**, *497*, 154963. [CrossRef]
45. Zhang, H.; Wu, F.; Huang, R.; Liu, X.; Zhang, Z.; Yao, T.; Zhang, Y.; Wu, Y. Symmetry Evolution Induced 2D Pt Single Atom Catalyst with High Density for Alkaline Hydrogen Oxidation. *Adv. Mater.* **2024**, *36*, 2404672. [CrossRef]
46. Zhao, X.; Huang, S.; Chen, Z.; Lu, C.; Han, S.; Ke, C.; Zhu, J.; Zhang, J.; Tranca, D.; Zhuang, X. Carbon nanosheets supporting Ni–N3S single-atom sites for efficient electrocatalytic CO_2 reduction. *Carbon* **2021**, *178*, 488–496. [CrossRef]
47. Fang, C.; Zhou, J.; Zhang, L.; Wan, W.; Ding, Y.; Sun, X. Synergy of dual-atom catalysts deviated from the scaling relationship for oxygen evolution reaction. *Nat. Commun.* **2023**, *14*, 4449. [CrossRef] [PubMed]
48. Zhang, N.; Gao, Y.; Ma, L.; Wang, Y.; Huang, L.; Wei, B.; Xue, Y.; Zhu, H.; Jiang, R. Single transition metal atom anchored on g-C3N4 as an electrocatalyst for nitrogen fixation: A computational study. *Int. J. Hydrogen Energy* **2023**, *48*, 7621–7631. [CrossRef]
49. Pang, Y.; Ding, Z.; Ma, A.; Fan, G.; Xu, H. Electroreduction of nitrate to ammonia on graphyne-based single-atom catalysts by combined density functional theory and machine learning study. *Sep. Purif. Technol.* **2025**, *354*, 129422. [CrossRef]
50. Mehmood, R.; Long, G.; Fan, W.; Li, M.; Liu, L.; Zhang, F. One dimensional nickel phosphide polymorphic heterostructure as carbon-free functional support loading single-atom iridium for promoted electrocatalytic water oxidation. *J. Energy Chem.* **2023**, *79*, 410–417. [CrossRef]
51. Feng, L.; Zhou, M.; He, D.; Yin, H.; Huang, Y.; Cao, L.; Fang, Y.; Chu, D.; Liu, Y.; Chen, H.; et al. Co-Zn single atoms anchored carbon nanotubes derived from anti-perovskite carbides for boosted hydrogen evolution and oxygen reduction reactions. *Chem. Eng. J.* **2024**, *496*, 154255. [CrossRef]
52. Liu, M.; Zhang, S.; Chen, M.; Zhou, S.; Wu, L. An isolated bimetallic Fe–Ru single-atom catalyst for efficient electrochemical nitrogen reduction. *J. Mater. Chem. A* **2023**, *11*, 14900–14910. [CrossRef]
53. Liu, Y.; Zhuang, Z.; Liu, Y.; Liu, N.; Li, Y.; Cheng, Y.; Yu, J.; Yu, R.; Wang, D.; Li, H. Shear-Strained Pd Single-Atom Electrocatalysts for Nitrate Reduction to Ammonia. *Angew. Chem. Int. Ed.* **2024**, *63*, e202411396.
54. Chen, W.; Luo, X.; Ling, S.; Zhou, Y.; Shen, B.; Slater, T.J.A.; Fernandes, J.A.; Lin, T.; Wang, J.; Shen, Y. Hemoglobin-derived Fe-Nx-S species supported by bamboo-shaped carbon nanotubes as efficient electrocatalysts for the oxygen evolution reaction. *Carbon* **2020**, *168*, 588–596. [CrossRef]
55. Narendra Kumar, A.V.; Muthu Prabhu, S.; Shin, W.S.; Yadav, K.K.; Ahn, Y.; Abdellattif, M.H.; Jeon, B.-H. Prospects of non-noble metal single atoms embedded in two-dimensional (2D) carbon and non-carbon-based structures in electrocatalytic applications. *Coord. Chem. Rev.* **2022**, *467*, 214613. [CrossRef]
56. He, J.; Chen, L.; Song, D.; Wang, C.; Yuan, Y.; Liu, Y.; Dong, K. From one-dimensional to three-dimensional, the criss-crossed fiber materials forge high-performance lithium-sulfur batteries. *Chem. Eng. J.* **2024**, *495*, 153126.

57. Tian, H.; Song, A.; Zhang, P.; Sun, K.; Wang, J.; Sun, B.; Fan, Q.; Shao, G.; Chen, C.; Liu, H.; et al. High Durability of Fe–N–C Single-Atom Catalysts with Carbon Vacancies toward the Oxygen Reduction Reaction in Alkaline Media. *Adv. Mater.* **2023**, *35*, 2210714. [CrossRef] [PubMed]
58. Tu, H.; Zhang, H.; Song, Y.; Liu, P.; Hou, Y.; Xu, B.; Liao, T.; Guo, J.; Sun, Z. Electronic Asymmetry Engineering of Fe–N–C Electrocatalyst via Adjacent Carbon Vacancy for Boosting Oxygen Reduction Reaction. *Adv. Sci.* **2023**, *10*, 2305194. [CrossRef] [PubMed]
59. Lyu, L.; Hu, X.; Lee, S.; Fan, W.; Kim, G.; Zhang, J.; Zhou, Z.; Kang, Y.-M. Oxygen Reduction Kinetics of Fe–N–C Single Atom Catalysts Boosted by Pyridinic N Vacancy for Temperature-Adaptive Zn–Air Batteries. *J. Am. Chem. Soc.* **2024**, *146*, 4803–4813. [CrossRef]
60. Cheng, Y.; Cao, B.; Xu, X.; Peng, L.; Liu, B.; He, J.; Zhang, J. Oxygen vacancy rich δ-MnO2 nanosheets encapsulating single cobalt atoms-anchored carbon nanotubes for efficient oxygen evolution. *Mater. Today Energy* **2024**, *40*, 101515. [CrossRef]
61. Yi, M.; Lv, S.; Yang, Q.; Lei, S.; Wang, H.; Huang, J.; Zhang, J. Ionic liquid meets ZnIn2S4: Synergistically tuning coordination environment of ZnIn2S4 grown on porous carbon by N, F doping and S-vacancies to load high concentration of single-atom Sb for efficient flexible Zn-Air batteries. *Appl. Catal. B Environ. Energy* **2025**, *361*, 124697. [CrossRef]
62. Zhang, L.; Jia, Y.; Gao, G.; Yan, X.; Chen, N.; Chen, J.; Soo, M.T.; Wood, B.; Yang, D.; Du, A.; et al. Graphene Defects Trap Atomic Ni Species for Hydrogen and Oxygen Evolution Reactions. *Chem* **2018**, *4*, 285–297. [CrossRef]
63. Yang, Q.; Liu, H.; Yuan, P.; Jia, Y.; Zhuang, L.; Zhang, H.; Yan, X.; Liu, G.; Zhao, Y.; Liu, J.; et al. Single Carbon Vacancy Traps Atomic Platinum for Hydrogen Evolution Catalysis. *J. Am. Chem. Soc.* **2022**, *144*, 2171–2178. [CrossRef] [PubMed]
64. Dao, V.; Choi, H.; Yadav, S.; Jiménez, J.D.; Kim, C.; Van Nguyen, T.; Chen, K.; Uthirakumar, P.; Van Le, Q.; Senanayake, S.D.; et al. LaCeOx coupled N-doped graphene/Ru single-atoms as a binary-site catalyst for efficient hydrogen evolution based on hydrogen spillover. *Appl. Catal. B Environ.* **2024**, *343*, 123452. [CrossRef]
65. An, B.; Zhou, J.; Zhu, Z.; Li, Y.; Wang, L.; Zhang, J. Uncovering the coordination effect on the Ni single-atom catalysts for CO_2 reduction including vacancy defect and non-vacancy defect structures. *Fuel* **2022**, *310*, 122472. [CrossRef]
66. Rong, X.; Wang, H.-J.; Lu, X.-L.; Si, R.; Lu, T.-B. Controlled Synthesis of a Vacancy-Defect Single-Atom Catalyst for Boosting CO_2 Electroreduction. *Angew. Chem. Int. Ed.* **2020**, *59*, 1961–1965. [CrossRef]
67. Dong, J.; Liu, Y.; Pei, J.; Li, H.; Ji, S.; Shi, L.; Zhang, Y.; Li, C.; Tang, C.; Liao, J.; et al. Continuous electroproduction of formate via CO_2 reduction on local symmetry-broken single-atom catalysts. *Nat. Commun.* **2023**, *14*, 6849. [CrossRef]
68. Choi, C.; Back, S.; Kim, N.-Y.; Lim, J.; Kim, Y.-H.; Jung, Y. Suppression of Hydrogen Evolution Reaction in Electrochemical N_2 Reduction Using Single-Atom Catalysts: A Computational Guideline. *ACS Catal.* **2018**, *8*, 7517–7525. [CrossRef]
69. Guo, X.; Gu, J.; Lin, S.; Zhang, S.; Chen, Z.; Huang, S. Tackling the Activity and Selectivity Challenges of Electrocatalysts toward the Nitrogen Reduction Reaction via Atomically Dispersed Biatom Catalysts. *J. Am. Chem. Soc.* **2020**, *142*, 5709–5721. [CrossRef]
70. Lee, C.H.; Pahari, S.; Barteau, M.A.; Kwon, J.S.-I. Exploring dynamics in single atom catalyst research: A comprehensive DFT-kMC study of nitrogen reduction reaction with focus on TM aggregation. *Appl. Catal. B Environ. Energy* **2024**, *358*, 124434. [CrossRef]
71. Shin, D.Y.; Lim, D.-H. DFT investigation into efficient transition metal single-atom catalysts supported on N-doped graphene for nitrate reduction reactions. *Chem. Eng. J.* **2023**, *468*, 143466. [CrossRef]
72. Xu, J.; Zhang, S.; Liu, H.; Liu, S.; Yuan, Y.; Meng, Y.; Wang, M.; Shen, C.; Peng, Q.; Chen, J.; et al. Breaking Local Charge Symmetry of Iron Single Atoms for Efficient Electrocatalytic Nitrate Reduction to Ammonia. *Angew. Chem. Int. Ed.* **2023**, *62*, e202308044. [CrossRef] [PubMed]
73. Chen, S.; Li, Y.; Bu, Z.; Yang, F.; Luo, J.; An, Q.; Zeng, Z.; Wang, J.; Deng, S. Boosting CO_2-to-CO conversion on a robust single-atom copper decorated carbon catalyst by enhancing intermediate binding strength. *J. Mater. Chem. A* **2021**, *9*, 1705–1712. [CrossRef]
74. Guo, Y.; Yao, S.; Xue, Y.; Hu, X.; Cui, H.; Zhou, Z. Nickel single-atom catalysts intrinsically promoted by fast pyrolysis for selective electroreduction of CO_2 into CO. *Appl. Catal. B Environ.* **2022**, *304*, 120997. [CrossRef]
75. Liu, S.; Li, C.; Zachman, M.J.; Zeng, Y.; Yu, H.; Li, B.; Wang, M.; Braaten, J.; Liu, J.; Meyer, H.M.; et al. Atomically dispersed iron sites with a nitrogen–carbon coating as highly active and durable oxygen reduction catalysts for fuel cells. *Nat. Energy* **2022**, *7*, 652–663. [CrossRef]
76. Rebarchik, M.; Bhandari, S.; Kropp, T.; Mavrikakis, M. Insights into the Oxygen Evolution Reaction on Graphene-Based Single-Atom Catalysts from First-Principles-Informed Microkinetic Modeling. *ACS Catal.* **2023**, *13*, 5225–5235. [CrossRef]
77. Zhang, X.; Xu, X.; Yao, S.; Hao, C.; Pan, C.; Xiang, X.; Tian, Z.Q.; Shen, P.K.; Shao, Z.; Jiang, S.P. Boosting Electrocatalytic Activity of Single Atom Catalysts Supported on Nitrogen-Doped Carbon through N Coordination Environment Engineering. *Small* **2022**, *18*, 2105329. [CrossRef]

78. Fei, H.; Dong, J.; Feng, Y.; Allen, C.S.; Wan, C.; Volosskiy, B.; Li, M.; Zhao, Z.; Wang, Y.; Sun, H.; et al. General synthesis and definitive structural identification of MN4C4 single-atom catalysts with tunable electrocatalytic activities. *Nat. Catal.* **2018**, *1*, 63–72. [CrossRef]
79. Zhang, L.; Yao, J.; Zhang, J.; He, W.; Li, Y.; Liang, L.; Liu, C.; Liu, H.; Hao, Q. Engineering Co and Ru dual-metal atoms on nitrogen-doped carbon as highly efficient bifunctional oxygen electrocatalysts. *Catal. Sci. Technol.* **2022**, *12*, 5435–5441. [CrossRef]
80. Tavakkoli, M.; Flahaut, E.; Peljo, P.; Sainio, J.; Davodi, F.; Lobiak, E.V.; Mustonen, K.; Kauppinen, E.I. Mesoporous Single-Atom-Doped Graphene–Carbon Nanotube Hybrid: Synthesis and Tunable Electrocatalytic Activity for Oxygen Evolution and Reduction Reactions. *ACS Catal.* **2020**, *10*, 4647–4658. [CrossRef]
81. Li, T.; Lu, T.; Li, X.; Xu, L.; Zhang, Y.; Tian, Z.; Yang, J.; Pang, H.; Tang, Y.; Xue, J. Atomically Dispersed Mo Sites Anchored on Multichannel Carbon Nanofibers toward Superior Electrocatalytic Hydrogen Evolution. *ACS Nano* **2021**, *15*, 20032–20041. [CrossRef]
82. Kuang, P.; Wang, Y.; Zhu, B.; Xia, F.; Tung, C.-W.; Wu, J.; Chen, H.M.; Yu, J. Pt Single Atoms Supported on N-Doped Mesoporous Hollow Carbon Spheres with Enhanced Electrocatalytic H2-Evolution Activity. *Adv. Mater.* **2021**, *33*, 2008599. [CrossRef]
83. Bala Musa, A.; Tabish, M.; Kumar, A.; Selvaraj, M.; Abubaker Khan, M.; Al-Shehri, B.M.; Arif, M.; Asim Mushtaq, M.; Ibraheem, S.; Slimani, Y.; et al. Microenvironment engineering of Fe-single-atomic-site with nitrogen coordination anchored on carbon nanotubes for boosting oxygen electrocatalysis in alkaline and acidic media. *Chem. Eng. J.* **2023**, *451*, 138684. [CrossRef]
84. Zheng, T.; Han, X.; Wang, J.; Xia, Z. Role of heteroatom-doping in enhancing catalytic activities and the stability of single-atom catalysts for oxygen reduction and oxygen evolution reactions. *Nanoscale* **2022**, *14*, 16286–16294. [CrossRef]
85. Yasin, G.; Ali, S.; Ibraheem, S.; Kumar, A.; Tabish, M.; Mushtaq, M.A.; Ajmal, S.; Arif, M.; Khan, M.A.; Saad, A.; et al. Simultaneously Engineering the Synergistic-Effects and Coordination-Environment of Dual-Single-Atomic Iron/Cobalt-sites as a Bifunctional Oxygen Electrocatalyst for Rechargeable Zinc-Air Batteries. *ACS Catal.* **2023**, *13*, 2313–2325. [CrossRef]
86. Zhou, S.; Chen, C.; Xia, J.; Li, L.; Qian, X.; Yin, F.-X.; He, G.; Chen, Q.; Chen, H. FeN4S1 Single-Atom Sites Anchored on Three-Dimensional Porous Carbon for Highly Efficient and Durable Oxygen Electrocatalysis. *ACS Nano* **2024**, *18*, 32995–33004. [CrossRef] [PubMed]
87. Chen, C.; Sun, Z.; Qin, G.; Wang, B.; Liu, M.; Liang, Q.; Li, X.; Pang, R.; Guo, Y.; Li, Y.; et al. Asymmetrically Coordinated Cu Dual-Atom-Sites Enables Selective CO_2 Electroreduction to Ethanol. *Adv. Mater.* **2024**, *36*, 2409797. [CrossRef] [PubMed]
88. Chen, Y.; Mao, J.; Zhou, H.; Xing, L.; Qiao, S.; Yuan, J.; Mei, B.; Wei, Z.; Zhao, S.; Tang, Y.; et al. Coordination Shell Dependent Activity of CuCo Diatomic Catalysts for Oxygen Reduction, Oxygen Evolution, and Hydrogen Evolution Reaction. *Adv. Funct. Mater.* **2024**, *34*, 2311664. [CrossRef]
89. Tao, X.; Liu, Y.; Lu, R.; Liu, J.; Wang, H.I.; Yang, J.; Bonn, M.; Müllen, K.; Zhou, Y. Phosphorus-Enhanced Bimetallic Single-Atom Catalysts for Hydrogen Evolution. *Adv. Energy Mater.* **2024**, 2404167. [CrossRef]
90. Yu, J.; Li, J.; Xu, C.-Y.; Li, Q.; Liu, Q.; Liu, J.; Chen, R.; Zhu, J.; Wang, J. Modulating the d-band centers by coordination environment regulation of single-atom Ni on porous carbon fibers for overall water splitting. *Nano Energy* **2022**, *98*, 107266. [CrossRef]
91. Zhou, Y.; Lu, R.; Tao, X.; Qiu, Z.; Chen, G.; Yang, J.; Zhao, Y.; Feng, X.; Müllen, K. Boosting Oxygen Electrocatalytic Activity of Fe–N–C Catalysts by Phosphorus Incorporation. *J. Am. Chem. Soc.* **2023**, *145*, 3647–3655. [CrossRef]
92. Li, S.-L.; Peng, M.; Song, Y.; Chen, Y.; Qiao, L.; Feng, Y.; Zhao, Y.; Gan, L.-Y. Screening transition metal and nonmetal atoms co-doped graphyne as efficient single-atom catalysts for nitrogen reduction. *Chem. Eng. J.* **2024**, *495*, 153275. [CrossRef]
93. Jiao, D.; Liu, Y.; Cai, Q.; Zhao, J. Coordination tunes the activity and selectivity of the nitrogen reduction reaction on single-atom iron catalysts: A computational study. *J. Mater. Chem. A* **2021**, *9*, 1240–1251. [CrossRef]
94. Han, L.; Hou, M.; Ou, P.; Cheng, H.; Ren, Z.; Liang, Z.; Boscoboinik, J.A.; Hunt, A.; Waluyo, I.; Zhang, S.; et al. Local Modulation of Single-Atomic Mn Sites for Enhanced Ambient Ammonia Electrosynthesis. *ACS Catal.* **2021**, *11*, 509–516. [CrossRef]
95. Shi, L.; Bi, S.; Qi, Y.; He, R.; Ren, K.; Zheng, L.; Wang, J.; Ning, G.; Ye, J. Anchoring Mo Single-Atom Sites on B/N Codoped Porous Carbon Nanotubes for Electrochemical Reduction of N_2 to NH_3. *ACS Catal.* **2022**, *12*, 7655–7663. [CrossRef]
96. Lu, X.; Wei, J.; Lin, H.; Li, Y.; Li, Y.-y. Boron Regulated Fe Single-Atom Structures for Electrocatalytic Nitrate Reduction to Ammonia. *ACS Appl. Nano Mater.* **2024**, *7*, 14654–14664. [CrossRef]
97. Lu, S.; Lou, F.; Zhao, Y.; Yu, Z. Regulating the coordination environment of single-atom catalysts for electrocatalytic CO_2 reduction. *J. Colloid Interface Sci.* **2023**, *646*, 301–310. [CrossRef] [PubMed]
98. Li, Z.; Wu, R.; Xiao, S.; Yang, Y.; Lai, L.; Chen, J.S.; Chen, Y. Axial chlorine coordinated iron-nitrogen-carbon single-atom catalysts for efficient electrochemical CO_2 reduction. *Chem. Eng. J.* **2022**, *430*, 132882. [CrossRef]
99. Ma, M.; Tang, Q. Axial coordination modification of M–N4 single-atom catalysts to regulate the electrocatalytic CO_2 reduction reaction. *J. Mater. Chem. C* **2022**, *10*, 15948–15956. [CrossRef]

100. Han, S.-G.; Ma, D.-D.; Zhou, S.-H.; Zhang, K.; Wei, W.-B.; Du, Y.; Wu, X.-T.; Xu, Q.; Zou, R.; Zhu, Q.-L. Fluorine-tuned single-atom catalysts with dense surface Ni-N4 sites on ultrathin carbon nanosheets for efficient CO_2 electroreduction. *Appl. Catal. B Environ.* **2021**, *283*, 119591. [CrossRef]
101. Sui, R.; Zhang, X.; Wang, X.; Wang, X.; Pei, J.; Zhang, Y.; Liu, X.; Chen, W.; Zhu, W.; Zhuang, Z. Silver based single atom catalyst with heteroatom coordination environment as high performance oxygen reduction reaction catalyst. *Nano Res.* **2022**, *15*, 7968–7975. [CrossRef]
102. Zhou, K.L.; Wang, Z.; Han, C.B.; Ke, X.; Wang, C.; Jin, Y.; Zhang, Q.; Liu, J.; Wang, H.; Yan, H. Platinum single-atom catalyst coupled with transition metal/metal oxide heterostructure for accelerating alkaline hydrogen evolution reaction. *Nat. Commun.* **2021**, *12*, 3783. [CrossRef] [PubMed]
103. Luo, M.; Wang, Z.; Li, Y.C.; Li, J.; Li, F.; Lum, Y.; Nam, D.-H.; Chen, B.; Wicks, J.; Xu, A.; et al. Hydroxide promotes carbon dioxide electroreduction to ethanol on copper via tuning of adsorbed hydrogen. *Nat. Commun.* **2019**, *10*, 5814. [CrossRef] [PubMed]
104. Zhu, E.; Sun, C.; Shi, C.; Yu, J.; Yang, X.; Xu, M. Isolated single-atom Fe-N4O1 catalytic site from a pre-oxidation strategy for efficient oxygen reduction reaction. *Chem. Eng. J.* **2023**, *463*, 142468. [CrossRef]
105. Zheng, J.; Yu, K.; Yuan, S.; Xiang, L.; Wang, K.; Jing, S.; Li, N. Coupling Ru single-atom and oxygen vacancy in Ru/SnO_2-x for efficient nitrogen reduction to ammonia via electronic synergistic effect. *J. Alloys Compd.* **2023**, *962*, 171028. [CrossRef]
106. Lu, F.; Yi, D.; Liu, S.; Zhan, F.; Zhou, B.; Gu, L.; Golberg, D.; Wang, X.; Yao, J. Engineering Platinum–Oxygen Dual Catalytic Sites via Charge Transfer towards Highly Efficient Hydrogen Evolution. *Angew. Chem. Int. Ed.* **2020**, *59*, 17712–17718. [CrossRef]
107. Wu, J.; Gao, J.; Lian, S.; Li, J.; Sun, K.; Zhao, S.; Kim, Y.D.; Ren, Y.; Zhang, M.; Liu, Q.; et al. Engineering the oxygen vacancies enables Ni single-atom catalyst for stable and efficient C-H activation. *Appl. Catal. B Environ.* **2022**, *314*, 121516. [CrossRef]
108. Yuan, C.-Z.; Wang, S.; San Hui, K.; Wang, K.; Li, J.; Gao, H.; Zha, C.; Zhang, X.; Dinh, D.A.; Wu, X.-L.; et al. In Situ Immobilizing Atomically Dispersed Ru on Oxygen-Defective Co_3O_4 for Efficient Oxygen Evolution. *ACS Catal.* **2023**, *13*, 2462–2471. [CrossRef]
109. Yao, Y.; Zhao, L.; Dai, J.; Wang, J.; Fang, C.; Zhan, G.; Zheng, Q.; Hou, W.; Zhang, L. Single Atom Ru Monolithic Electrode for Efficient Chlorine Evolution and Nitrate Reduction. *Angew. Chem. Int. Ed.* **2022**, *61*, e202208215. [CrossRef] [PubMed]
110. Wang, H.; Wen, N.; Wang, Y.; Jiao, X.; Xia, Y.; Chen, D. Boosting Electrochemical Reduction of CO_2 to Formate over Oxygen Vacancy Stabilized Copper–Tin Dual Single Atoms Catalysts. *Adv. Funct. Mater.* **2023**, *33*, 2303473. [CrossRef]
111. Sial, M.A.Z.G.; Guo, N.; Jalil, A.; Abbas, M.; Mateen, M.; Ullah, S.; Alam, U.; Bhat, Z.M.; Hussain, A.; Cai, X.; et al. Electrochemical C-C coupling mediated by novel Sn-SnO2 supported Cu single atoms: The case of CO_2 conversion to ethanol. *Chem. Eng. J.* **2024**, *489*, 151099. [CrossRef]
112. Kim, K.; Kim, C.; Bak, S.-M.; Nam, C.-Y.; Moon, J.H. Amorphous-crystalline transition-driven synthesis of Co single-atom catalysts on MoO3 for enhanced hydrogen evolution in acidic and alkaline media. *Chem. Eng. J.* **2024**, *488*, 150976. [CrossRef]
113. Wang, J.; Zhu, Y.; Zhong, X.; Hu, Z.; Huang, W.-H.; Pao, C.-W.; Cheng, H.; Alonso-Vante, N.; Ma, J. Universal synthesis strategy for preparation of transition metal oxide electrocatalysts doped with noble metal single atoms for oxygen evolution reaction. *Energy Adv.* **2024**, *3*, 2002–2012. [CrossRef]
114. Cheng, C.-C.; Ting, Y.-C.; Yen, F.-Y.; Li, G.-R.; Lee, C.-H.; Lee, K.-A.; Chang, S.-I.; Chen, H.-Y.T.; Lu, S.-Y. Synergistic Mo and W single atoms co-doped surface hydroxylated NiFe oxide as bifunctional electrocatalysts for overall water splitting. *Appl. Catal. B Environ. Energy* **2024**, *358*, 124356. [CrossRef]
115. Wang, Y.; Chen, Z.; Han, P.; Du, Y.; Gu, Z.; Xu, X.; Zheng, G. Single-Atomic Cu with Multiple Oxygen Vacancies on Ceria for Electrocatalytic CO_2 Reduction to CH4. *ACS Catal.* **2018**, *8*, 7113–7119. [CrossRef]
116. Zang, W.; Lee, J.; Tieu, P.; Yan, X.; Graham, G.W.; Tran, I.C.; Wang, P.; Christopher, P.; Pan, X. Distribution of Pt single atom coordination environments on anatase TiO_2 supports controls reactivity. *Nat. Commun.* **2024**, *15*, 998. [CrossRef]
117. Hu, B.; Sun, K.; Zhuang, Z.; Chen, Z.; Liu, S.; Cheong, W.-C.; Chen, C.; Hu, M.; Cao, X.; Ma, J.; et al. Distinct Crystal-Facet-Dependent Behaviors for Single-Atom Palladium-On-Ceria Catalysts: Enhanced Stabilization and Catalytic Properties. *Adv. Mater.* **2022**, *34*, 2107721. [CrossRef] [PubMed]
118. Chen, Y.; Wan, Q.; Cao, L.; Gao, Z.; Lin, J.; Li, L.; Pan, X.; Lin, S.; Wang, X.; Zhang, T. Facet-dependent electronic state of Pt single atoms anchoring on CeO2 nanocrystal for CO (preferential) oxidation. *J. Catal.* **2022**, *415*, 174–185. [CrossRef]
119. Tang, Y.; Wang, Y.-G.; Li, J. Theoretical Investigations of Pt1@CeO2 Single-Atom Catalyst for CO Oxidation. *J. Phys. Chem. C* **2017**, *121*, 11281–11289. [CrossRef]
120. Xu, J.; Zhang, C.; Liu, H.; Sun, J.; Xie, R.; Qiu, Y.; Lü, F.; Liu, Y.; Zhuo, L.; Liu, X.; et al. Amorphous MoOX-Stabilized single platinum atoms with ultrahigh mass activity for acidic hydrogen evolution. *Nano Energy* **2020**, *70*, 104529. [CrossRef]
121. Kyriakou, G.; Boucher, M.B.; Jewell, A.D.; Lewis, E.A.; Lawton, T.J.; Baber, A.E.; Tierney, H.L.; Flytzani-Stephanopoulos, M.; Sykes, E.C.H. Isolated Metal Atom Geometries as a Strategy for Selective Heterogeneous Hydrogenations. *Science* **2012**, *335*, 1209–1212. [CrossRef]

122. Cheng, X.; Wang, Y.; Lu, Y.; Zheng, L.; Sun, S.; Li, H.; Chen, G.; Zhang, J. Single-atom alloy with Pt-Co dual sites as an efficient electrocatalyst for oxygen reduction reaction. *Appl. Catal. B Environ.* **2022**, *306*, 121112. [CrossRef]
123. Cai, J.; Wei, Y.; Cao, A.; Huang, J.; Jiang, Z.; Lu, S.; Zang, S.-Q. Electrocatalytic nitrate-to-ammonia conversion with ~100% Faradaic efficiency via single-atom alloying. *Appl. Catal. B Environ.* **2022**, *316*, 121683. [CrossRef]
124. Li, X.; Shen, P.; Luo, Y.; Li, Y.; Guo, Y.; Zhang, H.; Chu, K. PdFe Single-Atom Alloy Metallene for N_2 Electroreduction. *Angew. Chem. Int. Ed.* **2022**, *61*, e202205923. [CrossRef] [PubMed]
125. Mao, J.; He, C.-T.; Pei, J.; Chen, W.; He, D.; He, Y.; Zhuang, Z.; Chen, C.; Peng, Q.; Wang, D.; et al. Accelerating water dissociation kinetics by isolating cobalt atoms into ruthenium lattice. *Nat. Commun.* **2018**, *9*, 4958. [CrossRef] [PubMed]
126. Wang, H.; Zhou, X.; Yu, T.; Lu, X.; Qian, L.; Liu, P.; Lei, P. Surface restructuring in AgCu single-atom alloy catalyst and self-enhanced selectivity toward CO_2 reduction. *Electrochim. Acta* **2022**, *426*, 140774. [CrossRef]
127. Gao, Q.; Yao, B.; Pillai, H.S.; Zang, W.; Han, X.; Liu, Y.; Yu, S.-W.; Yan, Z.; Min, B.; Zhang, S.; et al. Synthesis of core/shell nanocrystals with ordered intermetallic single-atom alloy layers for nitrate electroreduction to ammonia. *Nat. Synth.* **2023**, *2*, 624–634. [CrossRef]
128. Yao, Y.; Sulei, H.; Chen, W.; Huang, Z.-Q.; Wei, W.; Yao, T.; Liu, R.; Zang, K.; Wang, X.; Wu, G.; et al. Engineering the electronic structure of single atom Ru sites via compressive strain boosts acidic water oxidation electrocatalysis. In *Controllable Synthesis and Atomic Scale Regulation of Noble Metal Catalysts*; Springer Nature: London, UK, 2019; Volume 2, pp. 55–92.
129. Zhao, Y.; Liu, X.; Chen, D.; Liu, Z.; Yang, Q.; Lin, X.; Peng, M.; Liu, P.; Tan, Y. Atomic-level-designed copper atoms on hierarchically porous gold architectures for high-efficiency electrochemical CO_2 reduction. *Sci. China Mater.* **2021**, *64*, 1900–1909. [CrossRef]
130. Zhang, Y.; Chen, X.; Wang, W.; Yin, L.; Crittenden, J.C. Electrocatalytic nitrate reduction to ammonia on defective Au1Cu (111) single-atom alloys. *Appl. Catal. B Environ.* **2022**, *310*, 121346. [CrossRef]
131. Darby, M.T.; Stamatakis, M. Single-Atom Alloys for the Electrochemical Oxygen Reduction Reaction. *ChemPhysChem* **2021**, *22*, 499–508. [CrossRef]
132. Cheng, M.-J.; Clark, E.L.; Pham, H.H.; Bell, A.T.; Head-Gordon, M. Quantum Mechanical Screening of Single-Atom Bimetallic Alloys for the Selective Reduction of CO_2 to C1 Hydrocarbons. *ACS Catal.* **2016**, *6*, 7769–7777. [CrossRef]
133. Zheng, T.; Liu, C.; Guo, C.; Zhang, M.; Xu, L.; Jiang, Q.; Xue, W.; Li, H.; Li, A.; Pao, C.-W.; et al. Copper-catalysed exclusive CO_2 to pure formic acid conversion via single-atom alloying. *Nat. Nanotechnol.* **2021**, *16*, 1386–1393. [CrossRef]
134. Tong, Y.; Liu, J.; Wang, L.; Su, B.-J.; Wu, K.-H.; Juang, J.-Y.; Hou, F.; Yin, L.; Dou, S.X.; Liu, J.; et al. Carbon-Shielded Single-Atom Alloy Material Family for Multi-Functional Electrocatalysis. *Adv. Funct. Mater.* **2022**, *32*, 2205654. [CrossRef]
135. Zheng, G.; Li, Y.; Qian, X.; Yao, G.; Tian, Z.; Zhang, X.; Chen, L. High-Throughput Screening of a Single-Atom Alloy for Electroreduction of Dinitrogen to Ammonia. *ACS Appl. Mater. Interfaces* **2021**, *13*, 16336–16344. [CrossRef]
136. Li, L.; Qiu, H.; Zhu, Y.; Chen, G.; She, S.; Guo, X.; Li, H.; Liu, T.; Lin, Z.; Zhou, H.; et al. Atomic ruthenium modification of nickel-cobalt alloy for enhanced alkaline hydrogen evolution. *Appl. Catal. B Environ.* **2023**, *331*, 122710. [CrossRef]
137. Gao, D.; Yi, D.; Xia, J.; Yang, Y.; Wang, X. First-principles screening of Cu-based single-atom alloys for highly efficient electrocatalytic nitrogen reduction. *Mol. Catal.* **2024**, *555*, 113879. [CrossRef]
138. Pu, Y.; Chen, J.-L.; Zhao, J.-W.; Feng, L.; Zhu, J.; Jiang, X.; Li, W.-X.; Liu, J.-X. Nature of the Active Center for the Oxygen Reduction on Ag-Based Single-Atom Alloy Clusters. *JACS Au* **2024**, *4*, 2886–2895. [CrossRef] [PubMed]
139. Du, C.; Lu, S.; Wang, J.-a.; Wang, X.; Wang, M.; Fruehwald, H.M.; Wang, L.; Zhang, B.; Guo, T.; Mills, J.P.; et al. Selectively Reducing Nitrate into NH_3 in Neutral Media by PdCu Single-Atom Alloy Electrocatalysis. *ACS Catal.* **2023**, *13*, 10560–10569. [CrossRef]
140. Feng, Y.; An, W.; Wang, Z.; Wang, Y.; Men, Y.; Du, Y. Electrochemical CO_2 Reduction Reaction on M@Cu(211) Bimetallic Single-Atom Surface Alloys: Mechanism, Kinetics, and Catalyst Screening. *ACS Sustain. Chem. Eng.* **2020**, *8*, 210–222. [CrossRef]
141. Zhang, L.; Liu, H.; Liu, S.; Norouzi Banis, M.; Song, Z.; Li, J.; Yang, L.; Markiewicz, M.; Zhao, Y.; Li, R.; et al. Pt/Pd Single-Atom Alloys as Highly Active Electrochemical Catalysts and the Origin of Enhanced Activity. *ACS Catal.* **2019**, *9*, 9350–9358. [CrossRef]
142. Xi, S.; Zhao, P.; He, C.; Zhang, W. High-throughput screening of single-atom catalysts on 1 T-TMD for highly active and selective CO_2 reduction reaction: Computational and machine learning insights. *J. Catal.* **2024**, *436*, 115610. [CrossRef]
143. Ling, F.; Xia, W.; Li, L.; Zhou, X.; Luo, X.; Bu, Q.; Huang, J.; Liu, X.; Kang, W.; Zhou, M. Single Transition Metal Atom Bound to the Unconventional Phase of the MoS_2 Monolayer for Catalytic Oxygen Reduction Reaction: A First-Principles Study. *ACS Appl. Mater. Interfaces* **2021**, *13*, 17412–17419. [CrossRef]
144. Ran, N.; Song, E.; Wang, Y.; Zhou, Y.; Liu, J. Dynamic coordination transformation of active sites in single-atom MoS_2 catalysts for boosted oxygen evolution catalysis. *Energy Environ. Sci.* **2022**, *15*, 2071–2083. [CrossRef]
145. Pattengale, B.; Huang, Y.; Yan, X.; Yang, S.; Younan, S.; Hu, W.; Li, Z.; Lee, S.; Pan, X.; Gu, J.; et al. Dynamic evolution and reversibility of single-atom Ni(II) active site in 1T-MoS2 electrocatalysts for hydrogen evolution. *Nat. Commun.* **2020**, *11*, 4114. [CrossRef] [PubMed]

146. Xu, X.; Xu, H.; Cheng, D. Design of high-performance MoS2 edge supported single-metal atom bifunctional catalysts for overall water splitting via a simple equation. *Nanoscale* **2019**, *11*, 20228–20237. [CrossRef]
147. Ren, Y.; Sun, X.; Qi, K.; Zhao, Z. Single atom supported on MoS2 as efficient electrocatalysts for the CO_2 reduction reaction: A DFT study. *Appl. Surf. Sci.* **2022**, *602*, 154211. [CrossRef]
148. Yang, T.; Song, T.T.; Zhou, J.; Wang, S.; Chi, D.; Shen, L.; Yang, M.; Feng, Y.P. High-throughput screening of transition metal single atom catalysts anchored on molybdenum disulfide for nitrogen fixation. *Nano Energy* **2020**, *68*, 104304. [CrossRef]
149. Li, J.; Zhang, Y.; Liu, C.; Zheng, L.; Petit, E.; Qi, K.; Zhang, Y.; Wu, H.; Wang, W.; Tiberj, A.; et al. 3.4% Solar-to-Ammonia Efficiency from Nitrate Using Fe Single Atomic Catalyst Supported on MoS2 Nanosheets. *Adv. Funct. Mater.* **2022**, *32*, 2108316. [CrossRef]
150. Tursun, M.; Abdukayum, A.; Wu, C.; Wang, C. Screening WS2−based single−atom catalysts for electrocatalytic nitrate reduction to ammonia. *Int. J. Hydrogen Energy* **2024**, *73*, 183–190. [CrossRef]
151. Wang, J.; Fang, W.; Hu, Y.; Zhang, Y.; Dang, J.; Wu, Y.; Chen, B.; Zhao, H.; Li, Z. Single atom Ru doping 2H-MoS2 as highly efficient hydrogen evolution reaction electrocatalyst in a wide pH range. *Appl. Catal. B Environ.* **2021**, *298*, 120490. [CrossRef]
152. Zhang, Y.; Yang, R.; Li, H.; Zeng, Z. Boosting Electrocatalytic Reduction of CO_2 to HCOOH on Ni Single Atom Anchored WTe_2 Monolayer. *Small* **2022**, *18*, 2203759. [CrossRef] [PubMed]
153. Dang, Q.; Tang, S.; Liu, T.; Li, X.; Wang, X.; Zhong, W.; Luo, Y.; Jiang, J. Regulating Electronic Spin Moments of Single-Atom Catalyst Sites via Single-Atom Promoter Tuning on S-Vacancy MoS2 for Efficient Nitrogen Fixation. *J. Phys. Chem. Lett.* **2021**, *12*, 8355–8362. [CrossRef] [PubMed]
154. Yang, L.; Zhu, W. Accelerating electrocatalytic reduction of nitrate to ammonia by utilizing steric hindrance in single boron-decorated 2H/1T-MoS2: A theoretical insight. *Appl. Surf. Sci.* **2022**, *596*, 153624. [CrossRef]
155. Zhang, J.; Xu, X.; Yang, L.; Cheng, D.; Cao, D. Single-Atom Ru Doping Induced Phase Transition of MoS2 and S Vacancy for Hydrogen Evolution Reaction. *Small Methods* **2019**, *3*, 1900653. [CrossRef]
156. Zhu, T.; Gan, X.; Xiao, Z.; Dai, S.; Xiao, H.; Zhang, S.; Dong, S.; Zhao, H.; Wang, P. Single-atom dispersed Cu or Co on 2H-MoS2 monolayer for improving electrocatalytic activity of overall water splitting. *Surf. Interfaces* **2021**, *27*, 101538. [CrossRef]
157. Hoa, V.H.; Tran, D.T.; Prabhakaran, S.; Kim, D.H.; Hameed, N.; Wang, H.; Kim, N.H.; Lee, J.H. Ruthenium single atoms implanted continuous MoS2-Mo2C heterostructure for high-performance and stable water splitting. *Nano Energy* **2021**, *88*, 106277. [CrossRef]
158. Nie, Q.-J.; Ding, C.-C.; Liu, T.; Jin, W.; Hu, J.-S. Transition metal ions doped into Haeckelite-MoS2 as highly efficient and selective electrocatalyst for nitrogen reduction reaction. *Mater. Today Commun.* **2024**, *39*, 109423. [CrossRef]
159. Sun, T.; Tang, Z.; Zang, W.; Li, Z.; Li, J.; Li, Z.; Cao, L.; Dominic Rodriguez, J.S.; Mariano, C.O.M.; Xu, H.; et al. Ferromagnetic single-atom spin catalyst for boosting water splitting. *Nat. Nanotechnol.* **2023**, *18*, 763–771. [CrossRef]
160. Qi, K.; Cui, X.; Gu, L.; Yu, S.; Fan, X.; Luo, M.; Xu, S.; Li, N.; Zheng, L.; Zhang, Q.; et al. Single-atom cobalt array bound to distorted 1T MoS2 with ensemble effect for hydrogen evolution catalysis. *Nat. Commun.* **2019**, *10*, 5231. [CrossRef] [PubMed]
161. Pan, Y.; Wang, X.; Zhang, W.; Tang, L.; Mu, Z.; Liu, C.; Tian, B.; Fei, M.; Sun, Y.; Su, H.; et al. Boosting the performance of single-atom catalysts via external electric field polarization. *Nat. Commun.* **2022**, *13*, 3063. [CrossRef]
162. Aggarwal, P.; Sarkar, D.; Awasthi, K.; Menezes, P.W. Functional role of single-atom catalysts in electrocatalytic hydrogen evolution: Current developments and future challenges. *Coord. Chem. Rev.* **2022**, *452*, 214289. [CrossRef]
163. Kuznetsov, D.A.; Chen, Z.; Kumar, P.V.; Tsoukalou, A.; Kierzkowska, A.; Abdala, P.M.; Safonova, O.V.; Fedorov, A.; Müller, C.R. Single Site Cobalt Substitution in 2D Molybdenum Carbide (MXene) Enhances Catalytic Activity in the Hydrogen Evolution Reaction. *J. Am. Chem. Soc.* **2019**, *141*, 17809–17816. [CrossRef]
164. Kuznetsov, D.A.; Chen, Z.; Abdala, P.M.; Safonova, O.V.; Fedorov, A.; Müller, C.R. Single-Atom-Substituted Mo2CTx:Fe-Layered Carbide for Selective Oxygen Reduction to Hydrogen Peroxide: Tracking the Evolution of the MXene Phase. *J. Am. Chem. Soc.* **2021**, *143*, 5771–5778. [CrossRef] [PubMed]
165. Huang, B.; Li, N.; Ong, W.-J.; Zhou, N. Single atom-supported MXene: How single-atomic-site catalysts tune the high activity and selectivity of electrochemical nitrogen fixation. *J. Mater. Chem. A* **2019**, *7*, 27620–27631. [CrossRef]
166. Zhao, Q.; Zhang, C.; Hu, R.; Du, Z.; Gu, J.; Cui, Y.; Chen, X.; Xu, W.; Cheng, Z.; Li, S.; et al. Selective Etching Quaternary MAX Phase toward Single Atom Copper Immobilized MXene (Ti3C2Clx) for Efficient CO_2 Electroreduction to Methanol. *ACS Nano* **2021**, *15*, 4927–4936. [CrossRef]
167. Zhang, J.; Zhao, Y.; Guo, X.; Chen, C.; Dong, C.-L.; Liu, R.-S.; Han, C.-P.; Li, Y.; Gogotsi, Y.; Wang, G. Single platinum atoms immobilized on an MXene as an efficient catalyst for the hydrogen evolution reaction. *Nat. Catal.* **2018**, *1*, 985–992. [CrossRef]
168. Zhao, D.; Chen, Z.; Yang, W.; Liu, S.; Zhang, X.; Yu, Y.; Cheong, W.-C.; Zheng, L.; Ren, F.; Ying, G.; et al. MXene (Ti3C2) Vacancy-Confined Single-Atom Catalyst for Efficient Functionalization of CO_2. *J. Am. Chem. Soc.* **2019**, *141*, 4086–4093. [CrossRef]
169. Wan, J.; Wang, Y.; Tian, W.; Zhang, H.; Wang, Y. Single, double, and triple transition metal atoms embedded in defective V3C2O2 for nitrogen reduction reaction: A DFT study. *Appl. Surf. Sci.* **2021**, *569*, 151020. [CrossRef]

170. Gao, X.; Tse, E.C.M. Unraveling the Performance Descriptors for Designing Single-Atom Catalysts on Defective MXenes for Exclusive Nitrate-To-Ammonia Electrocatalytic Upcycling. *Small* **2024**, *20*, 2306311. [CrossRef]
171. Kan, D.; Wang, D.; Cheng, Y.; Lian, R.; Sun, B.; Chen, K.; Huo, W.; Wang, Y.; Chen, G.; Wei, Y. Designing of Efficient Bifunctional ORR/OER Pt Single-Atom Catalysts Based on O-Terminated MXenes by First-Principles Calculations. *ACS Appl. Mater. Interfaces* **2021**, *13*, 52508–52518. [CrossRef] [PubMed]
172. Ramalingam, V.; Varadhan, P.; Fu, H.-C.; Kim, H.; Zhang, D.; Chen, S.; Song, L.; Ma, D.; Wang, Y.; Alshareef, H.N.; et al. Heteroatom-Mediated Interactions between Ruthenium Single Atoms and an MXene Support for Efficient Hydrogen Evolution. *Adv. Mater.* **2019**, *31*, 1903841. [CrossRef]
173. Jiao, L.; Wan, G.; Zhang, R.; Zhou, H.; Yu, S.-H.; Jiang, H.-L. From Metal–Organic Frameworks to Single-Atom Fe Implanted N-doped Porous Carbons: Efficient Oxygen Reduction in Both Alkaline and Acidic Media. *Angew. Chem. Int. Ed.* **2018**, *57*, 8525–8529. [CrossRef]
174. Wei, Y.-S.; Sun, L.; Wang, M.; Hong, J.; Zou, L.; Liu, H.; Wang, Y.; Zhang, M.; Liu, Z.; Li, Y.; et al. Fabricating Dual-Atom Iron Catalysts for Efficient Oxygen Evolution Reaction: A Heteroatom Modulator Approach. *Angew. Chem. Int. Ed.* **2020**, *59*, 16013–16022. [CrossRef] [PubMed]
175. Yuan, S.; Pu, Z.; Zhou, H.; Yu, J.; Amiinu, I.S.; Zhu, J.; Liang, Q.; Yang, J.; He, D.; Hu, Z.; et al. A universal synthesis strategy for single atom dispersed cobalt/metal clusters heterostructure boosting hydrogen evolution catalysis at all pH values. *Nano Energy* **2019**, *59*, 472–480. [CrossRef]
176. Jiao, L.; Zhu, J.; Zhang, Y.; Yang, W.; Zhou, S.; Li, A.; Xie, C.; Zheng, X.; Zhou, W.; Yu, S.-H.; et al. Non-Bonding Interaction of Neighboring Fe and Ni Single-Atom Pairs on MOF-Derived N-Doped Carbon for Enhanced CO_2 Electroreduction. *J. Am. Chem. Soc.* **2021**, *143*, 19417–19424. [CrossRef]
177. Tao, H.; Choi, C.; Ding, L.-X.; Jiang, Z.; Han, Z.; Jia, M.; Fan, Q.; Gao, Y.; Wang, H.; Robertson, A.W.; et al. Nitrogen Fixation by Ru Single-Atom Electrocatalytic Reduction. *Chem* **2019**, *5*, 204–214. [CrossRef]
178. Zhang, W.-D.; Dong, H.; Zhou, L.; Xu, H.; Wang, H.-R.; Yan, X.; Jiang, Y.; Zhang, J.; Gu, Z.-G. Fe single-atom catalysts with pre-organized coordination structure for efficient electrochemical nitrate reduction to ammonia. *Appl. Catal. B Environ.* **2022**, *317*, 121750. [CrossRef]
179. Shankar, A.; Marimuthu, S.; Maduraiveeran, G. High-valent iron single-atom catalysts for improved overall water splitting via a reduced energy barrier and stabilization of the active center. *J. Mater. Chem. A* **2024**, *12*, 121–127. [CrossRef]
180. Zhang, Y.; Jiao, L.; Yang, W.; Xie, C.; Jiang, H.-L. Rational Fabrication of Low-Coordinate Single-Atom Ni Electrocatalysts by MOFs for Highly Selective CO_2 Reduction. *Angew. Chem. Int. Ed.* **2021**, *60*, 7607–7611. [CrossRef] [PubMed]
181. Chen, W.; Pei, J.; He, C.-T.; Wan, J.; Ren, H.; Wang, Y.; Dong, J.; Wu, K.; Cheong, W.-C.; Mao, J.; et al. Single Tungsten Atoms Supported on MOF-Derived N-Doped Carbon for Robust Electrochemical Hydrogen Evolution. *Adv. Mater.* **2018**, *30*, 1800396. [CrossRef]
182. Han, X.; Ling, X.; Yu, D.; Xie, D.; Li, L.; Peng, S.; Zhong, C.; Zhao, N.; Deng, Y.; Hu, W. Atomically Dispersed Binary Co-Ni Sites in Nitrogen-Doped Hollow Carbon Nanocubes for Reversible Oxygen Reduction and Evolution. *Adv. Mater.* **2019**, *31*, 1905622. [CrossRef]
183. Yan, C.; Li, H.; Ye, Y.; Wu, H.; Cai, F.; Si, R.; Xiao, J.; Miao, S.; Xie, S.; Yang, F.; et al. Coordinatively unsaturated nickel–nitrogen sites towards selective and high-rate CO_2 electroreduction. *Energy Environ. Sci.* **2018**, *11*, 1204–1210. [CrossRef]
184. Song, Z.; Zhu, Y.-N.; Liu, H.; Banis, M.N.; Zhang, L.; Li, J.; Doyle-Davis, K.; Li, R.; Sham, T.-K.; Yang, L.; et al. Engineering the Low Coordinated Pt Single Atom to Achieve the Superior Electrocatalytic Performance toward Oxygen Reduction. *Small* **2020**, *16*, 2003096. [CrossRef] [PubMed]
185. Pan, F.; Zhang, H.; Liu, K.; Cullen, D.; More, K.; Wang, M.; Feng, Z.; Wang, G.; Wu, G.; Li, Y. Unveiling Active Sites of CO_2 Reduction on Nitrogen-Coordinated and Atomically Dispersed Iron and Cobalt Catalysts. *ACS Catal.* **2018**, *8*, 3116–3122. [CrossRef]
186. Zhang, R.; Jiao, L.; Yang, W.; Wan, G.; Jiang, H.-L. Single-atom catalysts templated by metal–organic frameworks for electrochemical nitrogen reduction. *J. Mater. Chem. A* **2019**, *7*, 26371–26377. [CrossRef]

catalysts

MDPI

Review

Recent Advances in the Development of Nanocarbon-Based Electrocatalytic/Electrode Materials for Proton Exchange Membrane Fuel Cells: A Review

Adelina A. Zasypkina [1], Nataliya A. Ivanova [1], Dmitry D. Spasov [1,2], Ruslan M. Mensharapov [1], Matvey V. Sinyakov [1,3] and Sergey A. Grigoriev [1,2,4,5,*]

[1] Department of Electrochemical and Hydrogen Technologies, National Research Center "Kurchatov Institute", Akademika Kurchatova sq., 1, 123182 Moscow, Russia; adelinazasypkina@yandex.ru (A.A.Z.); ivanovana.1989@mail.ru (N.A.I.); spasovdd@outlook.com (D.D.S.); ruslan.mensharapov@gmail.com (R.M.M.); mmatveimatvei4@gmail.com (M.V.S.)

[2] Department of Chemistry and Electrochemical Energy, National Research University "Moscow Power Engineering Institute", Krasnokazarmennaya st., 14, 111250 Moscow, Russia

[3] Department of Isotope Technology and Hydrogen Energy, Institute of Modern Energetics and Nanotechnology, D. Mendeleev University of Chemical Technology of Russia, Miusskaya sq., 9, 125047 Moscow, Russia

[4] HySA Infrastructure Center of Competence, Faculty of Engineering, North-West University, Potchefstroom 2531, South Africa

[5] Laboratory for Aliphatic Organoboron Compounds, A.N. Nesmeyanov Institute of Organoelement Compounds, Russian Academy of Sciences, Vavilova st., 28, 119991 Moscow, Russia

* Correspondence: grigoryevsa@mpei.ru

Abstract: The global issue for proton exchange membrane fuel cell market development is a reduction in the device cost through an increase in efficiency of the oxygen reduction reaction occurring at the cathode and an extension of the service life of the electrochemical device. Losses in the fuel cell performance are due to various degradation mechanisms in the catalytic layers taking place under conditions of high electric potential, temperature, and humidity. This review is devoted to recent advances in the field of increasing the efficiency and durability of electrocatalysts and other electrode materials by introducing structured carbon components into their composition. The main synthesis methods, physicochemical and electrochemical properties of materials, and performance of devices on their basis are presented. The main correlations between the composition and properties of structured carbon electrode materials, which can provide successful solutions to the highlighted issues, are revealed.

Keywords: electrocatalyst; electrocatalytic layer; microporous layer; gas diffusion electrode; nanostructured carbon support; graphene-like materials; electrochemical performance; degradation; PEM fuel cell

1. Introduction

The commercial development of the proton exchange membrane fuel cell (PEMFC) market relies on overcoming several challenges. One significant limitation of fuel cell efficiency is the slow oxygen reduction reaction (ORR) that occurs at the cathode [1–3]. To accelerate the ORR and improve its efficiency, it is necessary to use platinum group catalysts; thus, pure platinum or platinum on a carbon support is used to increase the active catalyst surface [4]. Usually, carbon black is used as a platinum particle support due to its availability and low cost, as well as appropriate values of specific surface area and electrochemical characteristics [5–7].

It also remains important to reduce the degradation rate of membrane electrode assembly (MEA) components, particularly catalysts and their supports [8]. Catalyst degradation can occur either mechanically through membrane damage or through electrochemical reactions during the PEMFC operation. Thus, when water is generated in a fuel cell as a product of an electrochemical reaction, it stagnates in gas diffusion layers (GDLs), bipolar plates (BPs), or catalytic layers (CLs) [9]. This phenomenon negatively affects the mass transfer of reagents and, over time, leads directly to degradation of the above FC components and deterioration of the device output characteristics [10–12]. When humidity is too low, the proton exchange membrane changes its geometry and dimensions [13], leading to mechanical failure of the CL being in direct contact with the membrane [14,15].

Deterioration in FC performance is also caused by various mechanisms of catalytic layer degradation, which occur under the influence of electrical potential, elevated temperature, and humidity [16,17]. Carbon corrosion is the main cause of catalyst degradation [18]. When the potential reaches 0.207 V (relative to a standard hydrogen electrode), the oxidation process of the carbon support begins [19]. However, in low-temperature hydrogen FCs, this process is inhibited by the reduced temperature [20]. The process is accompanied by the spontaneous agglomeration of platinum particles and their recondensation, or Ostwald ripening (any increase in the number of catalyst particles leads to a decrease in the active surface and, consequently, a decrease in the electrochemical surface area (ESA)), as well as their separation and dissolution (Figure 1) [21–24]. In this case, platinum accelerates the carbon-oxidation reaction [25]. The degradation mainly occurs during the ORR taking place at the cathode, when the reduction intermediates and oxygen itself start affecting the catalytic layer [26].

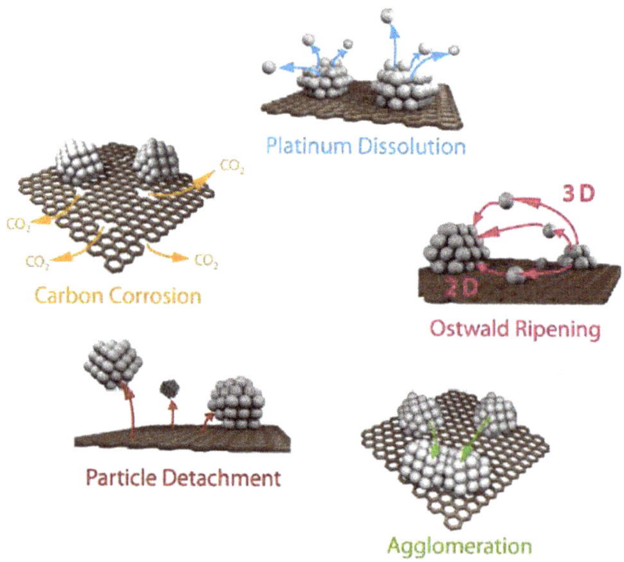

Figure 1. Simplified representation of suggested degradation mechanisms for platinum particles on a carbon support in fuel cells [22]. Copyright 2014, Beilstein Institute.

The most promising and investigated method for tackling the problem of CL degradation due to gradual oxidation of the carbon support is the use of structured materials as supports. Due to the original structure of graphite, it is possible to synthesize quite complex and diverse carbon nanomaterials on its basis. These include all types of graphene structures and their derivatives: carbon nanotubes (CNTs) and nanofibers (CNFs), graphene nanosheets, graphene oxides (GOs), reduced graphene oxides (rGOs), arrays of nanotubes, aligned and non-aligned, as well as graphene aerogels, honeycombs, spheres, and others [27–33].

Graphene and its various configurations possess excellent physical and chemical properties, including a large specific surface area, high electrical and thermal conductivity, durability, chemical resistance, and mechanical strength [34–36]. The use of complex carbon structures in CL leads to a more uniform distribution of catalytic particles on the surface and hence high electrochemical activity, gas permeability, and electrical conductivity [37–40]. At the same time, due to the appearance of micro- and macro-pores as a result of the specific packing of graphene particles, the water balance of FC is improved [30]. These materials are highly suitable and promising for use as electrocatalytic and electrode materials in fuel cell components, such as CLs, microporous layers (MPLs), and GDLs.

The current trends in the development of carbon nanomaterials technology are aimed at improving the methods for synthesizing materials with specified properties and expanding their applications, including electrochemical energy storage and conversion (supercapacitors, lithium-ion batteries, electrolysis systems (hydrogen evolution reaction (HER)), and fuel cells (hydrogen oxidation reaction (HOR) and ORR)). The growth in demand and applications of electrochemical current sources has led to a rapid development of research in the synthesis and properties of these structures, which have a favorable impact on the efficiency and durability of device components. The abundance of published materials in various research areas necessitates their analysis and systematization to address the challenges currently facing the PEMFC industry. This review will focus on these challenges and consider studies of the properties of electrocatalyst support materials. These properties include high conductivity, surface area, the presence of functional groups that improve metal deposition, and the durability of materials during operation. On the other hand, the decline in the number of studies on the properties of electrocatalysts highlights the problem of developing an effective catalytic layer that provides the hydrophilic/hydrophobic balance and optimal porosity for high transport efficiency. The main focus of this review is on systematizing properties and selecting the optimal material structure in terms of efficiency in the FC electrode (cathode).

2. Graphene Structures

2.1. Graphene Nanosheets

The structure of graphene is represented as one or more layers of sheets of one-atom-thick sp^2 hybridized carbon. Currently, there are many different techniques for making graphene, as each technique produces nanosheets with unique physical and chemical characteristics required in a particular field. Methods for producing graphene include physical or chemical layering, epitaxial growth through chemical vapor deposition (CVD), unzipping of carbon nanotubes through electrochemical, chemical, or physical methods, and reduction of sugars such as glucose or sucrose [38]. Graphene is the fundamental precursor of all graphitic carbon structures, including carbon nanotubes, nanofibers, spheres, and other nanostructures. The approximate thickness of one layer of graphene is 0.345 nm [27], which has a positive effect on the uniformity of the distribution of catalytic particles on it and the increase of active sites. Graphene nanosheets also have many different derivatives used as catalyst supports in PEMFCs, which are graphene oxides as well as graphene doped with heteroatoms.

For carbon materials, functional groups are important in the synthesis of platinum catalysts. They serve as adsorption sites for the nucleation of small Pt particles, resulting in a uniform distribution of Pt on the carbon support [41]. The capacitance of the double-layer electrode with platinum deposited on carbon depends on the surface morphology and the type of functional groups fixed in the support [42]. The smaller average Pt particle size is a result of the finer catalyst particle size, which indicates the existence of more anchoring sites on the graphene. This provides better dispersion of Pt particles [43]. Compared to spherical carbon particles, the planar structure of graphene facilitates the uptake of oxygen functional groups, mainly at the edges of graphene sheets [41].

While surface defects have been found to improve the activity of graphene-based electrocatalysts in ORR and enhance Pt dispersion, long-term observations have also revealed that these properties can result in carbon corrosion due to unsaturated valences and free s-electrons [44], as well as the adverse effects of highly concentrated s-electrons. Functional groups also decrease electrical conductivity and increase water affinity. This may make it more difficult to remove water from the fuel cell, which can lead to substrate corrosion and primarily Pt dissolution.

The addition of graphene nanosheets to the MPL improves the performance of PEMFC in both low- and high-humidity conditions [45]. The addition of graphene to MPL has been found to decrease membrane dehydration in low humidity conditions. Additionally, it has been observed to decrease liquid water saturation at the catalyst layer–GDL interface in high-humidity conditions [46]. This is due to the presence of a large number of micropores, which contribute to the retention of water needed to wet the membrane and electrolyte in the catalyst layer. Graphene exhibits a uniform distribution of micropores and small mesopores. This characteristic can enhance performance at high relative humidity and current densities by facilitating simultaneous capillary absorption of liquid water to reduce cathode flooding and improve the diffusion of reactants into the catalyst bed [47]. Graphene nanoplatelets exhibit dense horizontal stacking of graphene flakes in MPL, which leads to a reduction in pore volume. However, this effect results in a slight increase in the MPL wetting angle and therefore enhances hydrophobicity. In this case, there are fewer pathways for water removal from the surface, which hinders air flow towards the through-plane. As a result, through-plane permeability and efficiency decrease with increasing graphene content [45,48]. The electrical conductivity of graphene nanoplatelets is significantly affected by the number of graphene layers, lateral size, and material purity [49]. The resistivity measurements indicate that the inclusion of just 30% graphene in MPL significantly enhances the electronic conductivity of the layer, resulting in a 20% reduction in resistivity compared to a conventional carbon black-based MPL [45].

When comparing Pt catalysts based on graphene sheets and carbon black [50], this study found a three-fold increase in the ORR mass activity for graphene. This was attributed to higher charge transfer rates or lower binding energy with reactive adsorbates. The increase in ESA is due to the smaller average particle size of the Pt particles. This results in the presence of a disordered structure, more electroactive sites, and the ability to form more three-phase boundaries. Additionally, there are strong metal-support interactions and high dispersion of the nanoparticles [51–53]. Reducing the size of catalytic particles also leads to a higher utilization of said particles [50]. However, graphene-based catalysts generally exhibit superior degradation resistance and stability, which can be attributed to the presence of functional groups. Graphene may contain functional groups such as N, O, and S, depending on the synthesis or functionalization method used. During catalyst synthesis, active metal particles replace certain groups in the graphene structure. This significantly strengthens atomic bonding and improves catalyst degradation stability [27,41,46]. Electrochemical studies have shown that the catalytic activity and stability of graphene-based bimetallic catalysts are significantly enhanced by the addition of a secondary metal [54]. The initial decrease in activity is likely caused by the dissolution of subsurface and/or surface atoms of the second metal and/or Pt, followed by the re-formation of a stable number of electrons transferred per oxygen molecule for all catalysts [51].

Although graphene nanosheets have advantages, they can negatively affect MEA functioning. Therefore, when CVD graphene is used as a catalyst support deposited on GDLs, mass transfer limitations are observed at low polarization voltages when the current density suddenly decreases [55]. The high graphene content in the catalyst caused agglomeration, which characterizes the inefficiency of using such a layer to periodically remove water generated at the cathode [56]. The electrochemical performance deterioration is caused by mass transfer issues, which can be attributed to the presence of mesopores and densely stacked graphene sheets [45].

2.2. Graphene Oxide and Reduced Graphene Oxide

Graphene oxide is a single-layer sheet-like structure with functional groups. Oxygen groups such as hydroxyl and epoxy groups are located on the basal planes, while other groups, such as carboxy, carbonyl, phenol, etc., are mainly based on the leaf edges. These oxygen-containing groups and defects alter its physicochemical properties. GO is highly hydrophilic and disperses uniformly in water, making it an excellent electrode material [57]. However, the electrical conductivity of GO is lower than that of graphene because of the functional groups present on its surface [27].

Reduced graphene oxide is obtained in several steps, first from graphite to GO by thermal or chemical oxidation and then by reduction to reduced graphene oxide. Several methods exist for synthesizing GO, including the Brodie, Staudenmaier, and Hummers methods and their modified and improved variants [58,59].

The reduced graphene oxides have a more mesoporous structure with fewer micropores compared to graphene. They are aggregate friable aggregations of lamellar particles forming slit-like pores [60]. This is evidenced by characteristic sorption isotherms with a hysteresis loop, which is associated with capillary condensation occurring in mesopores, as well as with limiting absorption in the range of high P/P_0 (the ratio of pressure in the system to the condensation pressure) [61,62]. There is also a sufficient number of macropores, which play the role of gas transport channels in the structure. Thus, the specific porous structure of rGO with multiple mesopores and open and interconnected channels promotes rapid diffusion of fuel to the active Pt sites [62]. During the synthesis of metal particles onto the support, the doping process causes a decrease in surface area. However, the resulting catalysts maintain their original mesoporosity, which is a crucial aspect in the fabrication of catalysts for PEMFCs [62]. Gold nanoparticle catalysts supported on reduced graphene oxide (AuNP/rGO) with a high specific surface area and mesoporous structure, featuring a narrow and regular pore size distribution, have been successfully fabricated [63].

It is demonstrated in Ref [64] that rGO can serve as a capacitive material when incorporated into a catalytic layer along with platinum particles. Scanning electron microscopy (SEM) images of the rGO/Nafion layer on carbon paper revealed a homogeneous three-dimensional porous structure of rGO sheets (Figure 2a–c). The pore diameter of the rGO/Nafion layer was in the range of 0.7 to 9 μm, which was approximately 10–100 times longer than the free path length of O_2. This could facilitate the mass transfer of O_2. Elemental mapping images of the rGO/Nafion thin layer (Figure 2d–g) showed the presence of uniformly distributed S and F, which can be explained by the technique used to obtain the original graphene that underwent subsequent oxidation. After atomization of the catalytic ink, Pt/C/Nafion aggregated on the inner wall of the pores of the rGO sheets. Despite the inaccuracy of determining pore diameter from SEM images, the authors noted maintaining a porous structure while the pore diameter decreased.

The presence of functional groups on the surface of GO and rGO is clearly visible when GO is examined by Fourier-transform infrared spectroscopy (FTIR), the results of which are shown in Figure 3. In [65], functional groups related to graphite and rGO powders were investigated, from which the effect of oxidation on the properties of graphene can be determined. The graphite spectra in Figure 3 show a broad band in the 3000–3600 cm^{-1} region representing the valence vibration due to the presence of a hydroxyl group (O-H vibration). The spectrum shows typical peaks indicating oxygen-containing groups: 1919.77 cm^{-1} for C=O (carbonyl/carboxyl) and a broad band at 1000–1340 cm^{-1} for carboxy C-O, epoxy C-O, and alkoxy-C-O. The peak at 1619.81 cm^{-1} corresponds to the planar vibrations of the aromatic compound C=C, at 2112.61 cm^{-1} for C:O, and at 2271.16 cm^{-1} for CH. The IR spectra of rGO showed a significant decrease in the bands corresponding to the O-H and C:O groups, while the peaks corresponding to the epoxy and alkoxy C=O groups and C-H groups disappeared completely, as shown in Figure 3.

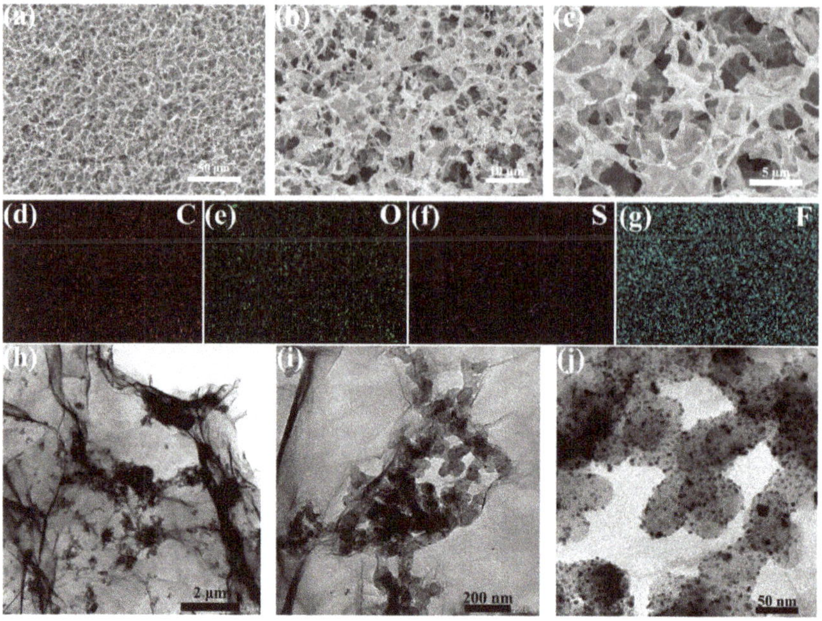

Figure 2. (**a–c**) SEM images of the rGO/Nafion layer. (**d–g**) Element mapping images of C, O, S, and F based on (**a**). (**h–j**) Transmission electron microscopy (TEM) images of the rGO sheets adsorbed Nafion and Pt/C particles in the rGO–NF–Pt electrode [64]. Copyright 2023, Elsevier.

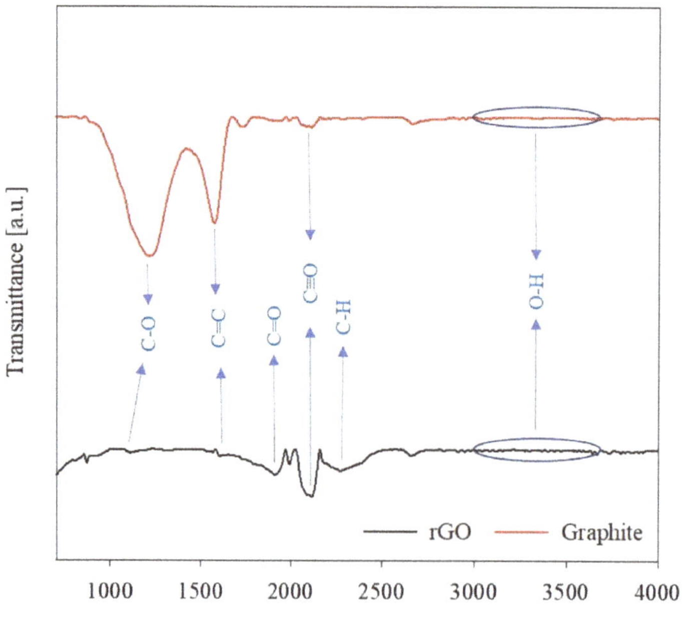

Figure 3. FTIR analysis related to graphite and rGO [65]. Copyright 2023, Springer.

The reduced graphene oxides have a structure between pure graphene and graphene oxide [66]. During chemical and thermal reduction, the quantity of oxygen-containing functional groups is considerably decreased. This is explained by the fact that the p-conjugated structure of graphene is essentially restored in the synthesized rGO, with the opening of epoxy rings having a significant influence on this situation. It has also been found that the thin nanolayer of rGO folds together to form a thick bulk structure due to the presence of strong van der Waals forces between each layer [67]. The reduction of oxygen-containing functional groups increases the hydrophobicity of rGO-based samples owing to the reduction in oxygen concentration [61], but due to the high porosity of these samples, this cannot affect the electrical conductivity [65].

Under reaction conditions, the spectra of the graphene-based catalysts indicate partial removal of major functional groups from GO compared to the original graphene oxide. The functional groups likely served as anchoring sites for metallic platinum ions during the reduction reaction, promoting the production of metal nanoparticles that are distributed on the rGO sheets [62]. The embedding of metal nanoparticles in the support structure and their interaction with it have a positive effect on the chemical and mechanical stability of the catalyst. This prevents the development of agglomeration processes and the migration of catalyst particles.

In [65], the effect of adding rGO and graphite to MPL by pressing on various properties of the GDL for which Toray carbon paper was used is studied. The study has demonstrated that MPLs based on rGO have a smooth and planar morphology as well as higher in-plane electrical resistivity compared to graphite-based MPLs. However, the rGO-based MPLs have lower roughness than the graphite-based MPLs. Replacing graphite with rGO increases the wetting angle, porosity, and permeability of the samples. And graphite-based MPLs have lower resistivity.

The catalytic activity of rGO-based catalysts is high due to the presence of a larger number of active sites that more effectively adsorb reactants during the catalytic ORR process [41]. The reduced graphene oxide itself can serve as electroactive sites and also reduce the ion diffusion pathway, which also results in high electrical conductivity and better electrical contacts [63].

The accelerated MEA degradation test results indicate that Pt/rGO exhibits nearly twice the degradation resistance compared to other materials [62]. Stronger electronic interactions between Pt nanoparticles and graphene support result in this effect. The metal-substrate interaction is strengthened, as shown by the negative shift of the Pt 4f binding energy in the X-ray photoelectron spectroscopy (XPS) spectra. This suggests that electrons are transferred from the carbon supports to Pt [39]. The resistance to degradation is also affected by the slower corrosion of the support, which is caused by fewer disordered sites for graphene derivatives.

The current density value in the electrooxidation of methanol using a multicomponent core-shell catalyst based on rGO is approximately four times higher than that of similar catalysts based on amorphous carbon support and approximately six times higher than that of Pt/C [68]. These results suggest that such an electrocatalyst exhibits high volumetric activity, high stability, and high resistance to CO poisoning, thus reducing the platinum loading in PEMFCs. The use of synthesized Pt/rGO material as a cathode catalyst increased the maximum specific power by up to 17% and significantly improved fuel cell performance, particularly at high current densities [39]. The efficiency of MEA depends on the catalyst activity and ESA, as well as the layer structure and thickness. These factors significantly affect the mass transfer of reactants and can cause a deterioration in load transfer resistance due to an imbalance in the three-phase boundary. It is important to maintain a balance at the three-phase boundary to ensure optimal efficiency [66]. To address this issue, it has been suggested that a mixture of graphene with either carbon black or CNT could be used [41]. The rGO–Vulcan mass ratio of 75:25 provides the highest ESA [69] as well as its lowest loss.

Although graphenes and their oxides have advantages as MEA components, such as high catalytic and electrochemical activity, stability, and degradation resistance, there is an issue with mass transfer of gas reagents and their products during cell operation, which significantly affects the efficiency of the PEMFC operation. Therefore, the formation of catalytic layers and MPL based on graphene and its oxides, which meet the requirements for structure and porosity and ensure efficient operation of the device, remains a significant challenge.

3. Carbon Nanotubes

Carbon nanotubes are used as supports for FC catalysts. They can be single-walled or multi-walled and can form arrays, including oriented ones, depending on the synthesis methods and application purposes. Several techniques are currently employed for the synthesis of CNTs, with the most prevalent being electric arc evaporation of graphite electrodes, laser ablation of graphite targets, and catalytic pyrolysis of gaseous hydrocarbons. CNTs are two-dimensional hexagonal tubular structures formed by a single layer of graphene with strong meshes of sp^2-hybridized atoms. CNTs can be 0.5–20 nm in diameter and up to hundreds of micrometers in length [70]. Due to their tubular structure, CNTs have a large specific surface area, which is one of the main advantages for the application of this support in PEMFCs [71]. In addition, the material's mechanical properties, such as its modulus of elasticity and hardness, are improved compared to those of other structured materials [72].

For single-walled CNTs (SWCNTs), each carbon atom carries a pair of π-electrons that do not participate in bond formation. These electrons form a conjugated π-electron cloud along the entire length of the CNT. Multi-walled carbon nanotubes (MWCNTs) consist of several concentric tubes of coiled graphene sheets arranged one inside the other, similar to a matryoshka doll. In certain instances, a multi-walled nanotube structure can be created by twisting a single graphene sheet around itself, resembling a sheet of parchment. For multi-walled CNTs, the carbon atoms within each layer are connected by strong covalent bonds. The multi-walled CNT layers are connected by weak van der Waals interactions, resulting in a layer-to-layer distance similar to that of graphite, approximately 3.4 Å.

The size of catalyst particles is a critical factor that determines the electrocatalytic activity and durability of the platinum electrocatalyst used in fuel cells. The electrochemical and catalytic activity of Pt nanoparticles synthesized on CNTs is improved due to their smaller size and uniform distribution. This is because the increased number of active sites and higher electrical conductivity promote rapid electron movement in the support [73–75]. The catalysts based on MWCNTs exhibited a positive shift in Pt $4f_{7/2}$ binding energies compared to crystalline Pt. The extent of this shift was found to be related to the size of the Pt nanoparticles on the MWCNTs; specifically, the smaller the nanoparticles, the more positive the shift [74]. This effect is reported to be attributed to initial state effects, such as changes in the coordination number of the atom or charge transfer effects [76]. Alternatively, the shift may be caused by end-state effects, such as relaxation effects [77]. This effect also makes it possible to significantly reduce the platinum catalyst loading. The use of nanotubes as a support for platinum particles enables high dispersibility, similar to that of graphene and its reduced oxide.

The synthesized CNTs initially exhibit various structural defects, including surface indentations, metal impurities, and amorphous carbon. These defects are evident in the XPS spectroscopy, which indicates the presence of sp^3-hybridized carbon [78]. The graphitization index (GI) indicates the amount of sp^2 carbon and is calculated from the interplanar distances obtained from X-ray diffraction analysis (XRD). Increasing the GI enhances the strength of p-sites (sp^2-hybridized carbon). These sites function to attach Pt particles to the support material and improve metal interactions with the support material [79]. The presence of structural defects in the grown MWCNTs is advantageous for achieving a more uniform deposition of Pt nanoparticles with reduced loading control due to the increased binding energy between MWCNTs and Pt nanoparticles and their good distribution [80]. The presence of C-O and C=O IR peaks in Pt/MWCNTs catalysts is attributed to their existence in the structure of carbon nanotubes used as a substrate for Pt nanostructures. Oxidized particles may form during the thermal decomposition of acetylene in the presence of trace amounts of oxygen in a CVD furnace [74].

To preserve the structure of CNTs during the synthesis of catalysts, an impregnation method is used. Acids and salts of metal precursors are most commonly used as precursors. Electrodes produced using this method exhibit greater stability as a result of the strong adhesion between the metal particles and the nanotubes [81,82].

The electrocatalytic activity of three-dimensional porous MWCNT electrodes was tested on a rotating disc electrode setup. At 0.9 V, the results indicate that Pt has a higher current in ORR per ESA and an extended kinetic current region, which remain stable after the durability stress test (DST). This phenomenon is also observed for ultra-low Pt loadings up to 14.2 $\mu g/cm^2$ [74]. A comparison of the performance of catalysts in PEMFC with 0.1 mg/cm^2 Pt loading on the cathode and of the impedance data showed that the Pt/CNT catalyst was the most active one [83]. Electrochemical measurements using cyclic voltammetry (CV) and linear sweep voltammetry (LSV) showed that Pt/CNT exhibited higher mass and specific activity compared to commercial Pt nanoparticles deposited on graphitized carbon (Pt/C) [84,85]. The selective activity towards HOR, improved by 30% compared to commercial samples, was the main indication of catalyst retention in SWCNTs [75]. It is also possible to increase catalytic activity by using bimetallic catalysts on MWCNT, for example, PtNi/MWCNT. The ORR mass activity of such a catalyst is 1.4 times higher than that of a conventional commercial electrocatalyst [86].

The use of CNTs instead of carbon black provides a high utilization rate of Pt in MEA [83] compared to the Pt/C catalyst. The higher use of platinum characterizes the higher number of Pt active sites and three-phase boundaries in the applied catalysts [79], and hence their high catalytic activity [87].

Carbon corrosion proceeds during a potential cycling test through the oxidation of functional groups of carbon support to CO_2 and CO and is enhanced in the presence of Pt nanoparticles [86]. CNTs in this case are less susceptible to corrosion than carbon black, even though the platinum particles are highly active. The main reason for the higher durability is the chemical inertness of the structured material [86]. ESA losses observed during DST are up to 40%, which is an excellent indicator of electrochemical resistance [79]. A study of the degradation properties of CNT-based catalysts under fuel cell operating conditions showed more stable behavior than with commercial Pt/C, including after 10,000 startup/shutdown cycles [84]. However, destruction of the electrode structure is observed due to the formation of larger pores in the catalytic layer with a significant thickness as a result of support corrosion [88]. The high thermal stability of these materials also enhances their degradation resistance. CNT-based catalyst synthesis can also significantly reduce electrode wear and improve long-term stability during potential cycling due to resistance to Pt agglomeration during ORR/HOR [75].

During tests on a fuel cell, a 15 μm thick CNT MPL was found to increase the peak power density of the device by 50.9% when air was supplied at the cathode and by 20.9% when pure oxygen was supplied at the cathode, compared to a conventional PEMFC [89]. When CNTs are used as catalyst support, the MEA also exhibits a high current density and electron transfer rate up to 50% higher than that of Pt/C [75,84,85]. Electrochemical impedance analysis revealed that the CNT MPL effectively reduced the charge transfer resistance. This improvement was attributed to the enhanced reaction kinetics and mass transfer through MPLs [86]. The MEA performance was significantly improved by adding a single-walled carbon nanotube nonwoven layer between the gas diffusion substrate and the hydrophobic MPL. The excellent electrical conductivity and rational pore structure of the material result in improved wettability and water-holding capacity of MEA, leading to enhanced performance under low humidity conditions [90]. Also, the porous structure of the electrode when CNTs are added to the electrode composition can improve the gas transfer and water drainage characteristics, which helps to improve the FC performance [91].

Vertically Aligned CNTs

Vertically aligned CNTs (VACNTs) are basically arrays of parallel-oriented nanotubes grown on a substrate (Figure 4b). In this case, nanotubes can be either single-walled or multi-walled. The synthesis of CNTs is carried out by the catalytic decomposition of hydrocarbons (for example, methane). The desired morphological and physical properties of nanotubes are achieved by applying different variations [92]. The performance of CNTs is also affected by a number of other factors, such as the material of the substrate used during synthesis, the tube growth catalyst used, and how it is applied [93–95]. The use of VACNTs as an electrocatalyst support greatly influences the degradation stability and performance of PEMFCs due to the peculiarity of their formation. The synthesis of aligned tubes directly on the substrate results in enhanced adhesion of the microporous and catalytic layers [96]. VACNTs possess a developed surface that facilitates the deposition of nanometer-sized platinum particles and prevents their agglomeration. This reduces the loading of expensive catalysts and increases their efficiency. When a film of aligned CNTs is used as an FC cathode, it increases the current density, performance, and electrocatalytic activity compared to a Pt/non-aligned CNT electrode [97]. The hydrophobicity of VACNTs ensures the prevention of water condensation on the cathode, resulting in long-term PEMFC stability achieved by their ability to control water balance and gas diffusion [98,99].

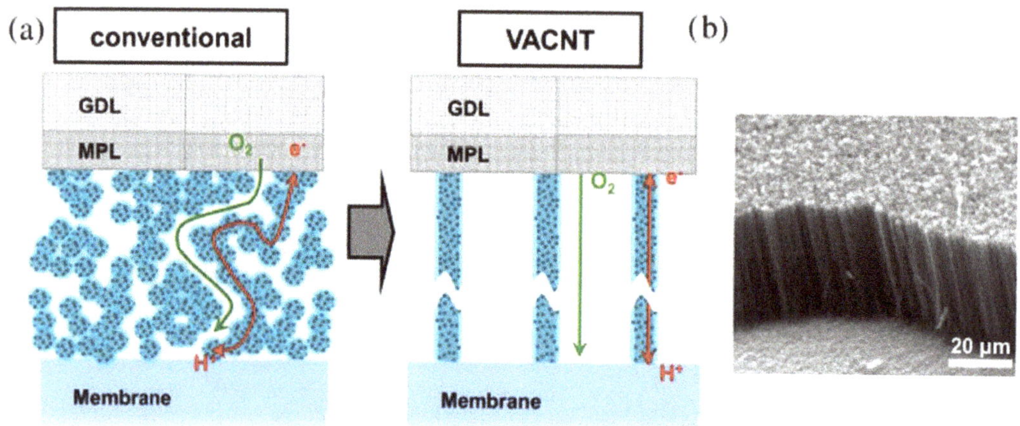

Figure 4. (**a**) Concept of VACNT electrodes; (**b**) SEM image of VACNTs [100]. Copyright 2014, Elsevier.

The formation of an electrode with MPL representing VACNTs is the most promising direction of development [100]. The concept of the electrode based on aligned nanotubes is shown in Figure 4. Here, nanotubes are grown vertically directly on the GDL, on which platinum catalytic particles are subsequently deposited. In this case, the diffusion of reagents due to the structure of the support is directed through the channels of the VACNT array with the catalyst, which enhances the chemisorption of gas reagents. The vertically aligned structure enhances the gas diffusivity, water transport, and effective utilization of Pt [100]. The platinum particle nanotube-based electrodes also provide improved proton and electron transport by creating an electrical contact between the substrate and CNTs, which greatly enhances the properties of the PEMFC device [99]. The thus-obtained carbon nanotube array is perpendicularly aligned along the entire length of the carbon fibers and covers the surface of the GDL. Fuel cell tests have demonstrated performance improvements of up to 30% [101].

The MPL structure for FC is also created by the direct growth of thin multi-walled VACNTs on a commercial GDL carbon fiber substrate without hydrophobic treatment [101].

In this case, it is possible to control the layer thickness by means of the layer-by-layer buildup of the CNT array. The formation of the near-surface catalytic layer based on MWCNTs as part of the microporous layer of the electrode is also carried out by methods that do not disturb the structure of the carbon support. When Pt-catalyst particles are attached to the top of the VACNT adjacent to the proton exchange membrane, the active region of the three-phase interface grows. Simultaneously, preserving the mass transport channels between the VACNTs without metal catalyst particles ensures the removal of reagents, which has a positive effect on the PEMFC performance [97]. It is possible to increase the three-phase interface between the catalyst, its support, and the PEM. The high peak power density of 1.61 W/cm^2 (H_2/O_2, 150 kPa) at an ultra-low Pt loading of 50 $\mu g/cm^2$ in this case is attributed to the presence of ordered hydrophobic and hydrophilic regions along the VACNTs. These regions ensure sufficient membrane wetting and promote rapid H_3O^+ transport, resulting in high Pt utilization, as well as facilitate mass transfer at high current densities [97].

Thus, CNTs, particularly vertically aligned nanotubes, are the most effective solution for addressing the issue of reagent mass transfer during MEA operation, which is due to the optimized structure of catalyst layers with mass transport channels and high adhesion to the membrane, or GDL.

4. Functionalization and Doping of Structured Materials

To enhance the adhesion between catalytic nanoparticles and the surface of the structured support and to increase the reactivity of CNTs by creating active sites that facilitate the distribution of catalyst nanoparticles, the support undergoes chemical or physical pre-modification [102–106]. In the case of chemical (or covalent) functionalization, new functional groups are attached to the carbon atoms [107]. Physical groups or atoms are attached through physical sorption [108], or surface defects are created. When functionalizing CNTs, one of the best options is to use a mixture of nitric and sulfuric acids [109]. The mechanism of action of such treatment is shown in Figure 5. This treatment is carried out to form various functional groups on the surface of nanotubes [110], such as carboxyl groups, thiol groups, etc.

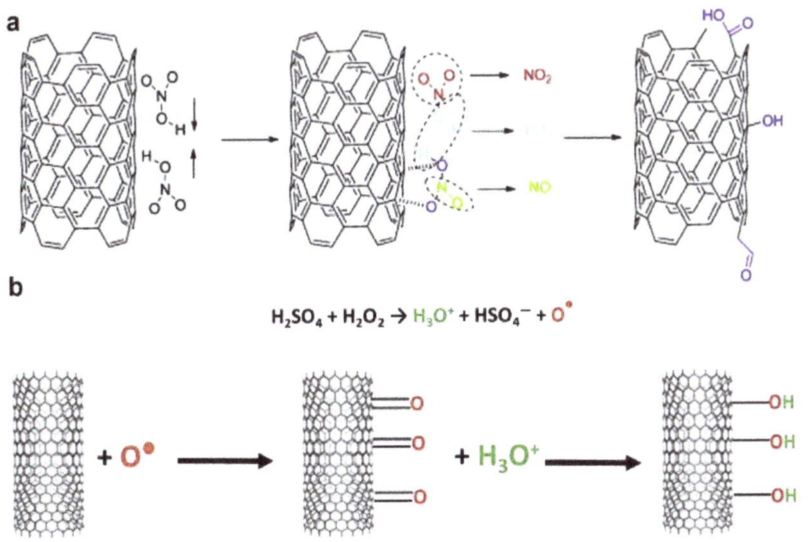

Figure 5. (**a**) Mechanism of oxidation of nitric acid; (**b**) mechanism of oxidation of piranha solution [104]. Copyright 2021, Elsevier.

In order to improve the physical and chemical properties of graphene materials, they are modified with various additives. Both heteroatoms, such as S [111], N [40,112–114], P [115,116], F [117], and various compounds, such as tin oxides [118,119], silicon oxides [120], and others [121], can serve as doping materials. These additives are introduced into the graphite structure by various chemical and physical methods and act as functional groups. Figure 6 shows the reaction schemes employed during the synthesis of modified CNTs by two distinct methods: adsorbed melamine polymerization (1) and melamine-formaldehyde resin polymerization on the CNT surface (2). As previously discussed, in catalyst support synthesis, catalytic nanoparticles are incorporated into the graphene lattice by replacing heteroatoms or attaching to functional groups that serve as adsorption sites. This enhances the interaction between the materials, resulting in better electrochemical activity and stability of the catalyst layer [112]. Also, the presence of doping components leads to an increase in the electrode–electrolyte interface, which contributes to lower resistance and improved conductivity and leads to an improvement in the overall performance of the fuel cell [115]. Heteroatom-doped graphenes and their derivatives can also exhibit catalytic properties comparable to classical Pt/C-type catalysts, which can significantly reduce the cost of FC due to the absence of expensive metals [122,123].

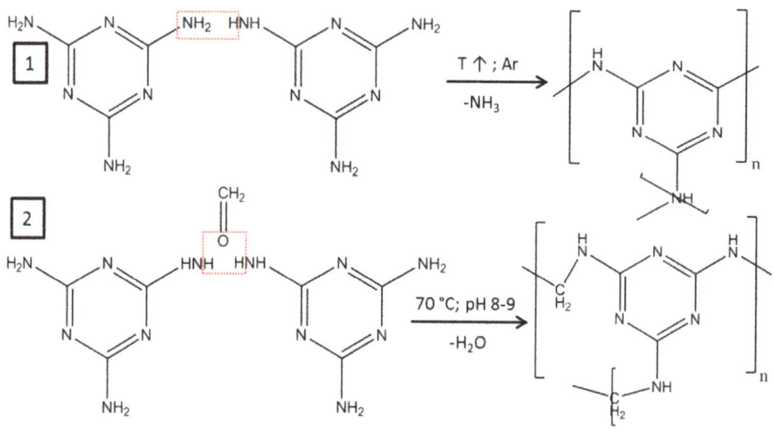

Figure 6. Synthesis scheme of nitrogen-containing polymers on the surface of CNTs. (**1**) Polymerization of melamine; (**2**) synthesis of a melamine–formaldehyde resin [83]. Copyright 2019, Springer.

CNTs, just like graphene, are doped with various heteroatoms to increase the durability of the catalytic layer. By using modifying compounds, it is possible to prevent direct contact between carbon and water, reducing the corrosion of carbon in electrochemical reactions [124]. Introducing modifying atoms directly into the carbon lattice results in increased durability and maximum power density of the PEMFC [125–127]. Modification of CNTs results in a reduction in the average size of Pt particles, which is due to an increase in the concentration of Pt ion anchoring sites during the synthesis of Pt/CNT catalysts. This, in turn, can lead to a higher utilization rate of the active component Pt observed in both the electrochemical cell and PEMFCs [50,83,124]. The effective doping also showed a reduction in excessively defective carbon atoms [78]. Nitrogen doping can increase catalytic activity, making this method attractive for ORR catalyst support [107,128,129].

5. Other Carbon Structures

5.1. Carbon Spheres

Carbon spheres are a type of carbon material that has a large surface area and a mesoporous structure [130–132]. These characteristics make them excellent for use as a support for platinum catalysts. The mesoporous structure of the carbon spheres provides a large surface area for the distribution of Pt nanoparticles, which increases the ESA of the catalyst. The porous nature of the carbon spheres also provides better access for reagent molecules to the Pt nanoparticles, resulting in improved catalytic and transport characteristics. Furthermore, carbon spheres exhibit excellent mechanical stability and thermal conductivity, rendering them a dependable and long-lasting support material for platinum catalysts. In general, mesoporous carbon spheres are promising materials for the development of highly efficient platinum catalysts for various applications.

Various parameters of carbon spheres, including their size, structure, and the number of shells, can be customized to meet specific needs. The design and creation of functional carbon spheres have garnered significant attention in the realm of research. As the demand for materials in energy conversion continues to grow, the fabrication of carbon spheres and their derivatives has become increasingly sophisticated, moving from simple to complex structures. This process involves modifying both the outer shell and the internal architecture of the spheres. As a result, a combination of three different methods (hard, soft, and self-templating) is utilized alongside a range of preparation techniques (such as selective etching, nanocasting, chemical vapor deposition, and spray pyrolysis) to achieve the desired outcomes [133–141].

Hard templating methods involve using rigid particles as templates to create hollow structures, which are removed after the desired shells are formed. For example, in the case of single-shelled carbon spheres, a carbonaceous precursor is coated on the template and then converted into a carbon shell through a carbonization process. The template is then selectively removed to create a carbon nanomaterial with internal void space. While removing hard templates can be a time-consuming process, it allows for the production of carbon spheres with a narrow size distribution and good morphological uniformity. Various templates, like silica, polymer, and metallic spheres, can be easily fabricated for this purpose. While hard templates require laborious removal by acid/alkaline etching or organic solvent dissolution, soft templating offers a more convenient and environmentally friendly approach through the self-assembly of precursor molecules with colloidal systems as templates, easily removed by pyrolysis or extraction. The self-templating method is an efficient way to prepare hollow carbon spheres without the need for additional templates. Solid or hollow polymer spheres can be directly synthesized and converted to hollow carbon spheres through self-assembly and calcination treatment.

In general, the choice of carbon sphere synthesis method for PEMFC will depend on the desired properties of the spheres, such as size, morphology, surface area, and pore structure. Each method has its own advantages and limitations. Careful optimization of synthesis parameters is crucial to achieving the desired performance in PEMFC.

One new approach is to develop a platinum catalyst without support, in which carbon spheres are used as a framework for the deposition of platinum particles and then removed. The electrocatalyst exhibits a surface area similar to that of commercial Pt/C, as determined by electrochemical studies. Additionally, based on DST results, it demonstrates twice the durability of Pt/C [130]. When using spheres as a support, their organized porous structure provides a combination of high electrocatalytic activity in ORR and high mass transfer efficiency. A well-designed porous structure and microstructure provide an optimized water/gas transport pathway into the fuel cell CLs [131].

It is also possible to synthesize a Pt catalyst using a new carbon support, a cup-shaped hollow carbon hemisphere (and an N-doped hemisphere) (carbon shell thickness ~ 4.6 nm). This morphology allows for good bonding of the broken shells and Pt loading on both sides. In addition, carbon has pores that are sized to match the size of the Pt particles, which results in their filling with Pt, as well as clogging the pores and, as a consequence, limiting the agglomeration of Pt, as determined by in situ XRD analysis. Both catalysts show high ESA (60–65 m^2/g) and cyclic durability (6000 cycles), which is superior to the commercial Pt/C catalyst. This enhanced durability under potential cycling can also be attributed to the pore size and limitations of Pt dissolution, as well as improved metal–substrate interaction [132].

5.2. Aerogels

Aerogels are a type of ultralight, highly porous material that is promising for use in PEMFC. Aerogels have a large surface area, which can provide a larger active area for chemical reactions to take place. This can improve the efficiency of the fuel cell and increase its output power. In addition, aerogels have good thermal stability and low thermal conductivity to maintain the operating temperature of the fuel cell and improve its overall performance. Furthermore, aerogels can be easily customized to obtain specific properties, such as pore size, surface area, and surface chemical composition, which can be optimized for use in PEMFCs. The process of preparing carbon gels consists of three main steps: polymerization, dehydration, and carbonization [142–149]. During the polymerization phase, hydrogels are formed through the combination and linking of molecules. Hydroxymethyl groups are added to resorcinol, which then undergoes condensation to form methylene and methylene–ether bridges. Crosslinking and agglomeration create the 3D hydrogel structure. Various alkali and acid catalysts are used in the addition and condensation reactions. Organic carbon aerogels can be synthesized using different aromatic aldehyde pairs and templating agents for specific properties.

Recently, graphene, carbon nanotubes, and biomass have been proposed as easier-to-use alternatives to traditional polymers. Various techniques, such as sol–gel, template, spacer support, self-support, and substrate-based methods, have been employed to create graphene aerogels. CNT-derived aerogels show potential as electrically conductive materials as they link together through van der Waals forces. However, directly forming CNT-based aerogels presents challenges like weak 3D structure, mechanical fragility, and limited flexibility. To address these issues, surfactants can be added to reduce CNT surface activity, and polymer additives can be used to strengthen the network.

As shown, the mesopore volume of the electrocatalyst on carbon aerogel by Brunauer–Emmett–Teller (BET) analysis is 2.7 times larger and the ESA is 2.2 times larger compared with that of conventional carbon black. Electrochemical durability experiments show greater stability of carbon aerogel compared to carbon black [150].

Using aerogel can increase the maximum power output of the fuel cell by 10% while maintaining its high performance and lowering potential losses for a wider current range. The carbon aerogel synthesized by the sol–gel method allows the platinum content in the electrocatalyst to be reduced to 4.2 wt.%. Platinum on carbon aerogel exhibits high catalytic activity in the ORR, with an ESA of 188 m^2/g_{Pt}. The high activity of the electrocatalyst on aerogel in the ORR is determined by its large mesoporosity, uniform distribution of dispersed Pt nanoparticles, and small size. The method used to prepare this electrocatalyst can be described as classical and relatively simple, making it suitable for mass production [151].

5.3. Other Structures

Carbon nanocages, nanohorns, nanocoils, uniform porous networks, and nano-onions are all viable carbon materials for PEMFCs [152–156].

The unique cellular structure of carbon nanocages can provide a large surface area and excellent conductivity, making them ideal for maximizing PEMFC efficiency and performance. The hollow interior of the carbon nanocages can provide improved mass transfer and accessibility of platinum sites, resulting in improved catalytic activity and stability of the electrocatalyst.

Carbon nanohorns, discovered in 1998, are interesting nanomaterials due to their special morphology and physicochemical properties. The carbon nanohorn structure consists of single graphene tubes with highly stressed tapered tips that aggregate into larger spherical angular superstructures with diameters of approximately 100 nm. The carbon nanohorns possess an unconventional structure, high purity and porosity, good electrical conductivity, and a large surface area, making them a promising electrocatalyst support. Like carbon nanotubes, carbon nanohorns are also insoluble in solvents, making it difficult to handle and use them effectively. Functionalization is a useful technique to improve the solubility of carbon nanohorns by oxidizing the tips, which have the most defects, increasing surface area and access to internal space. Covalent modification methods involve adding carboxylic groups to the tips of carbon nanohorns after oxidation or attaching organic molecules to the outer structure through cycloaddition reactions (Figure 7).

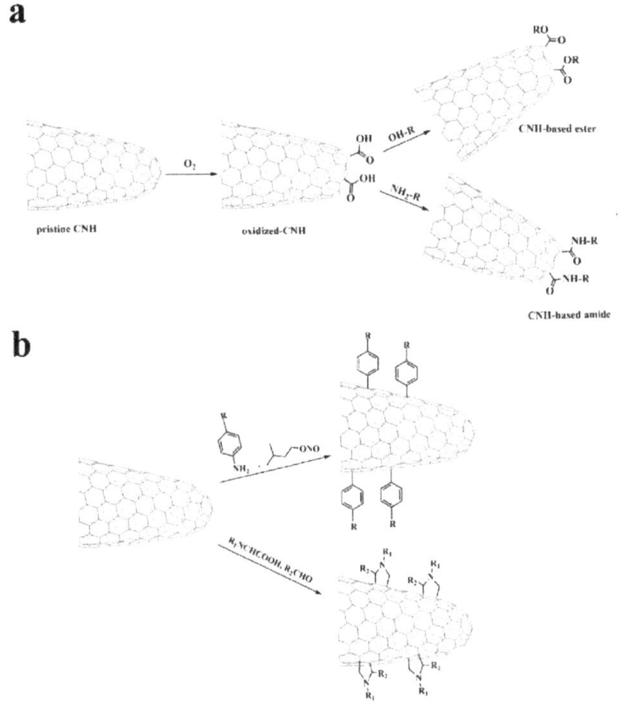

Figure 7. (**a**) Oxidation of carbon nanohorns and further post-functionalization of oxidized carbon nanohorns. (**b**) Functionalization of carbon nanohorns within situ-generated aryl diazonium salts and via the 1,3-dipolar cycloaddition reaction of in situ-generated azomethine ylides [153]. Copyright 2020, MDPI.

Carbon nanocoils are a unique form of carbon nanomaterial characterized by their distinctive helical structure, which resembles the miniature coils of a spring. They typically consist of sp^2-hybridized carbon atoms, which give them exceptional physical and chemical properties. Nanocoils have a large surface area, excellent electrical and thermal conductivity, and mechanical flexibility due to their helical morphology. These properties make them promising candidates for a variety of applications, including nanoelectronics, sensors, energy storage, and catalysis. In the context of energy applications, nanocoils have been studied for their potential use in fuel cells, supercapacitors, and batteries. Their large surface area and electrical conductivity could improve performance and efficiency [157,158].

Ordered homogeneous porous carbon networks are three-dimensional carbon structures with a well-defined and regular porous morphology. These networks have a large surface area, a uniform distribution of pore sizes, and interconnected porosity, making them ideal for various applications, such as energy storage, catalysis, and gas separation. The ordered and homogeneous nature of carbon networks provides improved mass transfer and reaction kinetics [159,160].

Carbon nano-onions are a unique form of carbon nanoparticles consisting of concentric graphite layers similar to those of an onion. They are formed by arc discharge, in which layers of graphene are forced to form spherical onion-like structures. Due to the unique structure of carbon nano-onions, they have outstanding properties such as high mechanical strength, excellent thermal and electrical conductivity, and a large surface area. These properties make them suitable for a variety of applications, including energy storage, nanocomposites, and biomedical applications.

The study of various nanostructured carbon materials is crucial for developing new types of electrodes that can enhance the performance, durability, and applicability of fuel cell technology. However, it is important to consider the economics and complexity of methods for synthesizing and scaling production of such carbon structures to improve the economic viability of fuel cell technology.

6. Summary and Conclusions

This review highlights the recent success of the use of carbon-based structures as electrodes, including PEMFC electrochemical performance. The use of structured supports shows promise in improving the efficiency and durability of electrodes for PEMFCs. Structured carbon materials, such as graphene, its oxides, and derivatives, including CNTs and their arrays, are the most promising for use in PEMFCs as electrocatalyst supports and additives in CL and GDL due to their well-developed surface and high degradation stability. Catalysts, on the other hand, exhibit improved electrocatalytic activity, stability, and degradation resistance due to the interaction between the support and active metal particles. This is achieved through the partial incorporation of platinum into the carbon material structure, resulting in increased dispersion of platinum particles during synthesis. However, the operation of MEAs with CL/GDL formed on the basis of structured carbon materials involves mass and gas transfer problems, which negatively affect the overall FC performance. The successful application of electrocatalytic material as a component of the PEMFC layer depends on its effectiveness as a support for the electrocatalyst, which must provide high values of electrical conductivity, dispersibility, and ESA of deposited particles, as well as their stability. Additionally, the ability to create a porous structure with sufficient values of mass transfer, reagent inlet/outlet characteristics, and hydrophilic/hydrophobic properties is essential.

Table 1 presents a summary of data from the recent literature that characterizes the main attributes of CLs based on structured graphene materials, as well as the performance of the PEMFC based on them. However, a deeper analysis of the literature reveals that there is insufficient data on the performance of PEMFCs due to the existing problems of reagent mass transfer. Thus, specific research aimed at improving the architecture of such electrodes to increase their specific capacity by improving mass transport makes this topic relevant [161].

It is important to note that, despite the similarity of the properties of the considered structures, the differences in their architecture and composition of surface groups and defects become determinative with regard to their application as electrode materials for PEMFC. Graphenes and their derivatives appear to be the most promising candidates for use as electrocatalyst support, but difficulties with the formation of an effective graphene-based layer have led to an almost complete absence of their testing in PEMFC, making the research of catalysts based on them a dead end. CNTs, in particular their arrays, appear to have greater potential. It is possible to form a catalytic layer of effective architecture with high diffusion characteristics, sufficient water balance, and improved power characteristics of the device. This issue is partially solved for vertically aligned carbon nanotubes due to their optimized structure, strong adhesion to GDL, and the existence of mass transport channels. The functionalization of this support and improvement of the properties of catalysts based on it may lead to even more successful results in terms of PEMFC technology. However, the task of optimizing the structure and porosity of graphene-like materials to enhance PEMFC efficiency remains relevant.

Table 1. A summary of data from the recent literature that characterizes the main attributes of CLs based on structured graphene materials, as well as the performance of the PEMFC based on them.

		Graphene Structures		**Carbon Nanotubes**		**Other Structures**
	Advantages	High dispersion of deposited platinum, ESA value, stability, and durability of catalysts Increasing FC output power when used as an additive		High catalyst stability and durability Ultra-low platinum loading High power output (both as a catalyst and as an electrode additive)		High dispersion of deposited platinum, ESA value, stability, and durability of catalysts Ultra-low platinum loading
	Disadvantages	Low values of mass transfer and diffusion of reagents in the layer Not applicable as a catalytic layer		Low dispersion of deposited platinum and ESA value of the catalysts		Complex fabrication technology Low FC output power No data
	Issue	Formation of the catalytic layer architecture		Functionalization to increase the dispersion of the active metal		Expanding research
	Parameters	Graphene nanosheets	Graphene oxides and reduced graphene oxides	Carbon nanotubes	Vertically aligned CNTs	**Other structures**
Electrocatalyst	ESA, m^2/g	80–100 [48,50]	20–50 [41,64,66]	10–40 [75,85,88]	~30 [98]	~200 [132,150]
Electrocatalyst	Degradation (ESA losses), %	~5	10–30	~30	~5	~35
Electrocatalyst	Composition	40 wt. % Pt/G	16.9 wt. % Pt/T-RGO	10 wt. % Pt/CNT	Pt-Co/VACNTs	40 wt. % Pt/broken hollow C spheres, 30 wt. % Pt/aerogels
Catalytic layer	CL architecture	-	40 wt. % Pt/rGO	Nafion ionomer, isopropanol, and 0.125 mg/cm^2 mMWCNT	Pt/VACNTs on cathode, VACNTs were used as MPL with a thickness of 4.6 μm	30 wt. % Pt/aerogels
Catalytic layer	Pt loading on anode/cathode, mg/cm^2	-	0.2/0.2 Pt/rGO on cathode Pt/C on anode	0.06/0.06 Pt/C (cathode) and PtRu/C (anode)	0.2/0.2	0.1/0.1
Catalytic layer	Maximum power density, W/cm^2	-	~0.62 [62]	~1.2 [91]	~1.4 [97]	~0.33 [150]
Electrode	Electrode architecture	Double-layer GDL with graphene MPL; MPL thickness 13–17 μm; mean pore size 162.41 nm	rGO was a capacitive material; pore diameter between 0.7 and 9 μm; a 60 wt. % Pt/C catalyst was applied to the GDL on top	Normal carbon black GDL without MPL and with 15 μm thick CNT sheet MPLs	VACNTs were used as MPLs with a thickness of ~40 μm; total porosity 56.3%	-
Electrode	Maximum power density, W/cm^2	~0.95 [48]	~0.75 [64]	~0.74 [89]	~1.41 [98]	-

In the field of other carbon structures, the question of the optimal balance between the increasing complexity of material fabrication technology and the success and reproducibility of the results obtained remains open. When using complex carbon nanostructures, such as spheres, it is crucial to evaluate the technological and economic efforts required against the benefits they provide in terms of catalyst properties and device performance.

Author Contributions: Conceptualization, A.A.Z. and N.A.I.; investigation, A.A.Z., N.A.I., D.D.S., R.M.M. and M.V.S.; resources, N.A.I.; writing—original draft preparation, A.A.Z., N.A.I. and D.D.S.; writing—review and editing, R.M.M. and S.A.G.; supervision, N.A.I.; project administration, N.A.I.; funding acquisition, N.A.I. All authors have read and agreed to the published version of the manuscript.

Funding: This work was carried out within the state assignment of the NRC "Kurchatov Institute".

Data Availability Statement: The data presented in this study are available on request from the corresponding authors.

Conflicts of Interest: The authors declare no conflicts of interest.

References

1. Mensharapov, R.M.; Ivanova, N.A.; Spasov, D.D.; Kukueva, E.V.; Zasypkina, A.A.; Seregina, E.A.; Grigoriev, S.A.; Fateev, V.N. Carbon-Supported Pt-SnO_2 Catalysts for Oxygen Reduction Reaction over a Wide Temperature Range: Rotating Disk Electrode Study. *Catalysts* **2021**, *11*, 1469. [CrossRef]
2. Zhao, H.; Xing, T.; Li, L.; Geng, X.; Guo, K.; Sun, C.; Zhou, W.; Yang, H.; Song, R.; An, B. Synthesis of Cobalt and Nitrogen Co-Doped Carbon Nanotubes and Its ORR Activity as the Catalyst Used in Hydrogen Fuel Cells. *Int. J. Hydrogen Energy* **2019**, *44*, 25180–25187. [CrossRef]
3. Stühmeier, B.M.; Pietsch, M.R.; Schwämmlein, J.N.; Gasteiger, H.A. Pressure and Temperature Dependence of the Hydrogen Oxidation and Evolution Reaction Kinetics on Pt Electrocatalysts via PEMFC-Based Hydrogen-Pump Measurements. *J. Electrochem. Soc.* **2021**, *168*, 064516. [CrossRef]
4. Harzer, G.S.; Orfanidi, A.; El-Sayed, H.; Madkikar, P.; Gasteiger, H.A. Tailoring Catalyst Morphology towards High Performance for Low Pt Loaded PEMFC Cathodes. *J. Electrochem. Soc.* **2018**, *165*, F770–F779. [CrossRef]
5. Mensharapov, R.M.; Spasov, D.D.; Ivanova, N.A.; Zasypkina, A.A.; Smirnov, S.A.; Grigoriev, S.A. Screening of Carbon-Supported Platinum Electrocatalysts Using Frumkin Adsorption Isotherms. *Inorganics* **2023**, *11*, 103. [CrossRef]
6. Spasov, D.D.; Ivanova, N.A.; Pushkarev, A.S.; Pushkareva, I.V.; Presnyakova, N.N.; Chumakov, R.G.; Presnyakov, M.Y.; Grigoriev, S.A.; Fateev, V.N. On the Influence of Composition and Structure of Carbon-Supported Pt-SnO2 Hetero-Clusters onto Their Electrocatalytic Activity and Durability in PEMFC. *Catalysts* **2019**, *9*, 803. [CrossRef]
7. Spasov, D.D.; Ivanova, N.A.; Mensharapov, R.M.; Zasypkina, A.A.; Seregina, E.A.; Grigoriev, S.A.; Fateev, V.N. Nanostructured Pt20/SiO_2^x/C Electrocatalysts for Water-Balance Stabilization in a Proton Exchange Membrane Fuel Cell. *Nanotechnologies Russ.* **2022**, *17*, 320–327. [CrossRef]
8. Ivanova, N.A.; Alekseeva, O.K.; Fateev, V.N.; Shapir, B.L.; Spasov, D.D.; Nikitin, S.M.; Presnyakov, M.Y.; Kolobylina, N.N.; Soloviev, M.A.; Mikhalev, A.I. Activity and Durability of Electrocatalytic Layers with Low Platinum Loading Prepared by Magnetron Sputtering onto Gas Diffusion Electrodes. *Int. J. Hydrogen Energy* **2019**, *44*, 29529–29536. [CrossRef]
9. Fedotov, A.A.; Grigoriev, S.A.; Millet, P.; Fateev, V.N. Plasma-Assisted Pt and Pt–Pd Nano-Particles Deposition on Carbon Carriers for Application in PEM Electrochemical Cells. *Int. J. Hydrogen Energy* **2013**, *38*, 8568–8574. [CrossRef]
10. Cai, Y.; Wu, D.; Sun, J.; Chen, B. The Effect of Cathode Channel Blockages on the Enhanced Mass Transfer and Performance of PEMFC. *Energy* **2021**, *222*, 119951. [CrossRef]
11. Jiao, D.; Jiao, K.; Zhong, S.; Du, Q. Investigations on Heat and Mass Transfer in Gas Diffusion Layers of PEMFC with a Gas–Liquid-Solid Coupled Model. *Appl. Energy* **2022**, *316*, 118996. [CrossRef]
12. Jeng, K.T.; Lee, S.F.; Tsai, G.F.; Wang, C.H. Oxygen Mass Transfer in PEM Fuel Cell Gas Diffusion Layers. *J. Power Sources* **2004**, *138*, 41–50. [CrossRef]
13. Mensharapov, R.; Ivanova, N.; Spasov, D.; Grigoriev, S.; Fateev, V. SAXS Investigation of the Effect of Freeze/Thaw Cycles on the Nanostructure of Nafion® Membranes. *Polymers* **2022**, *14*, 4395. [CrossRef] [PubMed]
14. Ivanova, N.A.; Spasov, D.D.; Grigoriev, S.A.; Kamyshinsky, R.A.; Peters, G.S.; Mensharapov, R.M.; Seregina, E.A.; Millet, P.; Fateev, V.N. On the Influence of Methanol Addition on the Performances of PEM Fuel Cells Operated at Subzero Temperatures. *Int. J. Hydrogen Energy* **2021**, *46*, 18116–18127. [CrossRef]
15. Rodgers, M.P.; Bonville, L.J.; Kunz, H.R.; Slattery, D.K.; Fenton, J.M. Fuel Cell Perfluorinated Sulfonic Acid Membrane Degradation Correlating Accelerated Stress Testing and Lifetime. *Chem. Rev.* **2012**, *112*, 6075–6103. [CrossRef] [PubMed]
16. Ren, P.; Pei, P.; Li, Y.; Wu, Z.; Chen, D.; Huang, S. Degradation Mechanisms of Proton Exchange Membrane Fuel Cell under Typical Automotive Operating Conditions. *Prog. Energy Combust. Sci.* **2020**, *80*, 100859. [CrossRef]
17. Jahnke, T.; Futter, G.; Latz, A.; Malkow, T.; Papakonstantinou, G.; Tsotridis, G.; Schott, P.; Gérard, M.; Quinaud, M.; Quiroga, M.; et al. Performance and Degradation of Proton Exchange Membrane Fuel Cells: State of the Art in Modeling from Atomistic to System Scale. *J. Power Sources* **2016**, *304*, 207–233. [CrossRef]
18. Kregar, A.; Kravos, A.; Katrašnik, T. Methodology for Evaluation of Contributions of Ostwald Ripening and Particle Agglomeration to Growth of Catalyst Particles in PEM Fuel Cells. *Fuel Cells* **2020**, *20*, 487–498. [CrossRef]
19. Fan, L.; Zhao, J.; Luo, X.; Tu, Z. Comparison of the Performance and Degradation Mechanism of PEMFC with Pt/C and Pt Black Catalyst. *Int. J. Hydrogen Energy* **2022**, *47*, 5418–5428. [CrossRef]
20. Urchaga, P.; Kadyk, T.; Rinaldo, S.G.; Pistono, A.O.; Hu, J.; Lee, W.; Richards, C.; Eikerling, M.H.; Rice, C.A. Catalyst Degradation in Fuel Cell Electrodes: Accelerated Stress Tests and Model-Based Analysis. *Electrochim. Acta* **2015**, *176*, 1500–1510. [CrossRef]
21. Li, C.; Zhu, L.; Song, L.; Zhu, G.; Zhang, Q.; Zhao, Y.; Gong, Q.; Sun, C.; Liu, Y.; Sasaki, K.; et al. Synergistic effect of polyaniline on stabilizing Pt nanoparticles in PEMFCs. *J. Mater. Chem. A* **2023**, *11*, 7756–7766. [CrossRef]
22. Meier, J.C.; Galeano, C.; Katsounaros, I.; Witte, J.; Bongard, H.J.; Topalov, A.A.; Baldizzone, C.; Mezzavilla, S.; Schüth, F.; Mayrhofer, K.J.J. Design Criteria for Stable Pt/C Fuel Cell Catalysts. *Beilstein J. Nanotechnol.* **2014**, *5*, 44–67. [CrossRef] [PubMed]

23. Liu, F.; Gao, Z.; Su, J.; Guo, L. An Experimental Investigation of the Effect of Platinum on the Corrosion of Cathode Carbon Support in a PEMFC. *ChemSusChem* **2022**, *12*, e202102726. [CrossRef] [PubMed]
24. Liu, P.; Yang, D.; Li, B.; Qu, T.; Ming, P.; Zhang, C.; Pan, X. Agglomeration behavior of carbon-supported platinum nanoparticles in catalyst ink: Modeling and experimental investigation. *J. Power Sources* **2024**, *602*, 234309. [CrossRef]
25. Ettingshausen, F.; Kleemann, J.; Marcu, A.; Toth, G.; Fuess, H.; Roth, C. Dissolution and Migration of Platinum in PEMFCs Investigated for Start/Stop Cycling and High Potential Degradation. *Fuel Cells* **2011**, *11*, 238–245. [CrossRef]
26. Pushkarev, A.S.; Pushkareva, I.V.; Ivanova, N.A.; du Preez, S.P.; Bessarabov, D.; Chumakov, R.G.; Stankevich, V.G.; Fateev, V.N.; Evdokimov, A.A.; Grigoriev, S.A. Pt/C and Pt/SnOx/C Catalysts for Ethanol Electrooxidation: Rotating Disk Electrode Study. *Catalysts* **2019**, *9*, 271. [CrossRef]
27. Das, B.; Das, S.; Tewary, S.; Bose, S.; Ghosh, S.; Ghosh, A. Graphene Nano Sheets for the Fuel Cell Applications. In *Advances in Nanosheets [Working Title]*; IntechOpen: London, UK, 2023.
28. Voznyakovskii, A.; Vozniakovskii, A.; Kidalov, S. New Way of Synthesis of Few-Layer Graphene Nanosheets by the Self Propagating High-Temperature Synthesis Method from Biopolymers. *Nanomaterials* **2022**, *12*, 657. [CrossRef] [PubMed]
29. Kozlova, M.; Butrim, S.; Solovyev, M.; Pushkarev, A.; Pushkareva, I.; Kalinichenko, V.; Akelkina, S.; Grigoriev, S. Structural and Electrochemical Characteristics of Platinum Nanoparticles Supported on Various Carbon Carriers. *C* **2022**, *8*, 14. [CrossRef]
30. Wang, K.; Zhou, T.; Cao, Z.; Yuan, Z.; He, H.; Fan, M.; Jiang, Z. Advanced 3D Ordered Electrodes for PEMFC Applications: From Structural Features and Fabrication Methods to the Controllable Design of Catalyst Layers. *Green Energy Environ.* **2023**, *in press*. S2468025723001462. [CrossRef]
31. Zasypkina, A.A.; Ivanova, N.A.; Spasov, D.D.; Mensharapov, R.M.; Alekseeva, O.K.; Vorobyeva, E.A.; Kukueva, E.V.; Fateev, V.N. Electrode with a Carbon Nanotube Array for a Proton Exchange Membrane Fuel Cell. *Inorganics* **2023**, *11*, 219. [CrossRef]
32. Verma, V.; Choudhury, S.R.; Rathour, V.; Choudhury, S.R.; Ganesan, V. Hollow Core Mesoporous Carbon Spheres as Catalyst Support for Improved Platinum Utilization in Phosphoric Acid Fuel Cells. *Microporous Mesoporous Mater.* **2024**, *367*, 113005. [CrossRef]
33. Mamat, M.S.; Grigoriev, S.A.; Dzhus, K.A.; Grant, D.M.; Walker, G.S. The Performance and Degradation of Pt Electrocatalysts on Novel Carbon Carriers for PEMFC Applications. *Int. J. Hydrogen Energy* **2010**, *35*, 7580–7587. [CrossRef]
34. Luo, C.; Xie, H.; Wang, Q.; Luo, G.; Liu, C. A Review of the Application and Performance of Carbon Nanotubes in Fuel Cells. *J. Nanomater.* **2015**, *2015*, 560392. [CrossRef]
35. Ortiz-Herrera, J.C.; Cruz-Martínez, H.; Solorza-Feria, O.; Medina, D.I. Recent Progress in Carbon Nanotubes Support Materials for Pt-Based Cathode Catalysts in PEM Fuel Cells. *Int. J. Hydrogen Energy* **2022**, *47*, 30213–30224. [CrossRef]
36. Su, H.; Hu, Y.H. Recent Advances in Graphene-based Materials for Fuel Cell Applications. *Energy Sci. Eng.* **2021**, *9*, 958–983. [CrossRef]
37. Sharma, S.; Pollet, B.G. Support Materials for PEMFC and DMFC Electrocatalysts—A Review. *J. Power Sources* **2012**, *208*, 96–119. [CrossRef]
38. Shaari, N.; Kamarudin, S.K. Graphene in Electrocatalyst and Proton Conductiong Membrane in Fuel Cell Applications: An Overview. *Renew. Sustain. Energy Rev.* **2017**, *69*, 862–870. [CrossRef]
39. Pushkareva, I.V.; Pushkarev, A.S.; Kalinichenko, V.N.; Chumakov, R.G.; Soloviev, M.A.; Liang, Y.; Millet, P.; Grigoriev, S.A. Reduced Graphene Oxide-Supported Pt-Based Catalysts for PEM Fuel Cells with Enhanced Activity and Stability. *Catalysts* **2021**, *11*, 256. [CrossRef]
40. Öztürk, A.; Bayrakçeken Yurtcan, A. Raw and Pyrolyzed (with and without Melamine) Graphene Nanoplatelets with Different Surface Areas as PEM Fuel Cell Catalyst Supports. *Carbon Lett.* **2021**, *31*, 1191–1214. [CrossRef]
41. Lagarteira, T.; Delgado, S.; Fernandes, C.; Azenha, C.; Mateos-Pedrero, C.; Mendes, A. The Role of Pt Loading on Reduced Graphene Oxide Support in the Polyol Synthesis of Catalysts for Oxygen Reduction Reaction. *Int. J. Hydrogen Energy* **2020**, *45*, 20594–20604. [CrossRef]
42. Smirnov, S.A.; Mensharapov, R.M.; Spasov, D.D.; Ivanova, N.A.; Grigoriev, S.A. Adsorption and Recombination of H+ and H3O+ on Graphene-Supported Pt1, Pt13, and Pt14 Nanoclusters: A First Principles Study. *Catalysts* **2024**, *14*, 108. [CrossRef]
43. Choi, S.M.; Seo, M.H.; Kim, H.J.; Kim, W.B. Synthesis of Surface-Functionalized Graphene Nanosheets with High Pt-Loadings and Their Applications to Methanol Electrooxidation. *Carbon* **2011**, *49*, 904–909. [CrossRef]
44. Tran, T.; Song, M.Y.; Kang, T.; Samdani, J.; Park, H.; Kim, H.; Jhung, S.H.; Yu, J. Iron Phosphide Incorporated into Iron-Treated Heteroatoms-Doped Porous Bio-Carbon as Efficient Electrocatalyst for the Oxygen Reduction Reaction. *ChemElectroChem* **2018**, *5*, 1944–1953. [CrossRef]
45. Lee, F.C.; Ismail, M.S.; Zhang, K.; Ingham, D.B.; Aldakheel, F.; Hughes, K.J.; Ma, L.; El-Kharouf, A.; Pourkashanian, M. Optimisation and Characterisation of Graphene-Based Microporous Layers for Polymer Electrolyte Membrane Fuel Cells. *Int. J. Hydrogen Energy* **2024**, *51*, 1311–1325. [CrossRef]
46. Chen, J.; Bailey, J.J.; Britnell, L.; Perez-Page, M.; Sahoo, M.; Zhang, Z.; Strudwick, A.; Hack, J.; Guo, Z.; Ji, Z.; et al. The Performance and Durability of High-Temperature Proton Exchange Membrane Fuel Cells Enhanced by Single-Layer Graphene. *Nano Energy* **2022**, *93*, 106829. [CrossRef]
47. Tanuma, T.; Kawamoto, M.; Kinoshita, S. Effect of Properties of Hydrophilic Microporous Layer (MPL) on PEFC Performance. *J. Electrochem. Soc.* **2017**, *164*, F499–F503. [CrossRef]

48. Ozden, A.; Shahgaldi, S.; Zhao, J.; Li, X.; Hamdullahpur, F. Assessment of Graphene as an Alternative Microporous Layer Material for Proton Exchange Membrane Fuel Cells. *Fuel* **2018**, *215*, 726–734. [CrossRef]
49. Fang, X.-Y.; Yu, X.-X.; Zheng, H.-M.; Jin, H.-B.; Wang, L.; Cao, M.-S. Temperature- and Thickness-Dependent Electrical Conductivity of Few-Layer Graphene and Graphene Nanosheets. *Phys. Lett. A* **2015**, *379*, 2245–2251. [CrossRef]
50. Lagarteira, T.; Delgado, S.; Garcia, G.P.; Ortiz, A.; Mendes, A. Oxygen Reduction Stability of Graphene-Supported Electrocatalyst: Electrochemical and Morphological Evidences. *Int. J. Hydrogen Energy* **2023**, *48*, 20901–20913. [CrossRef]
51. Daş, E.; Alkan Gürsel, S.; Bayrakçeken Yurtcan, A. Pt-Alloy Decorated Graphene as an Efficient Electrocatalyst for PEM Fuel Cell Reactions. *J. Supercrit. Fluids* **2020**, *165*, 104962. [CrossRef]
52. Liu, Y.; Li, Z.; Xu, S.; Xie, Y.; Ye, Y.; Zou, X.; Lin, S. Synthesis of Pt-Ni (Trace)/GNs Composite and Its Bi-Functional Electrocatalytic Properties for MOR and ORR. *J. Colloid Interface Sci.* **2019**, *554*, 640–649. [CrossRef] [PubMed]
53. Huo, L.; Liu, B.; Zhang, G.; Si, R.; Liu, J.; Zhang, J. 2D Layered Non-Precious Metal Mesoporous Electrocatalysts for Enhanced Oxygen Reduction Reaction. *J. Mater. Chem. A* **2017**, *5*, 4868–4878. [CrossRef]
54. Daş, E.; Gürsel, S.A.; Yurtcan, A.B. Simultaneously Deposited Pt-Alloy Nanoparticles over Graphene Nanoplatelets via Supercritical Carbon Dioxide Deposition for PEM Fuel Cells. *J. Alloys Compd.* **2021**, *874*, 159919. [CrossRef]
55. Boyaci San, F.G.; Dursun, S.; Yazici, M.S. PtCo on Continuous-phase Graphene as PEM Fuel Cell Catalyst. *Int. J. Energy Res.* **2021**, *45*, 1673–1684. [CrossRef]
56. Dursun, S.; Akay, R.G.; Yazici, M.S. CVD Graphene Supported Cobalt (II) Phthalocyanine as Cathode Electrocatalyst for PEM Fuel Cells. *Int. J. Hydrogen Energy* **2020**, *45*, 34837–34844. [CrossRef]
57. Chen, D.; Feng, H.; Li, J. Graphene Oxide: Preparation, Functionalization, and Electrochemical Applications. *Chem. Rev.* **2012**, *112*, 6027–6053. [CrossRef]
58. Hummers, W.S.; Offeman, R.E. Preparation of Graphitic Oxide. *J. Am. Chem. Soc.* **1958**, *80*, 1339. [CrossRef]
59. Marcano, D.C.; Kosynkin, D.V.; Berlin, J.M.; Sinitskii, A.; Sun, Z.; Slesarev, A.; Alemany, L.B.; Lu, W.; Tour, J.M. Improved Synthesis of Graphene Oxide. *ACS Nano* **2010**, *4*, 4806–4814. [CrossRef] [PubMed]
60. Grigoriev, S.A.; Fateev, V.N.; Pushkarev, A.S.; Pushkareva, I.V.; Ivanova, N.A.; Kalinichenko, V.N.; Yu Presnyakov, M.; Wei, X. Reduced Graphene Oxide and Its Modifications as Catalyst Supports and Catalyst Layer Modifiers for PEMFC. *Materials* **2018**, *11*, 1405. [CrossRef]
61. Oubraham, A.; Ion-Ebrasu, D.; Vasut, F.; Soare, A.; Sorlei, I.-S.; Marinoiu, A. Platinum-Functionalized Graphene Oxide: One-Pot Synthesis and Application as an Electrocatalyst. *Materials* **2023**, *16*, 1897. [CrossRef]
62. Marinoiu, A.; Carcadea, E.; Sacca, A.; Carbone, A.; Sisu, C.; Dogaru, A.; Raceanu, M.; Varlam, M. One-Step Synthesis of Graphene Supported Platinum Nanoparticles as Electrocatalyst for PEM Fuel Cells. *Int. J. Hydrogen Energy* **2021**, *46*, 12242–12253. [CrossRef]
63. Lazar, O.-A.; Marinoiu, A.; Raceanu, M.; Pantazi, A.; Mihai, G.; Varlam, M.; Enachescu, M. Reduced Graphene Oxide Decorated with Dispersed Gold Nanoparticles: Preparation, Characterization and Electrochemical Evaluation for Oxygen Reduction Reaction. *Energies* **2020**, *13*, 4307. [CrossRef]
64. Fu, X.; Wang, J.; Peng, F.; Wang, Y.; Hu, S.; Zhang, R.; Liu, Q. A Hybrid PEMFC/Supercapacitor Device with High Energy and Power Densities Based on Reduced Graphene Oxide/Nafion/Pt Electrode. *Int. J. Hydrogen Energy* **2023**, *48*, 16072–16082. [CrossRef]
65. Moradizadeh, L.; Yazdanpanah, P.; Karimi, G.; Paydar, M.H. Investigating the Role of Graphite and Reduced Graphene Oxide in the Fabrication of Microporous Layers for Proton Exchange Membrane Fuel Cells. *J. Mater. Sci.* **2023**, *58*, 12706–12723. [CrossRef]
66. Samancı, M.; Bayrakçeken Yurtcan, A. Chemically and Thermally Reduced Graphene Oxide Supported Pt Catalysts Prepared by Supercritical Deposition. *Int. J. Hydrogen Energy* **2022**, *47*, 19669–19689. [CrossRef]
67. Romero, A.; Lavin-Lopez, M.P.; Sanchez-Silva, L.; Valverde, J.L.; Paton-Carrero, A. Comparative Study of Different Scalable Routes to Synthesize Graphene Oxide and Reduced Graphene Oxide. *Mater. Chem. Phys.* **2018**, *203*, 284–292. [CrossRef]
68. Gomes, W.D.S.; Noce, R.D.; De Matos, T.D.S.; Andrade, F.V.; De Molfetta, F.A.; Iúdice De Souza, J.P. Cu@PtRu Core-Shell Nanostructured Electrocatalysts Anchored on Reduced Graphene Oxide Towards Methanol Oxidation. *Chem. Mater. Sci.* **2023**, *16*, 6508. [CrossRef]
69. Işıkel Şanlı, L.; Bayram, V.; Ghobadi, S.; Düzen, N.; Alkan Gürsel, S. Engineered Catalyst Layer Design with Graphene-Carbon Black Hybrid Supports for Enhanced Platinum Utilization in PEM Fuel Cell. *Int. J. Hydrogen Energy* **2017**, *42*, 1085–1092. [CrossRef]
70. Danafar, F.; Fakhru'l-Razi, A.; Salleh, M.A.M.; Biak, D.R.A. Fluidized Bed Catalytic Chemical Vapor Deposition Synthesis of Carbon Nanotubes—A Review. *Chem. Eng. J.* **2009**, *155*, 37–48. [CrossRef]
71. Esteves, L.M.; Oliveira, H.A.; Passos, F.B. Carbon Nanotubes as Catalyst Support in Chemical Vapor Deposition Reaction: A Review. *J. Ind. Eng. Chem.* **2018**, *65*, 1–12. [CrossRef]
72. Nezhad, E.Z.; Qu, X.; Musharavati, F.; Jaber, F.; Appleford, M.R.; Bae, S.; Uzun, K.; Struthers, M.; Chowdhury, M.E.H.; Khandakar, A. Effects of titanium and carbon nanotubes on nano/micromechanical properties of HA/TNT/CNT nanocomposites. *Appl. Surf. Sci.* **2021**, *538*, 148123. [CrossRef]
73. Devrim, Y.; Arıca, E.D. Multi-Walled Carbon Nanotubes Decorated by Platinum Catalyst for High Temperature PEM Fuel Cell. *Int. J. Hydrogen Energy* **2019**, *44*, 18951–18966. [CrossRef]
74. Pajootan, E.; Omanovic, S.; Coulombe, S. Controllable Dry Synthesis of Binder-Free Nanostructured Platinum Electrocatalysts Supported on Multi-Walled Carbon Nanotubes and Their Performance in the Oxygen Reduction Reaction. *Chem. Eng. J.* **2021**, *426*, 131706. [CrossRef]

75. Kim, T.; Kwon, Y.; Kwon, S.; Seo, J.G. Substrate Effect of Platinum-Decorated Carbon on Enhanced Hydrogen Oxidation in PEMFC. *ACS Omega* **2020**, *5*, 26902–26907. [CrossRef] [PubMed]
76. Aruna, I.; Mehta, B.R.; Malhotra, L.K.; Shivaprasad, S.M. Size Dependence of Core and Valence Binding Energies in Pd Nanoparticles: Interplay of Quantum Confinement and Coordination Reduction. *J. Appl. Phys.* **2008**, *104*, 064308. [CrossRef]
77. Uddin, A.S.M.I.; Yaqoob, U.; Hassan, K.; Chung, G.-S. Effects of Pt Shell Thickness on Self-Assembly Monolayer Pd@Pt Core-Shell Nanocrystals Based Hydrogen Sensing. *Int. J. Hydrogen Energy* **2016**, *41*, 15399–15410. [CrossRef]
78. Marbaniang, P.; Ingavale, S.; Karuppanan, P.; Swami, A.; Kakade, B. Rationale Approach of Nitrogen Doping at Defect Sites of Multiwalled Carbon Nanotubes: A Strategy for Oxygen Reduction Electrocatalysis. *Int. J. Hydrogen Energy* **2021**, *46*, 10268–10280. [CrossRef]
79. Devrim, Y.; Arıca, E.D. Investigation of the Effect of Graphitized Carbon Nanotube Catalyst Support for High Temperature PEM Fuel Cells. *Int. J. Hydrogen Energy* **2020**, *45*, 3609–3617. [CrossRef]
80. Li, B.; Wang, J.; Yuan, Y.; Ariga, H.; Takakusagi, S.; Asakura, K. Carbon Nanotube-Supported RuFe Bimetallic Nanoparticles as Efficient and Robust Catalysts for Aqueous-Phase Selective Hydrogenolysis of Glycerol to Glycols. *ACS Catal.* **2011**, *1*, 1521–1528. [CrossRef]
81. Haque, M.A.; Sulong, A.B.; Majlan, E.H.; Loh, K.S.; Husaini, T.; Rosli, R. Oxygen reduction reaction behaviours of carbon nanotubes supporting pt catalyst for proton exchange membrane fuel cell. *MJAS* **2019**, *23*, 147–154. [CrossRef]
82. Ivanova, N.A.; Spasov, D.D.; Zasypkina, A.A.; Alekseeva, O.K.; Kukueva, E.V.; Vorobyeva, E.A.; Kudinova, E.S.; Chumakov, R.G.; Millet, P.; Grigoriev, S.A. Comparison of the Performance and Durability of PEM Fuel Cells with Different Pt-Activated Microporous Layers. *Int. J. Hydrogen Energy* **2021**, *46*, 18093–18106. [CrossRef]
83. Gribov, E.N.; Kuznetsov, A.N.; Golovin, V.A.; Krasnikov, D.V.; Kuznetsov, V.L. Effect of Modification of Multi-Walled Carbon Nanotubes with Nitrogen-Containing Polymers on the Electrochemical Performance of Pt/CNT Catalysts in PEMFC. *Mater. Renew. Sustain. Energy* **2019**, *8*, 7. [CrossRef]
84. Huda, M.; Kawahara, T.; Park, J.-H.; Kawasumi, M.; Matsuo, Y. Single-Walled Carbon Nanotubes Supported Pt Electrocatalyst as a Cathode Catalyst of a Single Fuel Cell with High Durability against Start-up/Shut-down Potential Cycling. *ACS Appl. Energy Mater.* **2023**, *6*, 12226–12236. [CrossRef]
85. Pandia, R.M.; Berchmans, S. Electrochemically Activated Platinum Nanoparticles Trapped by Carbon Nanotubes for Methanol Oxidation Reaction. *J. Electroanal. Chem.* **2023**, *940*, 117492. [CrossRef]
86. Yao, Y.; Izumi, R.; Tsuda, T.; Oshima, Y.; Imanishi, A.; Oda, N.; Kuwabata, S. Platinum and PtNi Nanoparticle-Supported Multiwalled Carbon Nanotube Electrocatalysts Prepared by One-Pot Pyrolytic Synthesis with an Ionic Liquid. *ACS Appl. Energy Mater.* **2019**, *2*, 4865–4872. [CrossRef]
87. Mensharapov, R.M.; Ivanova, N.A.; Zasypkina, A.A.; Spasov, D.D.; Sinyakov, M.V.; Grigoriev, S.A.; Fateev, V.N. Model Study of CNT-Based PEMFCs' Electrocatalytic Layers. *Catalysts* **2022**, *12*, 1227. [CrossRef]
88. Kanninen, P.; Eriksson, B.; Davodi, F.; Buan, M.E.M.; Sorsa, O.; Kallio, T.; Lindström, R.W. Carbon Corrosion Properties and Performance of Multi-Walled Carbon Nanotube Support with and without Nitrogen-Functionalization in Fuel Cell Electrodes. *Electrochim. Acta* **2020**, *332*, 135384. [CrossRef]
89. Kim, J.; Kim, H.; Song, H.; Kim, D.; Kim, G.H.; Im, D.; Jeong, Y.; Park, T. Carbon Nanotube Sheet as a Microporous Layer for Proton Exchange Membrane Fuel Cells. *Energy* **2021**, *227*, 120459. [CrossRef]
90. Zhang, X.-F.; Liu, Y.-T.; Song, H.; Yao, T.-T.; Liu, Q.; Wu, G.-P. Single-Walled Carbon Nanotube Interlayer Modified Gas Diffusion Layers to Boost the Cell Performance of Self-Humidifying Proton Exchange Membrane Fuel Cells. *Int. J. Hydrogen Energy* **2023**, *48*, 30899–30908. [CrossRef]
91. Xi, J.; Meng, K.; Li, Y.; Wang, M.; Liao, Q.; Wei, Z.; Shao, M.; Wang, J. Performance Improvement of Ultra-Low Pt Proton Exchange Membrane Fuel Cell by Catalyst Layer Structure Optimization. *Chin. J. Chem. Eng.* **2022**, *41*, 473–479. [CrossRef]
92. Lee, J.H.; Kim, H.-S.; Yun, E.-T.; Ham, S.-Y.; Park, J.-H.; Ahn, C.H.; Lee, S.H.; Park, H.-D. Vertically Aligned Carbon Nanotube Membranes: Water Purification and Beyond. *Membranes* **2020**, *10*, 273. [CrossRef]
93. Zanin, H.; May, P.W.; Harniman, R.L.; Risbridger, T.; Corat, E.J.; Fermin, D.J. High Surface Area Diamond-like Carbon Electrodes Grown on Vertically Aligned Carbon Nanotubes. *Carbon* **2015**, *82*, 288–296. [CrossRef]
94. Cho, W.; Schulz, M.; Shanov, V. Growth and Characterization of Vertically Aligned Centimeter Long CNT Arrays. *Carbon* **2014**, *72*, 264–273. [CrossRef]
95. Kudinova, E.S.; Vorobyeva, E.A.; Ivanova, N.A.; Tishkin, V.V.; Alekseeva, O.K. A Magnetron Sputtering Method for the Application of the Ni Catalyst for the Synthesis Process of Carbon Nanotube Arrays. *Nanotechnologies Russ.* **2020**, *15*, 715–722. [CrossRef]
96. Ageev, O.A.; Blinov, Y.F.; Il'ina, M.V.; Il'in, O.I.; Smirnov, V.A.; Tsukanova, O.G. Study of Adhesion of Vertically Aligned Carbon Nanotubes to a Substrate by Atomic-Force Microscopy. *Phys. Solid State* **2016**, *58*, 309–314. [CrossRef]
97. Meng, Q.H.; Hao, C.; Yan, B.; Yang, B.; Liu, J.; Shen, P.K.; Tian, Z.Q. High-Performance Proton Exchange Membrane Fuel Cell with Ultra-Low Loading Pt on Vertically Aligned Carbon Nanotubes as Integrated Catalyst Layer. *J. Energy Chem.* **2022**, *71*, 497–506. [CrossRef]
98. Meng, X.; Deng, X.; Zhou, L.; Hu, B.; Tan, W.; Zhou, W.; Liu, M.; Shao, Z. A Highly Ordered Hydrophilic–Hydrophobic Janus Bi-Functional Layer with Ultralow Pt Loading and Fast Gas/Water Transport for Fuel Cells. *Energy Env. Mater.* **2021**, *4*, 126–133. [CrossRef]

99. Lu, C.; Shi, F.; Jin, J.; Peng, X. Study on the Properties of Vertical Carbon Nanotube Films Grown on Stainless Steel Bipolar Plates. *Materials* **2019**, *12*, 899. [CrossRef]
100. Murata, S.; Imanishi, M.; Hasegawa, S.; Namba, R. Vertically Aligned Carbon Nanotube Electrodes for High Current Density Operating Proton Exchange Membrane Fuel Cells. *J. Power Sources* **2014**, *253*, 104–113. [CrossRef]
101. Fontana, M.; Ramos, R.; Morin, A.; Dijon, J. Direct Growth of Carbon Nanotubes Forests on Carbon Fibers to Replace Microporous Layers in Proton Exchange Membrane Fuel Cells. *Carbon* **2021**, *172*, 762–771. [CrossRef]
102. Valdés-Madrigal, M.A.; Montejo-Alvaro, F.; Cernas-Ruiz, A.S.; Rojas-Chávez, H.; Román-Doval, R.; Cruz-Martinez, H.; Medina, D.I. Role of Defect Engineering and Surface Functionalization in the Design of Carbon Nanotube-Based Nitrogen Oxide Sensors. *IJMS* **2021**, *22*, 12968. [CrossRef] [PubMed]
103. Zheng, J.-S.; Zhang, X.-S.; Li, P.; Zhu, J.; Zhou, X.-G.; Yuan, W.-K. Effect of Carbon Nanofiber Microstructure on Oxygen Reduction Activity of Supported Palladium Electrocatalyst. *Electrochem. Commun.* **2007**, *9*, 895–900. [CrossRef]
104. Lavagna, L.; Nisticò, R.; Musso, S.; Pavese, M. Functionalization as a Way to Enhance Dispersion of Carbon Nanotubes in Matrices: A Review. *Mater. Today Chem.* **2021**, *20*, 100477. [CrossRef]
105. Balasubramanian, K.; Burghard, M. Chemically Functionalized Carbon Nanotubes. *Small* **2005**, *1*, 180–192. [CrossRef] [PubMed]
106. Banerjee, S.; Kahn, M.G.C.; Wong, S.S. Rational Chemical Strategies for Carbon Nanotube Functionalization. *Chem. A Eur. J* **2003**, *9*, 1898–1908. [CrossRef] [PubMed]
107. Samad, S.; Loh, K.S.; Wong, W.Y.; Lee, T.K.; Sunarso, J.; Chong, S.T.; Wan Daud, W.R. Carbon and Non-Carbon Support Materials for Platinum-Based Catalysts in Fuel Cells. *Int. J. Hydrogen Energy* **2018**, *43*, 7823–7854. [CrossRef]
108. Meng, L.; Fu, C.; Lu, Q. Advanced Technology for Functionalization of Carbon Nanotubes. *Prog. Nat. Sci.* **2009**, *19*, 801–810. [CrossRef]
109. Hernández-Fernández, P.; Montiel, M.; Ocón, P.; De La Fuente, J.L.G.; García-Rodríguez, S.; Rojas, S.; Fierro, J.L.G. Functionalization of Multi-Walled Carbon Nanotubes and Application as Supports for Electrocatalysts in Proton-Exchange Membrane Fuel Cell. *Appl. Catal. B Environ.* **2010**, *99*, 343–352. [CrossRef]
110. Chen, S.; Wei, Z.; Guo, L.; Ding, W.; Dong, L.; Shen, P.; Qi, X.; Li, L. Enhanced Dispersion and Durability of Pt Nanoparticles on a Thiolated CNT Support. *Chem. Commun.* **2011**, *47*, 10984. [CrossRef]
111. Shaheen Shah, S.; Abu Nayem, S.M.; Sultana, N.; Saleh Ahammad, A.J.; Aziz, M.A. Preparation of Sulfur-doped Carbon for Supercapacitor Applications: A Review. *ChemSusChem* **2022**, *15*, e202101282. [CrossRef]
112. Shirzadi Jahromi, H.; Saxena, S.; Sridhar, S.; Ghantasala, M.K.; Guda, R.; Rozhkova, E.A. Development of Nickel-ZIF-8 Doped Nitrogen Reduced Graphene Oxide Catalytic Materials for PEM Fuel Cell. In Proceedings of the ASME International Mechanical Engineering Congress and Exposition, New Orleans, LA, USA, 29 October 2023; Energy. American Society of Mechanical Engineers: New York, NY, USA, 2023; Volume 7, p. V007T08A045.
113. Bhaskaran, R.; Chetty, R. One-Pot Room Temperature Synthesis of Nitrogen-Doped Graphene and Its Application as Catalyst Support for ORR in PEMFCs. *ACS Appl. Energy Mater.* **2024**, *7*, 390–402. [CrossRef]
114. Lin, Y.-J.; Tsai, J.-E.; Huang, C.-C.; Chen, Y.-S.; Li, Y.-Y. Highly Porous Iron-Doped Nitrogen–Carbon Framework on Reduced Graphene Oxide as an Excellent Oxygen Reduction Catalyst for Proton-Exchange Membrane Fuel Cells. *ACS Appl. Energy Mater.* **2022**, *5*, 1822–1832. [CrossRef]
115. Nair, A.S.; Jafri, R.I. A Facile One-Step Microwave Synthesis of Pt Deposited on N & P Co-Doped Graphene Intercalated Carbon Black—An Efficient Cathode Electrocatalyst for PEM Fuel Cell. *Int. J. Hydrogen Energy* **2023**, *48*, 3653–3664. [CrossRef]
116. Han, C.; Chen, Z. Study on the Synergism of Thermal Transport and Electrochemical of PEMFC Based on N, P Co-Doped Graphene Substrate Electrode. *Energy* **2021**, *214*, 118808. [CrossRef]
117. Peera, S.G.; Menon, R.S.; Das, S.K.; Alfantazi, A.; Karuppasamy, K.; Liu, C.; Sahu, A.K. Oxygen Reduction Electrochemistry at F Doped Carbons: A Review on the Effect of Highly Polarized C-F Bonding in Catalysis and Stability of Fuel Cell Catalysts. *Coord. Chem. Rev.* **2024**, *500*, 215491. [CrossRef]
118. Spasov, D.D.; Ivanova, N.A.; Mensharapov, R.M.; Sinyakov, M.V.; Zasypkina, A.A.; Kukueva, E.V.; Trigub, A.L.; Kulikova, E.S.; Fateev, V.N. Study of the Cathode Pt-Electrocatalysts Based on Reduced Graphene Oxide with Pt-SnO2 Hetero-Clusters. *Inorganics* **2023**, *11*, 325. [CrossRef]
119. Ivanova, N.A.; Spasov, D.D.; Mensharapov, R.M.; Kukueva, E.V.; Zasypkina, A.A.; Fateev, V.N.; Grigoriev, S.A. Efficient and Stable Subzero Operation of a PEM Fuel Cell with a Composite Anode Using Hydrogen-Methanol Composition during Freeze/Thaw Cycles. *Int. J. Hydrogen Energy* **2023**, *48*, 11410–11420. [CrossRef]
120. Wang, R.; Li, X.; Li, H.; Wang, Q.; Wang, H.; Wang, W.; Kang, J.; Chang, Y.; Lei, Z. Highly Stable and Effective Pt/Carbon Nitride (CNx) Modified SiO2 Electrocatalyst for Oxygen Reduction Reaction. *Int. J. Hydrogen Energy* **2011**, *36*, 5775–5781. [CrossRef]
121. Ramírez-Palma, M.T.; Hernández-Padrón, G.; Gómez, J.M.; Rojas-González, F.; Castaño, V.M. NANOSTRUCTURED EPOXY-BASED ANTICORROSIVE COATINGS. *Surf. Rev. Lett.* **2020**, *27*, 1950202. [CrossRef]
122. Chowdury, M.S.K.; Cho, Y.J.; Park, S.B.; Lee, M.H.; Park, Y. Pt-Free Graphene Oxide-Hydrogen Membrane Fuel Cells (GOHMFCs). *Mater. Today Chem.* **2023**, *34*, 101770. [CrossRef]
123. Marinoiu, A.; Raceanu, M.; Carcadea, E.; Varlam, M. Nitrogen-Doped Graphene Oxide as Efficient Metal-Free Electrocatalyst in PEM Fuel Cells. *Nanomaterials* **2023**, *13*, 1233. [CrossRef] [PubMed]

124. Min, H.; Choi, J.-H.; Kang, H.E.; Kim, D.-J.; Yoon, Y.S. Enhanced Durability and Catalytic Performance of Pt–SnO_2 /Multi-Walled Carbon Nanotube with Shifted d-Band Center for Proton-Exchange Membrane Fuel Cells. *Small Struct.* **2024**, *5*, 2300407. [CrossRef]
125. Hoque, M.A.; Hassan, F.M.; Jauhar, A.M.; Jiang, G.; Pritzker, M.; Choi, J.-Y.; Knights, S.; Ye, S.; Chen, Z. Web-like 3D Architecture of Pt Nanowires and Sulfur-Doped Carbon Nanotube with Superior Electrocatalytic Performance. *ACS Sustain. Chem. Eng.* **2018**, *6*, 93–98. [CrossRef]
126. Lu, L.; Deng, H.; Zhao, Z.; Xu, B.; Sun, X. N-Doped Carbon Nanotubes Supported Pt Nanowire Catalysts for Proton Exchange Membrane Fuel Cells. *J. Power Sources* **2022**, *529*, 231229. [CrossRef]
127. Mardle, P.; Ji, X.; Wu, J.; Guan, S.; Dong, H.; Du, S. Thin Film Electrodes from Pt Nanorods Supported on Aligned N-CNTs for Proton Exchange Membrane Fuel Cells. *Appl. Catal. B Environ.* **2020**, *260*, 118031. [CrossRef]
128. Roudbari, M.N.; Ojani, R.; Raoof, J.B. Nitrogen Functionalized Carbon Nanotubes as a Support of Platinum Electrocatalysts for Performance Improvement of ORR Using Fuel Cell Cathodic Half-Cell. *Renew. Energy* **2020**, *159*, 1015–1028. [CrossRef]
129. Xu, Z.; Zhou, Z.; Li, B.; Wang, G.; Leu, P.W. Identification of Efficient Active Sites in Nitrogen-Doped Carbon Nanotubes for Oxygen Reduction Reaction. *J. Phys. Chem. C.* **2020**, *124*, 8689–8696. [CrossRef]
130. Dogan, D.C.; Cho, S.; Hwang, S.-M.; Kim, Y.-M.; Guim, H.; Yang, T.-H.; Park, S.-H.; Park, G.-G.; Yim, S.-D. Highly Durable Supportless Pt Hollow Spheres Designed for Enhanced Oxygen Transport in Cathode Catalyst Layers of Proton Exchange Membrane Fuel Cells. *ACS Appl. Mater. Interfaces* **2016**, *8*, 27730–27739. [CrossRef]
131. Shu, C.; Tan, Q.; Deng, C.; Du, W.; Gan, Z.; Liu, Y.; Fan, C.; Jin, H.; Tang, W.; Yang, X.; et al. Hierarchically Mesoporous Carbon Spheres Coated with a Single Atomic Fe–N–C Layer for Balancing Activity and Mass Transfer in Fuel Cells. *Carbon Energy* **2022**, *4*, 1–11. [CrossRef]
132. Mashindi, V.; Mente, P.; Mpofu, N.; Phaahlamohlaka, T.N.; Makgae, O.; Kirkland, A.I.; Forbes, R.; Ozoemena, K.I.; Levecque, P.B.; Coville, N.J. Platinum Supported on Pristine and Nitrogen-Doped Bowl-like Broken Hollow Carbon Spheres as Oxygen Reduction Reaction Catalysts. *J. Appl. Electrochem.* **2021**, *51*, 991–1008. [CrossRef]
133. Wang, Y.; Tan, G.; Yi, Q.; Fang, C.; Yi, R. N-Doped Hollow Carbon Sphere and Polyhedral Carbon Composite Supported Pt/Fe Nanoparticles as Highly Efficient Cathodic Catalysts of Proton-Exchange Membrane Fuel Cells. *ACS Appl. Energy Mater.* **2023**, *6*, 1228–1238. [CrossRef]
134. Panickar, R.; Sobhan, C.B.; Chakravorti, S. Chemical Vapor Deposition Synthesis of Carbon Spheres: Effects of Temperature and Hydrogen. *Vacuum* **2020**, *172*, 109108. [CrossRef]
135. Kızılduman, B.K.; Turhan, Y.; Doğan, M. Mesoporous Carbon Spheres Produced by Hydrothermal Carbonization from Rice Husk: Optimization, Characterization and Hydrogen Storage. *Adv. Powder Technol.* **2021**, *32*, 4222–4234. [CrossRef]
136. Xu, M.; Yu, Q.; Liu, Z.; Lv, J.; Lian, S.; Hu, B.; Mai, L.; Zhou, L. Tailoring Porous Carbon Spheres for Supercapacitors. *Nanoscale* **2018**, *10*, 21604–21616. [CrossRef] [PubMed]
137. Yu, S.; Dong, X.; Zhao, P.; Luo, Z.; Sun, Z.; Yang, X.; Li, Q.; Wang, L.; Zhang, Y.; Zhou, H. Decoupled Temperature and Pressure Hydrothermal Synthesis of Carbon Sub-Micron Spheres from Cellulose. *Nat. Commun.* **2022**, *13*, 3616. [CrossRef] [PubMed]
138. Mao, H.; Chen, X.; Huang, R.; Chen, M.; Yang, R.; Lan, P.; Zhou, M.; Zhang, F.; Yang, Y.; Zhou, X. Fast Preparation of Carbon Spheres from Enzymatic Hydrolysis Lignin: Effects of Hydrothermal Carbonization Conditions. *Sci. Rep.* **2018**, *8*, 9501. [CrossRef]
139. Yu, Q.; Xu, J.; Wu, C.; Zhang, J.; Guan, L. MnO_2 Nanofilms on Nitrogen-Doped Hollow Graphene Spheres as a High-Performance Electrocatalyst for Oxygen Reduction Reaction. *ACS Appl. Mater. Interfaces* **2016**, *8*, 35264–35269. [CrossRef] [PubMed]
140. Navas, D.; Fuentes, S.; Castro-Alvarez, A.; Chavez-Angel, E. Review on Sol-Gel Synthesis of Perovskite and Oxide Nanomaterials. *Gels* **2021**, *7*, 275. [CrossRef]
141. Karimi Estahbanati, M.R.; Feilizadeh, M.; Shokrollahi Yancheshmeh, M.; Iliuta, M.C. Effects of Carbon Nanotube and Carbon Sphere Templates in TiO_2 Composites for Photocatalytic Hydrogen Production. *Ind. Eng. Chem. Res.* **2019**, *58*, 2770–2783. [CrossRef]
142. Bokov, D.; Turki Jalil, A.; Chupradit, S.; Suksatan, W.; Javed Ansari, M.; Shewael, I.H.; Valiev, G.H.; Kianfar, E. Nanomaterial by Sol-Gel Method: Synthesis and Application. *Advances in Materials Science and Engineering* **2021**, *2021*, 5102014. [CrossRef]
143. Ebisike, K.; Okoronkwo, A.E.; Alaneme, K.K. Synthesis and Characterization of Chitosan–Silica Hybrid Aerogel Using Sol-Gel Method. *J. King Saud Univ. -Sci.* **2020**, *32*, 550–554. [CrossRef]
144. Ozouf, G.; Cognard, G.; Maillard, F.; Chatenet, M.; Guétaz, L.; Heitzmann, M.; Jacques, P.-A.; Beauger, C. Sb-Doped SnO_2 Aerogels Based Catalysts for Proton Exchange Membrane Fuel Cells: Pt Deposition Routes, Electrocatalytic Activity and Durability. *J. Electrochem. Soc.* **2018**, *165*, F3036. [CrossRef]
145. Shen, M.; Qi, J.; Gao, K.; Duan, C.; Liu, J.; Liu, Q.; Yang, H.; Ni, Y. Chemical Vapor Deposition Strategy for Inserting Atomic FeN4 Sites into 3D Porous Honeycomb Carbon Aerogels as Oxygen Reduction Reaction Catalysts in High-Performance Zn-Air Batteries. *Chem. Eng. J.* **2023**, *464*, 142719. [CrossRef]
146. Shaari, N.; Kamarudin, S.K. Current Status, Opportunities, and Challenges in Fuel Cell Catalytic Application of Aerogels. *Int. J. Energy Res.* **2019**, *43*, 2447–2467. [CrossRef]
147. Wang, Y.; Larsen, M.J.; Rojas, S.; Sougrati, M.-T.; Jaouen, F.; Ferrer, P.; Gianolio, D.; Berthon-Fabry, S. Influence of the Synthesis Parameters on the Proton Exchange Membrane Fuel Cells Performance of Fe–N–C Aerogel Catalysts. *J. Power Sources* **2021**, *514*, 230561. [CrossRef]

148. Alwin, S.; Sahaya Shajan, X. Aerogels: Promising Nanostructured Materials for Energy Conversion and Storage Applications. *Mater Renew Sustain. Energy* **2020**, *9*, 7. [CrossRef]
149. Öner, E.; Öztürk, A.; Yurtcan, A.B. Utilization of the Graphene Aerogel as PEM Fuel Cell Catalyst Support: Effect of Polypyrrole (PPy) and Polydimethylsiloxane (PDMS) Addition. *Int. J. Hydrog. Energy* **2020**, *45*, 34818–34836. [CrossRef]
150. Gu, K.; Kim, E.J.; Sharma, S.K.; Sharma, P.R.; Bliznakov, S.; Hsiao, B.S.; Rafailovich, M.H. Mesoporous Carbon Aerogel with Tunable Porosity as the Catalyst Support for Enhanced Proton-Exchange Membrane Fuel Cell Performance. *Mater. Today Energy* **2021**, *19*, 100560. [CrossRef]
151. Pasatoa, C.D.; Lawa, A.; Parka, J.; Kima, M.; Jina, P. The Synthesis of High Surface Area Carbon Aerogels. *J. Undergrad. Chem. Eng. Res* **2020**, *9*, 27–30.
152. Liu, X.; Tian, J.; Zhou, C.; Jiang, J.; Cheng, X.; Yang, L.; Wu, Q.; Wang, X.; Hu, Z. Constructing Membrane Electrodes of Low Pt Areal Loading with the New Support of N-Doped Carbon Nanocages for PEMFC. *FlatChem* **2023**, *40*, 100515. [CrossRef]
153. Kagkoura, A.; Tagmatarchis, N. Carbon Nanohorn-Based Electrocatalysts for Energy Conversion. *Nanomaterials* **2020**, *10*, 1407. [CrossRef] [PubMed]
154. Kausar, A. Polymer/Carbon Nanocoil Nanocomposite: Status and Future Directions. *Polym. -Plast. Technol. Mater.* **2021**, *60*, 816–829. [CrossRef]
155. Yeon, J.H.; Park, S.J.; Choi, I.; Choi, M. Generation of Carbon Nano-Onions by Laser Irradiation of Gaseous Hydrocarbons for High Durability Catalyst Support in Proton Exchange Membrane Fuel Cells. *J. Ind. Eng. Chem.* **2019**, *80*, 65–73. [CrossRef]
156. Yeon, J.H.; Choi, J.; Jang, S.; Choi, M. Hydrophilicity Control of Laser-Induced Amorphous Carbon-Encapsulated Carbon Nano-Onions and Their Application to Proton Exchange Membrane Fuel Cells under Low Humidity. *Carbon* **2021**, *184*, 910–922. [CrossRef]
157. Wu, J.; Sun, Y.-M.; Wu, Z.; Li, X.; Wang, N.; Tao, K.; Wang, G.P. Carbon Nanocoil-Based Fast-Response and Flexible Humidity Sensor for Multifunctional Applications. *ACS Appl. Mater. Interfaces* **2019**, *11*, 4242–4251. [CrossRef]
158. Zuo, X.; Zhao, Y.; Zhang, H.; Huang, H.; Zhou, C.; Cong, T.; Muhammad, J.; Yang, X.; Zhang, Y.; Fan, Z. Surface Modification of Helical Carbon Nanocoil (CNC) with N-Doped and Co-Anchored Carbon Layer for Efficient Microwave Absorption. *J. Colloid Interface Sci.* **2022**, *608*, 1894–1906. [CrossRef] [PubMed]
159. Ma, L.; Bi, Z.; Zhang, W.; Zhang, Z.; Xiao, Y.; Niu, H.; Huang, Y. Synthesis of a Three-Dimensional Interconnected Oxygen-, Boron-, Nitrogen-, and Phosphorus Tetratomic-Doped Porous Carbon Network as Electrode Material for the Construction of a Superior Flexible Supercapacitor. *ACS Appl. Mater. Interfaces* **2020**, *12*, 46170–46180. [CrossRef] [PubMed]
160. Wang, Y.; Luo, S.; Chen, M.; Wu, L. Uniformly Confined Germanium Quantum Dots in 3D Ordered Porous Carbon Framework for High-Performance Li-ion Battery. *Adv. Funct. Mater.* **2020**, *30*, 2000373. [CrossRef]
161. Sam, D.K.; Li, H.; Xu, Y.T.; Cao, Y. Porous Carbon Fabrication Techniques: A Review. *J. Ind. Eng. Chem.* **2024**, *in press*. [CrossRef]

catalysts

MDPI

Review

Investigation of Electrocatalytic Applications of Various Advanced Nanostructured Alloys—An Overview

Shashanka Rajendrachari [1,*], Gireesha R. Chalageri [2], Rayappa Shrinivas Mahale [3], Emre Altas [4], Yashwant Chapke [3] and Vinayak Adimule [5]

1 Department of Basic Sciences, School of Sciences and Humanities, SR University, Warangal 506371, India
2 School of Mechanical Engineering, KLE Technological University, Hubballi 580031, India; gireesha@kletech.ac.in
3 Department of Automation and Robotics, JSPMs Rajarshi Shahu College of Engineering, Pune 411033, India; rayappamahale@gmail.com (R.S.M.); yashchapke@gmail.com (Y.C.)
4 Department of Mechanical Engineering, Faculty of Engineering, Architecture and Design, Bartın University, Bartin 74110, Turkey; emrealtas@bartin.edu.tr
5 Department of Chemistry, Angadi Institute of Technology and Management (AITM), Savagaon Road, Belagavi 590009, India; chemistry@aitmbgm.ac.in
* Correspondence: shashankaic@gmail.com

Abstract: Cyclic voltammetry (CV) is one of the advanced techniques used to determine various bioactive molecules, organic dyes, pesticides, veterinary drugs, heavy metals, toxic chemicals, etc. To determine all the above analytes, one needs an electrocatalyst for their electrochemical redox reaction. Many researchers have reported the use of metal nanomaterials, metal oxide nanomaterials, metal–organic frameworks, surfactants, polymers, etc., as modifiers in carbon paste electrodes to enhance their current response, stability, sensitivity, and repeatability. But some of the emerging, cost-effective, and highly efficient electrocatalysts are advanced nanostructured alloy powders. These advanced alloys are used as a modifier to determine various bioactive analytes. These alloy-modified carbon paste electrodes (MCPEs) show excellent selectivity, sensitivity, and stability due to their extraordinary electrochemical properties, as the compositional elements of most of the alloys belong to d-block elements in the periodic table, and these transition elements are famous for their brilliant electrocatalytic properties. The present review article mainly focuses on the determination of dopamine, AA (AA), uric acid, methylene blue, methyl orange, Rhodamine B, and the L-Tyrosine amino acid by various alloys like stainless steel, high-entropy alloys, and shape-memory alloys and how these alloys could change the perception of metallurgists and electrochemists in the future. These alloys could be potential candidates for the development of various electrochemical sensors because of their high porosity and surface areas.

Keywords: mechanical alloying; nanostructured alloys; electrocatalysts; modifiers; carbon paste electrode

1. Introduction

Numerous medicinal and environmental applications depend on the quick and accurate detection of bioactive compounds. Conventional biomolecule detection techniques like spectroscopy and chromatography are frequently costly and time-consuming and necessitate intricate sample preparation [1,2]. We are in great need of analytical instruments that are highly sensitive, selective, stable, and robust in determining various bioactive compounds individually or simultaneously [3]. In analytical chemistry, medicine, and environmental monitoring, the creation of electrochemical sensors for the detection of bioactive compounds has grown in importance. High sensitivity, selectivity, and affordability, as well as the capacity for quick real-time analysis, are just a few benefits that these sensors provide [4]. Nowadays, electrochemical techniques like CV, electrochemical impedance spectroscopy, alternating current voltammetry, differential pulse voltammetry, photo-induced electrochemiluminescence, and electrochemiluminescence are highly efficient and effective techniques for determining bioactive compounds, organic dyes, surfactants, toxic chemicals, etc. [5]. Among these, CV is one of the popular electrochemical techniques performed by cycling the potential of working electrodes between two chosen values to determine the resultant current [6]. CV is a strong, sensitive, and adaptable technique for examining bioactive compounds in both research and clinical settings [7]. In domains including drug development, biomarker detection, and bioanalytical chemistry, it is essential to figure out their electrochemical characteristics, interactions, stability, and possible therapeutic uses. CV provides detailed information about the redox behavior of bioactive compounds, as well as their electrochemical reactivity, electron transfer kinetics, reaction mechanism, and quantitative and qualitative analysis [8].

Even though electrochemical sensors exhibit potential advantages, they still exhibit a few limitations in terms of selectivity, stability, sensitivity, and repeatability. To overcome these limitations, many researchers have used carbon paste electrodes (CPEs), which are stable, selective, economic, and consistent in detecting various bioactive compounds. Among the several electrochemical sensor platforms, carbon paste electrodes (CPEs) have widely attracted researchers due to their versatile properties, ease of preparation, and low manufacturing costs. Many researchers have used carbon paste electrodes (CPEs) to determine various bio-analytes using the CV technique [9–11], and it has been proven that these electrodes are highly stable, repeatable, sensible, and selective in nature. Further, these electrode properties can be improved by modifying CPEs with nanomaterials. These nanomaterials exhibit excellent values for surface area and surface energy, which can further improve the electrocatalytic properties of the electrode [12,13]. Many researchers have used various metal nanoparticles, metal oxide nanoparticles, and carbon-nanomaterial-modified CPEs [14–20], but very few researchers have reported the use of nanostructured alloys as modifiers.

The present review hypothesizes that the idea of adding nanostructured alloys like stainless steel, HEA, and shape-memory alloys into CPEs during the determination of various analytes will improve the sensitivity, selectivity, and stability of electrochemical sensors to a greater extent [21–24]. This could also overcome CPEs' limitations, like their poor conductivity, low stability, and sensitivity at low concentrations of the analyte, and improve their detection limits and quantification limits. These changes provide the sensors with the ability to identify a variety of bioactive compounds at low concentrations, which is crucial for tracking the levels of therapeutic drugs for the early diagnosis of illnesses, determining various toxic dyes present in water, etc. The main objective of the present review is to study the effects of modifiers, like nanostructured alloys on the electrochemical detection of various bioactive compounds, and also how they affect the sensitivity, selectivity, and stability of the electrodes at lower concentrations of the analyte.

2. Advantages of Using Nanostructured Alloys as Modifiers

One of the main advantages of using alloys as modifiers is their extraordinary stability; most alloys remain stable and do not undergo any undesirable electrochemical reactions, even with a wide range of potentials [25,26]. This will prevent the alloys from undergoing decomposition at higher operating voltages. One more characteristic of alloys is their electrode potential; the different elements present in an alloy can impart their electrode potential, resulting in a characteristic electrode potential that is obtained from all the elements present in an alloy. This electrode potential plays an important role in determining the electrocatalytic properties of alloys during the electrochemical redox reactions of the analytes [27–30]. In this review article, we reported the nanostructured alloys prepared from mechanical alloying. This method increases the surface area, and surface roughness of the alloys by refining their structures, and also improves their porosity ratios, thereby improving their electrocatalytic properties. The presence of different metals in an alloy results in heterogeneous surface characteristics, which have a strong influence on the electrochemical properties of the alloys [31–33]. On the other hand, surface roughness further improves the electrocatalytic properties of the alloy by creating reactive sites for the mobility of electrons.

3. Fabrication of Nanostructured Alloys by Ball-Milling Method

Recently, the ball-milling method has emerged as a well-established technique for fabricating nanostructured alloys, especially for producing fine powders of metallic alloys with improved qualities including strength, hardness, and wear resistance [34–36]. To create nano-sized grain formations, powders of various metals or elements are combined and processed under high energy conditions in a mechanical alloying process called ball milling. Figure 1 depicts the planetary ball mill present at Bartin University, Turkey.

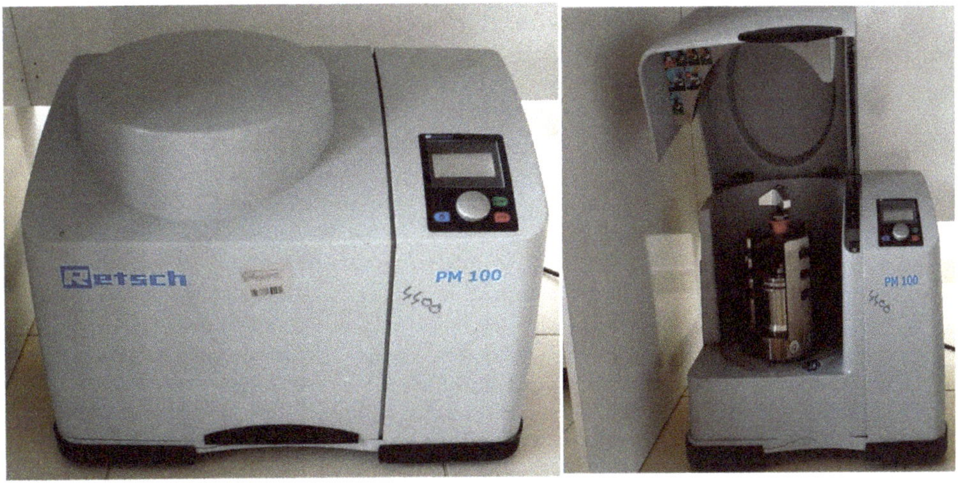

Figure 1. Retsch PM 100 planetary ball mill present in Bartin University, Turkey.

Ball milling is a technique composed of rotating cylindrical chambers loaded with steel or ceramic balls. The fundamental process involves adding the elemental powders to the grinding media at specific ball-to-powder weight ratios. The balls and powder collide as the mill turns, breaking up and reshaping the particles to produce finer powders. The impact and friction between the grinding balls and the particles provide the mechanical energy necessary for the ball-milling process [37]. A refined structure is produced as the particles are crushed, deformed, and welded by the balls as they tumble and revolve in the mill. As a result, nanoparticles are created, and solid-state processes can be triggered to create alloys [38,39]. For further details about ball milling, milling conditions, and how to prepare nanostructured alloys, refer to the published articles of the author of this review, Shashanka et al. [31,32,40–42]. The main advantages of using the mechanical alloying method are because of its low cost, scalability, versatility, improved properties, uniform homogeneity, control over the method, etc. Therefore, many researchers use mechanical alloying methods to prepare various advanced alloys. While preparing alloys by ball-milling method, one must take care of contamination, agglomeration, improper distribution of alloy, chances of wearing of machines, etc. In the present review article, we have discussed the electrocatalytic properties of DS, ferritic stainless steel, high-entropy alloys, and shape-memory alloys prepared by the ball-milling method.

4. Preparation of Nanostructured-Alloys-Modified Carbon Paste Electrodes (NsA-MCPE)

To prepare CPEs, we need graphite powder and a binder like paraffin oil or silicone oil. The proportion of binder to graphite powder is crucial [43,44]. The standard ratios of 70–80% graphite powder by weight and 20–30% binder, such as silicone oil, are mixed well in mortar and pestle for approximately 30 min [45,46]. Depending on the particular requirements of the experiment and the characteristics of the graphite powder, we can modify this ratio. To create a homogenous mixture, thoroughly mix the graphite powder and the binder material [47,48]. The carbon paste should resemble thick toothpaste in consistency and be uniformly smooth. Later, fill the carbon paste to the cavity present in the specially designed electrode, which acts as the working electrode. Make sure to fill it evenly to avoid air bubbles and ensure the paste is tightly packed. Leave a small part of the electrode body free of paste to attach the electrical contact. Carbon paste is packed properly and uniformly to avoid any formation of air bubbles in the electrode. To obtain a homogeneous smooth surface for electrochemical measurements, gently press the carbon paste electrode against a clean piece of paper towel or a level surface. To the surface of the CPE, attach a suitable conductive wire like copper for the electrical contact for the electrode. To avoid unwanted side reactions, make sure the copper wire is properly in contact with the carbon paste without touching the electrode's exposed surface [49,50]. Sometimes, curing of the electrode may require that, depending upon the type of binder used. Curing is a process of heating the electrode at a very low temperature to solidify the binder used to prepare the CPE; this will impart mechanical strength to the electrode. If we use mineral oil as a binder, then the CPE requires curing, and if we use silicone oil as a binder, then curing is not required. Curing of electrodes is required only when we use low-viscous binders to prepare the CPEs. After curing, we need to polish the surface of the electrode to make it smooth and uniform. On the other hand, NsA-MCPE is prepared by mixing different concentrations of the modifier (nanostructured alloys) in a CPE composition and grinding them with a mortar and pestle for approximately 30 min and follow the same steps as discussed above to prepare CPEs. Figure 2 depicts the CPE preparation process.

It is very important to pre-condition the electrode by applying multiple potential cycles in the electrochemical solution under study using cyclic voltammetry. This can help the electrode to perform consistently throughout measurements by eliminating contaminants or irregular side reactions because of them. Now, the electrode is ready to use for the electrochemical determination of various bioactive compounds by cyclic voltammetric method and placed in an analyte solution, along with a reference electrode and platinum wire counter electrode [51–53].

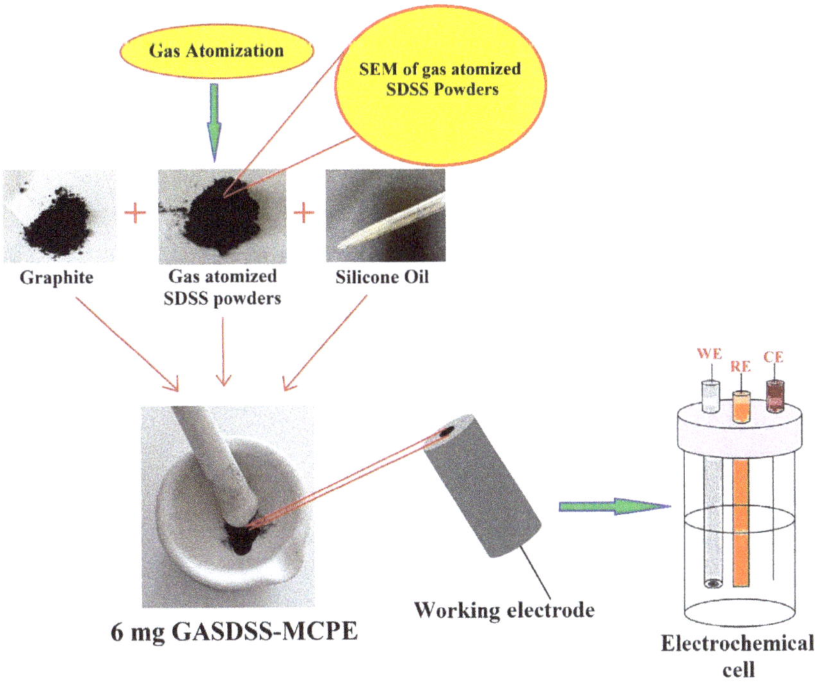

Figure 2. Preparation of CPE [28].

5. Duplex Stainless Steel-Modified Carbon Paste Electrode (DS-MCPE)

5.1. Introduction to Duplex Stainless Steel (DS)

DS is a popular grade of stainless steel, made up of nearly equal amounts of non-magnetic ferrite and magnetic austenite phases [54–56]. DS is appropriate for demanding applications requiring high strength, exceptional corrosion resistance, electrochemical properties, good weldability, high toughness, and enhanced fatigue resistance due to their dual-phase structure [57–59]. The marine, oil and gas, paper and pulp, chemical processing, and construction industries are among those that employ DS because of its exceptional strength, corrosion resistance, and affordability [60,61]. Recent applications of DSs are used as an electrocatalyst for various electrochemical applications [26–28].

5.2. Electrochemical Determination of Dopamine

Dopamine functions as a chemical messenger known as a neurotransmitter, which transmits messages to the brain through the central nervous system [62]. It is essential for numerous bodily processes, such as motivation, pleasure, happiness, mobility, and emotional response management. The ventral tegmental area (VTA) and the substantia nigra present in the brain manufacture dopamine. It is a component of the dopaminergic system, which affects mood, behavior, and even judgment [63,64]. Tyrosine, an amino acid, is the precursor for dopamine. It undergoes enzymatic processes to transform into L-DOPA and, subsequently, dopamine. When dopamine is produced, it attaches itself to target neurons' dopamine receptors to affect their activity. Deficiency of dopamine can affect the central nervous system and mental health, increase anxiety and depression, and cause attention deficit hyperactivity disorder (ADHD), Parkinson's disease, compulsive behavior, sleeping disorders, etc. [65–68]. Therefore, we must determine dopamine efficiently and effectively. The cyclic voltammetric method proved to be one of the best techniques used to determine dopamine. The chemical structure of dopamine is shown in Figure 3.

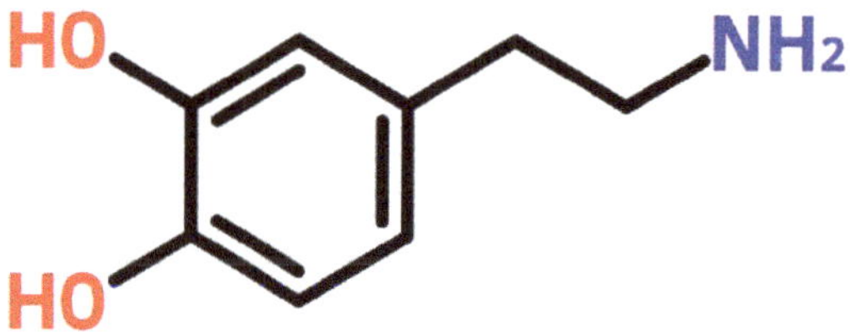

Figure 3. Chemical structure of dopamine [69].

In a reversible electrochemical process, dopamine is reduced and oxidized. Dopamine can usually be reduced back to its original form after being oxidized at the working electrode to produce dopamine quinone. Dopamine undergoes oxidation around +0.2 to +0.6 V to form dopamine quinone and, on the other hand, it will be reduced from dopamine quinone to dopamine at a potential of around +0.1 to +0.3 V. The resultant peak current of the electrochemical redox reaction is proportional to the concentration of dopamine present. The electrochemical determination of dopamine utilizing the DS-MCPE by cyclic voltammetry method was reported by Shashanka et al. [25]. The CV of a bare carbon paste electrode (BCPE) and 4 mg of DS-MCPE in 2 mM of DA at 100 mVs^{-1} and in a pH 7.2 phosphate buffer solution (PBS) is shown in Figure 4a. They have investigated one of the most crucial factors like modifier concentrations to determine the effectiveness of the electrode's performance as a dopamine sensor. To measure dopamine, they have utilized 2, 4, 6, and 8 mg of DS-MCPE; however, 4 mg of DS-MCPE has demonstrated the highest current response compared to the others. To investigate the effects of scan rate, dopamine concentrations, and pH, they utilized 4 mg of DS-MCPE. The corresponding CVs are displayed in Figure 4b, Figure 4c and Figure 4d, respectively.

As the scan rate is increased from 50 to 500 mVs^{-1}, the anodic and cathodic peak currents of dopamine increase during the electrochemical redox reaction due to direct electron transfers between the analyte and the electrode, as illustrated in Figure 4b. The potential difference between the anodic and cathodic peak potentials rises with the scan rate. Similarly, Figure 4c depicts that, as the concentration of dopamine increases from 2 to 3 mM, the anodic peak current (Ipa) also increases from 15.5 to 19 μA. This is because there are more dopamine molecules available at increasing concentrations, which causes Ipa to rise linearly. However, as can be seen in Figure 4d, pH tests suggest that dopamine's anodic peak potential (Epa) drops when the pH rises from 5.7 to 8. This is because protons (H^+) are frequently involved in the oxidation reaction and their concentration decreases as pH rises. This variation in proton concentration alters the electrochemical reaction's equilibrium and influences the voltage at which oxidation takes place. A basic explanation of the relationship between peak potential and pH is given by the Nernst equation; the shift is usually approximately 59 mV per pH unit, depending on the number of protons in the reaction. All these above results prove that DS-MCPE could be a potential candidate to use as an electrocatalyst to determine dopamine electrochemically using cyclic voltammetry.

5.3. Electrochemical Determination of AA

AA is a water-soluble and vital vitamin to human health, also referred to as vitamin C [70]. The molecular formula for this organic compound is $C_6H_8O_6$ and its chemical structure is shown in Figure 5. As a strong antioxidant, AA is essential for many physiological functions, such as collagen formation, bone, cartilage, and tooth maintenance, as well as immune system support [71,72]. Being an antioxidant, AA aids in scavenging free radicals and shielding the organism from oxidative damage. Vitamin E and other nutrients are also preserved as a result of this antioxidant action. Many fruits and vegetables, including bell peppers, oranges, strawberries, and kiwis, naturally contain AA [73,74]. It must be received through diet or supplements because humans are unable to produce it internally.

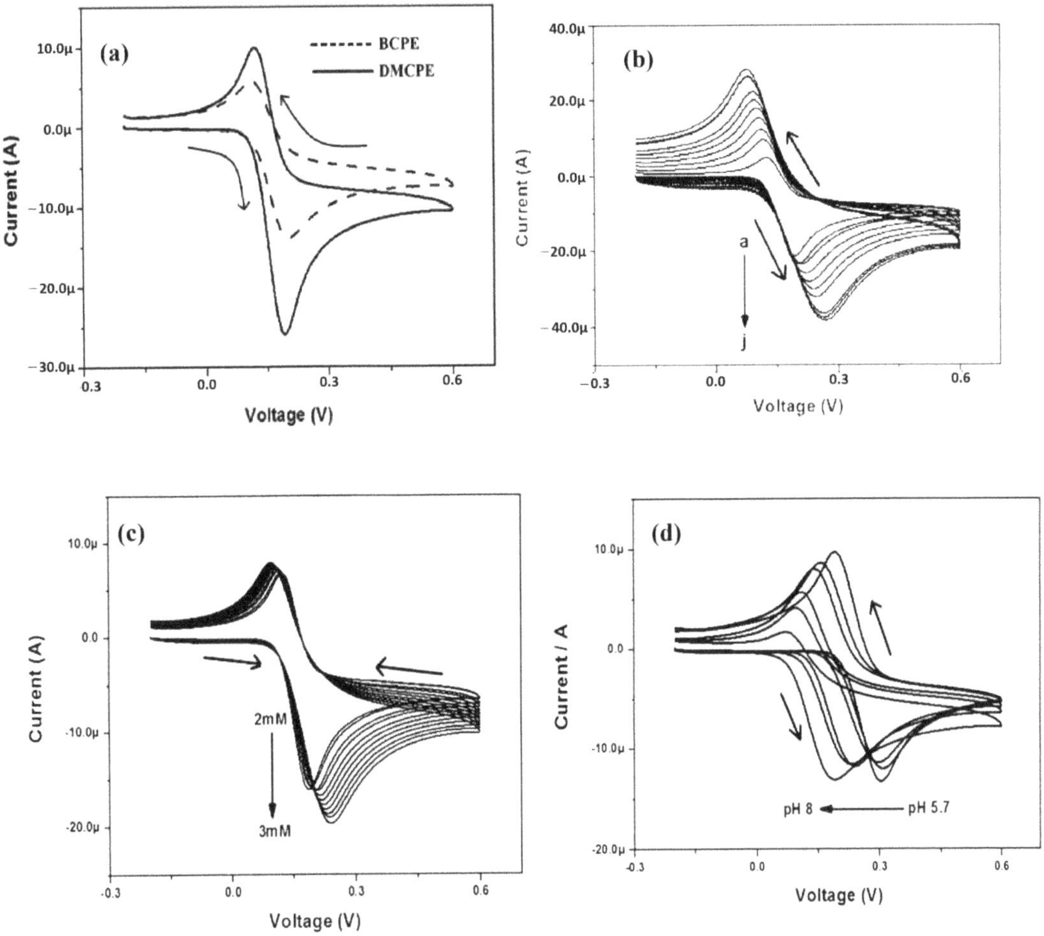

Figure 4. CV of (**a**) BCPE and DS MCPE, (**b**) scan rate investigations at (a = 50, b = 100, ..., j = 500 mV/s), (**c**) concentration variation of dopamine, (**d**) pH variation investigation on the redox reaction of dopamine using DS-MCPE [25].

AA is susceptible to air, heat, and light, and when it is exposed to air or UV radiation, it readily breaks down and loses its vitamin function. Because of this, it is frequently kept in cool, dark places and preserved in sealed containers. A typical dietary supplement of AA is used to prevent and cure vitamin C deficiency, which can result in scurvy, and its deficiency can also cause chronic illness, which causes malabsorption of nutrients [75,76]. Since the body cannot generate AA, it must be regularly consumed through diet or supplements to maintain good health [75–77]. Therefore, we need to frequently check for the AA and there are many techniques available to determine it. However, cyclic voltammetry is one of the most accurate ways to identify AA.

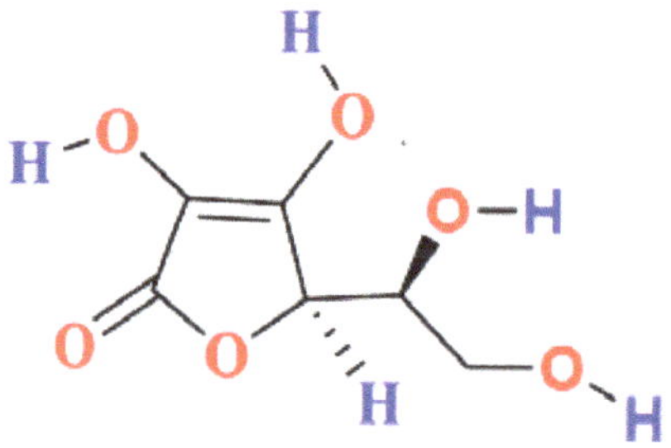

Figure 5. AA structure [77].

The use of mechanically alloyed nanostructured DS powders for electrochemical sensors was reported by Rayappa et al. [26]. The microstructure, lattice spacing, and SAED pattern of mechanically alloyed DS are shown in Figure 6.

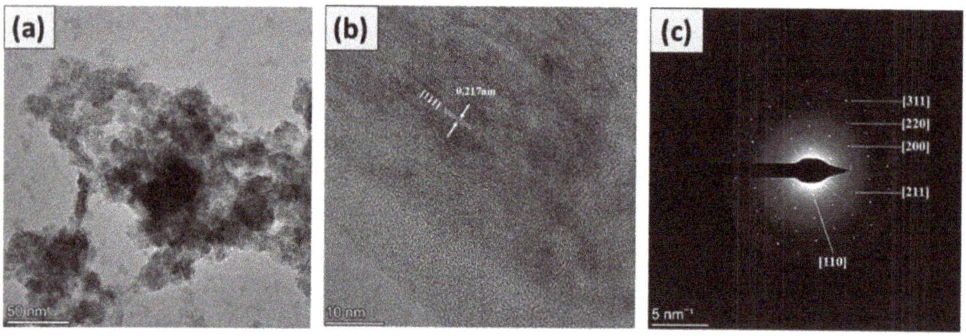

Figure 6. (**a**) TEM, (**b**) HRTEM, (**c**) SAED pattern of DS powder [26].

The fabricated 4 mg of DS-MCPE showed better current sensitivity to detect the AA in a PBS with a pH of 6.8 when compared to 2, 6, 8, and 10 mg of DS-MCPEs. It was determined that the DS-MCPE's active surface area was nearly twice as large as the BCPE's surface area. Additionally, they reported the limit of quantification (LOQ) and limit of detection (LOD), which were determined to be 0.688×10^{-8} M and 0.206×10^{-8} M, respectively. Additionally, they have reported the repeatability, reproducibility, and stability of the DS-MCPE and reported on the effects of scan rate, analyte concentration, pH fluctuations, and interference ions. The graph of various DS-MCPE concentrations, along with their corresponding Ipa and CV, is displayed in Figure 7a,b for 1 mM of AA at BCPE and 4 mg of DS-MCPE [26].

It is confirmed that, in Figure 7a, the Ipa of the 4 mg of DS-MCPE is around twice that of the BCPE. This demonstrates the significance of the modifier, which has a high current sensitivity and can identify the analytes even at low concentrations. The enhanced surface area of the electrode after the modification is the main reason for the higher current. The Ipa of BCPE and DS-MCPE differ from one another, as seen in Figure 7b. When electro-oxidizing 1 mM of AA, BCPE demonstrated an Ipa of 22.5 μA, while 4 mg of DS-MCPE recorded an Ipa of 37.2 μA. The increased surface area and reaction sites in the case of 4 mg of DS-MCPE are responsible for the increased Ipa. Similarly, Rayappa et al. [26]

also reported the CV of AA electrochemically oxidized at different pH and scan rates, and different analyte concentrations, as depicted in Figure 8a, Figure 8b, and Figure 8c, respectively. The pH studies validate the involvement of protons in the electrode response, which shows that increasing the pH from 6 to 6.8 causes Epa to move towards lower values. Participation of protons in the electrode reaction is confirmed by the pH studies, which show the shifting of Epa towards lower values with the increase in the pH from 6 to 6.8. Using the equation below, they have also determined that there are two protons and electrons involved in the electrochemical oxidation of AA:

$$\frac{\Delta E_p}{\Delta pH} = \frac{2.303mRT}{nF} \tag{1}$$

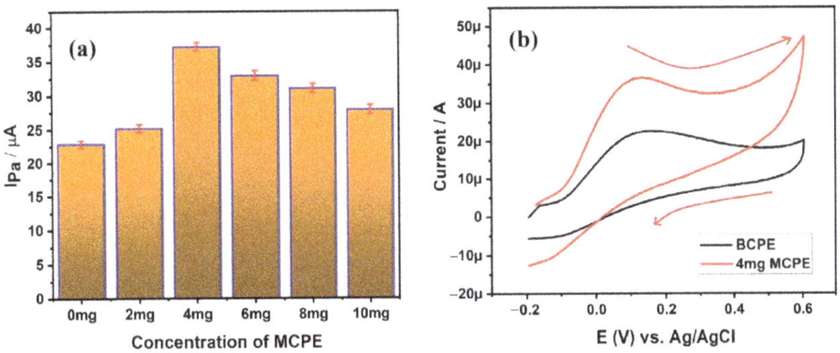

Figure 7. (**a**) DS-MCPE concentration graph with corresponding IPa, (**b**) AA CV at BCPE, and 4 mg of DS-MCPE [26].

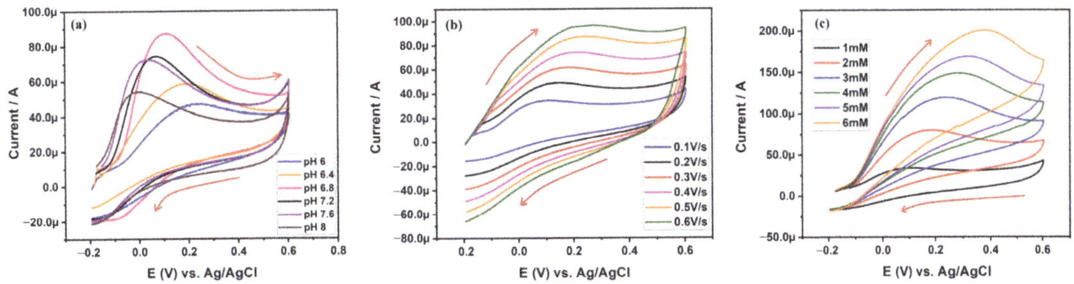

Figure 8. CV of electrochemically oxidized AA at different (**a**) pH PBS, (**b**) scan rates, (**c**) concentrations of AA [26].

Figure 8b demonstrates that raising the scan rate from 0.1 to 0.6 V/s during the electro-oxidation of AA increases the Ipa but decreases the Epa. It is well known that a higher scan rate results in a shorter period to maintain the same charge value, which raises the current. As a result, when AA is electro-oxidized, the anodic peak current value grows as the scan rate does [78]. Fast and direct electron transport between the analyte and the electrode surface is the outcome of this linear increase in Ipa. Similarly, Figure 8c reveals that an increase in the concentrations of the AA from 1 to 6 mM increases the Ipa linearly. Using the standard deviation of the blank CV and the slope obtained after plotting the Ipa vs. concentration of AA, Rayappa et al. [26] calculated LOD and LOQ and the values are 0.206×10^{-8} M and 0.688×10^{-8} M, respectively. Figure 9 shows the electro-oxidation of AA at a pH of 6.8 using DS-MCPE. They have also reported the effect of interfering ions during the determination of AA. They obtained variation of less than 4% in the electrochemical signal of Ipa and this implies that even when different interfering metal ions are present, the DS-MCPE remains extremely stable and selective.

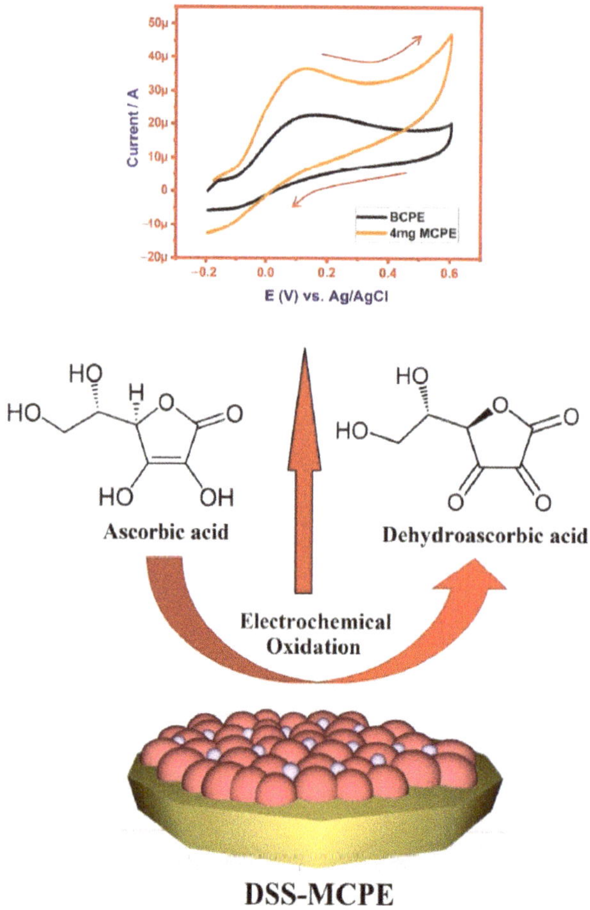

Figure 9. Possible scheme of electrochemical oxidation of AA under DS-MCPE [26].

The repeatability and reproducibility tests were performed five times each by switching the electrode and electrolyte after each run, respectively. The calculated relative standard deviations for reproducibility and repeatability were 3.02% and 2.15%, respectively. These findings suggest that the DS-MCPE is a more reliable and repeatable method for AA molecule detection. Additionally, they investigated the stability of the DS-MCPE by detecting AA in a pH of 6.8 PBS for 50 cycles. The first and final Ipa during the electro-oxidation of AA is recorded to determine the degree of stability. The calculated stability value of 96.98% was obtained and this demonstrates the outstanding stability of the DS-MCPE during the determination of AA.

5.4. Electrochemical Determination of MB (MB)

MB is a synthetic dye, that finds extensive use in both laboratory and medical settings. This dye has a characteristic blue color and is categorized as a thiazine dye chemically [79,80]. For almost a century, MB has been used for several purposes, including microscopy staining and medicinal uses [81,82]. Apart from its benefits, it has several drawbacks, particularly when used incorrectly or in conjunction with specific medical problems. The biggest worries are the possibility of toxicity at high dosages, serotonin syndrome when taken with certain drugs, and hemolysis in people with G6PD deficiency. Like any drug or substance, MB should be used with caution and under the supervision of medical specialists [83–85]. The chemical structure of MB is shown in Figure 10.

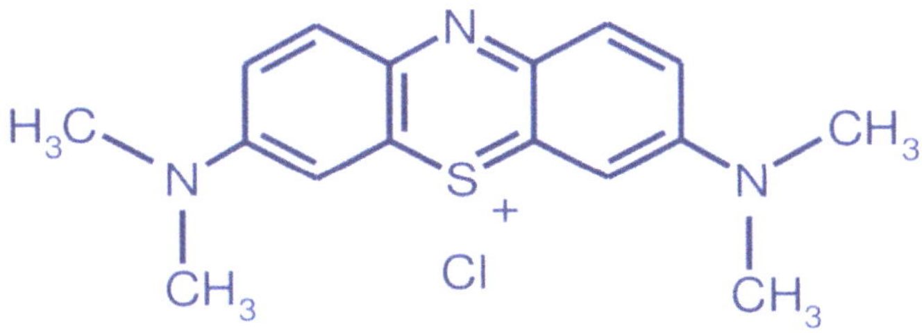

Figure 10. Chemical structure of MB [86].

MB is used as a coloring agent in a variety of sectors, such as textiles, leather, and paper. However, the significant MB dye discharge by these industries as wastewater into the surface and groundwater causes environmental issues, resulting in adverse effects on ecosystems and health issues [83–86]. The electrochemical oxidation of MB by cyclic voltammetry method is considered one of the best suitable methods. Because of its ease of use, sensitivity, low cost, and speed, this method provides detailed information about the electrochemical determination of MB dye, making it very useful in the biomedical sector [87–91].

Therefore, Rayappa et al. [27] determined the MB dye by cyclic voltammetry method using DS-MCPE. Figure 11a,b represents the Ipa of MB obtained at various concentrations of DS-MCPE and the CV of BCPE and DS-MCPE, respectively. From the figure, it is confirmed that 2 mg of DS-MCPE has shown excellent current response compared to BCPE and MCPE has shown more than 7 times the Ipa of BCPE. This confirms the importance of a modifier like DS powders [27].

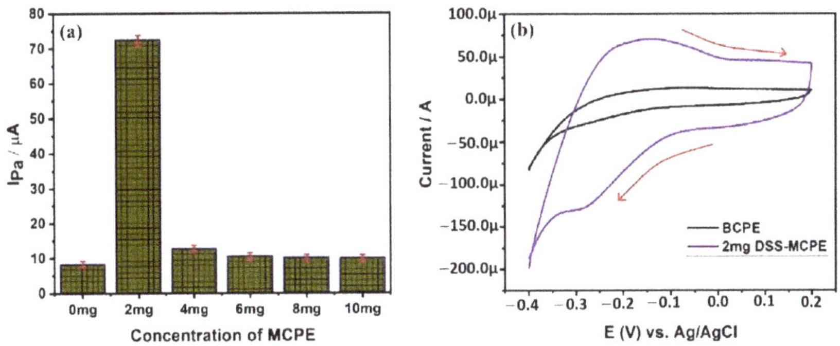

Figure 11. (**a**) Graph of Ipa at various concentrations of DS-MCPE, (**b**) CV of BCPE, and 2 mg of DS-MCPE [27].

On the other hand, the authors also reported the effect of pH variation, the effect of scan rate, and the effect of MB concentrations, and their respective CVs are depicted in Figure 12a–c. An increase in the scan rate and increase in the concentrations of MB increases the Ipa, as we discussed earlier. When the pH rises from 6 to 8, the Epa shifts to lower values. This indicates that the electrochemical oxidation of MB involves protons.

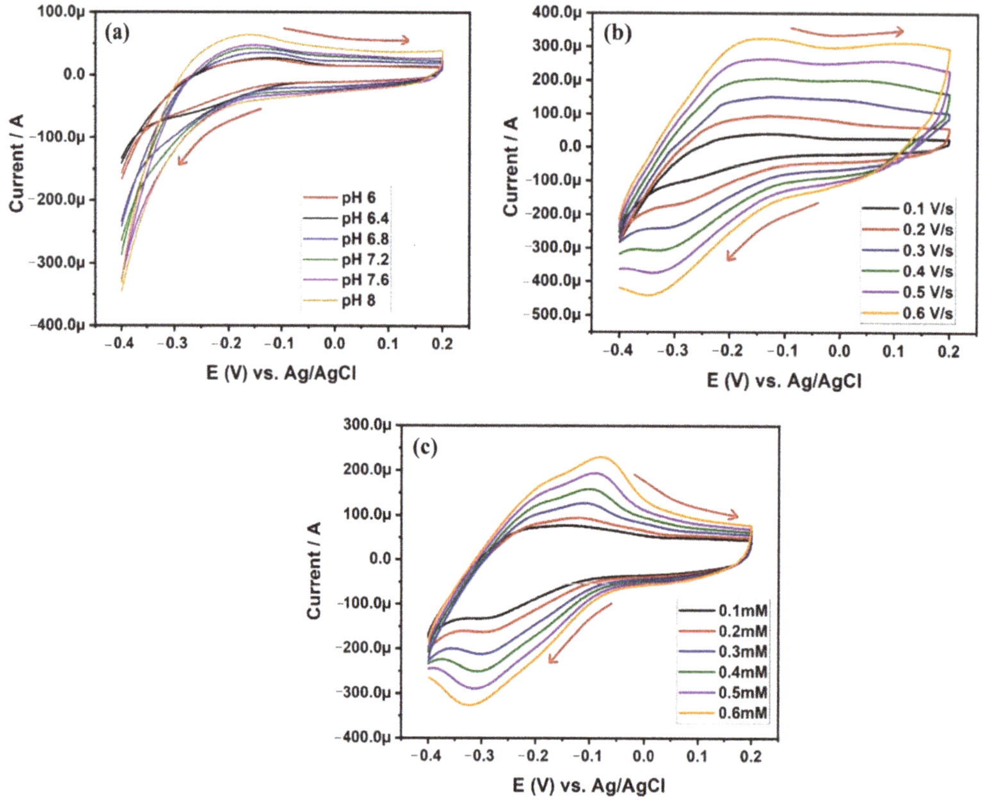

Figure 12. CV of electrochemically oxidized MB at different (**a**) PBS pH, (**b**) scan rates, (**c**) concentrations of MB, respectively, at DS-MCPE [27].

The calculated values of LOD and LOQ are 0.222×10^{-8} and 0.74×10^{-8} M, respectively [27]. Figure 13 shows the potential mechanism of the electrochemical reaction of MB using DS-MCPE. Several metal ions (Na^+, Cu^{2+}, Fe^{2+}, Mg^{2+}, Fe^{3+}, K^+) and bioactive compounds were used as the interferents and investigated up to what extent these interferents can affect the Ipa of MB. Notably, the Ipa and Epa of MB did not significantly change either positively or negatively, confirming DS-MCPE's strong selectivity. Furthermore, DS-MCPE showed exceptional stability even after 50 cycles of continuous 0.1 mM MB detection at pH 8, demonstrating its long-lasting effectiveness under electrochemical testing. Given the improved stability and sensitivity of the DS-MCPE electrode, it is a potential candidate for determining MB dye and can be incorporated into portable platforms for quick analysis. Future research will concentrate on improving device mobility and fabrication procedures and addressing pragmatic issues.

5.5. Electrochemical Determination of Rhodamine B (Rh B)

Rhodamine B is a synthetic dye that belongs to the Rhodamine family. It is frequently used in textile dyeing, biological research, and other industrial operations as a fluorescent tracer [92–94]. It has a fused tricyclic structure with nitrogen atoms and a benzene ring joined to a carboxyl group, as shown in Figure 14. When solid, it is a dark red or purple powder; when dissolved in water, it will show a pink color. Because of its fluorescent qualities, it can be used in a variety of processes, including DNA measurement, flow cytometry, and microscopy [95–97].

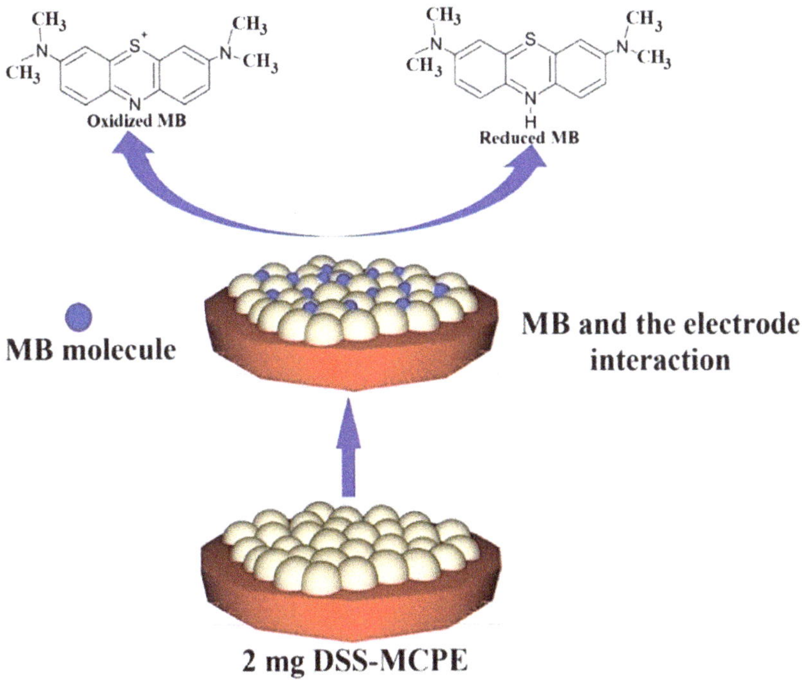

Figure 13. Possible scheme of electrochemical oxidation of MB at DS-MCPE [27].

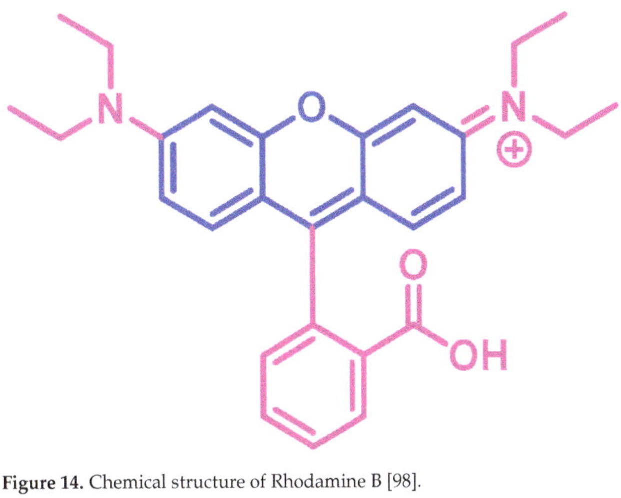

Figure 14. Chemical structure of Rhodamine B [98].

Rhodamine B is utilized extensively in science and industry, but its possible harmful effects on the environment and human health should not be disregarded. Long-term use of Rhodamine B can cause chronic diseases like kidney and liver damage, a person can develop vomiting and abdominal pain, and excessive use can affect the human reproduction system [92–96]. It is highly carcinogenic in nature and increases the risk of cancer. To reduce the possible danger, proper handling, safety precautions, and compliance with rules are crucial. Further study is required to completely comprehend the long-term effects on human health and the environment to appropriately assess and lessen the risks connected with its widespread use. Keeping in mind the adverse effects of Rhodamine B, we must develop a system that can determine Rhodamine B even at small concentrations.

Shashanka et al. [28] attempted to study their electrochemical properties using DS-MCPE using cyclic voltammetry technique and reported excellent results. Figure 15a,b represents the graph of Ipa reported at different concentrated DS-MCPEs and the CV of BCPE and DS-MCPE, respectively. According to the figure, 6 mg of DS-MCPE has shown more than 2 times the Ipa of BCPE because of increased surface area and reaction sites after modification. The authors have also calculated the active surface areas of 6 mg of DS-MCPE and BCPE and the values were found to be 0.302 and 0.761 cm^2, respectively. This shows the importance of the modification of the CPEs.

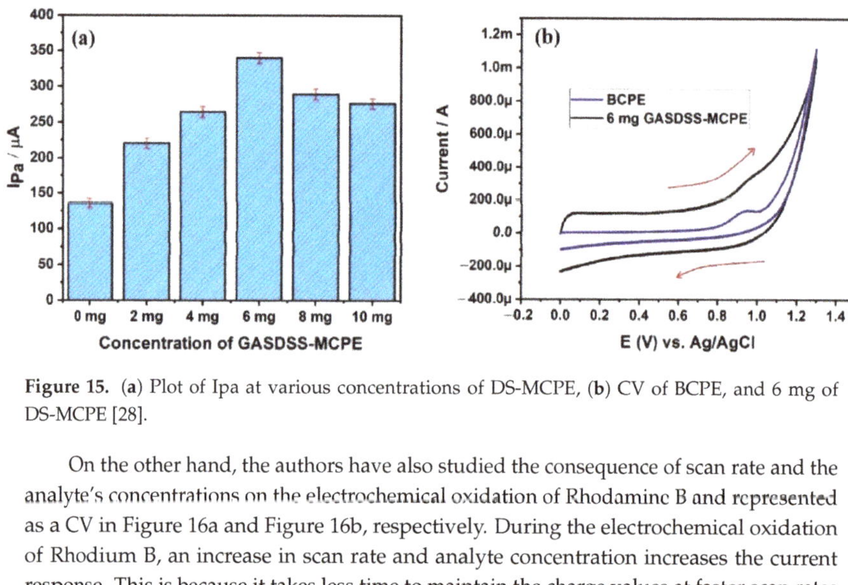

Figure 15. (**a**) Plot of Ipa at various concentrations of DS-MCPE, (**b**) CV of BCPE, and 6 mg of DS-MCPE [28].

On the other hand, the authors have also studied the consequence of scan rate and the analyte's concentrations on the electrochemical oxidation of Rhodamine B and represented as a CV in Figure 16a and Figure 16b, respectively. During the electrochemical oxidation of Rhodium B, an increase in scan rate and analyte concentration increases the current response. This is because it takes less time to maintain the charge values at faster scan rates and because the DS-MCPE surface and the analyte interact strongly, increasing electron mobility. For the 6 mg of DS-MCPE, the determined LOD and LOQ were 0.378 μM and 1.260 μM, respectively.

They have also reported that interference like vanillin, FA, uric acid, methylene blue, methyl orange, K^+, Fe^{2+}, and Mg^{2+} does not affect the current response of the 6 mg of DS-MCPE during the electrochemical oxidation of Rhodamine B. They have excellent stability, repeatability, and reproducibility results for the DS-MCPE, and, therefore, DS-MCPE is one the very important and potential electrodes that can be used to determine Rhodamine B in real samples. The electrochemical oxidation of Rhodamine B is shown in Figure 17.

5.6. Electrochemical Determination of Folic Acid (FA)

FA is essential for many body processes, particularly the creation and growth of cells. It is made artificially and can be found in supplements, fortified foods, and some natural foods such as fruits, vegetables, whole grains, and legumes [99,100]. The synthetic form of folate found naturally in food is called FA. During pregnancy, infancy, and adolescence, FA is crucial because the body needs it to manufacture DNA, make red blood cells, and maintain healthy cell division [101–104]. FA is a pteroylglutamic acid derivative, which consists of a pteridine ring, para-aminobenzoic acid, and glutamic acid, and its chemical structure is shown in Figure 18.

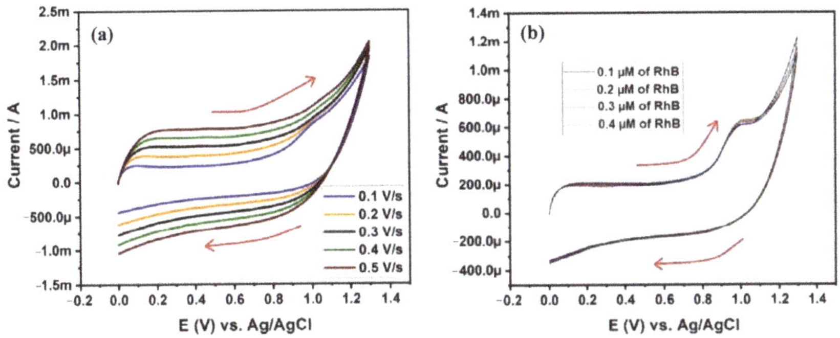

Figure 16. CV of Rhodamine B using 6 mg of DS-MCPE at different (**a**) scan rates and (**b**) concentrations of Rhodamine B analyte [28].

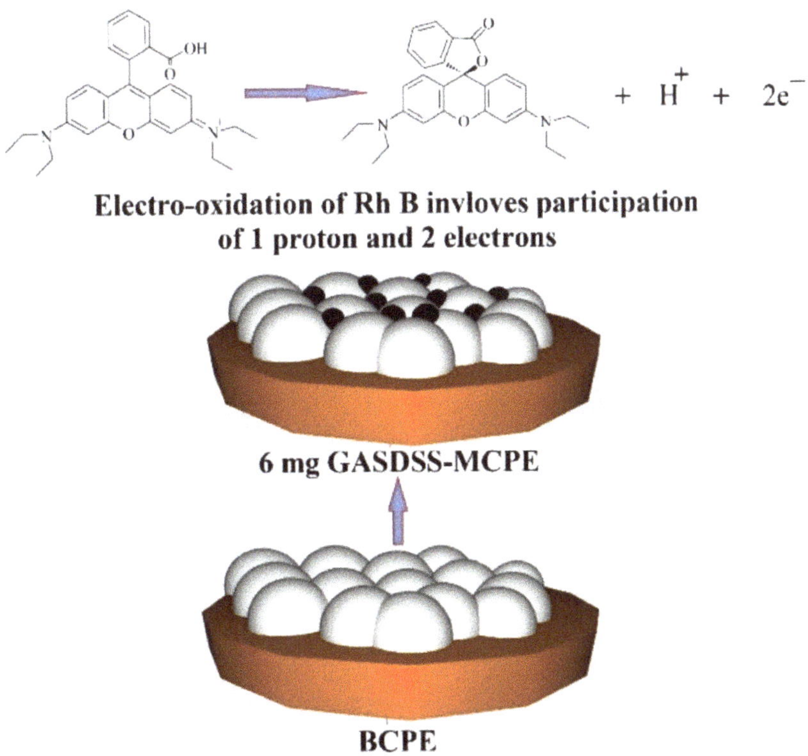

Figure 17. Possible scheme of electrochemical oxidation of Rhodamine B at DS-MCPE [28].

FA is very necessary to maintain cognitive function and mental health, and it reduces depression. Some studies also revealed that FA consumption can reduce the risk of colorectal and cervical cancer by promoting DNA stability and repair. Deficiency of FA during pregnancy can lead to spina bifida and anencephaly (improper growth of the spine and brain, respectively). Diseases like cardiovascular defects, immune system malfunction, dementia, depression, and macrocytic and megaloblastic anemia are caused because of FA deficiency. A crucial nutrient, FA has a wide range of effects on human health, especially in preventing birth abnormalities, promoting cellular function, and preserving general well-being. Its extensive use as a supplement and in food fortification highlights how crucial it is to contemporary healthcare and illness prevention [99–104]. Like any nutrient, though, balance is essential because both an excess and a shortage can cause health issues. Therefore, we need to determine the FA frequently, there are many analytical and electrochemical methods available to determine it, but cyclic voltammetry has proved to be one of the best techniques to determine FA.

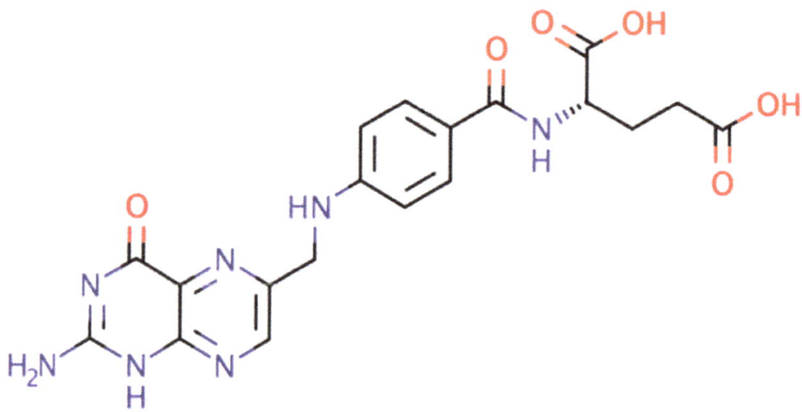

Figure 18. Chemical structure of FA [105].

Shashanka et al. [29] reported the determination of FA using DS-MCPE by cyclic voltammetry method. As we know, optimization of the experimental condition is very important to obtain good results. Authors fabricated DS-MCPE at different concentrations of DS powders as a modifier. Among, 2, 4, 6, and 8 mg of DS-MCPE, 4 mg of DS-MCPE has shown excellent current response during electrochemical oxidation of FA compared with the BCPE, as shown in the CV of Figure 19a. This is because of the increase in the surface area and the reactive sites due to the enhanced surface roughness after modifying the BCPE with 4 mg of DS powders. They have also studied the effect of scan rate and the effect of concentration variation of FA, and their respective CVs are shown in Figure 19b,c.

An increase in the scan rate and the concentration variation of FA increases the Ipa, respectively, due to the reduced time to keep the charge values at higher scan rates and the increased number of FA molecules at higher concentrations. This will create a strong electrochemical reaction between the DS-MCPE and the analyte FA, as a result of which the movement of electrons will increase, and thus increase the Ipa. On the other hand, an increase in pH from 5.7 to 8 will make Epa of FA shift towards negative potential. This confirms the participation of protons in the electrochemical oxidation of FA. It is possible to determine the number of electrons and protons participating in the electrochemical reaction of FA by investigating the effect of pH variation. As per the values mentioned in the article [29], we calculated the number of protons and electrons involved, and the values were found to be almost 2. At last, they have concluded that the fabricated FA sensor could be a potential sensor to diagnose FA deficiency diseases and they have reported all the electrode reactions are adsorption controlled.

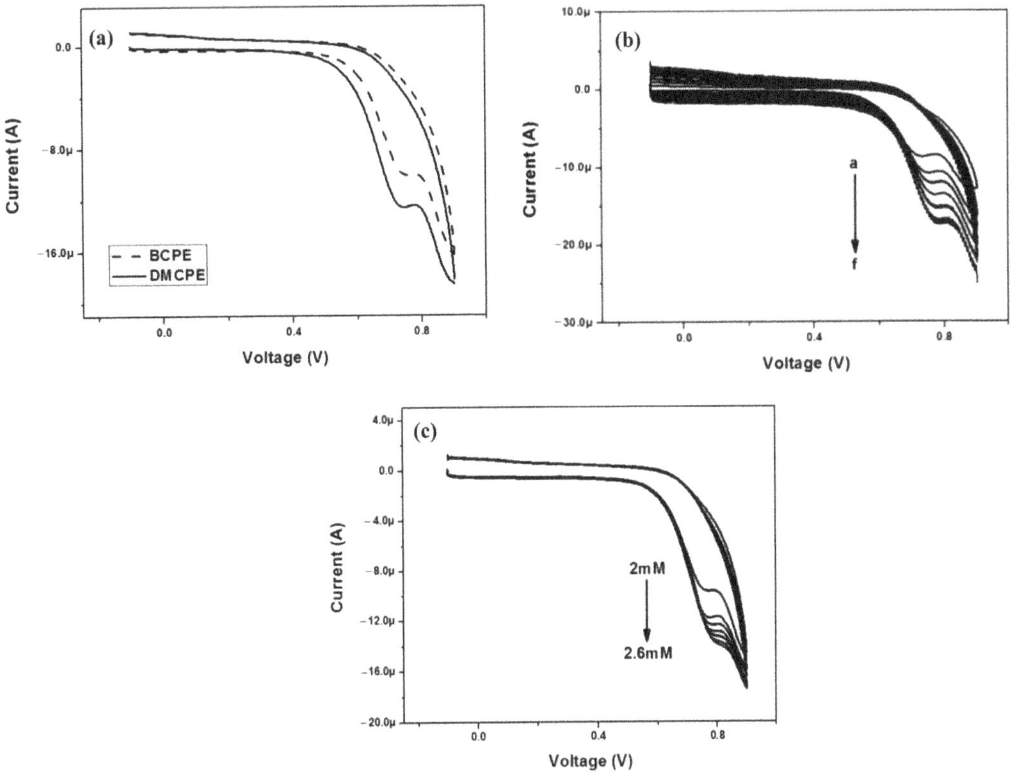

Figure 19. CVs obtained during the electrochemical oxidation of FA using (**a**) DS-MCPE and BCPE, (**b**) at different scan rates (a = 50, b = 100, . . ., f = 300 mV/s), (**c**) at different concentrations of FA [29].

6. High-Entropy-Alloy-Modified Carbon Paste Electrode (HEA-MCPE)

6.1. Introduction to High-Entropy Alloys (HEAs)

High-entropy alloys (HEAs) are the novel grades of modern alloys, which have attracted a lot of interest lately because of their special qualities and potential for a variety of uses. HEAs are made up of five or more major elements, each in large proportions (generally 5–35% by atomic percentage), in contrast to ordinary alloys, which are often made up of a single dominant element and a few alloying elements [106,107]. This produces a multi-phase, multi-configuration, extremely complex microstructure that combines good mechanical, thermal, chemical, and electrochemical properties [108–111].

The phrase "high entropy" describes the system's high configurational entropy, which is the outcome of mixing several different components. This opens the door to the creation of new solid-solution phases, including HCP, BCC, and FCC structures [112]. Despite the existence of many major elements, one of the main features of HEAs is their ability to create single-phase solid solutions. In contrast, secondary phases are frequently formed by extra components in typical alloys. Strength and hardness are boosted by the huge number of elements in HEAs, which stabilizes the solid solution phase at room temperature [113,114]. One of the popular, effective modern methods to prepare HEAs is mechanical alloying [110].

A graphical representation of preparing HEAs by the ball-milling method is depicted in Figure 20.

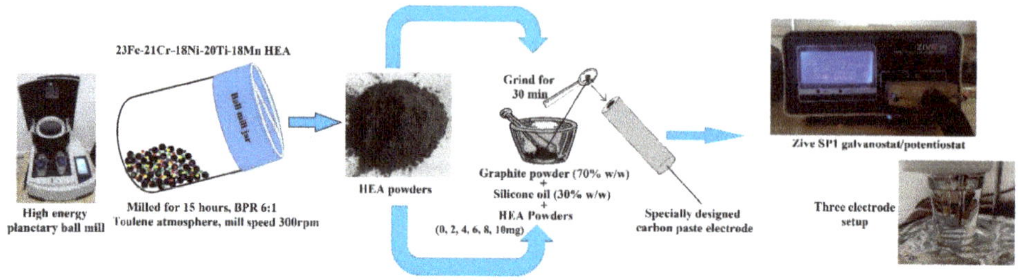

Figure 20. Graphical representation of ball milling of HEA and the fabrication of HEA-MCPE [30].

Figure 21a,b depicts the XRD and SEM images of ball-milled HEAs for 15 h from 0 h [30]. It has several advantages over traditional metallurgical methods. HEAs exhibit high mechanical strength, toughness, ductility, high resistance to corrosion, heat, and wear, and excellent magnetic, electrical, and electrochemical properties. Therefore, they are mainly used in aerospace, energy, coating, biomedical, defense, and automatic sectors. However, HEAs still have enormous potential to overcome many of the drawbacks of conventional alloys, like defining the phase diagrams, low cost, product designing, etc., and open the door to new technological advancements, as long as research and development continue. One of the recent novel applications of HEA is as an "electrocatalyst" to determine various dyes and bioactive compounds by cyclic voltammetry. Very few articles have been published about the electrocatalytic properties of HEAs to investigate redox reactions of various compounds, and we have reviewed a few of them and discussed how HEA plays an important role in acting as AA, methyl orange, and MB sensors.

6.2. Electrochemical Determination of AA

We discussed the importance, properties, applications, and effects of AA on human health earlier in this review article. We will discuss how HEA will help to determine the AA. Shashanka et al. [30] reported the excellent electrochemical properties of HEA powders and their potential use as an AA sensor. They have studied the effect of HEA as a modifier in the CPEs by varying the concentrations of the HEA during the electrochemical oxidation of AA. Among 2, 4, 6, 8, and 10 mg of modified HEA-MCPE, 8 mg of HEA-MCPE has shown an excellent current response, as shown in Figure 22a, and the CV of BCPE and HEA-MCPE in Figure 22b, respectively.

Using the Nernst equation, they determined the active surface area of HEA-MCPE and BCPE, which came out to be 0.0027 cm^2 and 0.0014 cm^2, respectively; this means that after the modification of CPE, its active surface area increased almost 2 times. This will increase the reactive sites and increase the mobility of electrons between the electrode and the AA. Therefore, always MCPEs will show a higher current response than BCPEs in most of the analytes. CVs obtained for the variation in pH and concentration of the AA are shown in Figure 23a and Figure 23b, respectively [30].

As per the figure, as the PBS pH increases from 6 to 8, the Epa is decreasing. On the other hand, concentration variation studies reveal that an increase in the concentrations of AA increases the Ipa linearly. At last, they have concluded that the manufactured HEA-MCPE has demonstrated excellent sensitivity and selectivity in identifying AA. In the medical field, the current electrode may be used as a sensor to diagnose disorders caused by AA deficiency. They have reported that the diffusion phenomenon controls the electrode processes that occur during the electrochemical oxidation of AA.

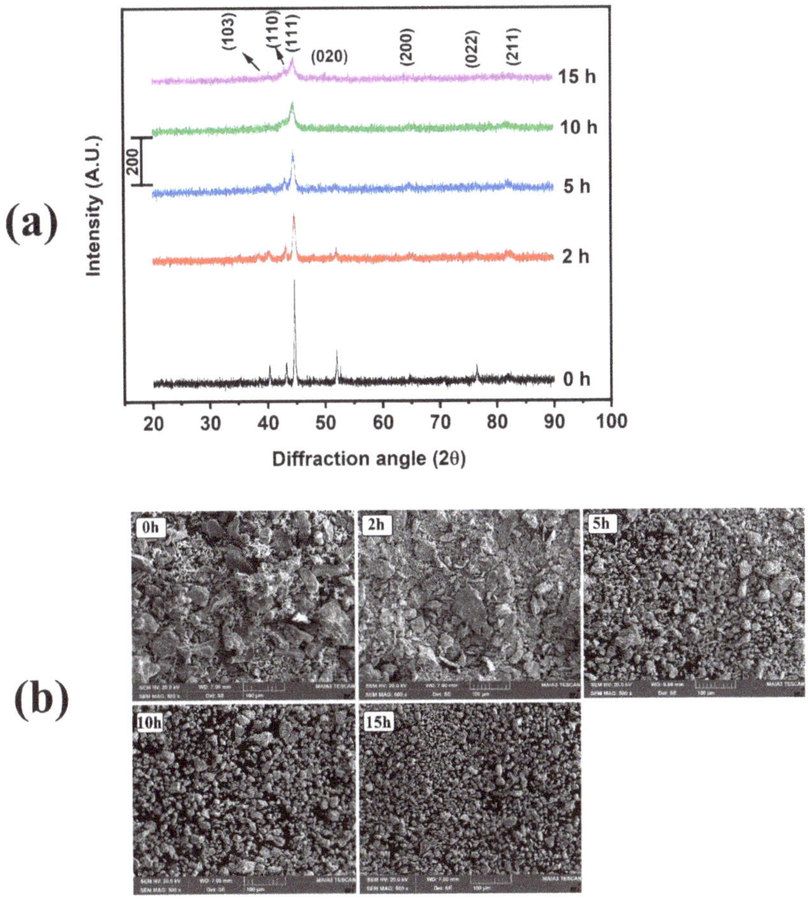

Figure 21. (a) XRD and (b) SEM of 0 to 15 h ball-milled HEA powders with regular time intervals [30].

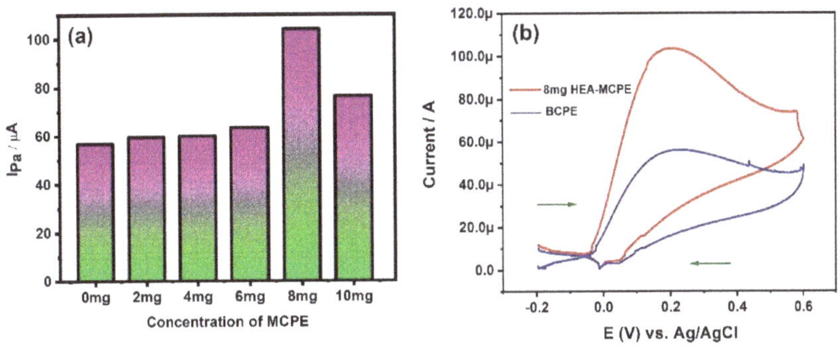

Figure 22. (a) Ipa at various concentrations of HEA-MCPE, (b) CV of BCPE, and 8 mg HEA-MCPE [30].

6.3. Electrochemical Determination of Methyl Orange

Shashanka et al. [31] have reported the use of HEA powders as a modifier in CPEs to determine methyl orange using CV. Figure 24 is the pictorial representation of the experiment carried out in the article [31].

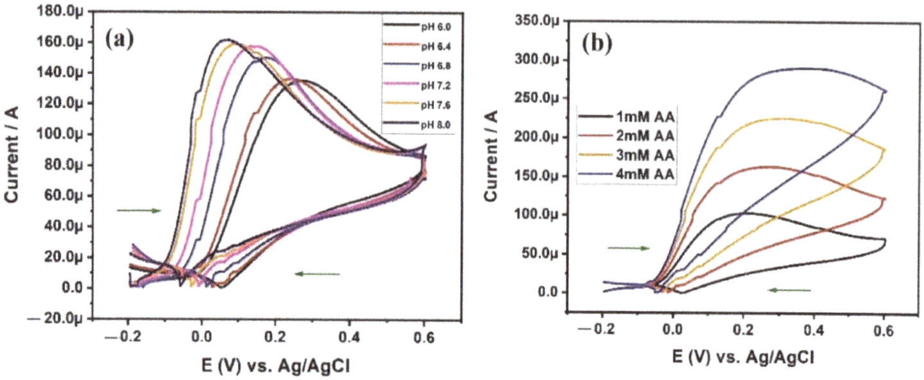

Figure 23. CVs electrochemical oxidation of AA using HEA-MCPE at different (**a**) pH of PBS and (**b**) concentrations of AA analyte [30].

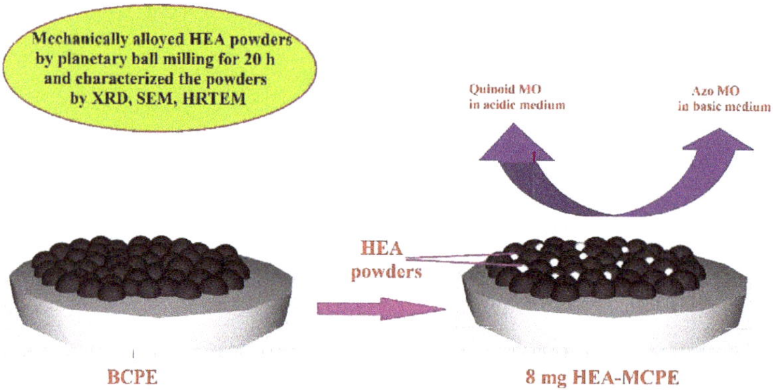

Figure 24. Pictorial representation of fabrication of HEA-MCPE and the electrochemical oxidation of methyl orange [31].

As seen in Figure 25a, they synthesized 2, 4, 6, 8, and 10 mg of HEA-MCPE and noted the corresponding Ipa achieved for the electrochemical oxidation of methyl orange. The graph indicates an increase in surface area and reactive sites up to 8 mg HEA-MCPE, followed by a decrease at 10 mg HEAMCPE. It also shows a rise in Ipa till 8 mg HEA-MCPE, after which it abruptly declines. Figure 25b depicts the over plot of CVs of a blank (without the analyte methyl orange), BCPE, and HEA-MCPE. There is an increase of 4 times the Ipa of 8 mg HEA-MCPE compared to BCPE. The pH studies revealed the participation of protons in the electrochemical oxidation of methyl orange and the calculated number of protons is 2.

The CV obtained for different pH solutions is shown in Figure 26a and its linear regression equation obtained is as follows: Ep (V) = 1.1452 − (0.060) pH (V/pH) (R^2 = 0.9980). The reported slope obtained after plotting a graph of Epa vs. pH is 0.0603 and the value is very close to the Nernst relation value of 0.059 V/pH. This confirms the participation of an equal number of electrons and protons during the electrochemical reaction. They have also performed the variation of the scan rate and the methyl orange and reported their effect on the Ipa of methyl orange, and their respective CVs are shown in Figure 26b,c. Using the Ipa values of Figure 26c, the authors have calculated the LOD and LOQ, and the values were found to be 1.14 nM and 3.8 nM, respectively. The investigations related to the effect of interferents, stability, reproducibility, and repeatability have shown excellent temperament of the HEA-MCPE. This HEA-MCPE can be a potential candidate in the future to detect methyl orange in wastewater, food products, etc. [31].

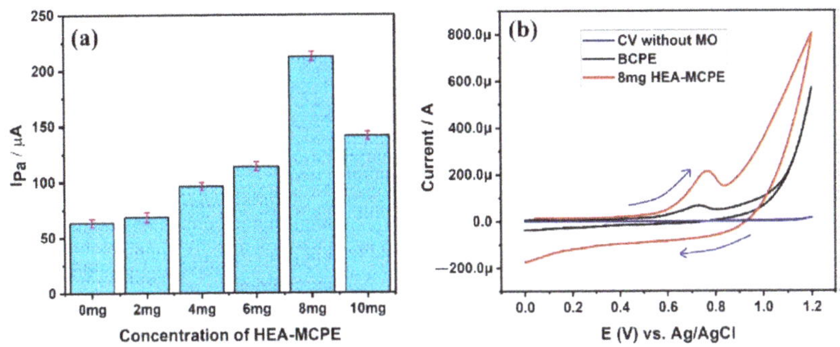

Figure 25. (**a**) Graph of respective Ipa reported at different concentrations of HEA-MCPE, (**b**) CV of a blank (without the analyte methyl orange), BCPE, and HEA-MCPE [31].

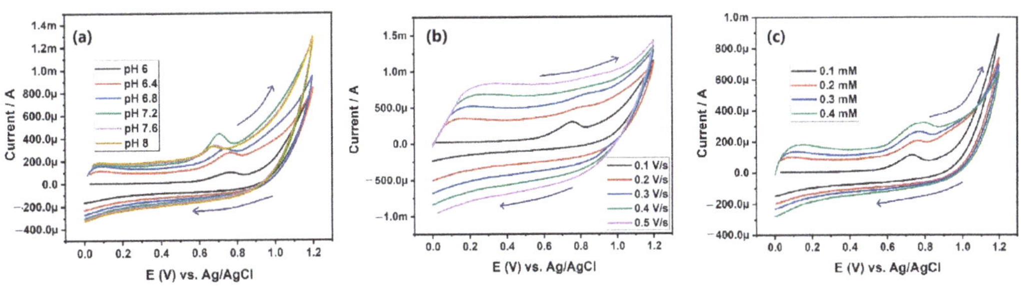

Figure 26. CVs of electrochemical oxidation of methyl orange using HEA-MCPE at different (**a**) PBS of pH, (**b**) scan rates, (**c**) concentrations of methyl orange analyte [31].

6.4. Electrochemical Determination of Methylene Blue

The accuracy, selectivity, sensitivity, repeatability, and reproducibility are further increased by modifying the CPE with ball-milled HEA powders. Shashanka et al. [32] prepared different concentrated HEA-MCPE (2, 4, 6, 8, and 10 mg) and recorded the respective Ipa during the electrochemical redox reaction of the MB and its graphical experimentation is as given in Figure 27 and the Ipa plotted the graph concerning different modifier concentration is as shown in Figure 28a. Among all, 4 mg HEA-MCPE has shown a maximum current response than others. Figure 28b indicates the CVs of BCPE and the 4 mg HEA-MCPE, and from the figure, it is confirmed that the Ipa of HEA-MCPE is 5 times more than the Ipa of BCPE. Therefore, they have used 4 mg HEA-MCPE for further studies.

Figure 29a depicts the CV of MB oxidized at different PBS of pH 6 to 7.6 at a scan rate of 100 V/s. Authors have plotted the Epa vs. different pH and their plot follows the linear regression equation as follows: Ep (V) = 0.5235 − (0.0651) pH(V/pH) (R^2 = 0.9938). This confirms that the electron mobility during the electrochemical redox reaction of MB mainly depends upon the protons. The slope obtained is 0.0651 and it is almost equal to the theoretical standard value of 0.059 V/pH. On the other hand, Figure 29b shows the CV obtained at different concentrations and the Ipa increases linearly. According to the results obtained during the scan rate variations, the mass transfer process is diffusion-controlled. They concluded that the prepared HEA-MCPE might be potentially used as an electrode to detect different dyes, medications, hazardous heavy metals, pesticides, etc. This will give metallurgists and mechanical engineers a new field of study to use different alloy powders for a range of electrocatalytic applications.

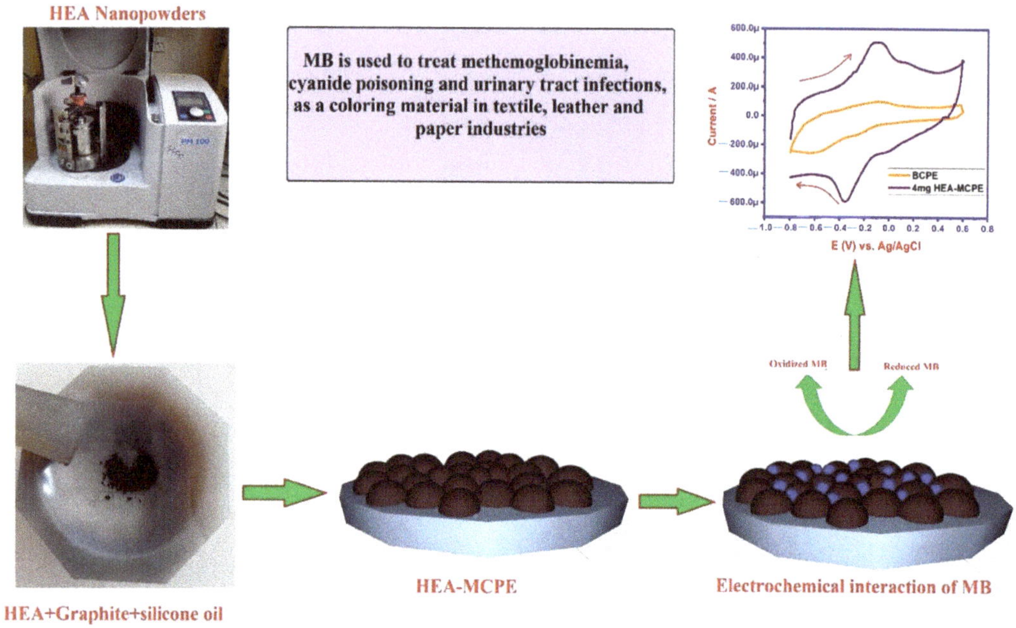

Figure 27. Pictorial representation of fabrication of HEA-MCPE and the electrochemical oxidation of MB [32].

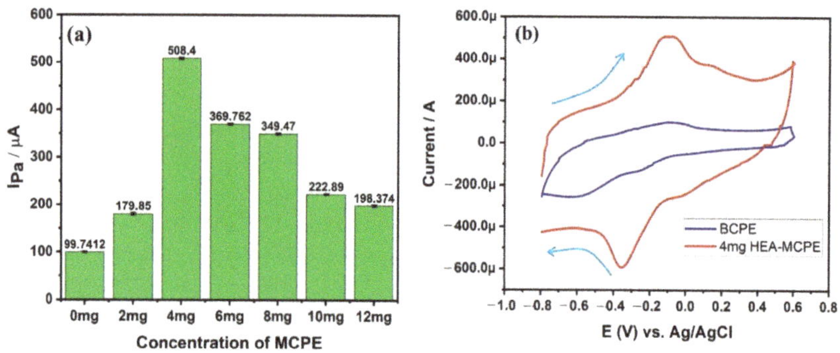

Figure 28. (**a**) Plot of Ipa vs. different concentrations of HEA-MCPE, (**b**) CV of BCPE, and 4 mg HEA-MCPE [32].

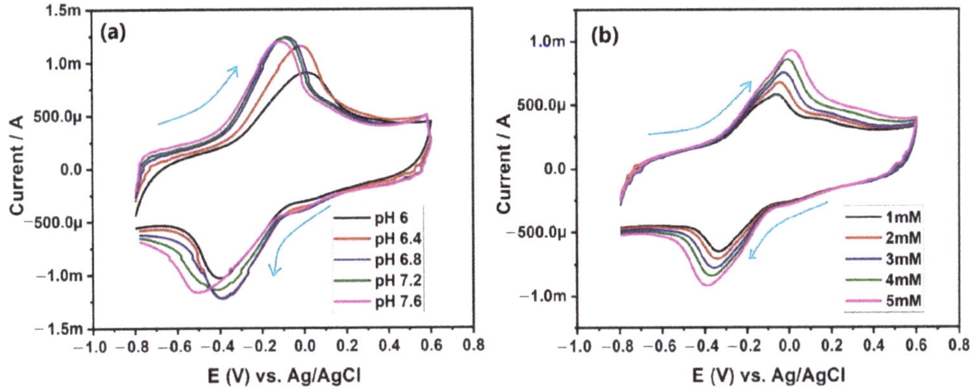

Figure 29. CVs of electrochemical redox reaction of MB under HEA-MCPE at different (**a**) PBS of pH, (**b**) concentrations of MB analyte [32].

7. Shape-Memory-Alloy-Modified Carbon Paste Electrode (SMA-MCPE)

7.1. Introduction to Shape-Memory Alloys (SMAs)

A special class of metallic materials known as shape-memory alloys (SMAs) can revert to a predetermined shape in response to a specific stimulus, like heat. They are categorized as "smart materials" because of their exceptional characteristics [115,116]. When exposed to stress or temperature fluctuations, SMAs go through a phase transformation that allows them to regain their former structure. One of the main characteristics of SMAs is this phenomenon, which is known as the shape-memory effect [117–120]. Although nickel–titanium (NiTi) alloys are widely used SMAs, other alloys, including copper–aluminum–nickel (Cu-Al-Ni) and iron-based SMAs, are also used as SMAs. NiTi alloys are also called nitinol and they undergo the reversible phase transition between a low-temperature and a high-temperature phase. The phase transformation mainly depends upon the temperature and they exhibit superelasticity, high damping capacity, excellent electrochemical properties, biocompatibility, fatigue resistance, etc. [116–120]. Therefore, they are used in stents, surgical instruments, orthodontics, actuators, automotive, aerospace, electronics, defense, microelectromechanical systems (MEMS), and electrochemical sensor applications.

7.2. Preparation of Shape-Memory Alloys (SMAs) by Mechanical Alloying

Shape-memory alloys can have their qualities improved and customized with the help of mechanical alloying. It makes it possible to precisely manipulate the microstructure and composition of the material, which can result in enhanced performance in a variety of demanding applications, superelasticity, and increased shape-memory effects [33]. If we control the basic ball-milling issues like oxidation and phase stability and perform milling optimization before the start of the actual milling, then we can fabricate high-quality nitinols. Mechanically alloyed nitinols possess excellent porosity, high surface area, and surface energy; therefore, nitinols could be potential electrocatalysts. If we disperse the elements like Hf, Cu, carbon nanotubes, and graphenes in small amounts to NiTi composition, then their properties will be improved further.

7.3. Electrochemical Determination of Dopamine

Shashanka et al. [33] investigated the electrocatalytic properties of ball-milled nitinols dispersed with Hf, and the graphical representation of the whole experiment starting from the milling to fabrication of NiTiHf-MCPE is shown in Figure 30. They have investigated the phases present in the alloy at different milling times from 0 to 20 h by XRD and it is given in Figure 31. As per the XRD, we believe 20 h of milling time is required to obtain highly refined SMA powders. Similarly, the progress of milling at different milling times is further investigated by SEM and EDS analysis as shown in Figure 32. The figure confirms the gradual decrease in the particle size from 0 to 20 h of ball milling with the irregular shape of the powders.

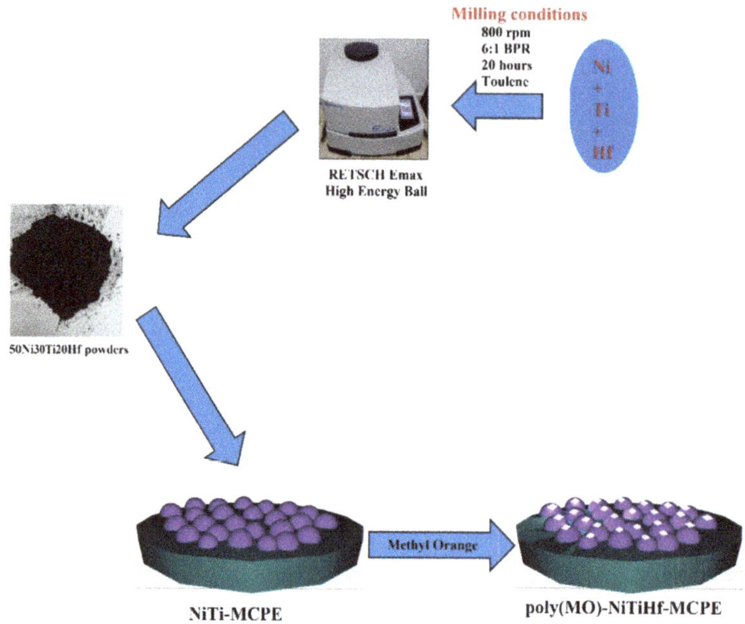

Figure 30. Pictorial representation of whole experiment starting from the milling to fabrication of NiTiHf-MCPE [33].

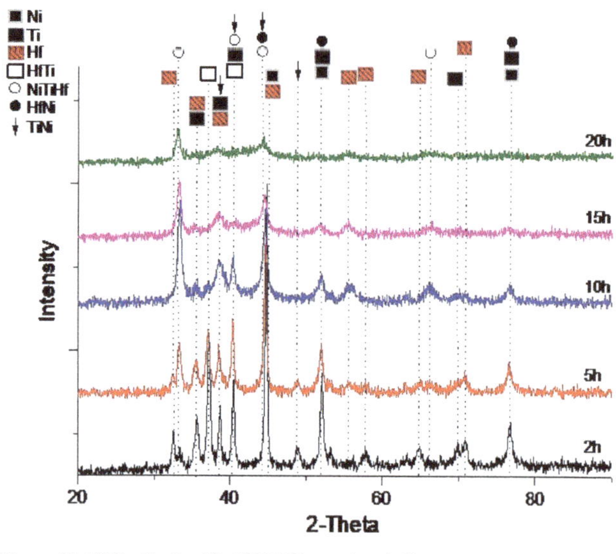

Figure 31. XRD of ball-milled NiTiHf powders [33].

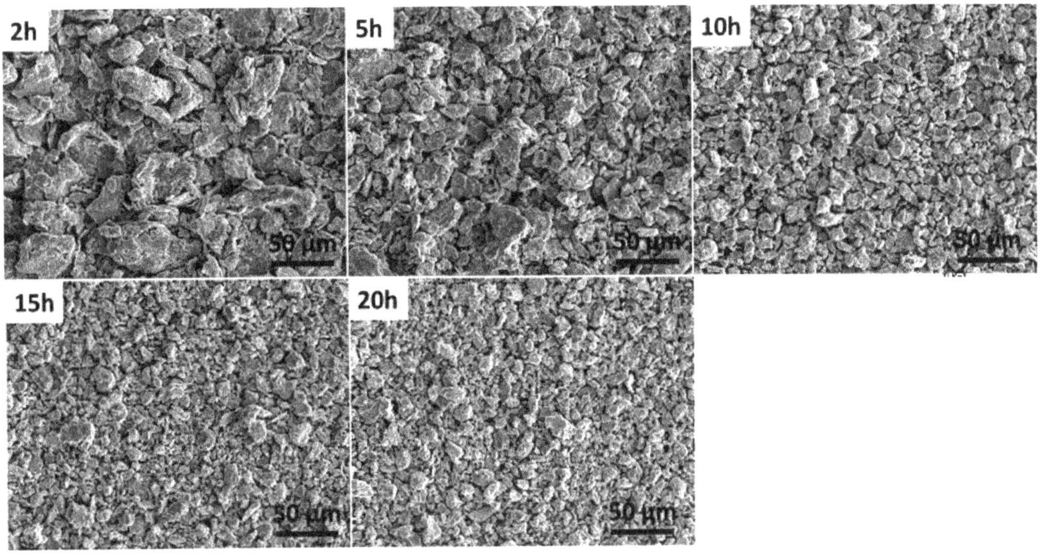

Figure 32. SEM and EDS analysis of ball-milled NiTiHf powders [33].

Initially, Shashanka et al. [33] investigated the electrocatalytic properties of NiTiHf-MCPE and found no remarkable increase in the current response; therefore, they electropolymerized the MCPE with methyl orange and studied their electrocatalytic properties in determining dopamine. After the process of electropolymerization, their current response increased from 40 to around 54 µA. Figure 33a shows the electropolymerized voltammograms after electropolymerized NiTiHf-MCPE with 1 mM methyl orange (MO) at a pH of 7 in 0.2 M phosphate buffer solution (PBS). For up to 20 cycles, the polymerization was carried out at a scan rate of 0.1 V/s and a potential range of 0.3 V to 1.2 V, respectively. It was verified by the voltammograms that a linear increase in Ipa could be seen as the number of cycles increased from 1 to 20; this is because a methyl orange monomer forms a polymer. The polymerization process enhances the NiTiHf-modified electrode's surface area, reactive sites, selectivity, and sensitivity in addition to raising the oxidation peak current. Figure 33b shows the CVs of electrochemical redox reactions of dopamine in BCPE, NiTiHf-MCPE, and poly(MO)-NiTiHf-MCPE, respectively. From the figure, it is confirmed that the electropolymerized MCPE has shown a maximum current response than the other two electrodes. Later authors calculated the active surface areas of BCPE, NiTiHf-MCPE, and poly(MO)-NiTiHf-MCPE using the Nernst equation, and the values were found to be 0.044, 0.089, and 0.098 cm^2, respectively. This shows the importance of modification and electropolymerization of the electrodes [33].

Figure 33c represents the effect of pH on the redox reaction of dopamine using poly(MO)-NiTiHf-MCPE, and it confirms the shifting of Epa to higher values linearly with the increase in the pH from 5 to 7.5. This confirms the participation of protons during the electrochemical redox reaction of dopamine. The linear regression equation for this study is as follows: Ep(V) = 0.1812 − (0.0591) pH (V/pH) (R^2 = 0.9984).

This linear regression equation demonstrates that during the redox reaction of dopamine, the electron mobility is mostly dependent on H+ [121]. The obtained slope of 0.0591 V/pH is extremely near to the conventional value of 0.059 V/pH. This demonstrates that the redox reaction of dopamine involves the same number of protons and electrons, which is determined to be two electrons and two protons. Figure 33d shows the CV of dopamine at different scan rates from 0.025 to 0.4 V/s at poly(MO)-NiTiHf-MCPE. It shows an increase in the IPA with the increase in scan rate. The possible electrochemical oxidation of dopamine on the surface of poly(MO)-NiTiHf-MCPE is shown in Figure 34. Authors have also studied the effect of interferents, repeatability, stability, and reproducibility of poly(MO)-NiTi–MCPE during the electrochemical redox reaction of dopamine and they have reported excellent results [33]. They have also determined dopamine and FA simultaneously using the same poly(MO)-NiTi–MCPE and concluded that this electrode could be a potential candidate to determine various bioactive molecules individually or simultaneously in real samples. Table 1 depicts the comparison of experimental results of different types of alloy-MCPE during the determination of various analytes.

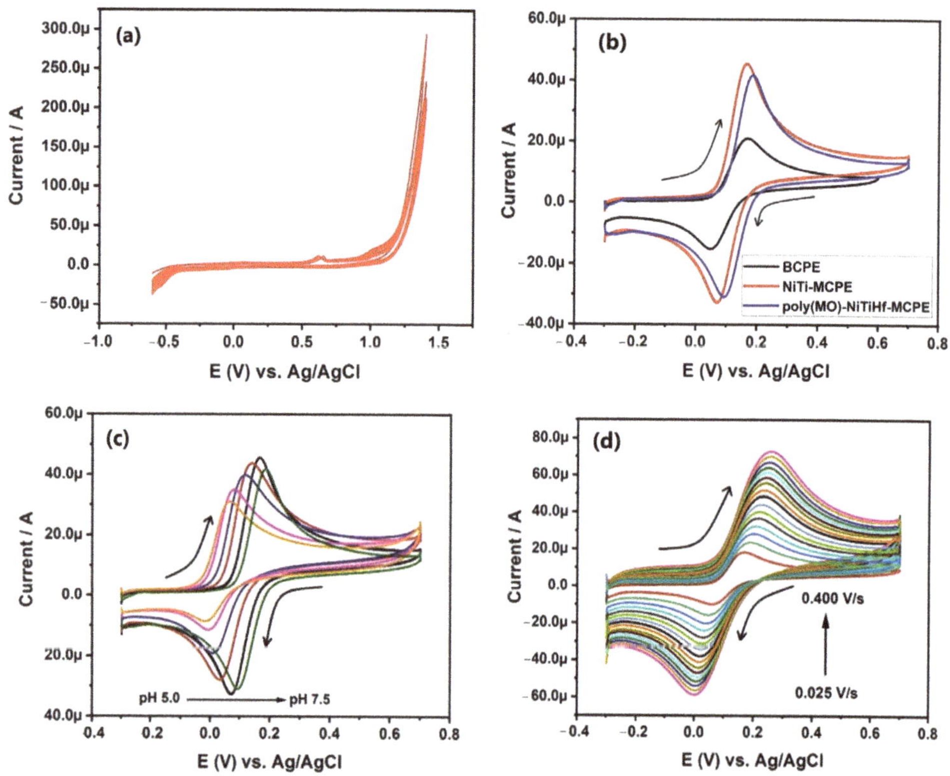

Figure 33. (**a**) CV after electropolymerized NiTiHf-MCPE with 1 mM methyl orange (MO) at a pH of 7 in 0.2 M phosphate buffer solution (PBS), (**b**) CV of electrochemical reaction of dopamine at BCPE, NiTi-MCPE, and poly(MO)-NiTiHf-MCPE, respectively; CVs of electrochemical redox reaction of dopamine using poly(MO)-NiTiHf-MCPE at different (**c**) PBS of pH, (**d**) scan rates [33].

Table 1. Comparison of experimental results of different types of alloy-MCPE during the determination of various analytes.

Modifiers Used	Electrochemical Technique	Analyte Determined	Important Findings	References
Fe-18Cr-13Ni (stainless steel)	CV	Folic Acid	Electrode reactions were adsorption controlled. Current sensitivity of 17.32 μA was recorded.	[29]

Table 1. *Cont.*

Modifiers Used	Electrochemical Technique	Analyte Determined	Important Findings	References
Duplex stainless steel	CV	Ascorbic Acid	The stainless steel showed good current sensitivity of 143.52 μA.	[25]
Duplex stainless steel	CV	Uric Acid	The stainless steel showed good current sensitivity of 19.36 μA.	[25]
Duplex stainless steel	CV	Dopamine	The stainless steel showed good current sensitivity of 25.61 μA.	[25]
Yttria dispersed Fe-18Cr-13Ni	CV	Ascorbic Acid	The stainless steel showed good current sensitivity of 370 mV.	[24]
Yttria dispersed duplex stainless steel	CV	Uric Acid	The stainless steel showed good current sensitivity of 31.01 μA.	[24]
Yttria dispersed duplex stainless steel	CV	Dopamine	The stainless steel showed good current sensitivity of 28.48 μA.	[24]
23Fe-21Cr-18Ni-20Ti-18Mn (HEA)	CV	Ascorbic Acid	For the concentration of 8 mg modifier, a maximum peak current of 104.07 μA was measured. For the high-entropy-alloy-modified carbon paste electrode and the bare carbon paste electrode, the active surface areas for the electron transfer process of ascorbic acid are calculated to be 0.0014 cm^2 and 0.0027 cm^2, respectively.	[30]
25Fe-19Cr-19Ni-18Ti-19Mn high-entropy alloy	CV	Methylene Blue	The anodic peak current of 508.4 μA was displayed by the 4 mg high-entropy-alloy-modified carbon paste electrode, while only 99.74 μA was displayed by the bare carbon paste electrode. This significant anodic peak current difference between the two different electrodes have demonstrated the value of the the modifier in enhancing the electrode sensor's sensitivity, robustness, and selectivity.	[32]
2507 super duplex stainless steel	CV	Ascorbic Acid	BCPE has shown an anodic peak current of 22.5 μA and 4 mg SDSS-MCPE has recorded 37.2 μA of anodic peak current during the electro-oxidation of 1 mM AA. LOD and LOQ were calculated to be 2.06 nM and 6.8 nM, respectively.	[26]
poly (methyl orange) shape memory alloy	CV	Dopamine	The calculated active surface area for BCPE, NiTiHf-MCPE, and the poly(MO)-NiTiHf-MCPE were found to be 0.044, 0.089, and 0.098 cm^2, respectively. LOD and LOQ were calculated to be 5.55 μM and 18.45 μM, respectively.	[33]

Table 1. *Cont.*

Modifiers Used	Electrochemical Technique	Analyte Determined	Important Findings	References
23Fe-21Cr-18Ni-20Ti-18Mn	CV	Methyl Orange	MO oxidizes at 700 mV. The calculated electrode surface area of BCPE and HEAMCPE was found to be 0.0546 and 0.4439 cm^2, respectively. The LOD obtained is 0.080 μM.	[31]

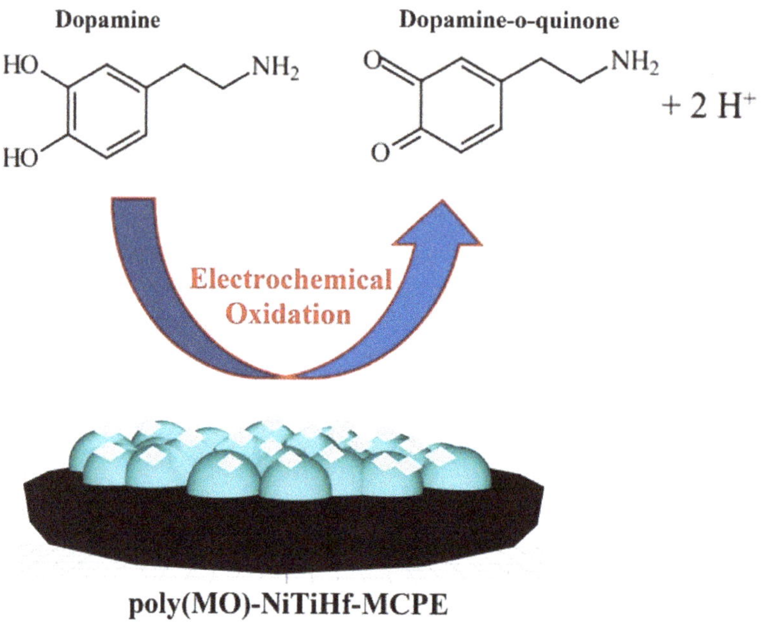

Figure 34. Possible electrochemical oxidation scheme of dopamine at poly(MO)-NiTiHf-MCPE [33].

8. Challenges and Future Perspectives

Although using nanostructured-alloy-modified CPEs for electrochemical sensing has several benefits, there are drawbacks as well:

- Reproducibility: Because of differences in how nanostructured alloys were prepared and how the nanoparticles were distributed in the past, it might be difficult to achieve similar performance across several sensors.
- Interference from Biological Matrices: Measurement accuracy may be impacted by the presence of interferents in complex biological samples, such as blood or urine. To reduce these interferences, sophisticated sensor designs and adjustments are required.
- Long-term Stability: Although nanostructured alloys are generally durable, oxidation and other causes may cause them to perform worse over time, which could compromise the sensors' long-term dependability.

By creating more reliable, repeatable electrode modification methods, improving sensor selectivity through sophisticated material design, and combining electrochemical sensors with portable devices for real-time applications, future research will probably concentrate on overcoming these obstacles.

9. Conclusions

Alloys are the perfect candidates for electrochemical sensing because of their exceptional conductivity and biocompatibility as electrocatalysts. Their catalytic qualities enable the oxidation or reduction of analytes at lower applied potentials, while their large surface area enables the effective adsorption of biomolecules. Selectivity and sensitivity towards particular analytes can be enhanced by the synergistic effects of alloys. Furthermore, the presence of nanostructured alloys on the electrode surface enhances the sensor's selectivity by enabling improved resolution of overlapping signals. These sensors are an important tool for environmental monitoring since they can identify minute levels of contaminants, harmful compounds, bioactive molecules, amino acids, and other substances in soil or water samples. The sensitivity of the sensor is raised by the significantly greater surface area that the nanoscale size of metal particles offers for the adsorption of target molecules. Because of their superior conductivity, these alloys—MCPE—produce more accurate and timely reactions by accelerating electron transport. They are also more durable and robust, which ensures that the sensors will perform effectively even under harsh circumstances and after extended use. Given the ongoing advancements in material science and sensor design, these modified electrodes are anticipated to play a key role in the development of next-generation diagnostic tools and environmental monitoring systems.

Author Contributions: Conceptualization, S.R. and V.A.; methodology, S.R. and R.S.M.; validation, Y.C. and G.R.C.; formal analysis, S.R. and R.S.M.; investigation, E.A. and V.A.; data curation, Y.C. and G.R.C.; writing—original draft preparation, S.R. and R.S.M.; writing—review and editing, S.R.; visualization, E.A., V.A. and Y.C.; supervision, S.R. All authors have read and agreed to the published version of the manuscript.

Funding: This research received no external funding.

Institutional Review Board Statement: Not applicable.

Informed Consent Statement: Not applicable.

Data Availability Statement: No new data were created or analyzed in this study. Data sharing is not applicable to this article.

Conflicts of Interest: The authors declare no conflicts of interest.

References

1. Yamaguchi, H.; Miyazaki, M. Enzyme-Immobilized Microfluidic Devices for Biomolecule Detection. *TrAC Trends Anal. Chem.* **2022**, *159*, 116908. [CrossRef]
2. Ziani, I.; Bouakline, H.; Guerraf, A.E.; Bachiri, A.E.; Fauconnier, N.M.-L.; Sher, F. Integrating AI and Advanced Spectroscopic Techniques for Precision Food Safety and Quality Control. *Trends Food Sci. Technol.* **2024**, *156*, 104850. [CrossRef]
3. Manjunatha, J.G. *Advances in Surfactant Biosensor and Sensor Technologies*; Elsevier: Amsterdam, The Netherlands, 2024. [CrossRef]
4. Charithra, M.M.; Manjunatha, J.G.G.; Hareesha, N.; Prinith, S.N.; Ravishankar, D.K.; Arpitha, H.J. An Overview of Recent Development in Carbon-Based Sensors for Neurotransmitter Detection. *Comb. Chem. High Throughput Screen.* **2023**, *26*, 2614–2624. [CrossRef] [PubMed]
5. Hareesha, N.; Manjunatha, J.G.; Girish, T.; ALOthman, Z.A. Analysis of Catechol as an Environmental Pollutant Based on Electropolymerised Methyl Orange Modified Carbon Paste Sensor. *Int. J. Environ. Anal. Chem.* **2023**, *104*, 9186–9198. [CrossRef]
6. Senthilkumar, N.; Babu, K.J.; Kumar, G.G.; Kim, A.R.; Yoo, D.J. Flexible Electrospun PVDF-HFP/NI/CO Membranes for Efficient and Highly Selective Enzyme Free Glucose Detection. *Ind. Eng. Chem. Res.* **2014**, *53*, 10347–10357. [CrossRef]
7. Tamilarasi, S.; Kumar, R.S.; Cho, K.-B.; Kim, C.-J.; Yoo, D.J. High-Performance Electrochemical Detection of Glucose in Human Blood Serum Using a Hierarchical NiO2 Nanostructure Supported on Phosphorus Doped Graphene. *Mater. Today Chem.* **2023**, *34*, 101765. [CrossRef]

8. Nadar, N.R.; Deepak, J.; Sharma, S.C.; Krushna, B.R.R.; Puneeth, N.; Sowjanya, R.; Varalakshmi, V.S.; Sahu, S.; Sargunam, B.; Nagabhushana, H.; et al. Graphene Oxide Based $Gd_2O_3:Eu^{3+}$ Nanocomposites: A Multifaceted Approach to Advanced Energy Storage and Bio Sensing Applications. *Inorg. Chem. Commun.* **2024**, *165*, 112515. [CrossRef]
9. Vaibhav, N.; Swamy, B.E.K.; Manjunatha, L.S.; Manjunatha, K.G.; Sharma, S.C. Electrochemical Determination of Uric Acid in Presence of FA Using Synthesized Cobalt Oxide Modified Carbon Paste Electrode. *Inorg. Chem. Commun.* **2024**, *165*, 112469. [CrossRef]
10. Nadar, N.R.; Deepak, J.; Sharma, S.C.; Krushna, B.R.R.; Puneeth, N.; Sowjanya, R.; Varalakshmi, V.S.; Sahu, S.; Sargunam, B.; Nagabhushana, H.; et al. Synergistic Doping Strategies Boosting Electrochemical Performance: GO-Y_2O_3: Eu^{3+}/Li^+ Nanocomposites for Supercapacitor and Biosensor Applications. *Inorg. Chem. Commun.* **2024**, *161*, 112397. [CrossRef]
11. Atta, N.F.; Gawad, S.A.A.; Galal, A.; Razik, A.A.; El-Gohary, A.R. Efficient Electrochemical Sensor for Determination of H_2O_2 in Human Serum Based on Nano Iron-nickel Alloy/Carbon Nanotubes/Ionic Liquid Crystal Composite. *J. Electroanal. Chem.* **2020**, *881*, 114953. [CrossRef]
12. Hernández-Saravia, L.P.; Martinez, T.; Llanos, J.; Bertotti, M. A Cu-NPG/SPE Sensor for Non-Enzymatic and Non-Invasive Electrochemical Glucose Detection. *Microchem. J.* **2020**, *160*, 105629. [CrossRef]
13. Hajmalek, S.; Jahani, S.; Foroughi, M.M. Simultaneous Voltammetric Determination of Tramadol and Paracetamol Exploiting Glassy Carbon Electrode Modified with $FeNi_3$ Nanoalloy in Biological and Pharmaceutical Media. *ChemistrySelect* **2021**, *6*, 8797–8808. [CrossRef]
14. Halfadji, A.; Naous, M.; Rajendrachari, S.; Ceylan, Y.; Ceylan, K.B.; Shekar, P.V.R. Effective Investigation of Electro-Catalytic, Photocatalytic, and Antimicrobial Properties of Porous CuO Nanoparticles Green Synthesized Using Leaves of Cupressocyparis Leylandii. *J. Mol. Struct.* **2023**, *1301*, 137318. [CrossRef]
15. Halfadji, A.; Naous, M.; Kharroubi, K.N.; Belmehdi, F.E.Z.; Rajendrachari, S. An Ultrasonic-Assisted Synthesis, Characterization, and Application of Nano-Fe_3O_4/TiO_2 as Nano-Catalyst for the Removal of Organic Dye by Like-Photo-Fenton Reactions. *Inorg. Chem. Commun.* **2023**, *158*, 111686. [CrossRef]
16. Halfadji, A.; Chougui, A.; Djeradi, R.; Ouabad, F.Z.; Aoudia, H.; Rajendrachari, S. TiO_2-Decorated by Nano-γ-Fe_2O_3 as a Catalyst for Efficient Photocatalytic Degradation of Orange G Dye under Eco-Friendly White LED Irradiation. *ACS Omega* **2023**, *8*, 39907–39916. [CrossRef]
17. Rajendrachari Shashanka, B.N.; M Avar, B.S.P.; Kumar, R.M.; Baek, K.H. Assessing the Food Quality Using Carbon Nanomaterial Based Electrodes by Voltammetric Techniques. *Biosensors* **2022**, *12*, 1173. [CrossRef]
18. Rajendrachari, S.; Jayaprakash, G.K.; Pandith, A.; Karaoglanli, A.C.; Uzun, O. Electrocatalytic Investigation by Improving the Charge Kinetics between Carbon Electrodes and Dopamine Using Bio-Synthesized CuO Nanoparticles. *Catalysts* **2022**, *12*, 994. [CrossRef]
19. Mahale, R.S.; Shamanth, V.K.; Hemanth, K.; Shashanka, R.; Sharath, P.C.; Sreekanth, N.V. Electrochemical Sensor Applications of Nanoparticle Modified Carbon Paste Electrodes to Detect Various Neurotransmitters: A Review. *Appl. Mech. Mater.* **2022**, *908*, 69–88. [CrossRef]
20. Shashanka, R.; Jayaprakash, G.K.; BG, P.; Kumar, M.; Swamy, B.E.K. Electrocatalytic Determination of AA Using a Green Synthesised Magnetite Nano-Flake Modified Carbon Paste Electrode by Cyclic Voltammetric Method. *Mater. Res. Innov.* **2021**, *26*, 229–239. [CrossRef]
21. Pavitra, V.; Praveen, B.M.; Nagaraju, G.; Shashanka, R. Energy Storage, Photocatalytic and Electrochemical Nitrite Sensing of Ultrasound-Assisted Stable TA_2O_5 Nanoparticles. *Top. Catal.* **2022**. [CrossRef]
22. Rajendrachari, S.; Ramakrishna, D. Functionalized Nanomaterial-Based Electrochemical Sensors: ASensitive Sensor Platform. In *Elsevier eBooks*; Woodhead Publishing: Cambridge, UK, 2022; pp. 3–25.
23. Rajendrachari, S.; Taslimi, P.; Karaoglanli, A.C.; Uzun, O.; Alp, E.; Jayaprakash, G.K. Photocatalytic Degradation of Rhodamine B (RhB) Dye in Waste Water and Enzymatic Inhibition Study Using Cauliflower Shaped ZnO Nanoparticles Synthesized by a Novel One-Pot Green Synthesis Method. *Arab. J. Chem.* **2021**, *14*, 103180. [CrossRef]
24. Shashanka, R.; Chaira, D.; Swamy, B.E.K. Fabrication of yttria dispersed duplex stainless steel electrode to determine dopamine, ascorbic and uric acid electrochemically by using cyclic voltammetry. *Int. J. Sci. Eng. Res.* **2016**, *7*, 1275–1285.
25. Shashanka, R.; Chaira, D.; Swamy, B.E.K. Electrochemical investigation of duplex stainless steel at carbon paste electrode and its application to the detection of dopamine, ascorbic and uric acid. *Int. J. Sci. Eng. Res.* **2015**, *6*, 1863–1871.
26. Mahale, R.S.; Vasanth, S.; Chikkegouda, S.P.; Rajendrachari, S.; Narsimhachary, D.; Basavegowda, N. Electrochemical Determination of AA by Mechanically Alloyed Super Duplex Stainless Steel Powders. *Metals* **2023**, *13*, 1430. [CrossRef]
27. Mahale, R.S.; Rajashekar, V.; Vasanth, S.; Chikkegowda, S.P.; Rajendrachari, S.; Mahesh, V. Fabrication of Mechanically Alloyed Super Duplex Stainless Steel Powder-Modified Carbon Paste Electrode for the Determination of MBby the Cyclic Voltammetry Technique. *ACS Omega* **2024**, *9*, 10660–10670. [CrossRef]

28. Rajendrachari, S.; Vinaykumar, R.; Katti, R.; Koujalagi, B.G.; Jayaprakash, G.K.; Ramkumar, R. Cyclic Voltammetric Determination of Rhodamine B Dye Using a Gas-Atomized Duplex Stainless Steel Powder-Modified Carbon Paste Electrode. *Top. Catal.* **2024**. [CrossRef]
29. Shashanka, R.; Chaira, D.; Swamy, B.E.K. Electrocatalytic Response of Duplex and Yittria Dispersed Duplex Stainless Steel Modified Carbon Paste Electrode in Detecting FA Using Cyclic Voltammetry. *Int. J. Electrochem. Sci.* **2015**, *10*, 5586–5598. [CrossRef]
30. Rajendrachari, S.; Adimule, V.; Gulen, M.; Khosravi, F.; Somashekharappa, K.K. Synthesis and Characterization of High Entropy Alloy 23Fe-21Cr-18Ni-20Ti-18Mn for Electrochemical Sensor Applications. *Materials* **2022**, *15*, 7591. [CrossRef]
31. Rajendrachari, S.; Mahale, R.S.; Adimule, V.; Gulen, M.; Mahesh, V.; Basavegowda, N. Fabrication of Mechanically Alloyed High Entropy Alloys (20-Fe-20Cr-20Ni-20Mn-20Ti) Modified Carbon Paste Electrode Sensor for the Cyclic Voltammetric Determination of Methyl Orange. *Inorg. Chem. Commun.* **2024**, *164*, 112428. [CrossRef]
32. Rajendrachari, S.; Adimule, V.M.; Jayaprakash, G.K.; Pandith, A. Electrochemical Oxidation of MB Dye in Wastewater Using Mechanically Alloyed High Entropy Alloy Modified Carbon Paste Electrode Using Cyclic Voltammetry. *Mater. Res. Express* **2023**, *10*, 054003. [CrossRef]
33. Rajendrachari, S.; Altaş, E.; Erdogan, A.; Küçük, Y.; Gök, M.S.; Khosravi, F. Electrochemical Determination of Dopamine by Poly (Methyl Orange) Shape Memory Alloy Modified Carbon Paste Electrode. *Inorg. Chem. Commun.* **2024**, *167*, 112826. [CrossRef]
34. Suryanarayana, C. Mechanical Alloying and Milling. *Prog. Mater. Sci.* **2001**, *46*, 1–184. [CrossRef]
35. Suryanarayana, C. Nanocrystalline Materials. *Int. Mater. Rev.* **1995**, *40*, 41–64. [CrossRef]
36. Suryanarayana, C.; Ivanov, E.; Boldyrev, V.V. The Science and Technology of Mechanical Alloying. *Mater. Sci. Eng. A* **2001**, *304–306*, 151–158. [CrossRef]
37. Prabhu, B.; Suryanarayana, C.; An, L.; Vaidyanathan, R. Synthesis and Characterization of High Volume Fraction Al–Al_2O_3 Nanocomposite Powders by High-Energy Milling. *Mater. Sci. Eng. A* **2006**, *425*, 192–200. [CrossRef]
38. Suryanarayana, C. Structure and Properties of Nanocrystalline Materials. *Bull. Mater. Sci.* **1994**, *17*, 307–346. [CrossRef]
39. Suryanarayana, C. Mechanical Alloying: A Novel Technique to Synthesize Advanced Materials. *Research* **2019**, *2019*, 4219812. [CrossRef]
40. Shashanka, R.; Chaira, D. Optimization of Milling Parameters for the Synthesis of Nano-Structured Duplex and Ferritic Stainless Steel Powders by High Energy Planetary Milling. *Powder Technol.* **2015**, *278*, 35–45. [CrossRef]
41. Chaira, D. Development of Nano-Structured Duplex and Ferritic Stainless Steels by Pulverisette Planetary Milling Followed by Pressureless Sintering. *Mater. Charact.* **2014**, *99*, 220–229. [CrossRef]
42. Chaira, D. Phase Transformation and Microstructure Study of Nano-Structured Austenitic and Ferritic Stainless Steel Powders Prepared by Planetary Milling. *Powder Technol.* **2014**, *259*, 125–136. [CrossRef]
43. Hocevar, S.; Ogorevc, B. Preparation and Characterization of Carbon Paste Micro-Electrode Based on Carbon Nano-Particles. *Talanta* **2007**, *74*, 405–411. [CrossRef] [PubMed]
44. Apetrei, C.; Apetrei, I.M.; De Saja, J.A.; Rodriguez-Mendez, M.L. Carbon Paste Electrodes Made from Different Carbonaceous Materials: Application in the Study of Antioxidants. *Sensors* **2011**, *11*, 1328–1344. [CrossRef] [PubMed]
45. Abbaspour, A. Chemically Modified Carbon Paste Electrode for Determination of Copper(II) by Potentiometric Method. *Talanta* **2002**, *56*, 91–96. [CrossRef] [PubMed]
46. Koirala, K.; Santos, J.H.; Tan, A.L.; Ali, M.A.; Mirza, A.H. Chemically Modified Carbon Paste Electrode for the Detection of Lead, Cadmium and Zinc Ions. *Sens. Rev.* **2016**, *36*, 339–346. [CrossRef]
47. Nesakumar, N.; Berchmans, S.; Alwarappan, S. Chemically Modified Carbon Based Electrodes for the Detection of Reduced Glutathione. *Sens. Actuators B Chem.* **2018**, *264*, 448–466. [CrossRef]
48. Ghassab, N.; Soleymanpour, A.; Shafaatian, B. Development of an Ultrasensitive Chemically Modified Carbon Paste Electrode for Selective Determination Trace Amount of Sulfate Ion. *Measurement* **2022**, *205*, 112231. [CrossRef]
49. Soleymanpour, A.; Shafaatian, B.; Mirfakhraei, H.S.; Rezaeifard, A. Development of a New Chemically Modified Carbon Paste Electrode for Selective Determination of Urinary and Serum Oxalate Concentration. *Talanta* **2013**, *116*, 427–433. [CrossRef]
50. Wang, H.; Qian, D.; Xiao, X.; Deng, C.; Liao, L.; Deng, J.; Lin, Y.-W. Preparation and Application of a Carbon Paste Electrode Modified with Multi-Walled Carbon Nanotubes and Boron-Embedded Molecularly Imprinted Composite Membranes. *Bioelectrochemistry* **2018**, *121*, 115–124. [CrossRef]
51. Hamidi, H.; Shams, E.; Yadollahi, B.; Esfahani, F.K. Fabrication of Bulk-Modified Carbon Paste Electrode Containing α-PW12O403−Polyanion Supported on Modified Silica Gel: Preparation, Electrochemistry and Electrocatalysis. *Talanta* **2007**, *74*, 909–914. [CrossRef]
52. Ali, T.A.; Mohamed, G.G. Determination of Mn(II) Ion by a Modified Carbon Paste Electrode Based on Multi-Walled Carbon Nanotubes (MWCNTs) in Different Water Samples. *Sens. Actuators B Chem.* **2014**, *202*, 699–707. [CrossRef]

53. Zaib, M.; Athar, M.M. Electrochemical Evaluation of Biologically Modified Carbon Paste Electrode with Cyclic Voltammetric Technique. *Port. Electrochim. Acta* **2019**, *37*, 285–294. [CrossRef]
54. Francis, R.; Byrne, G. Duplex Stainless Steels—Alloys for the 21st Century. *Metals* **2021**, *11*, 836. [CrossRef]
55. Straße, A.; Gumenyuk, A.; Rethmeier, M. Study on Duplex Stainless Steel Powder Compositions for the Coating of Thick Plates for Laser Beam Welding. *Adv. Eng. Mater.* **2022**, *24*, 2101327. [CrossRef]
56. Pezzato, L.; Calliari, I. Advances in Duplex Stainless Steels. *Materials* **2022**, *15*, 7132. [CrossRef]
57. Omiogbemi, I.M.-B.; Yawas, D.S.; Das, A.; Afolayan, M.O.; Dauda, E.T.; Kumar, R.; Gorja, S.R.; Chowdhury, S.G. Mechanical Properties and Corrosion Behaviour of Duplex Stainless Steel Weldment Using Novel Electrodes. *Sci. Rep.* **2022**, *12*, 22405. [CrossRef]
58. Alvarez-Armas, I. Duplex Stainless Steels: Brief History and Some Recent Alloys. *Recent Pat. Mech. Eng.* **2008**, *1*, 51–57. [CrossRef]
59. Han, Y.; Liu, Z.-H.; Wu, C.-B.; Zhao, Y.; Zu, G.-Q.; Zhu, W.-W.; Ran, X. A Short Review on the Role of Alloying Elements in Duplex Stainless Steels. *Tungsten* **2022**, *5*, 419–439. [CrossRef]
60. Besharatloo, H.; Carpio, M.; Cabrera, J.-M.; Mateo, A.M.; Fargas, G.; Wheeler, J.M.; Roa, J.J.; Llanes, L. Novel Mechanical Characterization of Austenite and Ferrite Phases within Duplex Stainless Steel. *Metals* **2020**, *10*, 1352. [CrossRef]
61. Duplex Stainless Steels. In *ASM International eBooks*; ASM: Almere, The Netherlands, 2008; pp. 91–107.
62. Olguín, H.J.; Guzmán, D.C.; García, E.H.; Mejía, G.B. The Role of Dopamine and Its Dysfunction as a Consequence of Oxidative Stress. *Oxidative Med. Cell. Longev.* **2015**, *2016*, 9730467. [CrossRef]
63. Speranza, L.; Di Porzio, U.; Viggiano, D.; De Donato, A.; Volpicelli, F. Dopamine: The Neuromodulator of Long-Term Synaptic Plasticity, Reward and Movement Control. *Cells* **2021**, *10*, 735. [CrossRef]
64. Franco, R.; Reyes-Resina, I.; Navarro, G. Dopamine in Health and Disease: Much More than a Neurotransmitter. *Biomedicines* **2021**, *9*, 109. [CrossRef] [PubMed]
65. Tian, Y. The Role of Dopamine Neurotransmitters in Neurological Diseases: New Sight. *Int. J. Mol. Sci.* **2024**, *25*, 7529. [CrossRef] [PubMed]
66. Meder, D.; Herz, D.M.; Rowe, J.B.; Lehéricy, S.; Siebner, H.R. The Role of Dopamine in the Brain-Lessons Learned from Parkinson's Disease. *NeuroImage* **2018**, *190*, 79–93. [CrossRef] [PubMed]
67. Latif, S.; Jahangeer, M.; Razia, D.M.; Ashiq, M.; Ghaffar, A.; Akram, M.; Allam, A.E.; Bouyahya, A.; Garipova, L.; Shariati, M.A.; et al. Dopamine in Parkinson's Disease. *Clin. Chim. Acta* **2021**, *522*, 114–126. [CrossRef]
68. Cramb, K.M.L.; Beccano-Kelly, D.; Cragg, S.J.; Wade-Martins, R. Impaired Dopamine Release in Parkinson's Disease. *Brain* **2023**, *146*, 3117–3132. [CrossRef]
69. Falamaş, A.; Petran, A.; Hada, A.-M.; Bende, A. Dopamine Photochemical Behaviour under UV Irradiation. *Int. J. Mol. Sci.* **2022**, *23*, 5483. [CrossRef]
70. Amare, M. Electrochemical Determination of AA in Pharmaceutical Tablets Using Carbon Paste Electrode. *Org. Med. Chem. Int. J.* **2019**, *8*, 124–132. [CrossRef]
71. Hamam, A.; Maiza, M. Voltammetric Determination of AA Using Cu Foil Electrode. *Int. J. Electrochem. Sci.* **2021**, *16*, 210426. [CrossRef]
72. Alemu, T.; Zelalem, B.; Amare, N. Voltammetric Determination of AA Content in Cabbage Using Anthraquinone Modified Carbon Paste Electrode. *J. Chem.* **2022**, *2022*, 7154170. [CrossRef]
73. Parkook, F.; Shahvandi, S.K.; Ghaedi, M.; Javadian, H.; Parkook, A. Determination of AA in Biological Samples Using an Electrochemical Sensor Modified with Au-Cu2O/MWCNTs Nanocomposite. *Diam. Relat. Mater.* **2024**, *144*, 110954. [CrossRef]
74. Rončević, I.Š.; Skroza, D.; Vrca, I.; Kondža, A.M.; Vladislavić, N. Development and Optimization of Electrochemical Method for Determination of Vitamin C. *Chemosensors* **2022**, *10*, 283. [CrossRef]
75. Gülen, M.; Taş, R.; Dünya, H.; Rajendrachari, S. Enlargement of the Surface Area of High-Entropy Alloys with Acid Treatment as Positive Electrode for High Specific Capacitance Supercapacitors. *ChemistryOpen* **2025**. [CrossRef] [PubMed]
76. Skrovankova, S.; Mlcek, J.; Sochor, J.; Baron, M.; Kynicky, J.; Jurikova, T. Determination of AA by Electrochemical Techniques and Other Methods. *Int. J. Electrochem. Sci.* **2015**, *10*, 2421–2431. [CrossRef]
77. Thbayh, D.K.; Reizer, E.; Kahaly, M.U.; Viskolcz, B.; Fiser, B. Antioxidant Potential of Santowhite as Synthetic and AA as Natural Polymer Additives. *Polymers* **2022**, *14*, 3518. [CrossRef]
78. Ramkumar, R.; Veerakumar, P.; Rajendrachari, S.; Dhakal, G.; Yun, J.; Shim, J.-J.; Kim, W.K. Copper Aerogel Frameworks—Electrochemical Detection of Dopamine and Catalytic Reduction of 4-Nitrophenol. *Microchem. J.* **2024**, *208*, 112486. [CrossRef]
79. Khan, I.; Saeed, K.; Zekker, I.; Zhang, B.; Hendi, A.H.; Ahmad, A.; Ahmad, S.; Zada, N.; Ahmad, H.; Shah, L.A.; et al. Review on Methylene Blue: Its Properties, Uses, Toxicity and Photodegradation. *Water* **2022**, *14*, 242. [CrossRef]
80. Bužga, M.; Machytka, E.; Dvořáčková, E.; Švagera, Z.; Stejskal, D.; Máca, J.; Král, J. Methylene Blue: A Controversial Diagnostic Acid and Medication? *Toxicol. Res.* **2022**, *11*, 711–717. [CrossRef]

81. Taldaev, A.; Terekhov, R.; Nikitin, I.; Melnik, E.; Kuzina, V.; Klochko, M.; Reshetov, I.; Shiryaev, A.; Loschenov, V.; Ramenskaya, G. MBin Anticancer Photodynamic Therapy: Systematic Review of Preclinical Studies. *Front. Pharmacol.* **2023**, *14*, 1264961. [CrossRef]
82. Xiong, Z.-M.; O'Donovan, M.; Sun, L.; Choi, J.Y.; Ren, M.; Cao, K. Anti-Aging Potentials of MBfor Human Skin Longevity. *Sci. Rep.* **2017**, *7*, 2475. [CrossRef]
83. Vasiljevic, Z.Z.; Dojcinovic, M.P.; Vujancevic, J.D.; Jankovic-Castvan, I.; Ognjanovic, M.; Tadic, N.B.; Stojadinovic, S.; Brankovic, G.O.; Nikolic, M.V. Photocatalytic Degradation of MBunder Natural Sunlight Using Iron Titanate Nanoparticles Prepared by a Modified Sol–Gel Method. *R. Soc. Open Sci.* **2020**, *7*, 200708. [CrossRef]
84. Fito, J.; Abewaa, M.; Mengistu, A.; Angassa, K.; Ambaye, A.D.; Moyo, W.; Nkambule, T. Adsorption of MBfrom Textile Industrial Wastewater Using Activated Carbon Developed from Rumex Abyssinicus Plant. *Sci. Rep.* **2023**, *13*, 5427. [CrossRef] [PubMed]
85. Oladoye, P.O.; Ajiboye, T.O.; Omotola, E.O.; Oyewola, O.J. MBDye: Toxicity and Potential Elimination Technology from Wastewater. *Results Eng.* **2022**, *16*, 100678. [CrossRef]
86. Modi, S.; Yadav, V.K.; Gacem, A.; Ali, I.H.; Dave, D.; Khan, S.H.; Yadav, K.K.; Rather, S.-U.; Ahn, Y.; Son, C.T.; et al. Recent and Emerging Trends in Remediation of MBDye from Wastewater by Using Zinc Oxide Nanoparticles. *Water* **2022**, *14*, 1749. [CrossRef]
87. Mulushewa, Z.; Dinbore, W.T.; Ayele, Y. Removal of MBfrom Textile Waste Water Using Kaolin and Zeolite-x Synthesized from Ethiopian Kaolin. *Environ. Anal. Health Toxicol.* **2021**, *36*, e2021007. [CrossRef]
88. Vutskits, L.; Briner, A.; Klauser, P.; Gascon, E.; Dayer, A.G.; Kiss, J.Z.; Muller, D.; Licker, M.J.; Morel, D.R. Adverse Effects of MBon the Central Nervous System. *Anesthesiology* **2008**, *108*, 684–692. [CrossRef]
89. Albert, M.; Lessin, M.S.; Gilchrist, B.F. Methylene Blue: Dangerous Dye for Neonates. *J. Pediatr. Surg.* **2003**, *38*, 1244–1245. [CrossRef]
90. Auerbach, S.S.; Bristol, D.W.; Peckham, J.C.; Travlos, G.S.; Hébert, C.D.; Chhabra, R.S. Toxicity and Carcinogenicity Studies of MBTrihydrate in F344N Rats and B6C3F1 Mice. *Food Chem. Toxicol.* **2009**, *48*, 169–177. [CrossRef]
91. Silva, M.; Caro, V.; Guzmán, C.; Perry, G.; Areche, C.; Cornejo, A. α-Synuclein and Tau, Two Targets for Dementia. *Stud. Nat. Prod. Chem.* **2020**, *67*, 1–25.
92. Saigl, Z.M. Various Adsorbents for Removal of Rhodamine B Dye: A Review. *Indones. J. Chem.* **2021**, *21*, 1039. [CrossRef]
93. Sharma, J.; Sharma, S.; Bhatt, U.; Soni, V. Toxic Effects of Rhodamine B on Antioxidant System and Photosynthesis of Hydrilla Verticillata. *J. Hazard. Mater. Lett.* **2022**, *3*, 100069. [CrossRef]
94. Al-Buriahi, A.K.; Al-Gheethi, A.A.; Kumar, P.S.; Mohamed, R.M.S.R.; Yusof, H.; Alshalif, A.F.; Khalifa, N.A. Elimination of Rhodamine B from Textile Wastewater Using Nanoparticle Photocatalysts: A Review for Sustainable Approaches. *Chemosphere* **2021**, *287*, 132162. [CrossRef] [PubMed]
95. Khubiev, O.M.; Egorov, A.R.; Semenkova, D.I.; Salokho, D.S.; Golubev, R.A.; Sikaona, N.D.; Lobanov, N.N.; Kritchenkov, I.S.; Tskhovrebov, A.G.; Kirichuk, A.A.; et al. Rhodamine B-Containing Chitosan-Based Films: Preparation, Luminescent, Antibacterial, and Antioxidant Properties. *Polymers* **2024**, *16*, 755. [CrossRef] [PubMed]
96. Chadelaud, T.; Zeghioud, H.; De La Garza, A.R.; Fuerte, O.; Benítez-Rico, A.; Revel, M.; Chávez-Miyauchi, T.E.; Djelal, H. Comparative Study of Rhodamine B Treatment: Assessing of Efficiency Processes and Ecotoxicity of By-Products. *Processes* **2023**, *11*, 2671. [CrossRef]
97. Kapoor, K.; Bala, M.; Choudhary, P.; Mridula, D.; Singh, R.K. Development of Method for Detection of Rhodamine B Dye in Chilli Powder (*Capsicum annuum* L.). *Asian J. Dairy Food Res.* **2022**, *43*, 539–543. [CrossRef]
98. Vizuete, G.; Santana-Romo, F.; Almeida-Naranjo, C.E. Virtual Screening of Fluorescent Heterocyclic Molecules and Advanced Oxidation Degradation of Rhodamine B in Synthetic Solutions. *Water* **2024**, *16*, 2141. [CrossRef]
99. Smith, A.D.; Kim, Y.-I.; Refsum, H. Is Folic Acid Good for Everyone? *Am. J. Clin. Nutr.* **2008**, *87*, 517–533. [CrossRef]
100. Shulpekova, Y.; Nechaev, V.; Kardasheva, S.; Sedova, A.; Kurbatova, A.; Bueverova, E.; Kopylov, A.; Malsagova, K.; Dlamini, J.C.; Ivashkin, V. The Concept of FA in Health and Disease. *Molecules* **2021**, *26*, 3731. [CrossRef]
101. Pamunuwa, G.; Nilakshi, H.; Rajapaksha, G.; Shakoor, F.; Karunaratne, D.N. Sensory and Physicochemical Properties and Stability of FA in a Pineapple Ready-to-Serve Beverage Fortified with Encapsulated FA. *J. Food Qual.* **2021**, *2021*, 9913884. [CrossRef]
102. Pertiwi, H.; Mahendra, M.Y.N.; Kamaludeen, J. FA: Sources, Chemistry, Absorption, Metabolism, Beneficial Effects on Poultry Performance and Health. *Vet. Med. Int.* **2022**, *2022*, 2163756. [CrossRef]
103. Ismail, S.; Eljazzar, S.; Ganji, V. Intended and Unintended Benefits of FA Fortification—A Narrative Review. *Foods* **2023**, *12*, 1612. [CrossRef]
104. Ajeigbe, K.; Aibangbee, K.; Saeed, S.; Ajeigbe, O.; Onifade, A. FA Protects and Heals Gastric Mucosa: Role of Acid Output, Inflammatory Cytokines, Angiogenic and Growth Factors. *J. Basic Appl. Zool.* **2022**, *83*, 15. [CrossRef]
105. Pérez-Carreón, K.; Martínez, L.M.; Videa, M.; Cruz-Angeles, J.; Gómez, J.; Ramírez, E. Effect of Basic Amino Acids on FA Solubility. *Pharmaceutics* **2023**, *15*, 2544. [CrossRef] [PubMed]

106. Krishna, S.A.; Noble, N.; Radhika, N.; Saleh, B. A Comprehensive Review on Advances in High Entropy Alloys: Fabrication and Surface Modification Methods, Properties, Applications, and Future Prospects. *J. Manuf. Process.* **2023**, *109*, 583–606. [CrossRef]
107. Kumar, A.; Singh, A.; Suhane, A. Mechanically Alloyed High Entropy Alloys: Existing Challenges and Opportunities. *J. Mater. Res. Technol.* **2022**, *17*, 2431–2456. [CrossRef]
108. Xie, X.; Li, N.; Liu, W.; Huang, S.; He, X.; Yu, Q.; Xiong, H.; Wang, E.; Hou, X. Research Progress of Refractory High Entropy Alloys: A Review. *Chin. J. Mech. Eng.* **2022**, *35*, 142. [CrossRef]
109. Wang, X.; Guo, W.; Fu, Y. High-Entropy Alloys: Emerging Materials for Advanced Functional Applications. *J. Mater. Chem. A* **2020**, *9*, 663–701. [CrossRef]
110. Rajendrachari, S. An Overview of High-Entropy Alloys Prepared by Mechanical Alloying Followed by the Characterization of Their Microstructure and Various Properties. *Alloys* **2022**, *1*, 116–132. [CrossRef]
111. Vaidya, M.; Muralikrishna, G.M.; Murty, B.S. High-Entropy Alloys by Mechanical Alloying: A Review. *J. Mater. Res.* **2019**, *34*, 664–686. [CrossRef]
112. Custodio, A.G.D.; Saha, G.C.; Aranas, C. Mechanical Alloying of AlCoCrFe and AlCoCrFeNi: Design and Experimental Evaluation of Medium- and High-Entropy Alloy Particles for Cold Spraying. *Powder Technol.* **2024**, *437*, 119556. [CrossRef]
113. Santiago-Andrades, L.; Vidal-Crespo, A.; Blázquez, J.S.; Ipus, J.J.; Conde, C.F. Mechanical Alloying as a Way to Produce Metastable Single-Phase High-Entropy Alloys beyond the Stability Criteria. *Nanomaterials* **2023**, *14*, 27. [CrossRef]
114. Kumar, A.; Singh, A.; Suhane, A. A Critical Review on Mechanically Alloyed High Entropy Alloys: Processing Challenges and Properties. *Mater. Res. Express* **2022**, *9*, 052001. [CrossRef]
115. Tsukamoto, H. Mechanical Behavior of Shape Memory Alloy Fiber Reinforced Aluminum Matrix Composites. *Mater. Today Commun.* **2021**, *29*, 102750. [CrossRef]
116. Balasubramanian, M.; Srimath, R.; Vignesh, L.; Rajesh, S. Application of Shape Memory Alloys in Engineering—A Review. *J. Phys. Conf. Ser.* **2021**, *2054*, 012078. [CrossRef]
117. Yavuzer, B.; Bicakci, Ü.; Simsek, D.; Ozyurek, D. Effect of Alloying Time and Heat Treatment on Microstructure and Tribological Properties of Mechanical Alloyed CU-AL-NI Shape Memory Alloy. *J. Mater. Eng. Perform.* **2023**, *33*, 11176–11187. [CrossRef]
118. Salwa, P.; Goryczka, T. Crystallization of Mechanically Alloyed NI50TI50 and TI50NI25CU25 Shape Memory Alloys. *J. Mater. Eng. Perform.* **2020**, *29*, 2848–2852. [CrossRef]
119. GöGebakan, M.; Soguksu, A.K.; Uzun, O.; Dogan, A. Production of Cu-Al-Ni Shape Memory Alloys by Mechanical Alloy. In *AIP Conference Proceedings*; American Institute of Physics: College Park, MD, USA, 2007; Volume 899, p. 654. [CrossRef]
120. Lee, T.-J.; Kim, W.-J. Effect of Severe Plastic Deformation and Post-Deformation Heat Treatment on the Microstructure and Superelastic Properties of TI-50.8 at.% Ni Alloy. *Materials* **2022**, *15*, 7822. [CrossRef]
121. Manjunatha, J.G.; Kanthappa, B.; Hareesha, N.; Raril, C.; Tighezza, A.M.; Albaqami, M.D. Enhanced Electrochemical Detection of Rutin Using Poly (Methyl Orange) Modified Carbon Paste Electrode as a Responsive Electrochemical Sensor. *Chem. Afr.* **2023**, *7*, 1141–1150. [CrossRef]

catalysts

MDPI

Review

Non-Platinum Group Metal Oxygen Reduction Catalysts for a Hydrogen Fuel Cell Cathode: A Mini-Review

Naomi Helsel and Pabitra Choudhury *

Chemical Engineering Department, New Mexico Tech, Socorro, NM 87801, USA; naomi.helsel@student.nmt.edu
* Correspondence: pabitra.choudhury@nmt.edu

Abstract: Although platinum-based catalysts are highly effective for the oxygen reduction reaction (ORR) in proton exchange membrane fuel cells (PEMFCs), their high cost and scarcity limit large-scale commercialization. As a result, platinum group metal-free catalysts—particularly Fe-N-C materials—have received increasing attention as promising alternatives. Despite significant progress, no platinum-group metal-free (PGM-free) catalyst has yet matched the performance and durability of commercial Pt/C in acidic media. Recent advances in synthesis strategies, however, have led to notable improvements in the activity, stability, and active site density of Fe-N-C catalysts. This review highlights key synthesis approaches, including pyrolysis, MOF-derived templates, and cascade anchoring, and discusses how these methods contribute to improved nitrogen coordination, electronic structure modulation, and active site engineering. The continued refinement of these strategies, alongside improved catalyst screening techniques, is essential for closing the performance gap and enabling the practical deployment of non-PGM catalysts in PEMFC technologies.

Keywords: ORR; PGM-Free Catalysis; PEMFC; Fe-N-C

1. Introduction

To limit the effect of climate change in the future, it is essential that major changes in sectors producing massive amounts of greenhouse gases are made, with the transportation sector being one of these areas. A total of 98% of the vehicles registered in the US in 2022 were gasoline vehicles [1]. These engines produce a vast amount of air pollution and require gasoline, which is fuel made from non-renewable crude oil and other petroleum liquids [2]. A typical internal combustion-powered vehicle emits about 4.6 metric tons of carbon dioxide per year, and 400 g of CO_2 per mile is emitted by a passenger vehicle [3]. The greenhouse effect is the process where heat is trapped near the Earth's surface due to greenhouse gases such as carbon dioxide and methane [4]. The greenhouse effect that greenhouse gases produce has led to a rapidly increasing surface temperature, which has caused sea levels to rise, extreme weather patterns, and habitat destruction [5]. Additionally, crude oil reserves are set to run out in approximately 50 years [6]. Carbon dioxide is at the highest levels ever recorded in the atmosphere and is a leading contributor to the climate change crisis. The reduction of emissions that would result from the reduced consumption of petroleum is integral to maintaining the integrity of our environment [7].

A fuel cell electric vehicle (FCEV) operating on hydrogen only emits water vapor and does not directly emit any CO_2 [3]. Since PEMFC-powered vehicles rely solely on hydrogen gas, the production method used to obtain it is the determining factor in the

CO_2-equivalent emissions of a PEMFC-powered vehicle [8]. Another limiting factor of PEMFCs is the amount of hydrogen available for use in the transportation sector and the limited infrastructure for refueling stations [9]. Increasing the amount of hydrogen produced annually would allow for the expansion of the emission-free vehicle market. The switch to emission-free vehicles would result in increased vehicle efficiency and reduce the sheer volume of petroleum used in the transportation sector. Competition with internal combustion engine vehicles lies in the cost, accessibility, and durability of PEMFC-powered vehicles [10].

Previous and current generations of commercial FCEVs use Pt/C and Pt-Co/C catalysts, respectively. As of 2021, the commercial Toyota Mirai Pt-Co/C catalyst reported a Pt loading of 0.33 $mg_{Pt}\ cm^{-2}$ for the cathode [11]. Pt-Co/C catalysts, while highly efficient under optimized conditions, are susceptible to performance losses due to carbon monoxide (CO) poisoning and ionomer interaction issues. Their effectiveness depends heavily on precise control of the Pt shell thickness, crystal structure, and the architecture of the carbon support, which must balance accessibility for reactants with protection against ionomer poisoning. Although innovations like mesoporous carbon supports (e.g., HSC-f) have improved Pt utilization and reduced loading requirements, these systems still face limitations related to CO tolerance and mass transport under certain conditions [12]. This reliance on high Pt loading is especially problematic in high-temperature PEMFCs (HT-PEMFCs), where loadings often exceed 1.0 mg Pt/cm^2 due to challenges like electrolyte flooding and anion adsorption, which further highlights the need for more CO-tolerant and cost-effective alternatives [13].

In contrast, Fe-N-C catalysts have emerged as a promising class of PGM-free materials for the oxygen reduction reaction (ORR) [14,15]. These catalysts typically feature atomically dispersed Fe-N_4 moieties coordinated within a nitrogen-doped carbon matrix. While they offer notable cost and performance advantages, challenges remain in achieving long-term durability in acidic PEMFC conditions, clearly identifying the active sites, and mitigating structural degradation over time. These structure–performance limitations are discussed in greater detail in the Synthesis Strategies Section of this review. These catalysts exhibit strong intrinsic tolerance to CO poisoning, making them more robust under practical fuel cell operating conditions where impurities are present, including trace CO potentially introduced through reformate use or ambient air exposure in real-world environments [16]. Moreover, Fe-N-C materials offer a significant cost advantage and can be synthesized with tunable porosity and active site distribution, supporting efficient mass transport and catalytic activity without the need for scarce and expensive noble metals [14]. This makes Fe-N-C catalysts an attractive and sustainable alternative to Pt-Co/C systems for future fuel cell applications.

This review provides a focused assessment of Fe-N-C catalysts specifically within the context of FCEV deployment, emphasizing their structural characteristics, synthesis strategies, and tolerance to air-derived impurities like CO. By directly comparing these materials to commercial Pt-based catalysts and highlighting recent advances in Fe-N-C performance under acidic PEMFC conditions, this work aims to bridge a critical gap in the literature and support the development of durable, cost-effective PGM-free alternatives for transportation applications.

2. Fundamental Aspects of Fuel Cells and ORR

Proton exchange membrane fuel cells (PEMFCs) directly convert the chemical energy in hydrogen to electricity with pure water and heat as byproducts. Specifically, a hydrogen fuel cell is an electrochemical cell that uses a spontaneous redox reaction to produce current.

The two sandwiched electrodes with an electrolyte, shown in Figure 1, are the anode and the cathode. All potentials reference the Standard Hydrogen Electrode (SHE), which is defined as 0 V under standard conditions (1 atm H_2, 1 M H^+, 25 °C). The half-cell and general redox reactions are as follows:

$$\begin{gathered}\text{Anodic Reaction (Hydrogen Oxidation Reaction)}\\ \mathrm{H_2} \rightarrow 2\mathrm{H}^+ + 2\mathrm{e}^- \quad E_a^o = 0\mathrm{V}\ (vs\mathrm{SHE})\\ \text{Cathodic Reaction (Oxygen Reduction Reaction)}\\ \mathrm{O_2} + 4\mathrm{H}^+ + 4\mathrm{e}^- \rightarrow 2\mathrm{H_2O} \quad E_c^o = 1.23\mathrm{V}\ (vs\mathrm{SHE})\\ \text{Overall Reaction}\\ 2\mathrm{H_2} + \mathrm{O_2} \rightarrow 2\mathrm{H_2O} \quad E^o = 1.23\mathrm{V}\ (vs\mathrm{SHE})\end{gathered}$$

where the standard electromotive power is 1.23 V, and E_a^o, and E_c^o are the anodic and cathodic standard electromotive force, respectively.

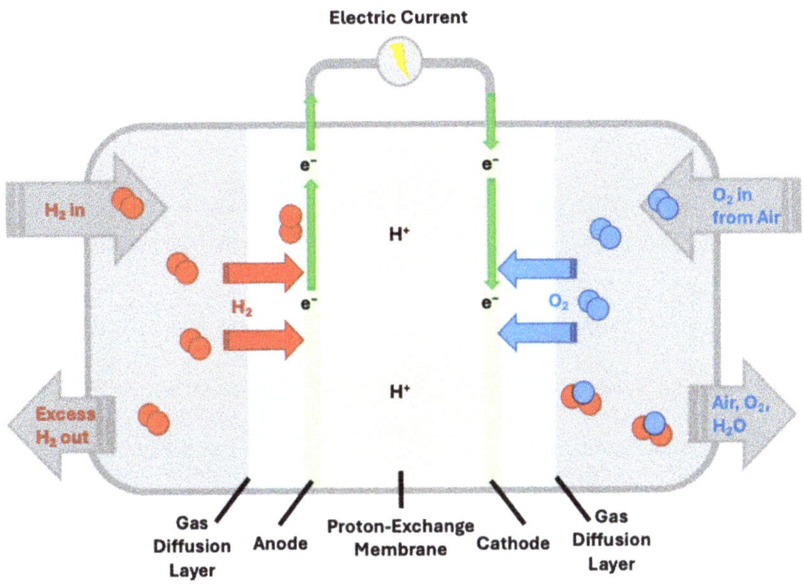

Figure 1. Diagram of a proton exchange membrane fuel cell.

Catalysts facilitate the overall reaction between hydrogen fuel and oxygen from the air to produce water vapor. Without a catalyst, the reaction would not be able to produce a usable amount of electricity [17]. As the cathode has sluggish reaction kinetics and catalyst stability issues, our focus will be on the cathode-side reaction (oxygen reduction reaction) and its catalyst.

Oxygen Reduction Reaction

With the oxygen molecule converting into water with an overall four-electron transfer, it is highly exothermic. However, since 1.23 V is required to enable this reaction, it is hard to enable this reaction with sluggish kinetics without a suitable catalyst. A catalyst is needed to lower the amount of potential required to make this reaction occur. The oxygen reduction reaction can occur in a direct four-electron transfer or a series of two-electron transfers and is shown below for acidic media [17]:

$$\text{Direct Pathway}$$
$$O_2 + 4H^+ + 4e^- \rightarrow 2H_2O \quad E^o = 1.23V\ (vsSHE)$$
$$\text{Series Pathway}$$
$$O_2 + 2H^+ + 2e^- \rightarrow H_2O_2 \quad E^o = 0.69V\ (vsSHE)$$
$$H_2O_2 + 2H^+ + 2e^- \rightarrow 2H_2O \quad E^o = 1.77V\ (vsSHE)$$

The direct pathway is the most favorable as it involves the breaking of the O-O bond of the oxygen molecule and provides the most released free energy. For complete oxygen reduction, the catalyst should be able to generate H_2O from O_2 through the direct pathway with minimal H_2O_2 production. The series pathway discharges nearly half of the free energy compared to the direct pathway because of the high energy required to break the O-O bond. The series pathway introduces harmful free radical species into the electrolyte while also providing lower energy conversion efficiency [17].

3. Synthesis Strategies and Material Design of Non-PGM ORR Catalysts

Expensive platinum-based catalysts are currently used on both electrodes of a proton exchange membrane fuel cell (PEMFC). On top of being expensive and precious, platinum-based catalysts are not considered perfect for both electrodes (the cathode and the anode). The cathode suffers from sluggish ORR kinetics, which leads to declining activity, methanol crossover, and CO poisoning effects [18–20]. The cathode reaction involves the reduction of an oxygen molecule and has been a major barrier to the development of economically enticing technology [19]. This poor ORR activity and low stability, coupled with the high price of platinum catalysts, limit the mass commercialization of PEM fuel cells with the current technology. Over the last few decades, much research has been undertaken to find a suitable alternative that offers the same, if not better, integrity and stability as the standard platinum-based catalysts [21]. Finding a suitable alternative to platinum-based catalysts has been a great challenge since no catalyst has currently surpassed platinum-based catalysts in integrity, stability, and oxygen reduction (ORR) performance [22,23]. Many platinum-free metal-based catalysts have been investigated to find a catalyst that can balance between ORR activity, electron conductivity, stability, and O_2 transport [24,25]. The most common categories of PGM-free catalysts for acidic ORR include iron–nitrogen–carbon (Fe-N-C) [26], carbon-based [27], and other molecular catalysts [28,29]. Studies showing Fe-N_4 moieties or Fe-N_4 coordination environments have been shown to have particularly high ORR activity, and well-defined Fe-N bonding sites serve as highly active and stable centers [30,31]. Beyond the density of Fe-N_x sites, their accessibility and exposure at the catalyst surface play a critical role in determining ORR activity. These structures are common across both Fe-N-C materials and molecular macrocycles, offering a strong link between heterogeneous and homogeneous catalyst design [32]. Q. Wang et al. demonstrated that lower catalyst loading led to increased H_2O_2 yield, indicating that inner Fe sites are less utilized, likely due to mass transport or site accessibility limitations [33]. This suggests that catalytic performance may be limited not by the presence of Fe-N_x moieties but by their availability at the reaction interface. In contrast, 2D catalysts, such as g-FePc [28], may suffer less from these limitations due to their atomically thin structure, which could facilitate greater exposure of active sites and more uniform utilization.

3.1. Iron–Nitrogen–Carbon (Fe-N-C) Catalysts

Among the transition metal–nitrogen–carbon catalysts, Fe-N-C catalysts have been shown to be the most promising for the oxygen reduction reaction due to their high activity and economic value [34]. Various synthesis methods of Fe-N-C composites can differ in metal loadings and active site densities. Some promising synthesis approaches include pyrolysis, Fe-doped MOF precursors, wet impregnation, ball-milling, cascade anchoring strategies, co-doping, and others. Fe-N-C synthesis strategies each offer trade-offs in activity, stability, and scalability. Pyrolysis is essential for Fe-N_x site formation, though it can also lead to the generation of inactive Fe species. In MOF-derived approaches, the MOF acts as a template that, upon pyrolysis, yields materials with high surface area and atomic dispersion; however, these methods can be complex and often result in low yields. Wet impregnation is scalable but may suffer from poor site uniformity. Emerging methods like cascade anchoring improve site accessibility but involve multi-step synthesis. Understanding these trade-offs is essential for the rational design of future Fe-N-C catalysts, particularly as the field moves toward optimizing site density, durability, and integration into practical MEA architectures.

Pyrolysis: Pyrolysis is one of the most common approaches for synthesizing Fe-N-C composites. This was employed by Ao et al., as shown in Figure 2a, where they first freeze-dried sodium chloride, dicyandiamide, D(+)-glucosamine hydrochloride, and anhydrous iron (III) chloride in a vacuum before heating to 550 °C for 3 h and annealing at 800 °C for another 2 h in an inert argon gas environment. Ao et al.'s synthesis yielded a Fe/N-functionalized 3D porous carbon network, and ultrafine Fe_4N nanoparticles were grown in the carbon framework during pyrolysis [35]. Another pyrolysis approach in which single Fe sites on N-doped porous carbon nanosheets were synthesized using a confined pyrolysis strategy was employed by Z. Yang et al., as shown in Figure 2b [36]. By utilizing a poly(ether imide) polymer as a carbon precursor and 1,10-phenanthroline (Phen) as a space isolation agent, they were able to promote the full Fe transformation to single Fe atoms without forming Fe nanoparticles. Specifically, cyanuric acid and melamine dispersed in ethanol were filtered and then pyrolyzed at 550 °C for 6 h to form C3N4 nanosheets, which were then freeze-dried and heated to 100 °C for 1 h with ferrous chloride, the poly(ether imide) polymer, and Phen to synthesize the final catalyst [36].

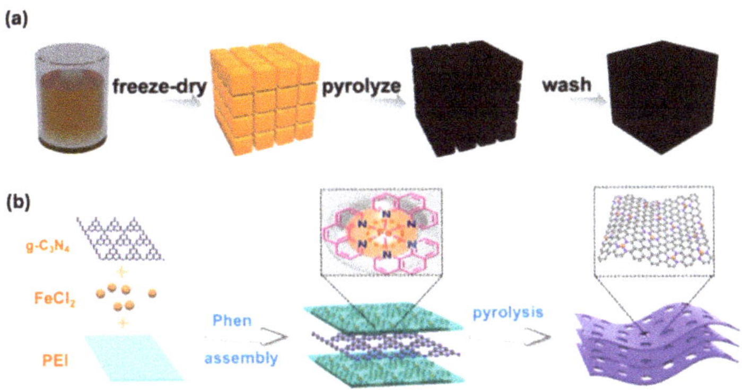

Figure 2. Fe-N-C catalyst preparation by (**a**) Ao et al. [35] (adapted with permission from Ref. [35]; copyright: 2013 Royal Society of Chemistry) and (**b**) Yang et al. [36] (adapted with permission from Ref. [36]; copyright: 2019 American Chemical Society).

While these two examples are discussed in detail, pyrolysis is also widely used in other synthesis strategies, including MOF-derived and support-assisted methods. MOF-derived catalysts, for instance, utilize metal–organic frameworks, like ZIF-8, as precursors, as mentioned in a separate section [37]. Upon pyrolysis, these frameworks decompose to form nitrogen-doped carbon matrices embedded with atomically dispersed Fe-N_4 sites [38,39]. This method benefits from the uniform distribution of metal centers and the inherent porosity of MOFs, which can enhance mass transport and active site accessibility [40]. Support-assisted strategies, on the other hand, involve impregnating nitrogen-rich carbon supports—such as carbon nanotubes or graphene—with iron precursors, followed by pyrolysis [41,42]. This approach allows for better control over the morphology and electronic properties of the resulting catalysts, as the characteristics of the support material can influence the dispersion and coordination environment of the active Fe sites [43]. Highlighting the distinctions between these pyrolysis-based methods—such as MOF-derived, support-assisted, and polymer-confined strategies—helps illustrate the wide range of structural control and active site engineering possible in Fe-N-C catalyst synthesis.

Cascade Anchoring Strategy: Traditional pyrolysis synthesis offers relatively low metal loadings (<5 at.%). To combat this low metal loading, strategies, such as cascade anchoring, can be employed to increase metal loading up to 12.1 wt.%. This was achieved by L. Zhao et al., who were able to produce a scalable synthesis with high metal-N_x loading. As shown in Figure 3, the metal ions are first chelated using glucose as a chelating agent before being anchored onto high-surface-area, oxygen-species-rich porous carbon supports and pyrolyzed at ~500 °C to allow for maximized protection of the metal ions. From their synthesis, they were able to determine that the chelation of metal ions, physical isolation of the chelate complex, and N-species binding at increased temperatures were key factors in achieving high-metal-loading single-atom catalysts (SACs) [44].

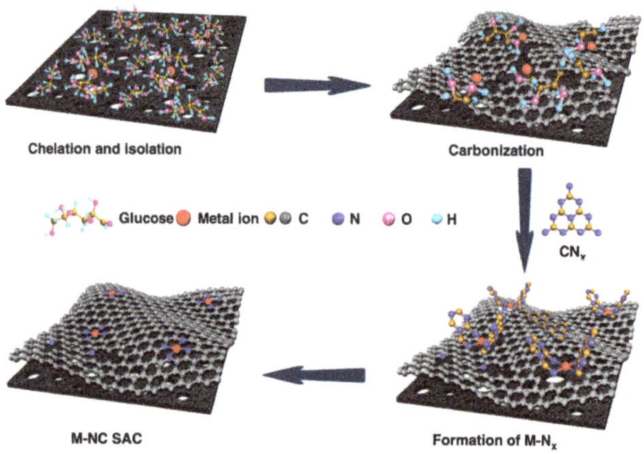

Figure 3. Cascade anchoring strategy employed by L. Zhao et al. to synthesize high-loading M-NC SACs [44]. Reprinted with permission from Ref. [44]. Copyright: 2019 Springer Nature. Licensed under CC BY 4.0 (http://creativecommons.org/licenses/by/4.0/ (accessed on 10 June 2025)).

Metal–Organic Frameworks: Hybrid crystalline materials, such as metal–organic frameworks (MOFs), have become some common precursors for metal–nitrogen–carbon complexes. Zeolitic imidazolate frameworks (ZIFs), which are a subclass of MOFs, contain N-filled ligands and tetrahedral coordination metal ions and are a type of coordination polymer. As shown in Figure 4, these ZIFs can be used as sacrificial templates to prepare nitrogen-doped carbon materials due to their nitrogen-rich composition. As employed by D. Zhao et al., metalated ZIFs were synthesized by heating ZnO, N-containing ligands, and tris-1,10-phenanthroline iron(II) perchlorate (TPI) (as seen in Figure 4a) [37]. After pyrolization at 1050 °C, acid washing, and a second 950 °C heat treatment, a final catalyst with a uniformly dispersed iron complex can be formed. Even though there is a lack of porosity in some of the precursors, the pyrolyzed ZIF catalysts are still able to have great surface areas. Ahluwalia et al. took a similar approach using a similar Fe-ZIF precursor and pyrolyzing at 1100 °C [45]. Their synthesis yielded a stable Fe-N-C catalyst with atomically dispersed Fe sites.

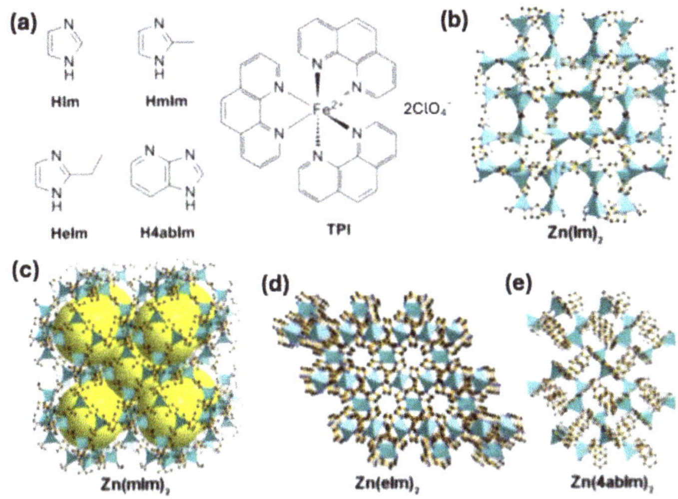

Figure 4. Chemical structures of (**a**) possible ligands and the iron additive (TPI) and the crystal structures of (**b**) $Zn(lm)_2$, (**c**) $Zn(mlm)_2$ (the yellow balls represent void), (**d**) $Zn(elm)_2$, and (**e**) $Zn(4ablm)_2$ [37]. Adapted with permission from Ref. [37]. Copyright: 2013 John Wiley and Sons.

As shown in Figure 5, another technique involving ZIFs and MOFs includes ball-milling. Zhan et al. used this approach to create bamboo-like carbon nanotubes (BLCNTs) with dispersed Fe-centered active sites (nitrogen-coordinated). Ball-milling NH_2-MIL-88B (iron source) and ZIF-8 (carbon source) precursors together destroys their crystal structure and uniformly mixes them. Zhan et al. controlled the proportion of precursors and utilized pyrolysis to control the morphology of the bamboo-like carbon nanotubes [46].

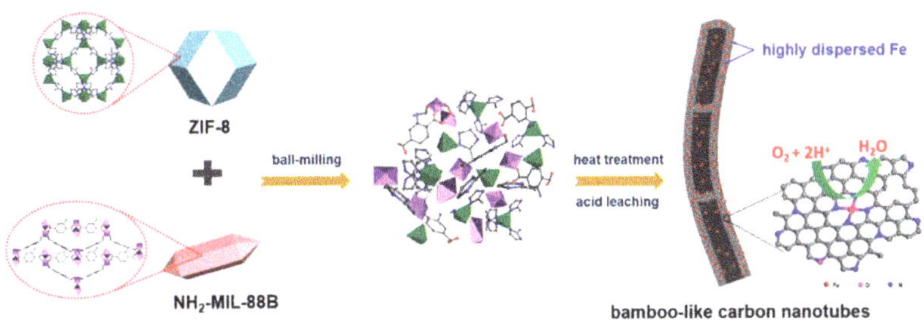

Figure 5. Schematic of the preparation of iron-containing bamboo-like carbon nanotubes [46]. Reprinted with permission from Ref. [46]. Copyright: 2020 American Chemical Society.

While MOFs, such as ZIFs, serve as sacrificial templates, they differ from traditional templates like silica. In silica-based templating, the silica is removed after pyrolysis, usually by acid etching, to generate porosity in the final material [47,48]. For example, Gianola et al. [47] highlighted how different silica removal methods, including acid-free approaches developed by Atanassov's group, influence catalyst porosity and performance. MOFs, on the other hand, are not removed but instead undergo a transformation during thermal treatment, forming part of the carbon-based framework and contributing directly to the structure and functionality of the resulting catalyst [49].

Secondary Metals Present in Fe-N-C: To change/tune the electronic properties of Fe-N-C catalysts, secondary metals or molecules are introduced into the carbon matrix. G. Yang et al. explored atomically dispersed Mn-N in Fe-N-C [50], and Jin et al. explored porous rod-like Bi-O structures [51]. A quick overview of their synthesis methods is shown in Figures 6 and 7. These will not be discussed in depth as they are similar to the synthesis methods discussed in the pyrolysis and metal–organic framework sections, with the addition of Mn and Bi precursors. In the case of Mn-N incorporation, Yang et al. reported that Mn atoms modulate the electronic environment around the Fe-N_4 active sites, resulting in a downshift of the d-band center of Fe. This downshift is described as electron delocalization from the Fe^{3+} centers, shifting their spin state from low spin ($t_{2g}5$ e_g0) to intermediate spin ($t_{2g}4$ e_g1), which enhances the overlap between Fe 3d orbitals and the π antibonding orbitals of O_2. Electronic tuning weakens the binding strength of oxygen intermediates, enhancing the overall ORR kinetics and improving catalytic activity and durability [50]. This type of electronic modulation is not unique to Mn; other secondary metals and Fe-N-C systems have also been shown to influence the spin state and d-band center of Fe, offering a versatile strategy to optimize catalytic performance [52].

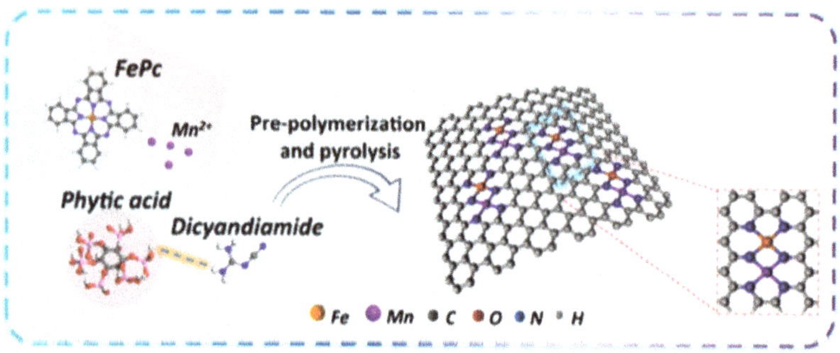

Figure 6. Synthesis procedure for Fe,Mn/N-C catalyst, as prepared by G. Yang et al. [50]. Figure adapted with permission from Ref. [50]. Copyright: 2021 Springer Nature. Licensed under CC BY 4.0 (http://creativecommons.org/licenses/by/4.0/ (accessed on 10 June 2025)).

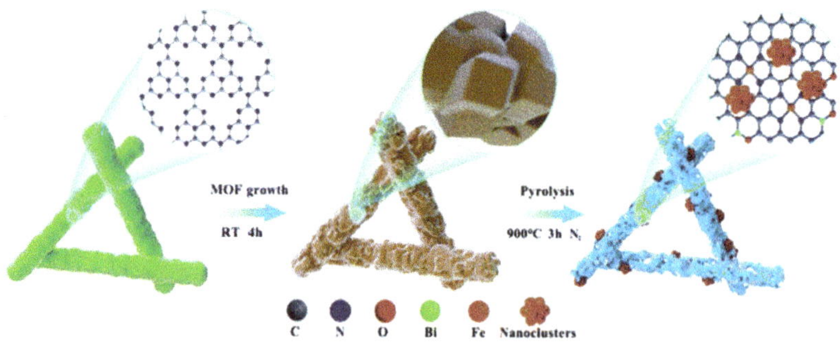

Figure 7. Synthesis procedure for Fe/Bi-RNC catalyst, as prepared by Jin et al. [51]. Reprinted with permission from Ref. [51]. Copyright: 2013 Royal Society of Chemistry.

3.2. Molecular Catalysts on Carbon Supports

Materials such as carbon matrices, carbon nanotubes, and graphene have also been studied independently for their application as cathode catalysts. Pristine graphene is considered a semiconductor with no bandgap due to its sp^2-bonded carbon atoms, which limits application in energy storage and catalysis [53–55]. Through covalent bonding with graphene, atoms such as fluorine (F), nitrogen (N), and boron (B) can be substitutionally doped into graphene to disrupt the sp^2 network and cause sp^3-defect regions [27]. This allows for the tuning of electronic properties and modification of surface chemistry. Fe-N-C complexes and macrocyclic compounds can be combined with carbon matrices, carbon nanotubes, and graphene to increase ORR activity and overall stability. Combining these materials in their pristine and modified (doped/exchanged) varieties has proven to be beneficial and favorable towards ORR activity and stability. Incorporating Fe-based macrocycles, like FePc, into conductive supports, such as graphene or CNTs, improves site dispersion and electron conductivity, leading to increases in onset potential by up to ~50 mV and 2–3× higher limiting current densities compared to unsupported counterparts [28,56–58]. In some cases, onset and half-wave potentials approach those of commercial Pt/C catalysts (e.g., ~0.92 V vs. RHE) [28], with enhanced stability, where over 90% of the initial activity is retained after 10,000 cycles in acidic media [59–61]. This approach also enables fine-tuning of the Fe coordination environment through ligand exchange or heteroatom doping of the carbon host, further enhancing ORR kinetics and promoting the desirable four-electron reduction pathway. First, doping of graphene and carbon nanotubes will be discussed before moving on to phthalocyanines and porphyrins that are commonly supported by these carbon-based structures as they prevent aggregation and can supply supplementary electron sites [28].

Doping of Graphene and Carbon Nanotubes: Dopants play a large part in the catalytic activity and electronic properties of carbon-based materials. Atomic-scale substitution can redistribute local strain within a catalyst particle, altering the energetic stability of surface configurations and potentially exposing more catalytically active sites. As shown by Husic et al., even a single dopant atom can trigger solid–solid transitions or shift structural preferences, leading to surface reconstructions that affect site accessibility and stability [62]. As mentioned in a previous section, sp^3-defect regions can be generated in the sp^2 network of graphene by introducing substitutional dopants, such as nitrogen (N) and boron (B). Though other atoms such as phosphorus (P) and sulfur (S) can also be doped into the graphene matrix, they are not as similar in size to carbon as N and B atoms and would disrupt the matrix to a much higher degree and deform the planar nature of graphene [27]. With relatively low doping percentages of <3 at.%, the electronic, magnetic, and optical properties in graphene can be greatly influenced due to the polarization that is created by the differences in electronegativity. Nitrogen and boron are common graphene dopants due to their similarity in size and electronegativity to carbon and have been studied intensively for various applications. Nitrogen-doped graphene has been successfully synthesized using thermal annealing, chemical vapor deposition (CVD), solvothermal synthesis, supercritical reaction, arc discharge, plasma, nitrogen bombardment, and ball-milling processes at doping concentrations between 0.6 and 16 at.% [27,52,63–65].

As shown in Figure 8a, nitrogen-doped graphene is made of various types of nitrogen in the graphene matrix, including pyridinic (sp^2-hybridized at edge sites), pyrrolic (sp^3-hybridized), and graphitic (sp^2-hybridized in the graphene matrix). Sun et al. mention that pyridinic and pyrrolic N allow for improved pseudo-capacitance and that graphitic N enhances conductivity (favoring electron transport during the charge/discharge process) [66]. The most common precursors for nitrogen-doped graphene include graphene or graphene oxide and a nitrogen-containing molecule such as NH_3. Boron-doped graphene for the application of ORR is slightly less common but does have a more energetically favorable formation energy of ~5.6 eV/atom (compared to 8.0 eV/atom for nitrogen-doping) [67]. Boron-doped graphene has been successfully synthesized using chemical vapor deposition (CVD), thermal annealing, thermal exfoliation, liquid process, arc discharge, and microwave plasma, and the Wurtz-type reductive coupling reaction processes dopant concentrations between 0.5 and 13.85 at.%. As shown in Figure 8b, two types of stable boron doping are possible in the boron-doped graphene. These include in-plane doping of boron (graphitic), which is sp^2-hybridized, and the formation of a "boron silane", which is where B atoms are located in a π-conjugated system [68]. Borane silane atoms have only been present in one synthesis method (CVD). The most common precursors for boron-doped graphene include graphene oxide and a boron-containing molecule, such as boron powder, BH_3-tetrahydrofuran (BH_3-THF), boron tribromide (BBr_3), and diborane (B_2H_6). The many available synthesis techniques make both N- and B-doped graphene a great candidate for the commercialization of fuel cells.

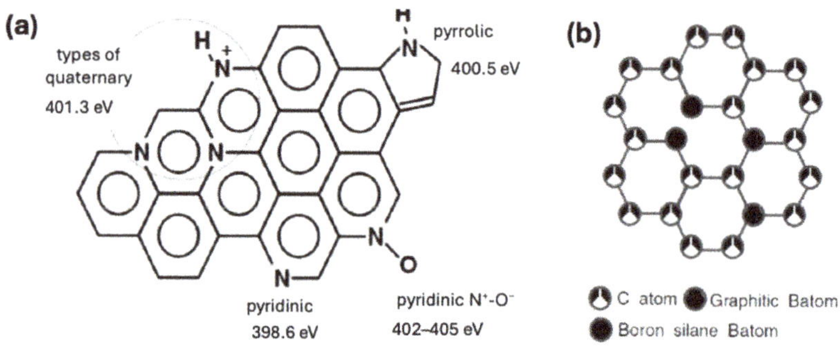

Figure 8. Schematic representations of the different dopant hybridizations of (**a**) nitrogen-doped graphene [69] (adapted with permission from Ref. [69]; copyright: 2009 Springer Nature) and (**b**) boron-doped graphene [70] (adapted with permission from Ref. [70]; copyright: 2012 John Wiley and Sons).

As shown in Figure 9, there are different types of carbon nanotube supports: single-walled, double-walled, and multi-walled nanotubes. The different types of carbon nanotubes behave differently from one another and provide support to other complexes deposited onto them. Most researchers purchase double- and multi-walled nanotubes and use them as received without any further purification step. However, in oxidative treatments, they can be subjected to 35% HNO_3 under reflux and heating conditions (100 °C) for 12 h. Single-walled carbon nanotubes can be produced using a laser ablation technique and can be purified to remove any amorphous carbon particles, which results in superficial oxidation [71].

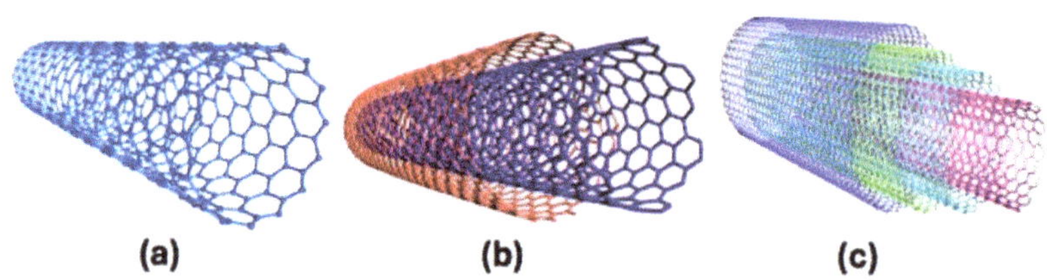

Figure 9. Schematic representation of (**a**) single-, (**b**) double-, and (**c**) multi-walled carbon nanotubes [72]. Adapted with permission from Ref. [72]. Copyright: 2021 Elsevier.

Iron Phthalocyanine on Carbon Nanotubes: As shown previously in Figure 9, carbon nanotubes can either be single-walled, double-walled, or multi-walled. However, catalyst preparation tends to be the same no matter the type of carbon nanotube (CNT). Morozan et al. were able to synthesize their catalysts by ultrasonicating a 1:2 w/w ratio of iron phthalocyanine, which was dissolved in THF and CNT powder, and then, the THF was evaporated off. Their proposed chemical representation of a hybrid catalyst for another phthalocyanine they studied is shown in Figure 10a [73]. Govan et al. took a slightly different approach by introducing eight cyano-ligand groups into their phthalocyanine, as shown in Figure 10b, in an attempt to tune their catalyst [74]. Their synthesis also included a physical deposition with sonication to ensure homogeneous dispersion in solution.

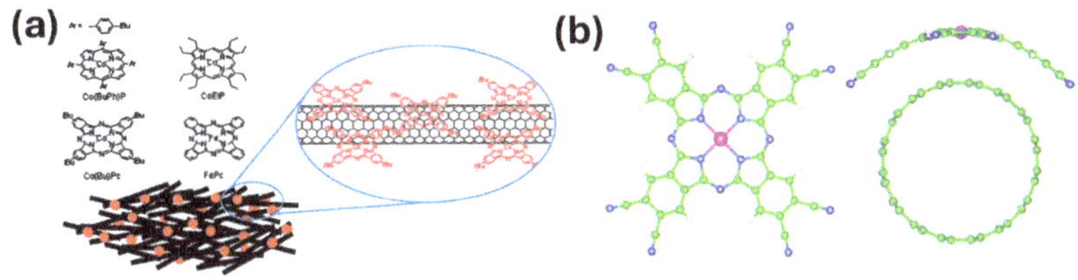

Figure 10. Schematic representations of (**a**) Morozan et al.'s catalyst [73] (reprinted with permission from Ref. [73]; copyright: 2011 Elsevier) and (**b**) Govan et al.'s catalyst [74] (figure adapted with permission from Ref. [74]; copyright: 2020 Elsevier; licensed under CC BY 4.0 (http://creativecommons.org/licenses/by/4.0/ (accessed on 10 June 2025)).

Iron Phthalocyanine on Reduced Graphene: Jiang et al. were able to synthesize iron phthalocyanine supported on chemically reduced graphene by first synthesizing graphene sheets through the partial reduction of graphene oxide, following the modified Hummers' method. Then, by having both iron phthalocyanine and graphene dispersed in N,N-dimethylformamide (DMF), they were able to incorporate iron phthalocyanine into the graphene by stirring for 6 h and sonicating [28]. This catalyst and doped-graphene variants were explored computationally by Helsel et al. and can be seen in Figure 11 [75–77].

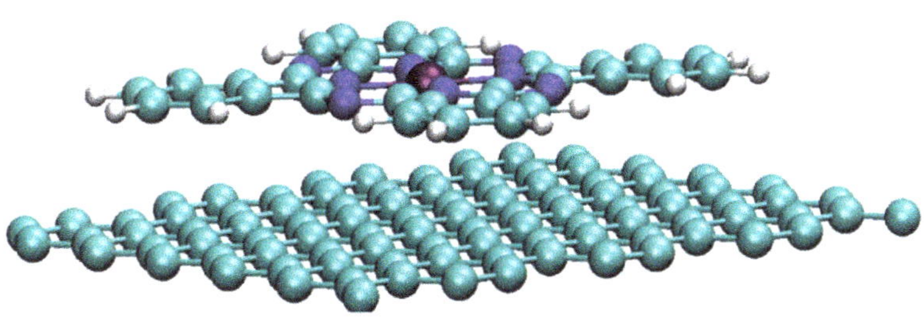

Figure 11. Schematic representation of the GFePc catalyst, as explored by Helsel et al. [77]. Blue, purple, cyan, and white represent N, Fe, C, and H, respectively. Reprinted with permission from Ref. [77]. Copyright: 2023 Elsevier.

Iron Polyporphyrin: Polyporphyrins can offer a high density of iron macrocyclic centers that are nitrogen-coordinated. Achieved successfully by Yuan et al., the overall synthesis and polymerization are displayed in Figure 12. After these steps were taken to synthesize the catalyst (PFeTTPP), it was thermally activated by being pyrolyzed from 600 to 1000 °C [29]. The simulated stacking of their catalyst is shown in Figure 12b.

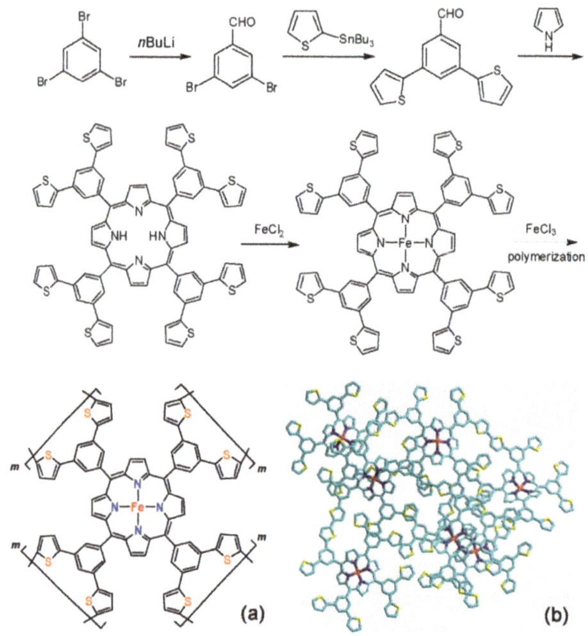

Figure 12. Synthesis steps and schematic representation of the (**a**) polymerization of PFeTTPP and (**b**) the simulated stacking of PFeTTPP employed by Yuan et al. [29]. Red, blue, light blue, yellow represents Fe, N, C, and S, respectively, with H not depicted for clarity. Reprinted with permission from Ref. [29]. Copyright: 2013 John Wiley and Sons.

4. Performance Evaluation of Non-PGM Catalysts

To be able to experimentally characterize a catalyst for fuel cell applications, electrochemical measurements and/or fuel cell tests must be conducted. Papers that involve purely computational work use a parameter called overpotential to quantify their catalyst's ORR activity. Research that involves mainly experimental work mostly utilizes a potentiostat, electrode rotator, and 3/4-electrode electrochemical cell to quantify parameters such as onset and half-wave potentials. If equipped with an available fuel cell, some researchers utilize actual fuel cell tests to obtain a better picture of the stability and practicality in actual applications.

4.1. Electrochemical Characterization Techniques

Electrochemical testing for fuel cell catalysts typically employs a three-electrode configuration using a rotating disk electrode (RDE) or rotating ring–disk electrode (RRDE), with the working electrode used to evaluate the catalyst. The working electrode is often glassy carbon, while RRDE setups include a surrounding ring, commonly platinum. However, for platinum-group metal-free (PGM-free) catalyst studies, platinum should be avoided as a counter electrode due to the risk of contamination through Pt dissolution and redeposition. Instead, graphite rods are preferred because they are chemically inert under typical electrochemical conditions, minimizing the risk of introducing unwanted species that might affect the catalyst's performance [78]. A reference electrode provides a stable potential (e.g., Ag/AgCl or saturated calomel), and reported values are commonly converted to the reversible hydrogen electrode (RHE) scale for consistency [79]. The three/four electrodes are arranged and then submerged in an acidic (<7 pH) O2-saturated electrolyte solution, which is typically 0.1 M $HClO_4$ for Fe-N-C catalysts. Fe-N-C catalysts are susceptible to poisoning by hydrochloric acid (HCl) as chloride ions strongly adsorb onto the Fe-N_x active sites, blocking oxygen access and significantly reducing ORR activity [80]. The studied mechanism, ORR, involves reducing oxygen. An O_2-saturated electrolyte solution allows for the reaction to occur without much resistance when certain voltages are applied.

Cyclic voltammetry (CV) is one of the most common electrochemical techniques in studying the reduction and oxidation processes of a particular electrode, catalyst, or species. In essence, cyclic voltammetry is where a range of potentials is swept in forward and reverse directions. CV reveals the cathodic peak potential and the half-wave potential, which can be calculated by taking the middle point between the cathodic and anodic peaks [81]. Linear sweep voltammetry (LSV) is another common electrochemical technique used where the potentials have a single forward sweep. Taking the intersection of two tangents on the non-faradaic zone and faradic zone in an LSV allows for the determination of the onset potential (E_{onset}) [82]. In order to study the kinetics and mechanisms involved in a glass carbon electrode, a method such as Koutecky–Levich analysis must be used. The analysis includes running multiple LSV sweeps at different rotation rates. This analysis reveals the number of electrons (n) involved in the reaction mechanism, or which ORR pathway the deposited catalyst takes [83]. Another key metric used to understand the activity of a catalyst is the Tafel slope. Tafel slopes help elucidate the kinetics of a catalyst by aiding the determination of the rate-limiting step, understanding its mass transport limitations, and the efficiency of the catalyst [84]. Due to the nature of this review, Tafel slopes will not be discussed, and reaction mechanisms can be deduced via the total number of electrons transferred from the Koutecky–Levich analysis. Half-wave potential alone does not provide a complete picture of ORR performance. Catalyst loading and mass activity (e.g., A mg^{-1} catalyst or A mg^{-1} Fe) are also critical metrics, especially when comparing selectivity and minimizing the effect of artifacts, such as peroxide re-reduction in high-loading films. Therefore, where possible, mass activity and loading information are highlighted in this review to provide more accurate benchmarking.

When utilizing a full membrane electrode assembly (MEA) with a prepared catalyst, the peak power density, along with the volumetric and gravimetric current densities, are considered key performance metrics. These values can be measured from the current-voltage polarization curves and power density tests. The values determined from the fuel cell test are more realistic than purely the electrochemical measurements; however, the cost, complexity, and time associated with operating a fuel cell, even in a lab setting, are more expensive than the general electrochemical cell setup used for preliminary catalyst screening [85].

Generally, to test the stability and durability of the studied catalyst on a lab scale, chronoamperometry is used over the course of a few hours to a few days by holding the electrode at a typical working potential for fuel cell operation, and then, current retention is measured over this time frame. It is generally directly compared to a control commercial Pt/C catalyst and can yield promising information regarding a catalyst's stability [34].

4.2. Comparison of ORR Activity

Table 1 shows a table of the catalysts discussed in the previous synthesis method section, along with a few key additions that were not discussed. The catalyst loadings are either reported directly from the reference or were estimated from the given drop-casted catalyst ink weights/volumes and estimated working electrode areas (denoted by ^ in Table 1). In the case of Fe-N-C-10/1-950 (BLCNTs) [46], the drop-casted volume was not specified; therefore, the loading could not be calculated from the given data.

Table 1. Comparison of loading, onset potential (E_{onset}), half-wave potential ($E_{1/2}$), and number of electrons (*n*) for various PGM-free catalysts. Loadings denoted with ^ are estimated from the given drop-casted catalyst ink weights/volumes and estimated working electrode areas. Entries in the number of electrons column with * denote that *n* and $\%H_2O_2$ formed was listed instead if present.

Environment	Catalyst	Loading (mg_{cat} cm^{-2})	Electrolyte	E_{onset} (V)	$E_{1/2}$ (V)	*n*	Ref.
Acidic	NGr	^ 0.71	0.1 M $HClO_4$	0.84	0.58	* (13.4% H_2O_2)	[86]
	B, N-Gr	^ 0.71	0.1 M $HClO_4$	0.86	0.61	* (1.8% H_2O_2)	
	Fe, Mn/N-C	0.1	0.1 M $HClO_4$	*	0.804	3.91	[50]
	Fe SAs/N-C	0.25	0.1 M $HClO_4$	0.95	0.798	*	[36]
	PFeTTPP-700	0.4	0.1 M $HClO_4$	0.93	0.73	3.96	[29]
	Fe-N-C-10/1-950	*	0.1 M $HClO_4$	0.96	0.78	~4	[46]
	Zn(mlm)$_2$TPIP	0.4	0.1 M $HClO_4$	0.902	0.76	3.9	[37]
	Zn(elm)$_2$TPIP	0.4	0.1 M $HClO_4$	0.914	0.78	3.8	
	Fe/Bi-RNC	^ 1.4	0.5 M H_2SO_4	0.899	0.97	3.93	[51]
	P*m*PDA-FeN$_x$/C (950 °C)	0.6	0.1 M H_2SO_4	0.94	0.82	* (<1% H_2O_2)	[33]
Alkaline	Fe SAs/N-C	0.25	0.1 M KOH	1.02	0.91	3.91	[36]
	Fe, Mn/N-C	0.1	0.1 M KOH	*	0.928	*	[50]
	FePc/a-MWCNTs	0.185	0.1 M NaOH	0.917	0.817	3.81	[73]
	OCNFePc-CNT	^ 0.2	0.1 M NaOH	0.95	*	3.8	[74]
	FeNCNs-800	0.36	0.1 M KOH	0.985	0.89	3.9	[35]
	Fe-NC SAC	0.6	0.1 M KOH	0.98	0.9	* (<3.5% H_2O_2)	[44]
	g-FePc	^ 0.113	0.1 M KOH	0.98	0.88	3.96	[28]

It is important to note that while the focus of this review is on Fe-N-C catalysts for PEMFC cathodes operating in acidic media, studies conducted on alkaline electrolytes, such as KOH, have contributed significantly to the understanding of ORR mechanisms and the development of active site structures. Therefore, select examples from alkaline studies are included for comparison and insight. Mass activity (i_m, A mg^{-1}) should be interpreted alongside catalyst loading (mg cm^{-2}) as both directly influence observed current density and selectivity, especially in systems like Fe-N-C, where thick films may mask peroxide production through re-reduction. However, mass activity values were not included in the comparison table, as they were not consistently reported across the reviewed studies and could not be reliably calculated in the absence of complete loading and kinetic current data.

Regardless of the preparation method, Table 1 shows that the main dominating pathway for Fe-N-C catalysts is the direct four-electron pathway, as most electron transfer numbers are around four, and there is minimal hydrogen peroxide (H_2O_2) formation in the catalysts, for which electron transfer numbers were not calculated. With 13.4% H_2O_2 formation, NGr (nitrogen-doped graphene) by itself seems to have a lower electron transfer number than other graphene. This could be due to nitrogen having one more valence electron than the carbon atoms in the graphene. Looking at carbon networks or materials that have iron and nitrogen built into them in a single catalyst, these materials have very promising half-wave and onset potentials. A high onset potential is indicative of a low overpotential required for actual fuel cell operation. In catalysts, such as Z. Yang et al.'s Fe SAs/N-C, where 3.5 wt.% single-iron-atom active sites are dispersed in a porous nitrogen-doped carbon matrix, the onset and half-wave potentials are exceptionally favorable, with the onset potential being over 1 V vs. RHE in alkaline media [36]. This could potentially mean that the space isolation agent they used to create the single-iron-atom active sites

(3.5 wt.%) within the carbon network has a larger effect than the increased metal loading found by the cascade anchoring strategy employed by L. Zhao et al. [44]. H. Jin et al. and Z. Yang et al. both modified their Fe-N-C catalysts with a secondary metal (Mn and Bi) to increase ORR activity [36,51]. Though the addition of the secondary metal improved the performance of both catalysts, neither catalyst performed better than the previously mentioned Fe SAs/N-C. There should be further studies conducted to investigate whether manganese or bismuth could be incorporated into the Fe SAs/N-C matrix to improve overpotential. Iron phthalocyanine on reduced graphene, studied by Y. Jiang et al., performed very well, even better than porous polyporphyrin with a high surface area [29]. Iron phthalocyanine is supported by Van der Waals dispersion forces, and it is known that the combination of iron phthalocyanine, supported by the reduced graphene substrate, allows for extra supplementary electron sites to be formed. Additionally, π- π interactions aid in preventing the aggregation of iron phthalocyanine, which aids in conductivity [28]. Experimental research should be conducted to tune the graphene substrate, like substrate doping (with N and B atoms), as mentioned previously, to potentially increase the overall ORR activity of the catalyst. The carbon nanotube (CNT) catalysts discussed in this review either have iron phthalocyanine physically adsorbed on the CNTs through Van der Waals dispersion forces or iron highly dispersed throughout the CNTs. A. Morozan et al. discussed the use of iron phthalocyanine deposited on single-, double-, and multi-walled carbon nanotubes, in which the acid-treated multi-walled CNTs displayed the best performance [73]. When looking at Govan et al.'s similar experiment with iron phthalocyanine that has eight hydrogens substituted with cyano groups deposited on single-walled CNTs, this material offers a higher onset potential than even iron phthalocyanine on multi-walled CNTs (the single-walled CNTs performed worse) [74]. This increase in ORR activity could potentially mean that the extra cyano groups on the ends of the phthalocyanine are aiding the charge transfer to make the reaction easier to complete. Overall, it seems that catalysts that had single iron atoms or Fe-N_4 -coordinated molecules performed better than just uniformly dispersing Fe in a carbon network, even at much higher loading percentages.

Although Fe-N-C catalysts often exhibit excellent ORR activity in alkaline media, their performance typically declines under acidic conditions due to harsher operating environments and proton-rich electrolytes, which lead to lower onset and half-wave potentials. Protonation of pyridinic-N species and carbon support corrosion are the main factors in this activity decline [87]. In acidic media, the commercial Pt/C catalyst has an approximate onset potential of 0.9 V and a half-wave potential of 0.8–0.85 V vs. SHE [88]. Though no Fe-N-C catalyst has yet surpassed Pt/C in terms of activity and stability due to the harsh operating conditions of PEMFCs, the gap in performance for this class of catalysts has narrowed significantly in the last few years. The facile synthesis methods described in this review have enabled greater active site dispersion and the modulation of electronic properties through nitrogen coordination and secondary metal co-doping, thereby enhancing the structural and catalytic performance of Fe-N-C catalysts.

While most Fe-N-C studies focus on half-cell measurements, such as RDE or RRDE, to evaluate intrinsic ORR activity, a smaller number extend testing to full MEA fuel cell configurations. As summarized in Table 2, studies on catalysts, such as PFeTTPP-700, Zn(mlm)$_2$TPIP, Zn(elm)$_2$TPIP, Fe SAs/N-C, and Fe-N-C-10/1-950, report polarization curves and peak power densities under H_2-O_2 conditions, with loadings ranging from 1.5 to 4.0 mg_{cat} cm^{-2}. These studies also report or enable the estimation of volumetric current densities as a more device-relevant metric that normalizes currents by both geometric area and catalyst layer thickness. For instance, the Fe-N-C-10/1-950 catalyst achieves an estimated volumetric current density of ~0.85 A cm^{-3} at 0.7 V and a peak power density of

770 mW cm^{-2} at approximately 0.48 V. While some values are directly reported and others are estimated from polarization plots, this comparison highlights that several Fe-N-C catalysts now approach the performance range of commercial Pt-based systems, particularly when evaluated by space-constrained metrics relevant to practical device integration. For context, leading commercial catalysts, such as Pt/C and Pt-Co/C, have achieved peak power densities of up to 0.84 W cm^{-2} at 0.67 V (LANL, using 0.1 mg_{Pt} cm^{-2} loading under H_2/air, 80 °C, 100% RH, 150 kPa) and 0.94–1.0 W cm^{-2} with Pt-Co alloy catalysts under similar conditions, as reported by General Motors and other groups. These values reflect the current benchmarks for DOE target-aligned PEMFC performance using precious-metal-based electrodes [89]. While Fe-N-C catalysts still fall short of these absolute performance levels, the gap has narrowed significantly, particularly in terms of volumetric current density and power output per unit loading. These results suggest that with continued improvements in site accessibility, conductivity, and catalyst layer architecture, non-precious metal catalysts are becoming increasingly viable for practical PEMFC applications.

Table 2. A comparison of the H_2/O_2 fuel cell test performance of the studied Fe-N-C catalysts in various media. Peak power densities and volumetric current densities denoted with ^ are estimated from polarization powder density plots from the referenced publications. The current densities marked with # denote that the volumetric current density was not given, just the normal current density.

Catalyst	Loading (mg_{cat} cm^{-2})	Volumetric Current Density (A cm^{-3})	Peak Power Density (mW cm^{-2})	Ref.
PFeTTPP-700	4	20.2 @ 0.8 V	730 @ 0.4 V	[29]
Zn(mlm)$_2$TPIP	2.2	67 @ 0.8 V	620 @ 0.43 V	[37]
Zn(elm)$_2$TPIP	2.2	88.1 @ 0.8 V	500 @ 0.34 V	
Fe SAs/N-C	1.5	# (^ 1.5 A cm^{-2} @ 0.89 V)	750 @ ^ 0.5 V	[36]
Fe-N-C-10/1-950	4	# (0.85 A cm^{-2} @ 0.7 V)	770 @ ^ 0.48 V	[46]

4.3. Future Perspectives and Challenges

Recent advances in Fe-N-C catalysts for hydrogen fuel cells have led to significant progress in multiple areas. These include high oxygen reduction reaction (ORR) activity approaching that of Pt/C ($E_{1/2}$ = ~0.8 V), high power density in MEAs (e.g., >0.7 A cm^{-2} at 0.7 V), and site densities exceeding 10^{19} sites $gram^{-1}$ [90]. The development of atomically dispersed Fe-N_x active sites and the use of 2D or porous supports, such as g-FePc and BLCNTs, have further improved active site utilization [28,46]. Additionally, volumetric current density has emerged as an increasingly relevant performance metric, reflecting improvements in catalyst layer architecture and practical device integration.

Despite these achievements, several critical challenges remain. The key issues include limited long-term stability under fuel cell operating conditions, acid leaching of Fe sites, and lower performance under H_2/air compared to H_2/O_2 conditions. The ability to accurately characterize which Fe-N_x sites are electrochemically active and how accessible they are in a working electrode remains a major limitation to rational catalyst design. Trade-offs also persist between maximizing site density and ensuring sufficient mass transport, particularly in thicker catalyst layers. In addition, kinetic limitations in O-O bond cleavage and the initial electron transfer step in the ORR continue to hinder performance relative to Pt-based benchmarks. Among these, stability remains the most significant barrier to practical deployment. While a few recent studies, such as Fe-N-C-10/1-950 retaining 70% of its current density after 24 h in an MEA, have shown encouraging durability, most Fe-N-C catalysts still undergo substantial degradation over time [46]. The causes are multifaceted: acid-induced demetalation of Fe-N_x sites, carbon corrosion at high potentials, and mechanical or chemical collapse of the porous structure during prolonged operation. The harsh acidic environment and voltage cycling during fuel cell start-up/shutdown further accelerate these degradation pathways. Moreover, Fe-N_x sites may leach or transform into inactive species, especially in the presence of H_2O_2 intermediates.

Efforts to address these issues, such as stronger Fe anchoring, sacrificial scavengers for reactive oxygen species, and graphitized supports, have shown partial success, but durability still lags far behind that of commercial Pt/C. Moving forward, research should place greater emphasis on in situ diagnostics to track catalyst evolution under working conditions and standardized durability testing protocols that enable fair comparison across studies. Ultimately, achieving a commercially viable performance will require integrated advances in both materials design and electrode engineering, including attention to ionomer distribution, water management, and catalyst layer morphology. An overview of the key challenges and progress made in Fe-N-C catalysts can be found in Figure 13.

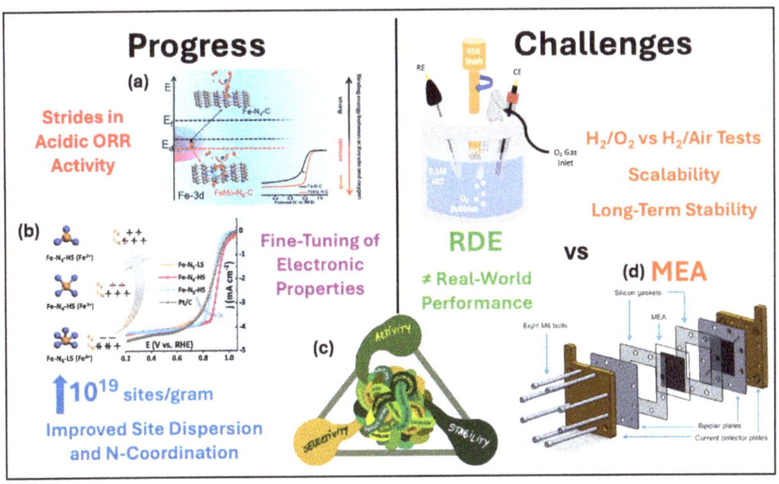

Figure 13. Overview of key challenges and progress made in Fe-N-C catalysts for a fuel cell cathode. (**a**) Dual metal Fe-Mo atom sites regulating Fe-N_4 moiety [91] (adapted with permission from Ref. [91]; copyright: 2022 American Chemical Society). (**b**) Manipulation of Fe-N_x changing spin resulting in excellent ORR activity [92] (adapted with permission from Ref. [92]; copyright: 2023 Elsevier). (**c**) Gordian knot of electrocatalysis emphasizing activity, selectivity, and stability [93] (adapted with permission from Ref. [93]; copyright: 2020 John Wiley and Sons; licensed under CC BY 4.0 [http://creativecommons.org/licenses/by/4.0/ (accessed on 10 June 2025)]). (**d**) Schematic of single-membrane electrode assembly [94] (adapted with permission from Ref. [94]; copyright: 2016 Elsevier).

5. Conclusions

Fe-N-C catalysts have emerged as the most promising non-precious alternatives to Pt/C for the oxygen reduction reaction in proton exchange membrane fuel cells. While still inferior to the commercial Pt/C and Pt-Co/C catalysts, recent advances in synthesis strategies have allowed for significant activity improvements in the durability and activity of Fe-N-C catalysts in acidic conditions. Pyrolysis, metal–organic framework (MOF)-derived structures, wet impregnation, and cascade anchoring have enabled improved site dispersion, enhanced nitrogen coordination, and the fine-tuning of electronic properties through heteroatom co-doping. Allowing for the increase in active site density and overall catalyst stability, Fe-N-C catalysts are now closer to practical use in fuel cells. Long-term stability and scalable production of these catalysts remain a key challenge in non-PGM fuel cell research. Continued investigation into the relationship between catalyst structure, electronic properties, and performance is necessary to fully bridge the gap to commercial viability. Achieving commercially viable stability will require coordinated progress in both molecular design and electrode-level engineering. Strategies such as graphitized supports, stronger Fe-N bonding, and protective architectures show promise but remain insufficient under realistic MEA operating conditions. Future work should prioritize operando diagnostics and standardized durability protocols to better understand degradation pathways and guide rational improvements.

Funding: This research was funded by U.S. Department of Energy, Office of Science, Office of Basic Energy Sciences, with grant number DE-SC0024595. Additionally, thanks to the New Mexico Space Grant Consortium (NMSGC) and the State Legislative Fund, New Mexico, for financial support.

Data Availability Statement: No new data were created or analyzed in this study. Data sharing is not applicable to this article.

Conflicts of Interest: The authors declare no competing financial interests.

References

1. Alternative Fuels Data Center: Vehicle Registration Counts by State. 2022. Available online: https://afdc.energy.gov (accessed on 7 June 2025).
2. EIA (U.S. Energy Information Administration). Gasoline Explained. 2016. Available online: https://www.eia.gov/energyexplained/gasoline/ (accessed on 7 June 2025).
3. United States Environmental Protection Agency. *Greenhouse Gas Emissions from a Typical Passenger Vehicle*; US EPA: Washington, DC, USA, 2022.
4. NASA. What Is the Greenhouse Effect? In *Global Climate Change: Vital Signs of the Planet*; NASA: Washington, DC, USA, 2019.
5. The Causes of Climate Change. In *Climate Change: Vital Signs of the Planet*; NASA: Washington, DC, USA, 2019.
6. Kuo, G. *When Fossil Fuels Run Out, What Then?* MAHB: Stanford, CA, USA, 2019.
7. Lindsey, R. Climate Change: Atmospheric Carbon Dioxide. 2023. Available online: https://Climate.gov (accessed on 7 June 2025).
8. Dulău, L.-I. CO_2 Emissions of Battery Electric Vehicles and Hydrogen Fuel Cell Vehicles. *Clean Technol.* **2023**, *5*, 696–712. [CrossRef]
9. Kim, J.W.; Boo, K.J.; Cho, J.H.; Moon, I. 1-Key challenges in the development of an infrastructure for hydrogen production, delivery, storage and use. In *Advances in Hydrogen Production, Storage and Distribution*; Basile, A., Iulianelli, A., Eds.; Woodhead Publishing: Sawston, UK, 2014; pp. 3–31.
10. Thompson, S.T.; James, B.D.; Huya-Kouadio, J.M.; Houchins, C.; DeSantis, D.A.; Ahluwalia, R.; Wilson, A.R.; Kleen, G.; Papageorgopoulos, D. Direct hydrogen fuel cell electric vehicle cost analysis: System and high-volume manufacturing description, validation, and outlook. *J. Power Sources* **2018**, *399*, 304–313. [CrossRef]
11. Wang, C.; Spendelow, J.S. Recent developments in Pt–Co catalysts for proton-exchange membrane fuel cells. *Curr. Opin. Electrochem.* **2021**, *28*, 100715. [CrossRef]
12. Yarlagadda, V.; Carpenter, M.K.; Moylan, T.E.; Kukreja, R.S.; Koestner, R.; Gu, W.; Thompson, L.; Kongkanand, A. Boosting Fuel Cell Performance with Accessible Carbon Mesopores. *ACS Energy Lett.* **2018**, *3*, 618–621. [CrossRef]
13. Müller-Hülstede, J.; Uhlig, L.M.; Schmies, H.; Schonvogel, D.; Meyer, Q.; Nie, Y.; Zhao, C.; Vidakovic, J.; Wagner, P. Towards the Reduction of Pt Loading in High Temperature Proton Exchange Membrane Fuel Cells—Effect of Fe−N−C in Pt-Alloy Cathodes. *ChemSusChem* **2023**, *16*, e202202046. [CrossRef]
14. Chen, L.; Wan, X.; Zhao, X.; Li, W.; Liu, X.; Zheng, L.; Liu, Q.; Yu, R.; Shui, J. Spatial porosity design of Fe-N-C catalysts for high power density PEM fuel cells and detection of water saturation of the catalyst layer by a microwave method. *J. Mater. Chem. A* **2022**, *10*, 7764–7772. [CrossRef]
15. Asset, T.; Atanassov, P. Iron-Nitrogen-Carbon Catalysts for Proton Exchange Membrane Fuel Cells. *Joule* **2020**, *4*, 33–44. [CrossRef]
16. Birry, L.; Zagal, J.H.; Dodelet, J.-P. Does CO poison Fe-based catalysts for ORR? *Electrochem. Commun.* **2010**, *12*, 628–631. [CrossRef]
17. Kumar, A.; Zhang, Y.; Liu, W.; Sun, X. The chemistry, recent advancements and activity descriptors for macrocycles based electrocatalysts in oxygen reduction reaction. *Coord. Chem. Rev.* **2020**, *402*, 213047. [CrossRef]
18. Winter, M.; Brodd, R.J. What Are Batteries, Fuel Cells, and Supercapacitors? *Chem. Rev.* **2004**, *104*, 4245–4270. [CrossRef]

19. Wang, X.X.; Swihart, M.T.; Wu, G. Achievements, challenges and perspectives on cathode catalysts in proton exchange membrane fuel cells for transportation. *Nat. Catal.* **2019**, *2*, 578–589. [CrossRef]
20. Rabis, A.; Rodriguez, P.; Schmidt, T.J. Electrocatalysis for Polymer Electrolyte Fuel Cells: Recent Achievements and Future Challenges. *ACS Catal.* **2012**, *2*, 864–890. [CrossRef]
21. Vaca, S. *Toyota Introduces Second-Generation Mirai Fuel Cell Electric Vehicle as Design and Technology Flagship Sedan*; Toyota USA Newsroom: Plano, TX, USA, 2020.
22. Bezerra, C.W.B.; Zhang, L.; Lee, K.; Liu, H.; Marques, A.L.B.; Marques, E.P.; Wang, H.; Zhang, J. A review of Fe-N/C and Co–N/C catalysts for the oxygen reduction reaction. *Electrochim. Acta* **2008**, *53*, 4937–4951. [CrossRef]
23. Proietti, E.; Jaouen, F.; Lefèvre, M.; Larouche, N.; Tian, J.; Herranz, J.; Dodelet, J.-P. Iron-based cathode catalyst with enhanced power density in polymer electrolyte membrane fuel cells. *Nat. Commun.* **2011**, *2*, 416. [CrossRef]
24. Wang, X.; Li, Z.; Qu, Y.; Yuan, T.; Wang, W.; Wu, Y.; Li, Y. Review of Metal Catalysts for Oxygen Reduction Reaction: From Nanoscale Engineering to Atomic Design. *Chem* **2019**, *5*, 1486–1511. [CrossRef]
25. Shen, S.; Chen, J.; Yan, X.; Cheng, X.; Zhao, L.; Ren, Z.; Li, L.; Zhang, J. Insights into properties of non-precious metal catalyst (NPMC)-based catalyst layer for proton exchange membrane fuel cells. *J. Power Sources* **2021**, *496*, 229817. [CrossRef]
26. Liu, J.; Li, E.; Ruan, M.; Song, P.; Xu, W. Recent Progress on Fe/N/C Electrocatalysts for the Oxygen Reduction Reaction in Fuel Cells. *Catalysts* **2015**, *5*, 1167–1192. [CrossRef]
27. Duan, J.; Chen, S.; Jaroniec, M.; Qiao, S.Z. Heteroatom-Doped Graphene-Based Materials for Energy-Relevant Electrocatalytic Processes. *ACS Catal.* **2015**, *5*, 5207–5234. [CrossRef]
28. Jiang, Y.; Lu, Y.; Lv, X.; Han, D.; Zhang, Q.; Niu, L.; Chen, W. Enhanced Catalytic Performance of Pt-Free Iron Phthalocyanine by Graphene Support for Efficient Oxygen Reduction Reaction. *ACS Catal.* **2013**, *3*, 1263–1271. [CrossRef]
29. Yuan, S.; Shui, J.-L.; Grabstanowicz, L.; Chen, C.; Commet, S.; Reprogle, B.; Xu, T.; Yu, L.; Liu, D.-J. A Highly Active and Support-Free Oxygen Reduction Catalyst Prepared from Ultrahigh-Surface-Area Porous Polyporphyrin. *Angew. Chem.* **2013**, *52*, 8349–8353. [CrossRef]
30. Liu, K.; Fu, J.; Lin, Y.; Luo, T.; Ni, G.; Li, H.; Lin, Z.; Liu, M. Insights into the activity of single-atom Fe-N-C catalysts for oxygen reduction reaction. *Nat. Commun.* **2022**, *13*, 2075. [CrossRef] [PubMed]
31. Zhou, Y.; Lu, R.; Tao, X.; Qiu, Z.; Chen, G.; Yang, J.; Zhao, Y.; Feng, X.; Müllen, K. Boosting Oxygen Electrocatalytic Activity of Fe-N-C Catalysts by Phosphorus Incorporation. *J. Am. Chem. Soc.* **2023**, *145*, 3647–3655. [CrossRef] [PubMed]
32. Marshall-Roth, T.; Libretto, N.J.; Wrobel, A.T.; Anderton, K.J.; Pegis, M.L.; Ricke, N.D.; Voorhis, T.V.; Miller, J.T.; Surendranath, Y. A pyridinic Fe-N_4 macrocycle models the active sites in Fe/N-doped carbon electrocatalysts. *Nat. Commun.* **2020**, *11*, 5283. [CrossRef] [PubMed]
33. Wang, Q.; Zhou, Z.-Y.; Lai, Y.-J.; You, Y.; Liu, J.-G.; Wu, X.-L.; Terefe, E.; Chen, C.; Song, L.; Rauf, M.; et al. Phenylenediamine-Based FeNx/C Catalyst with High Activity for Oxygen Reduction in Acid Medium and Its Active-Site Probing. *J. Am. Chem. Soc.* **2014**, *136*, 10882–10885. [CrossRef]
34. Lefèvre, M.; Proietti, E.; Jaouen, F.; Dodelet, J.-P. Iron-Based Catalysts with Improved Oxygen Reduction Activity in Polymer Electrolyte Fuel Cells. *Science* **2009**, *324*, 71–74. [CrossRef]
35. Ao, X.; Zhang, W.; Li, Z.; Lv, L.; Ruan, Y.; Wu, H.-H.; Chiang, W.-H.; Wang, C.; Liu, M.; Xiao Cheng, Z. Unraveling the high-activity nature of Fe-N-C electrocatalysts for the oxygen reduction reaction: The extraordinary synergy between Fe–N_4 and Fe_4N. *J. Mater. Chem. A* **2019**, *7*, 11792–11801. [CrossRef]
36. Yang, Z.; Wang, Y.; Zhu, M.; Li, Z.; Chen, W.; Wei, W.-C.; Yuan, T.; Qu, Y.; Xu, Q.; Zhao, C.; et al. Boosting Oxygen Reduction Catalysis with Fe–N_4 Sites Decorated Porous Carbons toward Fuel Cells. *Acs Catal.* **2019**, *9*, 2158–2163. [CrossRef]
37. Zhao, D.; Shui, J.-L.; Grabstanowicz, L.R.; Chen, C.; Commet, S.M.; Xu, T.; Lu, J.; Liu, D.-J. Highly Efficient Non-Precious Metal Electrocatalysts Prepared from One-Pot Synthesized Zeolitic Imidazolate Frameworks. *Adv. Mater.* **2014**, *26*, 1093–1097. [CrossRef]
38. Bates, J.S.; Khamespanah, F.; Cullen, D.A.; Al-Omari, A.A.; Hopkins, M.N.; Martinez, J.J.; Root, T.W.; Stahl, S.S. Molecular Catalyst Synthesis Strategies to Prepare Atomically Dispersed Fe-N-C Heterogeneous Catalysts. *J. Am. Chem. Soc.* **2022**, *144*, 18797–18802. [CrossRef]
39. Yin, S.; Yi, H.; Liu, M.; Yang, J.; Yang, S.; Zhang, B.-W.; Chen, L.; Cheng, X.; Huang, H.; Huang, R.; et al. An in situ exploration of how Fe/N/C oxygen reduction catalysts evolve during synthesis under pyrolytic conditions. *Nat. Commun.* **2024**, *15*, 6229. [CrossRef]
40. Huang, Y.; Chen, Y.; Xu, M.; Ly, A.; Gili, A.; Murphy, E.; Asset, T.; Liu, Y.; De Andrade, V.; Segre, C.U.; et al. Catalysts by pyrolysis: Transforming metal-organic frameworks (MOFs) precursors into metal-nitrogen-carbon (M-N-C) materials. *Mater. Today* **2023**, *69*, 66–78. [CrossRef]

41. Malik, W.; Victoria Tafoya, J.P.; Doszczeczko, S.; Jorge Sobrido, A.B.; Skoulou, V.K.; Boa, A.N.; Zhang, Q.; Ramirez Reina, T.; Volpe, R. Synthesis of a Graphene-Encapsulated Fe3C/Fe Catalyst Supported on Sporopollenin Exine Capsules and Its Use for the Reverse Water–Gas Shift Reaction. *ACS Sustain. Chem. Eng.* **2023**, *11*, 15795–15807. [CrossRef] [PubMed]
42. Zhang, M.; Ma, Z.; Song, H. Preparation and Application of Fe-N Co-Doped GNR@CNT Cathode Oxygen Reduction Reaction Catalyst in Microbial Fuel Cells. *Nanomaterials* **2021**, *11*, 377. [CrossRef]
43. Ma, Q.; Jin, H.; Zhu, J.; Li, Z.; Xu, H.; Liu, B.; Zhang, Z.; Ma, J.; Mu, S. Stabilizing Fe-N-C Catalysts as Model for Oxygen Reduction Reaction. *Adv. Sci.* **2021**, *8*, 2102209. [CrossRef]
44. Zhao, L.; Zhang, Y.; Huang, L.-B.; Liu, X.-Z.; Zhang, Q.-H.; He, C.; Wu, Z.-Y.; Zhang, L.-J.; Wu, J.; Yang, W.; et al. Cascade anchoring strategy for general mass production of high-loading single-atomic metal-nitrogen catalysts. *Nat. Commun.* **2019**, *10*, 1278. [CrossRef]
45. Ahluwalia, R.K.; Wang, X.; Osmieri, L.; Peng, J.; Cankur, C.; Park, J.; Myers, D.J.; Hyun Kyu, C.; Neyerlin, K.C. Stability of Atomically Dispersed Fe-N-C ORR Catalyst in Polymer Electrolyte Fuel Cell Environment. *J. Electrochem. Soc.* **2021**, *168*, 024513. [CrossRef]
46. Zhan, Y.; Xie, F.; Zhang, H.; Jin, Y.; Meng, H.; Chen, J.; Sun, X. Highly Dispersed Nonprecious Metal Catalyst for Oxygen Reduction Reaction in Proton Exchange Membrane Fuel Cells. *ACS Appl. Mater. Interfaces* **2020**, *12*, 17481–17491. [CrossRef]
47. Gianola, G.; Cosenza, A.; Roiron, C.; Pirri, C.F.; Specchia, S.; Atanassov, P.; Zeng, J. Effect of silica leaching treatment during template-assisted synthesis on the performance of FeNC catalysts for oxygen reduction reaction. *Electrochim. Acta* **2025**, *525*, 146085. [CrossRef]
48. Muhyuddin, M.; Mostoni, S.; Honig, H.C.; Mirizzi, L.; Elbaz, L.; Scotti, R.; D'Arienzo, M.; Santoro, C. Enhancing Electrocatalysis: Engineering the Fe–N_x–C Electrocatalyst for Oxygen Reduction Reaction Using Fe-Functionalized Silica Hard Templates. *ACS Appl. Energy Mater.* **2024**, *7*, 11691–11702. [CrossRef]
49. Radwan, A.; Jin, H.; He, D.; Mu, S. Design Engineering, Synthesis Protocols, and Energy Applications of MOF-Derived Electrocatalysts. *Nano-Micro Lett.* **2021**, *13*, 132. [CrossRef]
50. Yang, G.; Zhu, J.; Yuan, P.; Hu, Y.; Qu, G.; Lu, B.-A.; Xue, X.; Yin, H.; Cheng, W.; Jung An, C.; et al. Regulating Fe-spin state by atomically dispersed Mn-N in Fe-N-C catalysts with high oxygen reduction activity. *Nat. Commun.* **2021**, *12*, 1734. [CrossRef] [PubMed]
51. Jin, H.; Zhu, J.; Yu, R.; Li, W.; Ji, P.; Liang, L.; Liu, B.; Hu, C.; He, D.; Mu, S. Tuning the Fe–N4 sites by introducing Bi–O bonds in a Fe-N-C system for promoting the oxygen reduction reaction. *J. Mater. Chem. A* **2022**, *10*, 664–671. [CrossRef]
52. Luo, F.; Roy, A.; Sougrati, M.T.; Khan, A.; Cullen, D.A.; Wang, X.; Primbs, M.; Zitolo, A.; Jaouen, F.; Strasser, P. Structural and Reactivity Effects of Secondary Metal Doping into Iron-Nitrogen-Carbon Catalysts for Oxygen Electroreduction. *J. Am. Chem. Soc.* **2023**, *145*, 14737–14747. [CrossRef] [PubMed]
53. Wu, J.; Pisula, W.; Müllen, K. Graphenes as Potential Material for Electronics. *Chem. Rev.* **2007**, *107*, 718–747. [CrossRef]
54. Stoller, M.D.; Park, S.; Zhu, Y.; An, J.; Ruoff, R.S. Graphene-Based Ultracapacitors. *Nano Lett.* **2008**, *8*, 3498–3502. [CrossRef]
55. Sahoo, N.G.; Pan, Y.; Li, L.; Chan, S.H. Graphene-Based Materials for Energy Conversion. *Adv. Mater.* **2012**, *24*, 4203–4210. [CrossRef]
56. Higgins, D.; Zamani, P.; Yu, A.; Chen, Z. The application of graphene and its composites in oxygen reduction electrocatalysis: A perspective and review of recent progress. *Energy Environ. Sci.* **2016**, *9*, 357–390. [CrossRef]
57. Park, J.S.; Chang, D.W. Iron Phthalocyanine/Graphene Composites as Promising Electrocatalysts for the Oxygen Reduction Reaction. *Energies* **2020**, *13*, 4073. [CrossRef]
58. Koh, K.H.; Noh, S.H.; Kim, T.-H.; Lee, W.J.; Yi, S.-C.; Han, T.H. A graphene quantum dot/phthalocyanine conjugate: A synergistic catalyst for the oxygen reduction reaction. *RSC Adv.* **2017**, *7*, 26113–26119. [CrossRef]
59. Tian, H.; Song, A.; Zhang, P.; Sun, K.; Wang, J.; Sun, B.; Fan, Q.; Shao, G.; Chen, C.; Liu, H.; et al. High Durability of Fe-N-C Single-Atom Catalysts with Carbon Vacancies toward the Oxygen Reduction Reaction in Alkaline Media. *Adv. Mater.* **2023**, *35*, 2210714. [CrossRef]
60. Mahmood, A.; Xie, N.; Zhao, B.; Zhong, L.; Zhang, Y.; Niu, L. Optimizing Surface N-Doping of Fe-N-C Catalysts Derived from Fe/Melamine-Decorated Polyaniline for Oxygen Reduction Electrocatalysis. *Adv. Mater. Interfaces* **2021**, *8*, 2100197. [CrossRef]
61. Chung, H.T.; Cullen, D.A.; Higgins, D.; Sneed, B.T.; Holby, E.F.; More, K.L.; Zelenay, P. Direct atomic-level insight into the active sites of a high-performance PGM-free ORR catalyst. *Science* **2017**, *357*, 479–484. [CrossRef] [PubMed]
62. Husic, B.E.; Schebarchov, D.; Wales, D.J. Impurity effects on solid–solid transitions in atomic clusters. *Nanoscale* **2016**, *8*, 18326–18340. [CrossRef] [PubMed]
63. Long, D.; Li, W.; Ling, L.; Miyawaki, J.; Mochida, I.; Yoon, S.-H. Preparation of Nitrogen-Doped Graphene Sheets by a Combined Chemical and Hydrothermal Reduction of Graphene Oxide. *Langmuir* **2010**, *26*, 16096–16102. [CrossRef]

64. Qian, W.; Cui, X.; Hao, R.; Hou, Y.; Zhang, Z. Facile Preparation of Nitrogen-Doped Few-Layer Graphene via Supercritical Reaction. *ACS Appl. Mater. Interfaces* **2011**, *3*, 2259–2264. [CrossRef]
65. Shao, Y.; Zhang, S.; Engelhard, M.H.; Li, G.; Shao, G.; Wang, Y.; Liu, J.; Aksay, I.A.; Lin, Y. Nitrogen-doped graphene and its electrochemical applications. *J. Mater. Chem.* **2010**, *20*, 7491–7496. [CrossRef]
66. Sun, Z.; Yan, Z.; Yao, J.; Beitler, E.; Zhu, Y.; Tour, J.M. Growth of graphene from solid carbon sources. *Nature* **2010**, *468*, 549–552. [CrossRef]
67. Tang, Y.-B.; Yin, L.-C.; Yang, Y.; Bo, X.-H.; Cao, Y.-L.; Wang, H.-E.; Zhang, W.-J.; Bello, I.; Lee, S.-T.; Cheng, H.-M.; et al. Tunable Band Gaps and p-Type Transport Properties of Boron-Doped Graphenes by Controllable Ion Doping Using Reactive Microwave Plasma. *ACS Nano* **2012**, *6*, 1970–1978. [CrossRef]
68. Yu, S.; Zheng, W.; Wang, C.; Jiang, Q. Nitrogen/Boron Doping Position Dependence of the Electronic Properties of a Triangular Graphene. *ACS Nano* **2010**, *4*, 7619–7629. [CrossRef]
69. Biddinger, E.J.; von Deak, D.; Ozkan, U.S. Nitrogen-Containing Carbon Nanostructures as Oxygen-Reduction Catalysts. *Top. Catal.* **2009**, *52*, 1566–1574. [CrossRef]
70. Li, X.; Fan, L.; Li, Z.; Wang, K.; Zhong, M.; Wei, J.; Wu, D.; Zhu, H. Boron Doping of Graphene for Graphene–Silicon p–n Junction Solar Cells. *Adv. Energy Mater.* **2012**, *2*, 425–429. [CrossRef]
71. Palacin, T.; Khanh, H.L.; Jousselme, B.; Jegou, P.; Filoramo, A.; Ehli, C.; Guldi, D.M.; Campidelli, S. Efficient Functionalization of Carbon Nanotubes with Porphyrin Dendrons via Click Chemistry. *J. Am. Chem. Soc.* **2009**, *131*, 15394–15402. [CrossRef] [PubMed]
72. Sobamowo, M.G.; Akanmu, J.O.; Adeleye, O.A.; Akingbade, S.A.; Yinusa, A.A. Coupled effects of magnetic field, number of walls, geometric imperfection, temperature change, and boundary conditions on nonlocal nonlinear vibration of carbon nanotubes resting on elastic foundations. *Forces Mech.* **2021**, *3*, 100010. [CrossRef]
73. Morozan, A.; Campidelli, S.; Filoramo, A.; Jousselme, B.; Palacin, S. Catalytic activity of cobalt and iron phthalocyanines or porphyrins supported on different carbon nanotubes towards oxygen reduction reaction. *Carbon* **2011**, *49*, 4839–4847. [CrossRef]
74. Govan, J.; Abarca, G.; Aliaga, C.; Sanhueza, B.; Orellana, W.; Cárdenas-Jirón, G.; Zagal, J.H.; Tasca, F. Influence of cyano substituents on the electron density and catalytic activity towards the oxygen reduction reaction for iron phthalocyanine. The case for Fe(II) 2,3,9,10,16,17,23,24-octa(cyano)phthalocyanine. *Electrochem. Commun.* **2020**, *118*, 106784. [CrossRef]
75. Helsel, N.; Choudhury, P. Regulating Electronic Descriptors for the Enhanced ORR Activity of FePc-Functionalized Graphene via Substrate Doping and/or Ligand Exchange: A Theoretical Study. *J. Phys. Chem. C* **2022**, *126*, 4458–4471. [CrossRef]
76. Helsel, N.; Choudhury, P. Enhancing the Stability of a Pt-Free ORR Catalyst via Reaction Intermediates. *Adv. Mater. Interfaces* **2023**, *10*, 2202132. [CrossRef]
77. Helsel, N.; Choudhury, P. Investigation of bifunctionality of FePc-functionalized graphene for enhanced ORR/OER activity. *Mol. Catal.* **2023**, *545*, 113213. [CrossRef]
78. Creel, E.B.; Lyu, X.; McCool, G.; Ouimet, R.J.; Serov, A. Protocol for Screening Water Oxidation or Reduction Electrocatalyst Activity in a Three-Electrode Cell for Alkaline Exchange Membrane Electrolysis. *Front. Energy Res.* **2022**, *10*, 871604. [CrossRef]
79. Wang, L.; Lee, C.-Y.; Schmuki, P. Solar water splitting: Preserving the beneficial small feature size in porous α-Fe2O3 photoelectrodes during annealing. *J. Mater. Chem. A* **2013**, *1*, 212–215. [CrossRef]
80. Hu, Y.; Jensen, J.O.; Pan, C.; Cleemann, L.N.; Shypunov, I.; Li, Q. Immunity of the Fe-N-C catalysts to electrolyte adsorption: Phosphate but not perchloric anions. *Appl. Catal. B Environ.* **2018**, *234*, 357–364. [CrossRef]
81. Elgrishi, N.; Rountree, K.J.; McCarthy, B.D.; Rountree, E.S.; Eisenhart, T.T.; Dempsey, J.L. A Practical Beginner's Guide to Cyclic Voltammetry. *J. Chem. Educ.* **2018**, *95*, 197–206. [CrossRef]
82. Lu, G.; Yang, H.; Zhu, Y.; Huggins, T.; Ren, Z.J.; Liu, Z.; Zhang, W. Synthesis of a conjugated porous Co(ii) porphyrinylene–ethynylene framework through alkyne metathesis and its catalytic activity study. *J. Mater. Chem. A* **2015**, *3*, 4954–4959. [CrossRef]
83. Treimer, S.; Tang, A.; Johnson, D.C. A Consideration of the Application of Koutecký-Levich Plots in the Diagnoses of Charge-Transfer Mechanisms at Rotated Disk Electrodes. *Electroanalysis* **2002**, *14*, 165–171. [CrossRef]
84. Gasteiger, H.A.; Kocha, S.S.; Sompalli, B.; Wagner, F.T. Activity benchmarks and requirements for Pt, Pt-alloy, and non-Pt oxygen reduction catalysts for PEMFCs. *Appl. Catal. B Environ.* **2005**, *56*, 9–35. [CrossRef]
85. Riasse, R.; Lafforgue, C.; Vandenberghe, F.; Micoud, F.; Morin, A.; Arenz, M.; Durst, J.; Chatenet, M. Benchmarking proton exchange membrane fuel cell cathode catalyst at high current density: A comparison between the rotating disk electrode, the gas diffusion electrode and differential cell. *J. Power Sources* **2023**, *556*, 232491. [CrossRef]
86. Choi, C.H.; Chung, M.W.; Kwon, H.C.; Park, S.H.; Woo, S.I. B, N- and P, N-doped graphene as highly active catalysts for oxygen reduction reactions in acidic media. *J. Mater. Chem. A* **2013**, *1*, 3694. [CrossRef]
87. Lian, Y.; Xu, J.; Zhou, W.; Lin, Y.; Bai, J. Research Progress on Atomically Dispersed Fe-N-C Catalysts for the Oxygen Reduction Reaction. *Molecules* **2024**, *29*, 771. [CrossRef]

88. Zhang, Y.; Li, J.; Peng, Q.; Yang, P.; Fu, Q.; Zhu, X.; Liao, Q. Toward an objective performance evaluation of commercial Pt/C electrocatalysts for oxygen reduction: Effect of catalyst loading. *Electrochim. Acta* **2022**, *429*, 140953. [CrossRef]
89. Yang, G.; Lee, C.; Qiao, X.; Babu, S.K.; Martinez, U.; Spendelow, J.S. Advanced Electrode Structures for Proton Exchange Membrane Fuel Cells: Current Status and Path Forward. *Electrochem. Energy Rev.* **2024**, *7*, 9. [CrossRef]
90. Yin, S.; Chen, L.; Yang, J.; Cheng, X.; Zeng, H.; Hong, Y.; Huang, H.; Kuai, X.; Lin, Y.; Huang, R.; et al. A Fe-NC electrocatalyst boosted by trace bromide ions with high performance in proton exchange membrane fuel cells. *Nat. Commun.* **2024**, *15*, 7489. [CrossRef] [PubMed]
91. Zhu, P.; Xiong, X.; Wang, X.; Ye, C.; Li, J.; Sun, W.; Sun, X.; Jiang, J.; Zhuang, Z.; Wang, D.; et al. Regulating the FeN4 Moiety by Constructing Fe–Mo Dual-Metal Atom Sites for Efficient Electrochemical Oxygen Reduction. *Nano Lett.* **2022**, *22*, 9507–9515. [CrossRef] [PubMed]
92. Xue, D.; Yuan, P.; Jiang, S.; Wei, Y.; Zhou, Y.; Dong, C.-L.; Yan, W.; Mu, S.; Zhang, J.-N. Altering the spin state of Fe-N-C through ligand field modulation of single-atom sites boosts the oxygen reduction reaction. *Nano Energy* **2023**, *105*, 108020. [CrossRef]
93. Masa, J.; Andronescu, C.; Schuhmann, W. Electrocatalysis as the Nexus for Sustainable Renewable Energy: The Gordian Knot of Activity, Stability, and Selectivity. *Angew. Chem. Int. Ed.* **2020**, *59*, 15298–15312. [CrossRef]
94. Pollet, B.G.; Franco, A.A.; Su, H.; Liang, H.; Pasupathi, S. 1-Proton exchange membrane fuel cells. In *Compendium of Hydrogen Energy*; Barbir, F., Basile, A., Veziroğlu, T.N., Eds.; Woodhead Publishing: Oxford, UK, 2016; pp. 3–56.

catalysts

MDPI

Review

Oxygen Reduction Reactions of Catalysts with Asymmetric Atomic Structures: Mechanisms, Applications, and Challenges

Hengxing Qiu, Shilong Wen, Qiuju Fu * and Xuebo Zhao *

Shandong Provincial Key Laboratory of Chemical Energy Storage and Novel Cell Technology, Advanced Materials Institute, School of Materials Science and Engineering, Qilu University of Technology (Shandong Academy of Sciences), Jinan 250353, China; qiuhengxing6211@163.com (H.Q.); wenslt@163.com (S.W.)
* Correspondence: fuqiuju@qlu.edu.cn (Q.F.); zhaoxuebo@qlu.edu.cn (X.Z.)

Abstract: Asymmetric-atomic-structure catalysts can modulate the interactions between active sites and intermediates through their unique electronic filling states and asymmetric charge distribution, breaking the linear relationship between adsorption energy and activity, thereby enhancing the catalytic performance of the oxygen reduction reaction (ORR). By introducing heteroelements, vacancies, or clusters into symmetric-atomic-structure catalysts (e.g., M-N_4), asymmetric configurations (such as M-N_x, M-N_x-S/B/O, etc.) can be formed. These modifications substantially alter their internal structure, trigger charge redistribution, and create asymmetric sites to reduce reaction energy barriers, effectively regulating the adsorption strength of oxygen intermediates and significantly improving ORR performance. This review systematically summarizes recent advancements in asymmetric-atomic-structure catalysts for ORR, elucidating the intrinsic "structure–performance–application" relationships to provide theoretical guidance for developing high-performance asymmetric atomic catalysts. First, the ORR mechanisms, including the two-electron and four-electron pathways, are introduced. Furthermore, strategies to modulate catalyst selectivity and activity through doping with metallic/nonmetallic elements or introducing defects are discussed. Finally, prospects for asymmetric-atomic-structure catalysts in next-generation energy storage and conversion technologies are outlined, offering novel insights to overcome current ORR performance bottlenecks.

Keywords: asymmetric atomic structure; oxygen reduction reaction; electrocatalysts; mechanism

1. Introduction

The depletion of fossil fuels and the associated energy demands and environmental issues have inevitably spurred interest in developing new energy sources and energy storage/conversion technologies, including hydrogen fuel cells, zinc–air batteries, and others [1–9]. In these devices, the oxygen reduction reaction (ORR) plays a pivotal role. The ORR, a core reaction in fuel cells and metal–air batteries, can proceed via two distinct pathways in flow cells depending on the number of electrons transferred during electrocatalysis: the two-electron pathway, generating hydrogen peroxide (H_2O_2) as an intermediate (O_2 + $2H^+ + 2e^- \rightarrow H_2O_2$) [10,11], or the four-electron pathway, producing water (H_2O) as the final product ($O_2 + 4H^+ + 4e^- \rightarrow 2H_2O$) [12]. Despite being an exothermic reaction, practical applications of ORR still face significant challenges in energy utilization efficiency [13]. The high bond energy of O-O leads to a high activation energy barrier, resulting in sluggish reaction kinetics [14]; the actual operating potential is lower than the theoretical value, necessitating additional energy input and compromising energy efficiency [15]; the inherent complexity of the two-electron and four-electron transfer pathways severely limits energy conversion efficiency and device longevity [16]. Furthermore, catalyst instability in acidic or alkaline electrolytes, where they are prone to degradation or poisoning [17–19], poses additional critical barriers. These challenges collectively hinder the large-scale commercialization of ORR-based technologies. Platinum (Pt) is widely recognized as the most effective and commonly used ORR catalyst [20]. Although nanotechnology has paved the way for advancements in Pt catalysts—such as precise control of size and morphology to create Pt nanowires, exposure of high-index facets in Pt nanoparticles, and alloying strategies to form nanocrystals, nanosheets, and nanocages [21,22]—these approaches have mitigated catalytic activity as a limiting factor for ORR. However, the low atomic utilization efficiency of Pt (where only surface atoms participate in reactions, with less than 20% of internal atoms being utilized) poses significant challenges to the cost and performance of fuel cells [23–25]. These issues mean that the high cost and limited durability of Pt catalysts remain major obstacles to their widespread commercial application [26,27]. Consequently, the development of non-precious-metal catalyst materials with cost-effectiveness, abundant raw material availability, high ORR activity, excellent stability, and well-defined selectivity has become an imperative objective. In recent years, non-precious-metal symmetric atomic catalysts (e.g., M-N_4 sites with highly symmetric coordination structures) [28–31] have attracted significant attention due to their cost-effectiveness, high stability, universal synthesis of symmetric configurations, highly ordered active site distribution, and modifiable ligand environments. However, their performance remains constrained by low active site density, insufficient exposure efficiency, imbalanced adsorption strength of intermediates, and metal leaching in acidic media. This has facilitated the emergence of asymmetric atomic engineering strategies as an innovative paradigm for ORR catalyst design [32]. The core principle of asymmetric atomic engineering lies in constructing heterogeneous coordination microenvironments, facilitating precise modulation of the geometric and electronic structures of active centers. This approach breaks the performance limitations of conventional catalysts, fostering synergistic enhancement of activity, selectivity, and stability.

Since the early 21st century, conventional ORR catalysts have predominantly manifested highly symmetric coordination environments [33] (e.g., the planar square symmetry of Fe-N_4). The homogeneous electron distribution in symmetric structures facilitates uniform adsorption and activation of O_2, while modulation of intermediate adsorption energies helps prevent over-adsorption-induced active site poisoning. However, limitations such as restricted electronic structure tunability, corrosion susceptibility in acidic media, and low conductivity have impeded breakthroughs in catalyst stability, activity, and selectivity. To date, significant theoretical advancements have been achieved in understanding asymmetric-atomic-structure catalysts for the ORR. The core challenge of both $2e^-$ and $4e^-$ORR pathways lies in balancing the kinetics of intermediate adsorption and desorption [34]. In $2e^-$ORR, asymmetric coordination lowers the d-band center position of the active metal atom toward the Fermi level, weakening *OOH adsorption and enabling its direct desorption to produce H_2O_2. Conversely, for $4e^-$ORR, elevating the d-band center strengthens the adsorption of intermediates (*O, *OH), while asymmetric coordination promotes O_2 dissociation to achieve complete reduction to H_2O. The adsorption configuration of O_2 critically dictates pathway selectivity: end-on adsorption preserves the O-O bond (favoring $2e^-$ORR), whereas side-on adsorption cleaves it (driving $4e^-$ORR) [35]. Asymmetric catalysts optimize O_2 adsorption modes by tailoring the electronic structure of metal centers through heterogeneous coordination environments. Synergistic effects between elements further elevate ORR performance [36]. For instance, one element activates O_2 via adsorption, while another facilitates the desorption of oxygenated intermediates, ensuring efficient reaction kinetics. Moreover, it has been reported that the choice of carbon supports profoundly impacts catalytic activity [37]. The porous structure of carbon governs reactant/product transport efficiency: hierarchical porous hollow carbon spheres accelerate mass transfer due to their high specific surface area and open channels [38], while graphitic carbon enhances conductivity, reduces charge transfer resistance, and improves ORR kinetics. Additionally, 3D porous carbon supports provide abundant anchoring sites to disperse active centers, maximizing active site exposure [39]. These insights collectively establish a robust framework for designing high-performance asymmetric ORR catalysts.

Although asymmetric atom-structured catalysts have made progress in oxygen reduction reaction, they still have problems like poor structural stability, weak anti-poisoning ability, and high demand for structural control precision. Also, there is a lack of research on the mechanism of asymmetric atom-structured catalysts in two-electron and four-electron pathways (Figure 1) and summaries on activity enhancement. In this review, we outline the mechanism and application of ORR in the two-electron and four-electron pathways of asymmetric-atomic-structure catalysts, and provide feasible strategies for the rational design of asymmetric-atomic-structure catalysts. This paper introduces the oxygen reduction reaction mechanisms of the two-electron pathway and the four-electron pathway. In addition, this review synthesizes the reasons for the improvement of catalyst activity, including reasonably regulating the charge coordination environment, defect initiation, and equilibrium adsorption energy through element doping. Finally, by comprehensively looking forward to the development prospects of asymmetric-atomic-structure catalysts, some design principles of the catalysts are proposed to provide new design paradigms for the subsequent generation of asymmetric-atomic-structure oxygen reduction reaction catalysts.

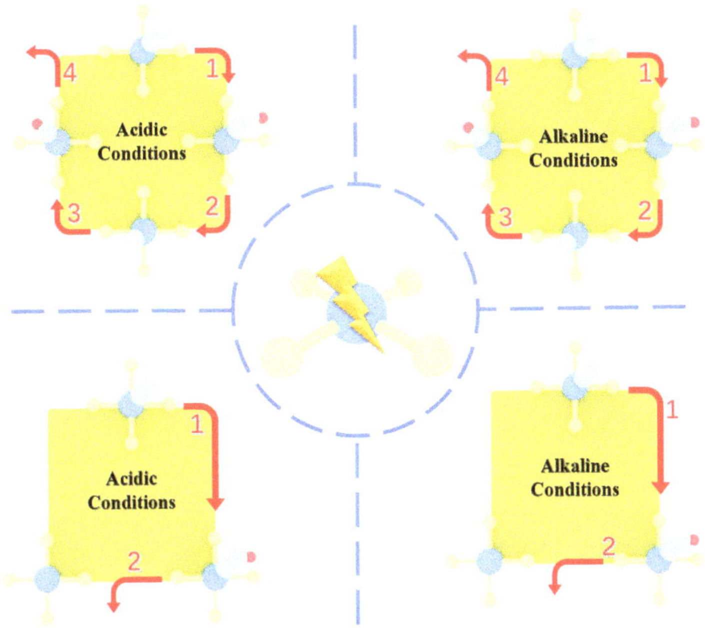

Figure 1. The two-electron and four-electron pathways of the oxygen reduction reaction under acidic and alkaline conditions.

2. Pathway and Mechanism of Oxygen Reduction Reaction

2.1. Two-Electron Pathway and Mechanism of Oxygen Reduction Reaction

The $2e^-$ORR pathway involves the catalytic reduction of molecular oxygen through a $2e^-$ transfer process, producing hydrogen peroxide (H_2O_2) or superoxide radicals (O_2^-) as characteristic products. In this process, the adsorption mode of O_2 molecules on the catalyst surface (end-on or side-on) critically determines the reaction pathway. End-on adsorption, where O_2 binds to the catalyst surface (e.g., Pt) via a single oxygen atom, may instigate O-O bond cleavage, favoring the $4e^-$ pathway [40]. Conversely, side-on adsorption, characterized by parallel alignment of O_2 on surfaces (e.g., carbon-based materials), preserves the O-O bond and promotes the $2e^-$ pathway [41]. Mechanistically, adsorbed O_2 undergoes proton-coupled electron transfer (PCET) to form intermediates such as *OOH (in acidic media) or $*O_2^-$ (in alkaline media), contingent upon pH. These intermediates subsequently accept a second electron and combine with protons, reducing to H_2O_2 (acidic) or HO_2^- (alkaline). Although less energy-efficient than the $4e^-$ route (direct H_2O formation), the $2e^-$ pathway is indispensable in applications like H_2O_2 synthesis, biomedical systems, and wastewater treatment [42–44].

The $2e^-$ pathway is fundamentally dictated by the partial reduction of O_2 molecules through the acceptance of two electrons, yielding H_2O_2 as the primary product. Its selectivity hinges critically on the catalyst's adsorption strength toward the key intermediate (*OOH). Weak adsorption promotes H_2O_2 desorption, suppressing further progression to the $4e^-$ pathway, while strong adsorption drives complete O_2 reduction via the $4e^-$ route. Thus, refining the adsorption energy of *OOH through heteroatom doping—which modifies the electronic coordination configuration of active sites—is a pivotal strategy to enhance $2e^-$ selectivity (Figure 2a–c) [45]. Additionally, introducing heteroatoms into a catalyst's stable lattice generates electron-rich or electron-deficient regions, creating vacancy defects [46]. In asymmetric atomic structures, such defects alter the local electron density and charge distribution, fine-tuning the catalyst's O_2 adsorption energy and activation capacity. Defects further regulate the adsorption strength of critical intermediates (e.g., *OOH) on the catalyst surface (Figure 2d–g) [47]. An idealized adsorption strength stabilizes intermediates, facilitating subsequent electron transfer steps without premature desorption or over-adsorption that could divert the reaction toward the $4e^-$ pathway [48]. When a catalyst contains a single metal element as the active center, incorporating non metallic heteroatoms disrupts the symmetric coordination structure. The electronegativity mismatch between the metal and nonmetal induces electronic redistribution or a downshift of electron orbitals at the metal center, enhancing ORR activity [49]. Moreover, nonmetallic doping reduces the metal center's adsorption strength for oxygen-containing intermediates, promoting their conversion to H_2O_2 [50].

Introducing nonmetallic elements such as N and O into the catalyst breaks the symmetry of the catalyst structure. This alters orbital hybridization, lowers the d-band center of the central atom, and modifies the electronic structure. Consequently, it weakens the adsorption strength of intermediates along the two-electron pathway, shifting the adsorption strength towards favoring desorption. Furthermore, catalysts featuring bimetallic asymmetric coordination promote the two-electron pathway due to their unique side-on adsorption configuration. Additionally, defect-reconstructed catalysts, such as those with dissolved cation vacancies, alter the oxidation state of the central atom. This optimizes the d-band, lowers the adsorption energy of intermediates, and steers the reaction towards the two-electron pathway.

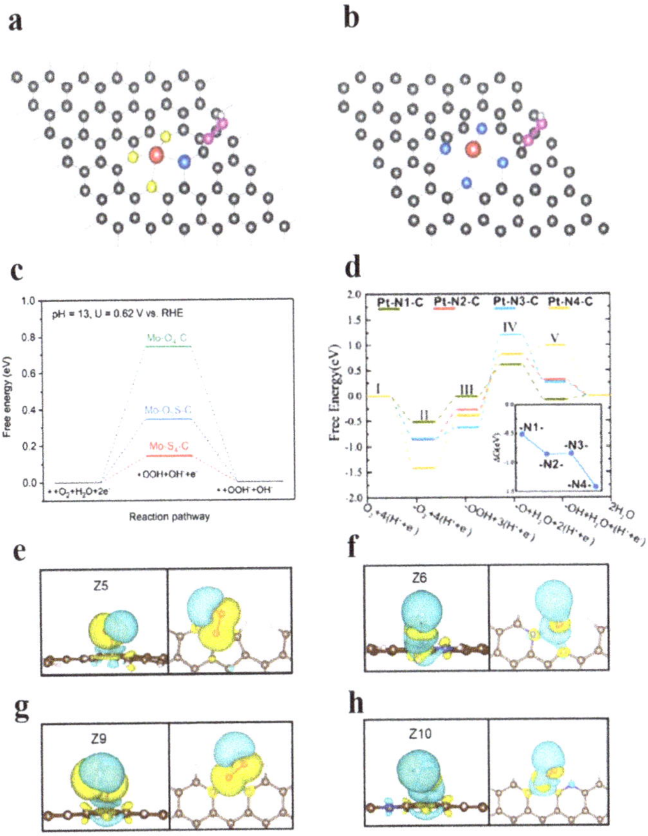

Figure 2. (**a**,**b**) Atomic configuration of adsorption of *OOH on Mo-O_3S-C and Mo-S_4-C. (**c**) Free energy diagrams of 2e$^-$ORR on the three studied substrates at the reaction equilibrium potential. (**d**) Free energy diagrams of different Pt-N_x-C catalysts. (**a**–**d**) Reproduced with permission [45]. Copyright 2023, Elsevier. (**e**) Non-doped pentagonal defect (Z5). (**f**) Doped Gr-N pentagonal defect (Z6). (**g**) Doped graphene edge (Z9). (**h**) Doped graphene edge (Z10). (**e**–**h**) Reproduced with permission [47]. Copyright 2023, American Chemical Society.

2.2. Four-Electron Pathway and Mechanism of Oxygen Reduction Reaction

The 4e$^-$ORR pathway is pivotal for attaining high-efficiency energy conversion, with water (H_2O, acidic media) or hydroxide ions (OH^-, alkaline media) as final products, offering high energy efficiency and environmental compatibility. A critical step in the 4e$^-$ pathway is the cleavage of the O-O bond in molecular oxygen, which typically serves as the rate-determining step (RDS).

The 4e$^-$ORR pathway revolves around the complete cleavage of the O-O bond in molecular oxygen (O_2), achieving full reduction to H_2O [51]. In this process, O_2 molecules adsorb onto the catalyst surface via chemisorption, where the O-O bond is stretched as electrons transfer from the catalyst to O_2's antibonding *π orbitals, leading to direct dissociation into two O atoms or stepwise reduction via PCET to form *OOH. These oxygen atoms sequentially accept electrons and protons, ultimately generating H_2O (in acidic media) or OH^- (in alkaline media). The complete reduction of O_2 through 4e$^-$ transfer maximizes energy release, offering superior energy efficiency. The selectivity of the 4e$^-$ORR pathway is governed by the adsorption strength of the *OOH intermediate, which dictates reaction pathway bifurcation.

The electronic structure and charge distribution of the catalyst modulate the active site's interaction with intermediates, making heteroatom doping a key strategy to tailor the local coordination environment of the central active site and augment $4e^-$ selectivity (Figure 3a–e) [52]. Additionally, defect engineering introduces vacancies near active sites, disrupting the periodic atomic arrangement and altering the central atom's electronic distribution to favor the $4e^-$ pathway. Finally, embedding metal clusters into the catalyst structure leverages synergistic effects between clusters and single atoms to adjust the d-band center position of the metal center (Figure 3f,g). This balances the adsorption capacity for oxygen-containing intermediates, facilitating multi-step electron transfer in ORR [53]. Clusters also enhance the structural stability of asymmetric atomic configurations, preventing aggregation or leaching of single atoms during catalysis and prolonging catalyst durability [54].

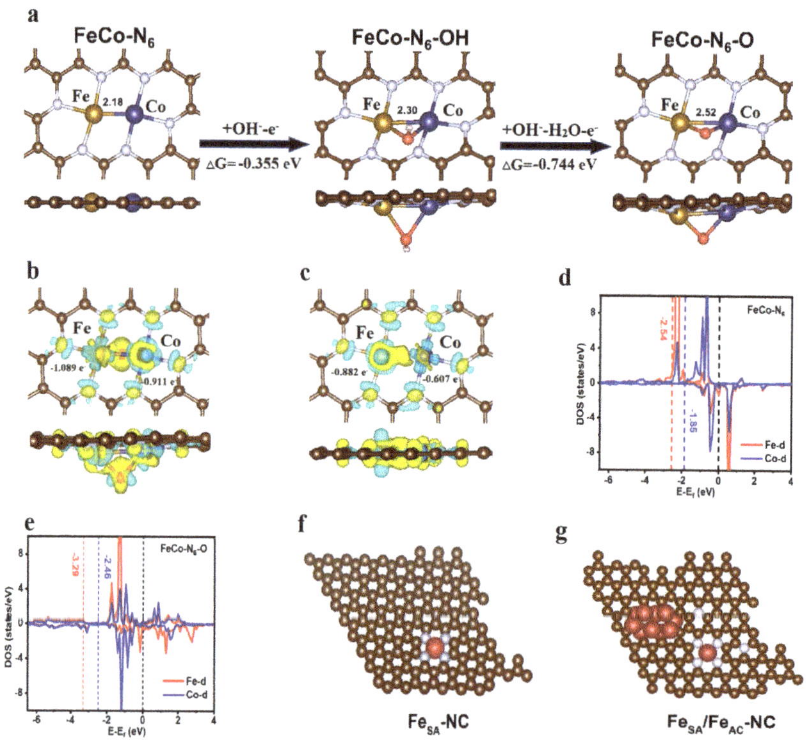

Figure 3. (**a**) Schematic diagram of the formation of the FeCo-N_6-O structure from FeCo-N_6 through two consecutive steps of OH^- electrochemical oxidation and the corresponding change in free energy. (**b**,**c**) The charge density difference plots of FeCo-N_6 and FeCo-N_6-O. (**d**,**e**) The PDOS of FeCo$-N_6$ and FeCo$-N_6-$O, respectively. The vertical red and blue lines in the bottom panel define the d-band center of Fe and Co atoms, respectively. (**a**–**e**) Reproduced with permission [52]. Copyright 2024, American Chemical Society. (**f**,**g**) Structural models of Fe_{SA}-NC and Fe_{SA}/Fe_{AC}-NC. (**f**,**g**) Reproduced with permission [53]. Copyright 2024, Elsevier BV.

Introducing nonmetallic elements such as B and S into the catalyst to form an asymmetric structure lowers the oxidation state of the central active atom and alters the electron distribution in its d-orbitals. This simultaneously enhances O_2 adsorption while accelerating the cleavage of the O-O bond. The co-coordination of multiple nonmetallic elements modifies the coordination structure of the central metal atom, thereby influencing the hybridization mode of its atomic orbitals. The synergistic effect between the nonmetallic elements and the metal elements promotes the conversion of *O to *OH, thereby biasing the oxygen reduction reaction (ORR) towards the four-electron pathway.

2.3. Research on Reaction Pathways

2.3.1. Rotating Ring–Disk Electrode Technique

The Rotating Ring–Disk Electrode (RRDE) technique is an advanced research method combining hydrodynamic regulation and electrochemical detection, widely used to investigate electrocatalytic mechanisms (e.g., ORR) [55]. The operational mechanism of the RRDE relies on forced convection generated by electrode rotation and selective detection of intermediates at the ring electrode, enabling high-sensitivity real-time dynamic analysis of ORR pathways. In RRDE measurements, the ORR occurs at the disk electrode, while intermediate products are selectively oxidized at the ring electrode. By analyzing the ratio of ring current to disk current, the H_2O_2 yield and electron transfer number can be calculated. For instance, Cui et al. [56] documented that the prepared BIM-Co_2Zn_8-500 catalyst achieved a H_2O_2 selectivity of 92.11% and an electron transfer number of 2.15 in the 0.05–0.65 V vs. RHE range, highlighting its exceptional selectivity for the $2e^-$ORR pathway (Figure 4a). Wang et al. [57] demonstrated that the C@PVI-(NCTPP)Fe-800 catalyst exhibited a H_2O_2 yield below 2% and an electron transfer number greater than 3.95 within the 0.2–0.8 V vs. RHE range (Figure 4b), confirming its predominance in the efficient $4e^-$ORR pathway. The Co-N_5 catalyst constructed by Zhang et al. [58] exhibited a high H_2O_2 selectivity of up to 92.5% in 1 M KOH within the potential window of 0.3–0.7 V vs. RHE using the RRDE technique, with the electron transfer number (n) stably maintained between 2.0 and 2.3. This confirms that the dominant reaction pathway is two-electron oxygen reduction. The Fe_{SA}-Fe_3C/NC catalyst synthesized by Huang et al. [59] consistently showed an H_2O_2 yield below 5.2% across a wide potential window of 0.2–0.8 V vs. RHE. The electron transfer number calculated via RRDE was n ≈ 4.0, further confirming that the ORR process proceeds efficiently via the four-electron pathway.

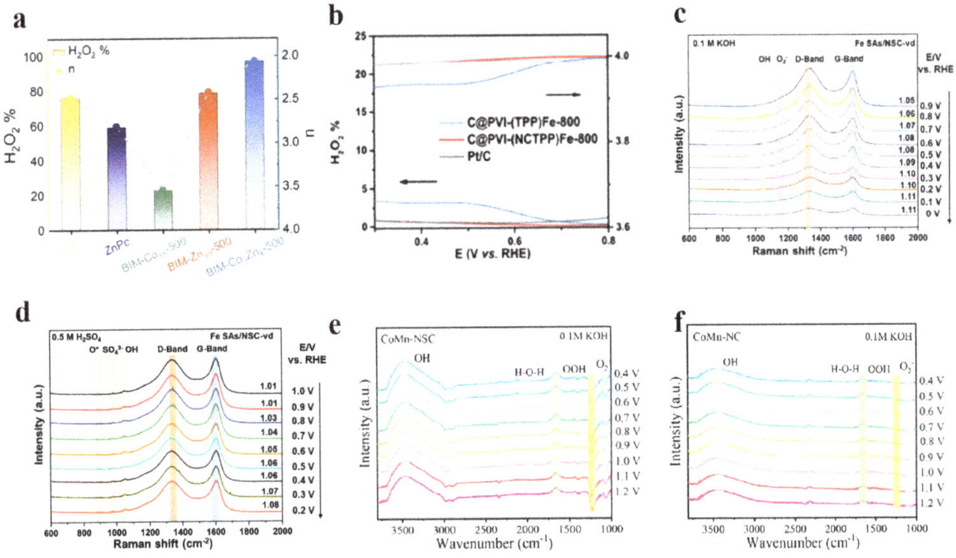

Figure 4. Research method of the oxygen reduction reaction pathway. (**a**) The RRDE technique was used to test the H_2O_2 yield and electron transfer of the C@PVI-(NCTPP)Fe-800 catalyst. (**a**) Reproduced with permission [56]. Copyright 2025, Elsevier BV. (**b**) The RRDE technique was used to test the H_2O_2

yield and the number of electron transfers of the BIM-Co_2Zn_8-500 catalyst. (**b**) Reproduced with permission [57]. Copyright 2024, American Chemical Society. In situ Raman spectra of the Fe/NSC-vd catalyst under the conditions of (**c**) 0.1M KOH and (**d**) 0.5M H_2SO_4. (**c**,**d**) Reproduced with permission [60]. Copyright 2023, Wiley-Blackwell. In situ attenuated total reflection surface-enhanced infrared absorption spectra of (**e**) the CoMn-NSC catalyst and (**f**) the CoMn-NC catalyst under 0.1 M KOH conditions. (**e**,**f**) Reproduced with permission [61]. Copyright 2025, Wiley-VCH Verlag.

2.3.2. In Situ Raman Spectroscopy

In situ Raman spectroscopy is a real-time monitoring method that integrates Raman spectroscopy with in situ reaction conditions, enabling dynamic surveillance of molecular structures, chemical bond changes, and reaction intermediate formation under actual reaction environments, providing direct evidence for elucidating ORR mechanisms [62]. Raman spectroscopy is predicated upon the inelastic scattering effect, detecting frequency shifts caused by interactions between incident light and molecular vibrational modes to reveal molecular vibrations, rotations, and lattice vibrations. In situ Raman spectroscopy couples Raman spectrometers with reaction systems (e.g., electrochemical cells) via specially designed reaction cells or probes, allowing real-time monitoring of structural changes during reactions. The "in situ" feature ensures real-time observation without disrupting the sample's native state or environment. For example, Pan et al. [60] demonstrated that for the Fe/NSC-vd catalyst in 0.1 M KOH, in situ Raman spectra revealed rapid generation and transformation of *OOH intermediates as the potential decreased, while the protonation process of O_2^- was slower (Figure 4c). This suggests that Fe-N coordination structures likely serve as the primary active centers for O_2 adsorption and activation under alkaline conditions. In contrast, in 0.5 M H_2SO_4, in situ Raman spectra showed slower conversion of O_2 to O_2^-, while *OOH intermediate formation and transformation remained pronounced, indicating that Fe-S coordination structures may dominate as active centers for pivotal ORR steps in acidic media (Figure 4d). These findings highlight dynamic structural changes and intermediate adsorption behaviors during the $4e^-$ORR pathway of the Fe/NSC-vd catalyst under divergent electrolytes, confirming distinct reaction mechanisms in alkaline and acidic conditions.

2.3.3. In Situ ATR-SEIRAS

In situ attenuated total reflection surface-enhanced infrared absorption spectroscopy (ATR-SEIRAS) is a highly sensitive surface analysis technique combining attenuated total reflection (ATR) and surface-enhanced infrared absorption spectroscopy (SEIRAS) [63]. It enables real-time monitoring of molecular vibrational information of surface-adsorbed species in practical reaction environments (e.g., electrochemical interfaces, catalytic reactions). The surface enhancement effect grants this technique ultra-high sensitivity for detecting surface species, allowing identification of trace amounts of low-concentration substances and even sub-monolayer-level surface-selective detection, surpassing the limitations of conventional infrared spectroscopy. The in situ monitoring capability permits real-time observation in the actual ORR environment without complex sample pretreatment or labeling, capturing transient intermediates and kinetic information during reactions. With surface specificity, high sensitivity, and environmental compatibility, in situ ATR-SEIRAS has become one of the most effective techniques for elucidating molecular behavior at solid/liquid/gas interfaces. Exemplifying this, Gao et al. [61] analyzed reaction intermediates in the CoMn-NSC catalyst in a 0.1M KOH electrolyte using in situ ATR-SEIRAS: the O-O stretching vibration at 1240 cm^{-1} indicated the presence of superoxide intermediates (O_2^-), the HOH bending vibration at 1638 cm^{-1} reflected interactions between water molecules and intermediates, and the broad peak at 3000–3550 cm^{-1} corresponded to O-H stretching vibrations, confirming hydroxyl adsorption (Figure 4e,f). By tracking potential-dependent spectral changes, the generation of oxygen-containing intermediates and transformation of superoxide intermediates were monitored, validating the CoMn-NSC catalyst's efficacy in promoting the $4e^-$ ORR pathway.

2.3.4. Density Functional Theory

Density functional theory (DFT) calculations, a quantum mechanical computational method, have become a core tool in materials simulation by efficiently solving electron density [64]. The Hohenberg–Kohn theorem, one of the cornerstones of DFT, states that the ground-state energy of a system is a unique functional of electron density, meaning all ground-state properties are derivable from a given electron density distribution. The Kohn–Sham equations, another theoretical foundation of DFT, incorporate the exchange–correlation energy functional, which describes the exchange and correlation interactions between electrons. Accurate calculation of this energy is critical to DFT precision. Although approximative treatments of the exchange–correlation functional introduce errors, DFT demonstrates significant value in predicting material properties and guiding experiments. In ORR studies, DFT is applied to analyze the geometric and electronic structures of catalytic sites [65]. For example, Zhao et al. [66] revealed that high-shell heteroatoms in the $FeCoN_6$ catalyst regulate charge distribution through DFT combined with machine learning, significantly reducing overpotential. Jia et al. [67] synthesized the O-PtCo-FeNC catalyst and used DFT to demonstrate that Fe sites lower the d-band center of Pt, weakening intermediate adsorption energy and enhancing ORR activity and stability. DFT calculations also determine reaction mechanisms and pathways by analyzing intermediate adsorption and free energy landscapes. Liu et al. [68] constructed four structural models for the Fe_{SA}/N,S-PHLC catalyst via DFT, highlighting the correlation between the FeN_3SOH/Pyridine N model and oxygen intermediates. By calculating Gibbs free energy diagrams for different models at U = 1.23 V, the FeN_3SOH/Pyridine N model exhibited relatively low overpotential during *OOH transformation, favoring O-OH bond cleavage and promoting the $4e^-$ ORR pathway.

DFT has become an indispensable tool in ORR research, enabling comprehensive analysis from atomic-level mechanisms to macroscopic performance prediction. It deepens understanding of ORR mechanisms, accelerates the development of low-cost, high-performance catalysts, and drives advancements in technologies such as fuel cells and metal–air batteries.

2.3.5. X-Ray Absorption Near-Edge Structure

X-ray absorption near-edge structure (XANES) is a cutting-edge spectroscopic technique based on synchrotron radiation sources. It analyzes the local electronic structure and geometric environment of materials by detecting changes in the probability of core electron excitation to unoccupied orbitals for specific elements. Its working principle involves measuring the X-ray absorption coefficient of a specific element in the region near its absorption edge (typically from 10–20 eV below the edge to 30–50 eV above the edge). When the energy of the incident X-ray photon reaches a level sufficient to excite a core electron of the specific element into an unoccupied state (the conduction band or molecular orbital), the absorption coefficient exhibits a sharp rise (the absorption edge) [69]. As revealed by Co K-edge XANES spectroscopy analysis of the ZIF-ZC91@$Co(OH)_2$-V_{Co} catalyst synthesized by Zhang et al. [70], the Co K-edge absorption edge shifts towards higher energy. This indicates an increase in the Co oxidation state from +2 to nearly +3. This valence state elevation arises because the cation vacancies (V_{Co}) induce electron loss from surrounding Co atoms, triggering an electronic restructuring. This optimization enhances the catalyst's adsorption capability for the *OOH intermediate, consequently improving H_2O_2 selectivity. Analysis of both the Cu K-edge and L-edge XANES for the Cu-S_1N_3 asymmetric structure catalyst synthesized by Jiang et al. [71] revealed that the Cu L-edge position lies between those of CuPc and CuS. Combined with the K-edge absorption edge position, this indicates an average Cu oxidation state of +1.97. During the oxygen reduction reaction (ORR) process, the absorption edge shifted towards lower energy and the white line intensity decreased, confirming a reduction in the Cu oxidation state from +2 to +1. This demonstrates that the low-valent Cu^+ species serves as the ORR active center. The results confirm the existence of C–S and N–Cu bonds, providing evidence for the direct coordination of S to Cu.

2.3.6. Fourier Transform Infrared Spectroscopy

Fourier Transform Infrared Spectroscopy (FTIR) is a powerful technique used to obtain the infrared absorption or emission spectra of solid, liquid, or gas samples. It investigates the molecular structure, chemical bonds, and functional groups of substances based on their selective absorption of infrared radiation. The working principle of FTIR involves using an interferometer to generate an interferogram that contains information across all infrared frequencies. This interferogram passes through the sample. A detector then receives the interferogram signal modified by the sample's absorption. Subsequently, a computer performs a Fourier transform on this complex interferogram signal, converting it into the conventional infrared absorption spectrum. In situ Fourier Transform Infrared Spectroscopy (In situ FTIR) is an advanced technique that enables real-time monitoring of dynamic structural changes in materials under authentic reaction conditions (such as specific temperature, pressure, atmosphere, or electrochemical environments). This is achieved by integrating a specialized reaction cell (e.g., high-temperature/high-pressure cell, electrochemical cell, catalytic reaction cell) into the FTIR optical path, allowing the infrared beam to pass directly through the sample while it is in its operational state. By rapidly acquiring time-resolved interferograms and performing real-time Fourier transforms, instantaneous spectra revealing molecular bond vibrations/rotations during the reaction are obtained. This allows for the capture of real-time information on intermediates, surface-adsorbed species, and reaction pathways [72]. For the ZnCo-MTF catalyst prepared by Liu et al. [73], Fourier Transform Infrared Spectroscopy (FTIR) revealed that the precursor material ZnCo-ZIF exhibited characteristic peaks of 2-methylimidazole (e.g., C=N stretching vibration, C-H bending vibration, etc.). During the reaction process, the characteristic peaks of 2-methylimidazole gradually diminished, while those of 1,2,3-triazole progressively intensified (e.g., N-H/N-N vibrations, triazole ring skeletal vibrations). The final product, ZnCo-MTF, displayed only the strong characteristic peaks of 1,2,3-triazole, with the 2-methylimidazole peaks completely absent. This demonstrates that the FTIR spectrum of the final product matches the characteristic peaks of the pure 1,2,3-triazole ligand, indicating the complete replacement of 2-methylimidazole by 1,2,3-triazole within the ZnCo-MTF framework, resulting in a new metal–organic framework (MOF) structure. This transformation provides the structural foundation for the subsequently enhanced catalytic performance (high H_2O_2 production rate/selectivity). During the synthesis of the Fe-N_4-SM catalyst constructed by Shen et al. [74], the precursor containing the defect-prone ligand 5-ATZ exhibited a Zn-N vibration peak at 423.0 cm^{-1}. The displacement of this Zn-N vibration peak, coupled with the preservation of the main framework characteristics, confirmed the successful synthesis of the dual-ligand ZTIF precursor. Furthermore, this analysis revealed that the thermal instability of 5-ATZ underpins the subsequent selective C-N bond cleavage and serves as the structural basis for atomic strain modulation.

3. Control of Oxygen Reduction Reaction Pathways

3.1. Control of Two-Electron Pathways

It is a reasonable choice to introduce one or more nonmetallic elements instead of the nonmetallic elements in the original structure to form an asymmetric atomic structure in order to control the tendency of the catalyst to follow the two-electron pathway in the reaction, so as to improve the catalytic efficacy of $2e^-$ORR [75].

The incorporation of heteroelements can create catalysts with asymmetric configurations, which induces uneven charge distribution at the central active sites and modifies electron orbital states. As elucidated by Ivan Parkin et al. [76], a surface engineering strategy was employed to synthesize atomically dispersed Co on N-doped carbon black, forming an asymmetric coordination configuration (Co-C/N/O). The asymmetric structure facilitates hybridization between Co 3d orbitals and C/N/O p-orbitals, which lowers the d-band center of Co, modulates the electronic structure, and weakens the adsorption strength of intermediates (e.g., *OOH) (Figure 5a,b), thereby preventing over-strong adsorption that would drive the $4e^-$ pathway. Oxygen atoms act as electron-withdrawing sites from Co, while N and C serve as electron-donating components, establishing a charge gradient that stabilizes *OOH adsorption. DFT calculations reveal a positive correlation between the vertical displacement of Co atoms from the graphene plane and the *OOH adsorption energy, highlighting the potential for geometric configuration optimization to fine-tune catalytic activity. Similarly, Zhang et al. [58] synthesized a Co-N_5 catalyst by precisely controlling the coordination number of Co atoms through the thermal stability differences of precursors (Figure 5c). The introduction of an additional N atom in the Co-N_5 configuration induces uneven charge distribution and electronic rearrangement, breaking the linear scaling relations (LSRs) intrinsic to conventional symmetric Co-N_4 catalysts and thereby resolving the activity–selectivity trade-off. DFT calculations demonstrate that the altered d-orbital occupancy of Co optimizes the adsorption energy of *OOH intermediates, preventing both O-O bond cleavage ($4e^-$ pathway) and insufficient O_2 activation (blocked $2e^-$ pathway). In situ Raman spectroscopy reveals that Co-N_5 sites stabilize *OOH intermediates and slow their consumption rate. KSCN poisoning experiments further confirm that atomically dispersed Co serves as the catalytically active center, with changes in d-orbital states being the key factor for enhanced performance.

Additionally, the incorporation of heteroatoms modulates the adsorption capability of the central active atom toward oxygen-containing intermediates, tailoring the adsorption strength to favor *OOH desorption. Gao et al. [77] synthesized an asymmetrically coordinated Zn-N_3O catalyst by adsorbing Zn ions onto polypyrrole (PPy) to achieve an elevated Zn loading of 11.34 wt%, forming Zn-N/O bonds (Figure 5d). The introduction of O-atom coordination withdraws electrons from Zn, increasing its oxidation state and redistributing its electron cloud density. This induces a contraction of Zn's d-orbital electron density, which strengthens the interaction between Zn's d-orbitals and the antibonding orbitals of *OOH, thereby diminishing the O-O bond dissociation energy barrier and optimizing the adsorption energy of Zn toward *OOH intermediates to favor the $2e^-$ORR pathway. DFT calculations reveal that the density of states (DOS) near the Fermi level in the Zn-N_3O structure significantly increases, indicating enhanced hybridization between Zn's d-orbitals and O's p-orbitals, which improves intermediate adsorption. Zn K-edge EXAFS spectroscopy further confirms structural modulation by O-coordination, showing altered coordination numbers and bond lengths between Zn and N/O atoms. Additionally, Han et al. [78] developed a high-coordination Co-N_5 moiety by precisely regulating the coordination environment of Co atoms, leveraging adjacent electron-withdrawing epoxy groups to construct a Co-N_5-O-C catalyst. By adjusting the number of N atoms in the first coordination shell of Co single-atom catalysts (SACs) or partially substituting them with O atoms, Co-N_xO_y SACs were formed. This modification alters the charge distribution and d-band center of the metal atom, achieving an optimized electronic structure that shifts the reaction pathway toward the $2e^-$ORR. The high-coordination N atoms modulate Co's d-band center, fine-tuning the adsorption energy of the *OOH intermediate. DFT calculations demonstrate that the Co-N_5 moiety exhibits a reduced overpotential for the $2e^-$ORR compared to the Co-N_4 configuration, thereby enhancing pathway selectivity toward H_2O_2 production.

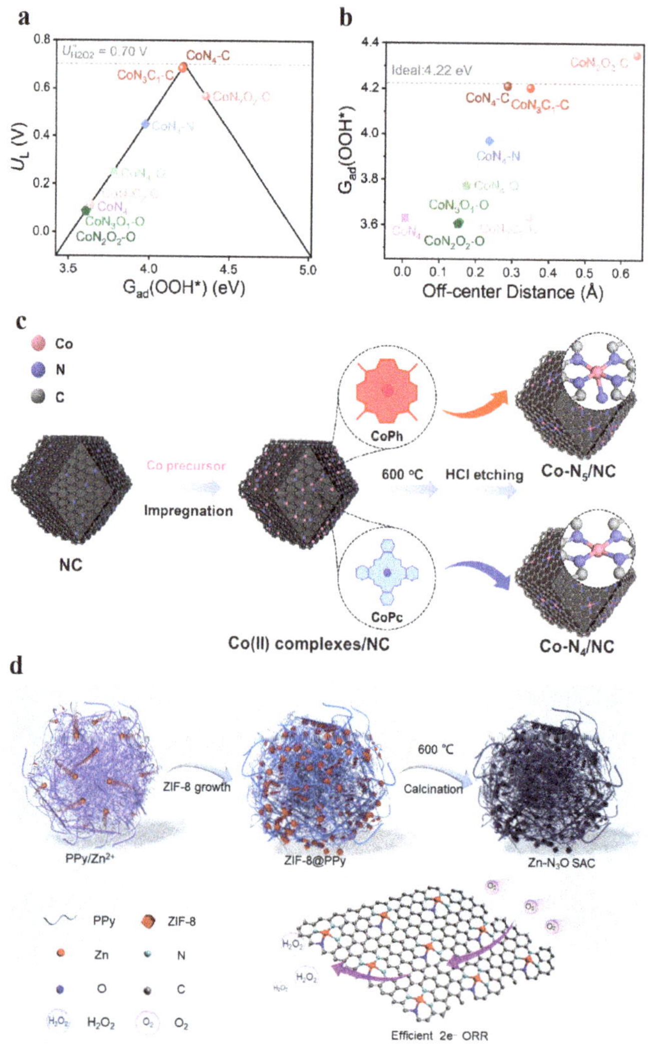

Figure 5. (**a**) The calculated ORR activity volcano plot. (**b**) The relationship between the distance of Co atoms from the center and the adsorption energy of *OOH. (**a**,**b**) Reproduced with permission [76]. Copyright 2024, Springer Nature. (**c**) Schematic diagram of the synthesis of Co-N_5/NC and Co-N_4/NC catalysts. (**c**) Reproduced with permission [58]. Copyright 2025, John Wiley and Sons Ltd. (**d**) Schematic diagram of the synthesis of Zn-N_3O-SAC. (**d**) Reproduced with permission [77]. Copyright 2022, Elsevier.

Bimetallic or multimetallic-doped ORR catalysts typically orchestrate the adsorption strength of intermediates (*OOH) to achieve a balance where oxygen-containing intermediates neither undergo over-adsorption (leading to further reduction via the $4e^-$ pathway) nor under-adsorption (causing sluggish kinetics), thereby avoiding O-O bond cleavage [79,80]. In such catalysts, the complementary electronic properties of the two metals synergize: one metal adsorbs and activates O_2 to form *OOH, while the other weakens *OOH adsorption, enabling its desorption to generate H_2O_2. Additionally, differences in bimetallic coordination alter the electronic distribution of active sites, enhancing selectivity for the $2e^-$ pathway. The asymmetric coordination environment further induces hybrid orbital energy matching with O_2's $*\pi$ orbitals, facilitating single-electron transfer to produce superoxide radicals (O_2^-), which subsequently undergo selective reduction to H_2O_2. This dual-metal synergy not only modulates intermediate adsorption but also tailors electronic interactions to stabilize the $2e^-$ pathway. Liu et al. [73] fabricated a hollow nanocubic-structured catalyst (ZnCo-MTF) using ZnCo-ZIF as a precursor and 1,2,3-triazole as a ligand. As shown in Figure 6a,b, the synergistic interaction between Zn and Co induces a unique side-on adsorption configuration for O_2: one oxygen atom weakly binds to Zn, while the other strongly interacts with Co. This binding mode effectively polarizes the O_2 molecule, lowering the O-O bond dissociation barrier and promoting desorption of the *OOH intermediate, thereby selectively driving the $2e^-$ pathway to generate H_2O_2. The coordination between 1,2,3-triazole ligands and Co forms *d-pπ delocalized orbitals**, increasing the occupation of Co-N antibonding orbitals and weakening Co-N bonds to activate the Co center. Simultaneously, Zn incorporation optimizes the electron density of Co via charge transfer, enhancing O_2 adsorption capability. DFT calculations reveal that the DOS of ZnCo-MTF spans the Fermi level, signifying superior conductivity compared to monometallic systems, which accelerates electron transfer kinetics.

Strategies to disrupt structural symmetry in catalysts extend beyond metal/nonmetal doping and include defect engineering to reconstruct the original framework. For instance, Zhang et al. [70] synthesized a ZIF-ZC91@Co(OH)$_2$-V_{Co} catalyst via electrochemical self-optimization reconstruction of ZIF-ZC91, where partial Zn^{2+} leaching generates cobalt cation vacancies. These vacancies, formed through selective Zn dissolution, modulate the catalyst's electronic structure and surface morphology. The presence of cation vacancies optimizes the adsorption strength of *OOH intermediates (Figure 6c), suppresses O-O bond cleavage, and promotes the $2e^-$ORR pathway. Mechanistically, the vacancies alter the oxidation state of Co and refine its d-band characteristics, thereby reducing the adsorption energy of *OOH. DFT calculations confirm that the energy barrier for the $2e^-$ pathway is lower than that of the $4e^-$ route, effectively inhibiting O-O bond dissociation.

Notably, incorporating heteroelements into catalyst structures not only induces disparate charge distribution at the central atom but also creates defects near the active sites. These defects generate localized electron-rich regions or charge-asymmetric environments, stabilizing the *OOH intermediate (precursor to H_2O_2) and preventing its further reduction to *O or *OH, thereby blocking the $4e^-$ORR pathway. As demonstrated by Zhang et al. [81], a coaxial Co single-atom catalyst (Co_{SA}-N-C/CNTs) was synthesized via an SCVD strategy. Concentrated nitric acid oxidation was employed to introduce surface defects (e.g., carbon vacancies) and oxygen-containing functional groups on CNTs, forming defective CNTs (d-CNTs). These defects provide abundant anchoring sites, enabling the stable immobilization of Co atoms as single-atom sites on the carbon substrate to form high-density Co-N_x active centers. The defects alter the electronic state of the carbon substrate, modulating the electron density of Co-N_x sites and thereby tuning the adsorption strength of intermediates (*OOH) (Figure 6d,e). DFT calculations reveal that the ΔG*OOH of Co-N_x sites approaches the apex of the volcano plot, indicating that defect-induced electronic optimization achieves near-optimal adsorption energy for enhanced 2e$^-$ORR selectivity. Additionally, the defect-engineered Co_{SA}-N-C/CNTs exhibit a microporous structure, where the pores and surface roughness (resulting from defects) accelerate the diffusion of reactants (O_2) and products (H_2O_2), while exposing more active sites to enhance catalytic performance.

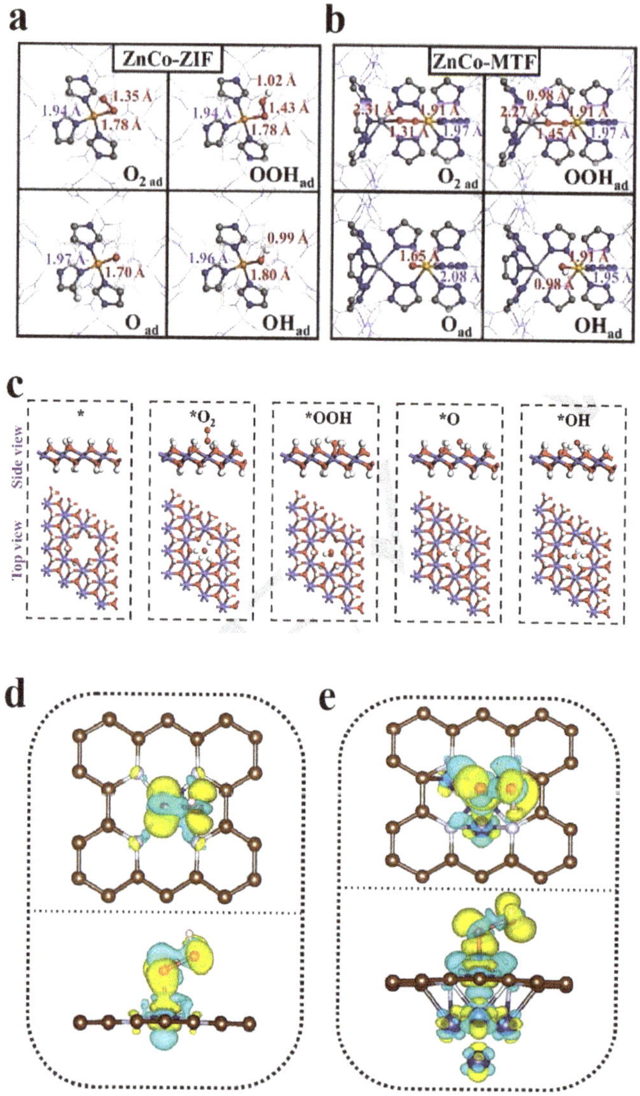

Figure 6. (**a**,**b**) Snapshots of the intermediate structures on the surfaces of ZnCo-ZIF (001) and ZnCo-MTF (001). (**a**,**b**) Reproduced with permission [73]. Copyright 2023, John Wiley and Sons Ltd. (**c**) Adsorption configurations of different intermediates in the two-electron and four-electron transfer ORR pathways on medium-cationic-vacancy catalysts. "*" represent the intermediate. (**c**) Reproduced with permission [70]. Copyright 2025, John Wiley and Sons Ltd. (**d**,**e**) The difference in charge density of Co_{SA}-N-C and Co@Co-N-C before and after their interaction with *OOH. (**d**,**e**) Reproduced with permission [81]. Copyright 2024, Wiley-VCH Verlag.

3.2. Control of Four-Electron Paths

3.2.1. Incorporation Forms of Nonmetallic Elements

To modulate the adsorption energy of central active sites, nonmetallic elements are incorporated into inherently symmetric and stable catalysts, constituting asymmetric atomic structures [82]. Due to the higher electronegativity of nonmetallic elements compared to metallic counterparts, their incorporation alters the electron density distribution of the central active element. The nonmetallic dopant withdraws electrons from neighboring atoms (e.g., carbon or the central metal atom), creating electron-deficient regions. This polarization effect strengthens O_2 adsorption and activation, lowers the free energy barrier for oxygen adsorption reactions, and consequently augments ORR activity. For instance, Ding et al. [83] synthesized a single-atom Co catalyst with N/B asymmetric coordination (Co-N_3B) via a pyrolysis method and anchored it on boron–nitrogen Co-doped graphene (BNG) to form Co-N_3B active sites (Figure 7a). The incorporation of B atoms, owing to their low electronegativity, facilitates electron transfer to the Co center, reducing its oxidation state and increasing its electron density. This electronic modulation enhances Co's ability to adsorb and activate oxygen intermediates (e.g., *O, *OOH). Projected density of states (PDOS) analysis from DFT reveals stronger d-p orbital hybridization between Co's d-orbitals and O's p-orbitals near the Fermi level in Co-N_3B, indicating enhanced orbital overlap that promotes the 4e^-ORR pathway. Attenuated total reflection surface-enhanced infrared absorption spectroscopy (ATR-SEIRAS) experiments demonstrate that the Co-BNG catalyst exhibits stronger adsorption capacity for oxygen intermediates (e.g., *O_2^- and *OOH) and maintains excellent stability during prolonged reaction cycles. Similarly, Zhang et al. [84] fabricated a Co-N_3-C/CNT catalyst via a bidentate pre-coordination-assisted pyrolysis strategy. DFT calculations elucidate how the asymmetric low-coordination structure of Co-N_3 disrupts the symmetric charge distribution of Co-N_4. The reduced coordination (Co-N_3) induces asymmetric electron density distribution at the Co atom (Figure 7b–d), enhancing oxygen activation capabilities and optimizing the adsorption/desorption equilibrium of ORR intermediates. The Co-N_3 configuration exhibits lower energy barriers during ORR, particularly in the critical *O-to-*OH conversion step, demonstrating superior intrinsic catalytic activity compared to symmetric Co-N_4 counterparts. Furthermore, Yin et al. [85] successfully constructed a CoN_3S-PC catalyst by anchoring single Co atoms on porous carbon (PC) and subsequently doping S atoms into the first coordination shell of the Co active sites, forming an asymmetric CoN_3S coordination structure. The incorporation of S atoms induces electronic rearrangement and geometric distortion at the Co active center, disrupting its charge symmetry and enhancing interactions between Co 3d orbitals and oxygen molecules/intermediates (Figure 7e,f). This strengthens O_2 adsorption at the Co sites and accelerates O-O bond cleavage. DFT calculations reveal that S doping modifies the electronic distribution of Co's d-orbitals, promoting O_2 adsorption/activation while suppressing H_2O_2 generation. Concurrently, the CoN_3S sites exhibit a reduced thermodynamic energy barrier and kinetic superiority for the 4e^-ORR pathway, ensuring efficient and selective oxygen reduction.

The coordination of nonmetallic elements directly with the central active site increases the electron density peripherally to the active atom, thereby influencing the adsorption energy of intermediates. Jiang et al. [71] designed an asymmetrically coordinated Cu-S_1N_3 single-atom active site via an atomic interface engineering strategy, constructing a single-copper-atom catalyst (S-Cu-ISA/SNC) based on MOF-derived hierarchically porous carbon. The introduction of S atoms disrupts the conventional symmetric CuN_4 coordination, forming an asymmetric Cu-S_1N_3 active site that modifies the electronic structure around the central Cu atom. The relatively lower electronegativity of S atoms reduces the loss of valence electrons from Cu atoms, altering the local electron density distribution (Figure 8a) and thereby optimizing the adsorption free energy of ORR intermediates (e.g., *OOH, *O, *OH) (Figure 8b) while diminishing the reaction energy barrier. In situ soft X-ray absorption near-edge structure (XANES) tests and in situ FTIR spectroscopy elucidated the electronic and atomic structural evolution of Cu sites during ORR. DFT calculations, the Koutecky–Levich (K-L) equation, and RRDE tests collectively validated the catalyst's performance, confirming the presence of low-valent Cu active centers, an electron transfer number ($n \approx 4$), and a low H_2O_2 yield (<4%). Yang et al. [86] synthesized an FeN_3S/Cmeso catalyst by embedding single Fe atoms into a mesoporous carbon matrix and introducing S atoms to directly coordinate with the Fe center, forming an FeN3S coordination structure. The direct coordination of S atoms to the Fe center disrupts the symmetry of the FeN_4 configuration, increasing the electron density around Fe (Figure 8c). The incorporation of S lowers the d-band center of Fe, weakening the orbital overlap between active sites and oxygenated intermediates (e.g., *OH), thereby optimizing their adsorption energy (Figure 8d). DFT calculations demonstrate that the S-induced coordination enhances the electron cloud density of Fe atoms, which reduces the orbital overlap between active sites and oxygen-containing intermediates. This attenuated adsorption strength facilitates intermediate desorption and lowers the energy barrier during the oxygen reduction process.

The co-coordination of multiple nonmetallic elements modifies the coordination number and coordination structure of the central metal atom, thereby influencing the hybridization mode of metal atomic orbitals [87]. Different coordination structures exhibit distinct geometric configurations and symmetries, which constrain or guide the direction and extent of hybridization in metal atomic orbitals. Following multi-nonmetallic element coordination, charge transfer occurs between the nonmetallic elements and the central metal atom. Nonmetallic elements tend to attract electrons, causing a partial transfer of electron density from the metal atom to the nonmetallic ligands and a reduction in the electron cloud density of the metal atom. This charge redistribution modifies the electronic configuration of the metal atom's d-orbitals, affecting the formation and characteristics of its hybridized orbitals. Additionally, highly electronegative nonmetallic elements can lower the d-band center of metal sites through electronic modulation, geometric restructuring, and synergistic catalytic effects. This weakens the adsorption strength toward oxygen-containing intermediates (such as *OOH, *O, and *OH), bringing the adsorption energy closer to the peak of the "volcano curve". Such optimization of reaction kinetics significantly promotes the $4e^-$ ORR pathway while suppressing H_2O_2 generation routes, such as reducing the probability of *OOH protonation to form H_2O_2. As reported by Liu et al. [88], an asymmetric nitrogen–oxygen-coordinated tin single-atom catalyst (Sn-N/O-C) with Sn atoms as active centers was developed via a simple one-step pyrolysis strategy, forming asymmetric SnN_2O sites through the co-coordination of N and O atoms. The N and O co-coordination disrupted the symmetry of the Sn sites, optimized the p-orbital electron distribution of Sn, and induced strong hybridization between the Sn 5p orbitals and the O_2 2p orbitals. This altered the theoretical d-band center of Sn, promoting electron transfer and accelerating O_2 activation and reduction (Figure 8e,f). DFT calculations revealed that the Sn atoms with broken symmetric coordination facilitated charge transfer and O_2 activation. RRDE measurements demonstrated a hydrogen peroxide (H_2O_2) yield of <5% in alkaline media and <3% in acidic media, with an electron transfer number close to 4, indicating the dominance of the $4e^-$ pathway. Similarly, Liu et al. [89] introduced O atoms into the first coordination shell of Fe and incorporated adjacent defects, inducing a transition of Fe from a low-spin state to a medium-spin state and forming a Fe-N_3O coordination structure, thereby synthesizing a single-atom iron catalyst (Fe-N/O-C). The coordination of N and O atoms in Fe-N/O-C redistributes the charge, increases the electron density at the Fe sites, and significantly weakens the adsorption of oxygen-containing species such as O_2, mitigating over-adsorption of O_2. Theoretical calculations reveal that the d-band center of Fe-N/O-C is lower than that of Fe-N-C, leading to weakened adsorption of intermediates. The formation of adsorbed *OOH from O_2 becomes the new rate-determining step (replacing the original *OH desorption), with a reduced energy barrier. In situ Raman spectroscopy further confirms this shift in the PDS. RRDE tests demonstrate an electron transfer number close to 4 and an extremely low H_2O_2 yield (<5%), indicating that a highly efficient 4e$^-$ pathway dominates the reaction.

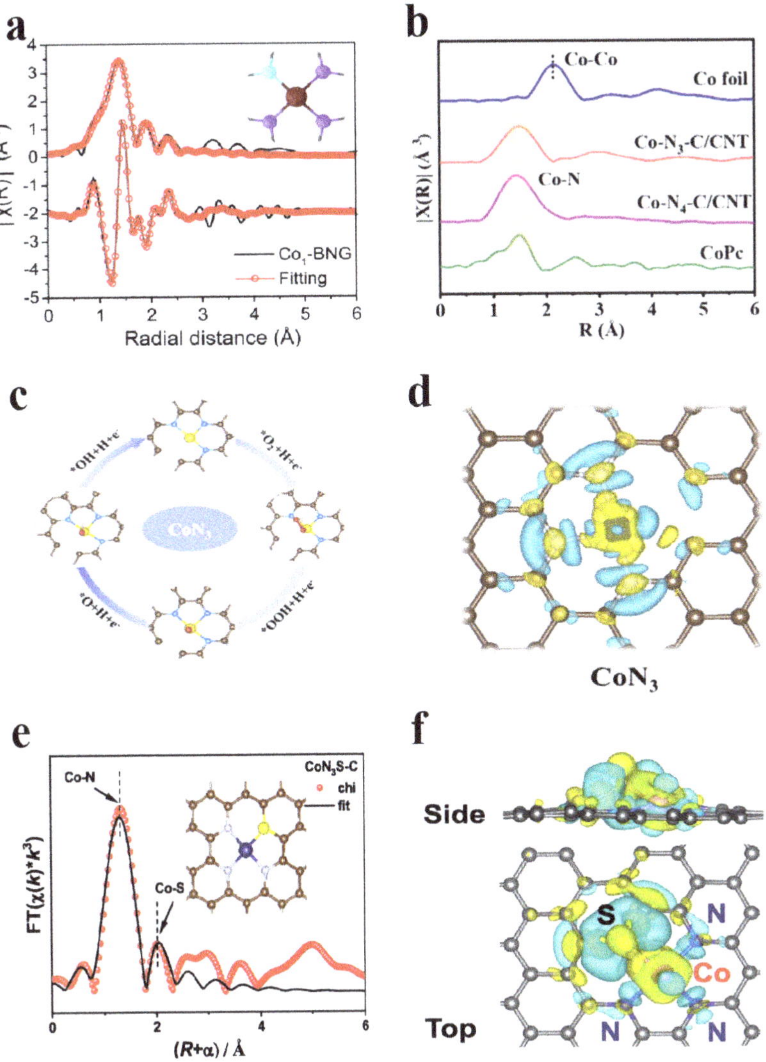

Figure 7. (**a**) The first shell fitting of the EXAFS spectra of Co_1-BNG. (**a**) Reproduced with permission [83]. Copyright 2024, Wiley-VCH Verlag. (**b**) FT-EXAFS of Co-N_3-C/CNT, Co-N_4-C/CNT and reference materials. (**c**) Optimized structures of *O_2, *OOH, *O and *OH on Co-N_3. (**d**) The charge density difference of Co-N_3. (**b–d**) Reproduced with permission [84]. Copyright 2025, Elsevier BV. (**e**) FT-EXAFS fitting curve of the CoN_3S-PC-1.4 catalyst. (**f**) The charge density difference at the CoN_3S site. (**e**,**f**) Reproduced with permission [85]. Copyright 2024, Elsevier BV.

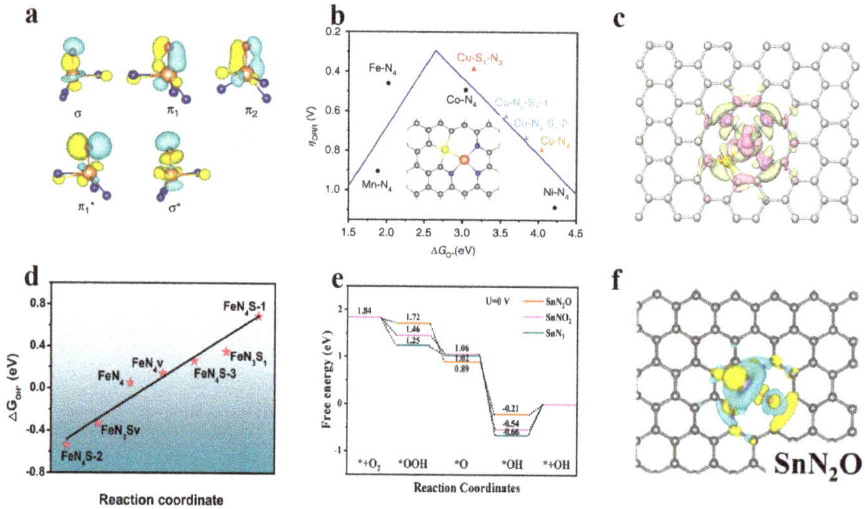

Figure 8. (**a**) The molecular orbital of *O adsorbed on Cu-S_1N_3 is in S-Cu-ISA/SNC. (**b**) ORR overpotential (ηORR) as a function of the free adsorption energy (ΔG*O) of *O on different Cu central groups. (**a**,**b**) Reproduced with permission [71]. Copyright 2020, Springer Nature. (**c**) Charge density difference plot of FeN_3S_1. (**d**) Absorption energy of *OH intermediates at different active sites. (**c**,**d**) Reproduced with permission [86]. Copyright 2024, Elsevier Ltd. (**e**) The variation of Gibbs free energy of the ORR pathway at different Sn sites under a potential bias voltage of U = 0 V. (**f**) The charge density difference plot and relative charge state of Sn. (**e**,**f**) Reproduced with permission [88]. Copyright 2024, Wiley-VCH Verlag.

3.2.2. Co-Incorporation Forms of Nonmetallic and Metallic Elements

The introduction of metal elements beyond the central active site disrupts the local charge symmetry of the original catalyst structure and redistributes the electron cloud density of the active centers [90]. The incorporation of additional metal elements also narrows the energy band of the central active metal sites, enhancing electron mobility. This enables faster electron transfer during catalytic reactions, thereby accelerating the ORR rate. The synergistic interplay of multiple metal elements modulates the adsorption strength of active sites toward oxygen-containing intermediates (e.g., *O, *OOH), ensuring effective adsorption of these intermediates on the catalyst's surface while preventing overly strong adsorption that impedes desorption. This equilibrium promotes the $4e^-$ transfer pathway. Additionally, multimetal systems provide diverse active sites that act synergistically and collectively participate in the ORR process [80]. As reported by Shang et al. [91], a three-step pyrolysis strategy was employed to anchor Se atoms and other metal atoms (e.g., Fe) onto a C_2N substrate, forming SeN_2-MN_2 active sites, thereby synthesizing an asymmetric Se-based diatomic SeM-C_2N catalyst (Figure 9a). Among these, SeFe-C_2N demonstrated optimal ORR performance. The heteronuclear Se atom induces short-range modulation to polarize the charge distribution of the Fe atom, reducing its oxidation state, enhancing the co-adsorption capacity for intermediates (e.g., *O and *OH), and forming heteronuclear diatomic sites, with diatomic sites accounting for 82% of the active centers. The introduction of Se increases the electron density at the Fe atom, creating a localized polarized electric field that lowers the ΔG value of the rate-determining step. This moderates the adsorption strength of intermediates, accelerates O_2 activation and electron transfer, and achieves more efficient conversion of O_2 to H_2O. DFT calculations integrated with operando XAS techniques confirmed that Se sites adsorb *OOH, while the Fe-Se dual sites co-adsorb *O, synergistically promoting the *O→*OH conversion. The K-L equation, fitted from LSV curves at varying rotation speeds, yielded an electron transfer number of n ≈ 4. RRDE measurements directly determined a H_2O_2 yield of <4%, and the calculated electron transfer number (based on ring and disk currents) ranged from n = 3.97 to 4.05, conclusively confirming the dominance of the $4e^-$ pathway. As reported by Chang et al. [92], a Fe-Co dual-site N-doped porous carbon catalyst (Fe,Co/N-C) was synthesized using a host–guest strategy. By adjusting the precursor ratio of $FeCl_3$ to Co-ZIF (optimal at Fe/Co = 1:1), the Fe,Co/N-C catalyst incorporated Fe-Co dual sites that provided more favorable active sites for O-O bond activation. Compared to single-metal sites, the dual sites significantly reduced the energy barrier for O-O bond cleavage, thereby enhancing catalytic activity. DFT calculations confirmed that upon O_2 adsorption on the Fe-Co dual sites, the O-O bond was stretched, and the cleavage energy barrier decreased to 0.25 eV (vs. 0.65 eV for a single Fe site), facilitating the four-electron pathway. K-L equation analysis and RRDE tests indicated an electron transfer number of n ≈ 3.96 and a H_2O_2 yield below 1.17%.

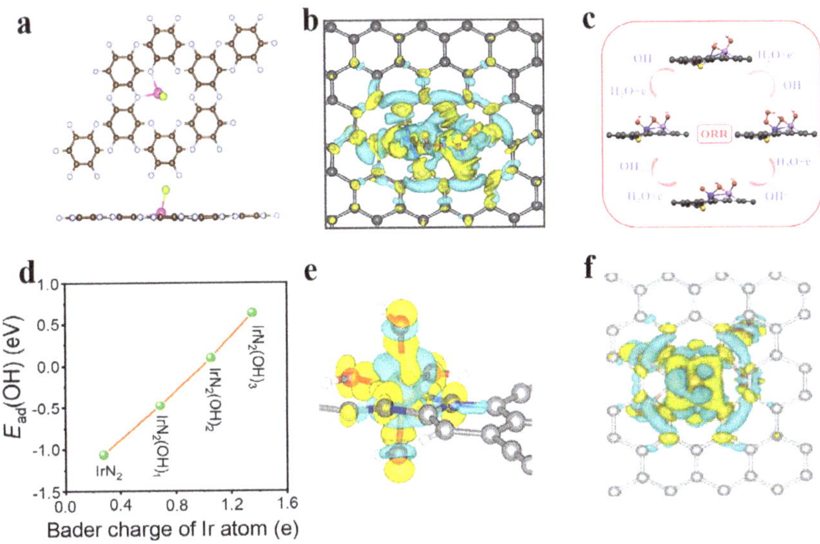

Figure 9. (**a**) Atomic configuration of FeSe-C_2N. (**a**) Reproduced with permission [91]. Copyright 2025, Springer Nature. (**b**) The differential charge density map of CoN_2S-MnN_3-2OH. (**c**) Schematic diagram of the ORR process on the CoN_2S-MnN_3-2OH surface. (**b**,**c**) Reproduced with permission [61]. Copyright 2025, Wiley-VCH Verlag. (**d**) Correlation between Bader charge and the adsorption energy of hydroxyl (OH) groups of Ir atoms in IrO_2, $IrO_2(OH)_1$, $IrO_2(OH)_2$ and $IrO_2(OH)_3$. (**e**) The charge density difference of $IrO_2(OH)_4$. (**d**,**e**) Reproduced with permission [93]. Copyright 2023, Elsevier BV. (**f**) The charge density of Fe_{SA}-Fe_3C/NC. (**f**) Reproduced with permission [59]. Copyright 2025, John Wiley and Sons Ltd.

Incorporating low-electronegativity nonmetallic elements into catalysts possessing dual-metal sites can further modulate the distribution of surface electron cloud density and alter the spin state of metal sites. The interaction between low-electronegativity nonmetallic elements and metal sites enhances the electronic properties of the metal centers. For example, in Fe-based catalysts, the introduction of Si induces a transition of Fe ions from a low-spin state to a medium-spin state. Fe sites in the medium-spin state exhibit higher reactivity, improved oxygen adsorption capacity, and stronger interactions with oxygen-containing intermediates (e.g., *O and *OOH), thereby accelerating the ORR rate [94]. Additionally, low-electronegativity nonmetallic elements can form isolation layers during high-temperature treatments, preventing the aggregation of metal sites into nanoparticles and maintaining the dispersion and high activity of active sites. Finally, these elements tend to donate electrons to the dual-metal sites, causing an upward shift in the d-band center, which fortifies the adsorption of ORR intermediates (*OOH, *OH). As reported by Gao et al. [61], an adsorption–pyrolysis process was utilized to anchor adjacent S,N-coordinated Co atoms and N-coordinated Mn atoms (forming CoN_2S-MnN_3) onto a nitrogen-doped carbon substrate, constructing CoN_2S-MnN_3-2OH active sites and synthesizing the CoMn-NSC catalyst. In this structure, S atoms preferentially coordinate with Co to form CoN_2S, while Mn retains its N-coordinated MnN_3 configuration. The strong interactions between Co and Mn atoms, coupled with multi-electron filling effects induced by asymmetric charge distribution (Figure 9b), enhance orbital hybridization between Co and O_2, lower the O-O bond cleavage energy barrier, optimize adsorption energies of reaction intermediates, and improve electron transfer efficiency through dual-atom synergy. Here, Co acts as the primary active site for O_2 adsorption, while Mn serves as an auxiliary site that modulates the local electronic environment. This adjustment shifts the d-band center of Co closer to the Fermi level, strengthening its adsorption and activation capabilities for O_2, thereby boosting overall catalytic activity (Figure 9c). DFT calculations revealed that the d-band center of Co in the asymmetric structure enhances hybridization with the $*\pi$ orbitals of O_2, improves antibonding electron filling efficiency, and optimizes adsorption strength. K-L equation analysis of polarization curves and RRDE tests confirmed an electron transfer number of $n \approx 4$ and a H_2O_2 yield below 5%. In situ Raman spectroscopy and in situ ATR-SEIRAS further validated the efficient adsorption, activation, and reduction of O_2 on the CoMn-NSC catalyst during the reaction. Su et al. [95] employed a heteroatomic pair cooperative modulation strategy to alter the coordination environment of Fe sites by introducing Co and B heteroatomic pairs, thereby synthesizing the Fe-B-Co/NC catalyst. The introduction of Co atoms achieved electronic synergy between Fe and Co, modulating the d-band center position of Fe and balancing the competitive relationships between RDS during the oxygen reduction reaction. Concurrently, asymmetric coordination transformed Fe from an intermediate-spin state ($t^5{}_{2g}e^1{}_g$) to a high-spin state ($t^4{}_{2g}e^2{}_g$). The high-spin Fe sites enhanced orbital hybridization with O_2, promoting O-O bond cleavage to directly generate *O intermediates (rather than accumulating *OOH), thereby enabling rapid four-electron reaction kinetics and improving oxygen reduction efficiency. K-L equation and RRDE tests confirmed that the Fe-B-Co/NC catalyst exhibited an average electron transfer number of 3.87 within the 0.5 V–0.65 V potential range, indicating a $4e^-$ pathway during ORR with a H_2O_2 yield below 6.9%. In situ EIS and SR-FTIR techniques confirmed that Fe sites accelerate O-O bond cleavage in *OOH to rapidly form *O intermediates. In 0.1 M KOH, the Fe-B-Co/NC catalyst demonstrated a half-wave potential ($E_{1/2}$) of 0.891 V and diffusion-limited current density (Jd) of 5.9 mA cm^{-2}, surpassing commercial Pt/C (0.851 V, 5.19 mA cm^{-2}). During 50 h stability testing, 95% of the initial current density was retained.

3.2.3. Other Incorporation Forms (Functional Groups, Defect Sites, Metallic Clusters, etc.)

Similar to $2e^-$ORR, strategies to augment $4e^-$ORR selectivity extend beyond doping with solely metallic or nonmetallic elements. Functional groups (e.g., OH^-) can also be incorporated to coordinate with central active atoms, or combined with elements (e.g., S) for multi-coordination, optimizing the electronic structure of the central atom [96]. Additionally, carbon frameworks with pronounced edge effects (e.g., HsGDY) reconfigure the edge morphology of the catalyst structure, inducing unique electronic states at the edges.

This effectively modulates the material's band structure, enhancing pathway selectivity. Constructing asymmetric anchoring sites at the edges of the carbon substrate strengthens the bonding between metal atoms and the carbon matrix, reducing metal aggregation and leaching during reactions. As documented by Hou et al. [93], iridium single-atom catalysts (Ir-N-HsGDY) were synthesized by anchoring Ir atoms at the sp-N and pyridinic N sites of hydrogen-substituted graphdiyne (HsGDY) and coordinating them with hydroxyl groups (OH^-), forming $IrN_2(OH)_3$ active sites. The introduction of OH^- ligands broadened the Ir d-band and displaced its center downward, enhancing orbital hybridization between Ir and O/N, optimizing the electronic structure (Figure 9d,e), weakening the adsorption strength of intermediates (e.g., *OH), and reducing reaction energy barriers. The macroporous structure of HsGDY provided space for OH^- incorporation, while its hierarchical design effectively isolated Ir precursors to prevent aggregation and ensure single-atom dispersion. DFT calculations confirmed that the $IrN_2(OH)_3$ active site exhibited the lowest ORR overpotential (0.54 eV), significantly outperforming unmodified IrN_2 (2.24 eV). When applied to PEMFCs, the Ir-N-HsGDY catalyst achieved a half-wave potential ($E_{1/2}$) of 0.82 V vs. RHE in acidic media (0.1 M $HClO_4$), comparable to commercial Pt/C with superior methanol tolerance. In alkaline media (0.1 M KOH), it demonstrated a half-wave potential ($E_{1/2}$) of 0.89 V vs. RHE within the potential window above 0.78 V, significantly outperforming commercial Pt/C, and exhibited only an 18 mV negative shift in $E_{1/2}$ after 5000 CV cycles, indicating high durability. Lv et al. [97] synthesized the Mn-N-HsGDY catalyst by asymmetrically anchoring Mn atoms as single atoms at edge sites of the HsGDY carbon substrate, incorporating two nitrogen species (sp-N and pyridinic N) and OH^- ligands to form asymmetric $MnN_2(OH)_3$ active sites, thereby disrupting the electronic state of conventional symmetric M-N_4 configurations. The strong coordination between sp-N (high electronegativity) and Mn, combined with the electron-withdrawing effect of OH^- ligands, synergistically drove the Mn center to a high oxidation state (+3 to +4), altered the electronic structure of Mn atoms, and reduced the desorption energy barrier of the intermediate *OH. DFT calculations elucidated that the asymmetric coordination of the $MnN_2(OH)_3$ site weakened *OH adsorption, lowering the RDS energy barrier to 0.502 eV, significantly outperforming symmetric sites (e.g., MnN_4). RRDE-derived electron transfer numbers and K-L equation analysis of LSV curve slopes at different rotation rates both yielded an average electron transfer number (n) close to 4, confirming the dominance of the $4e^-$ ORR pathway.

Additionally, etching defects (e.g., vacancies) adjacent to the central atom that disrupt the original symmetric electronic distribution can also enhance $4e^-$ ORR performance [98]. Defects alter the electron cloud distribution around the central active atom, modifying electron delocalization, thereby influencing the catalyst's electronic structure. This creates an unsaturated coordination environment for atoms at defect sites, providing additional active sites for ORR [99]. Zhang et al. [100] prepared an atomically dispersed Fe-N-C catalyst modified with proximal carbon vacancies (Fe-N_4-V_c) through pyrolysis combined with urea etching to selectively cleave C-N bonds near Fe-N_4 sites, forming carbon vacancies (Vc). In the Fe-N_4-V_c structure, the introduced adjacent carbon vacancies induced redistribution of electron density in Fe-N_4, lowered the oxidation state of Fe (coexistence of Fe^{2+} and Fe^{3+} with an average oxidation state of +2.6), and generated asymmetric electronic states. This caused an upward shift of the Fe d-band center, reduced electron occupancy in antibonding orbitals, optimized adsorption behaviors of O_2 and intermediates, lowered reaction free energy barriers, and thereby enhanced ORR activity. DFT calculations demonstrated that Vc significantly reduced the adsorption energy of O_2 (from -2.27 eV to -1.66 eV) and optimized the adsorption free energy of intermediates (e.g., *OH). Among ORR free energy diagrams of distinct structures, Fe-N_4-V_c exhibited the smallest RDS energy barrier and the highest activity. Shen et al. [74] synthesized the Fe-N_4-SM catalyst by selectively cleaving C-N bonds around Fe-N_4 sites to construct an asymmetric atomic strain environment, thereby modulating the electronic structure of the Fe center. The selective cleavage of C-N bonds adjusted the asymmetric atomic strain environment at Fe sites, perturbed the symmetric electronic structure of Fe-N_4, modulated the d-band broadening and d-orbital electron distribution of the Fe center, and consequently altered the adsorption strength between the catalyst and ORR intermediates, enabling precise regulation of catalytic activity. RRDE tests and K-L equation analysis revealed an average electron transfer number of ~3.98 and H_2O_2 yield below 1%, confirming the dominance of the four-electron pathway. In situ Raman spectroscopy detected vibration peaks of *O_2 and *OOH intermediates, supported by theoretical models, further validating the four-electron mechanism. DFT calculations demonstrated that the RDS of Fe-N4-SM shifted from *OH desorption (energy barrier 0.83 eV) to the first electron transfer step (energy barrier 0.25 eV).

When functional groups (e.g., OH^-) and defect vacancies are co-incorporated into a catalyst structure, they may exhibit a synergistic effect to jointly enhance the ORR performance. For example, Zhan et al. [101] employed a microenvironment engineering strategy to anchor Fe single atoms on a defective N-doped carbon (DNC) substrate by regulating micromorphology and interfacial microenvironments, synthesizing the Fe/DNC catalyst. Carbon vacancies reshaped the electronic properties of FeN_4(OH) sites via inductive effects, optimizing hybridization between Fe 3d and O_2 2p orbitals, balancing energy barriers for *OOH formation and *OH reduction, and regulating charge distribution at Fe sites through electron induction to promote O_2 adsorption and intermediate desorption (e.g., *OH), thereby accelerating reaction kinetics. The introduction of axial hydroxyl (OH^-) ligands disrupted the symmetry of FeN_4, enhancing adsorption capability for oxygen intermediates and reducing kinetic barriers. RRDE analysis revealed extremely low H_2O_2 yields (0.35% in alkaline, 6% in acidic, and <5% in neutral conditions) and an electron transfer number close to 4 (≈4.06), confirming the ascendancy of the $4e^-$ pathway and validating high ORR efficiency/selectivity. DFT simulations identified FeN_4(OH) as the true active center, with the defective model (FeN_4(OH)-A) exhibiting the lowest RDS energy barrier (0.43 eV), outperforming symmetric FeN_4(OH) (0.64 eV).

Finally, co-incorporating metal active atoms and metal clusters into the catalyst structure can also generate asymmetric atomic configurations [102]. The presence of metal clusters modifies the charge distribution and local coordination environment surrounding metal active sites, facilitating O_2 molecule access and adsorption at active sites while stabilizing the adsorption of subsequent reaction intermediates, thereby promoting $4e^-$ORR. Additionally, clusters and asymmetric atomic structures collectively create multi-active sites, enabling synergistic catalytic effects. Huang et al. [59] successfully synthesized the Fe_{SA}-Fe_3C/NC catalyst by encapsulating Fe atoms and Fe_3C clusters on nitrogen-doped carbon. The incorporation of Fe_3C clusters disrupted the symmetric electronic structure of FeN_4, depressed the d-band center of Fe atomic sites, thereby altering the adsorption strength of oxygen intermediates on Fe-N_4-C sites, reduced the adsorption energy of *OOH intermediates (0.38 eV), accelerated O-O bond cleavage, and promoted the $4e^-$ORR pathway. Additionally, the high specific surface area (443 m^2/g) and mesoporous structure (4 nm pore size) of the nitrogen-doped carbon enhanced active site exposure and mass transfer efficiency. The synergistic interaction between atomically dispersed Fe-N_4 sites and Fe_3C clusters provided an efficient catalytic platform for ORR. DFT calculations demonstrated that the introduction of Fe_3C clusters increased the electron cloud density of Fe d-orbitals (Figure 9f), augmenting O_2 adsorption and activation while weakening *OH adsorption strength, favoring *OH desorption. This optimized the adsorption/desorption behavior of intermediates, reduced reaction energy barriers, and improved ORR kinetics and activity. In situ infrared spectroscopy confirmed that *OOH intermediate formation was the rate-determining step, and the FeSA-Fe_3C/NC catalyst effectively desorbed *OOH intermediates, thereby lowering reaction barriers and facilitating ORR progression.

4. Application of Oxygen Reduction Reaction

4.1. Device for Producing H_2O_2

Currently, the traditional anthraquinone process remains the dominant technology for H_2O_2 production due to its maturity and high capacity. This method hinges upon the hydrogenation–oxidation cycle of 2-ethylanthraquinone over palladium catalysts (hydrogenation to form hydroanthraquinone, followed by oxidation to regenerate anthraquinone and release H_2O_2), integrated with extraction and purification for continuous production. It offers advantages such as mature technology, high productivity (annual production exceeding 100,000 tons per single unit), and high product purity. However, it faces environmental and economic challenges, including dependence on fossil-fuel-derived hydrogen, high energy consumption, solvent leakage, catalyst costs, and byproduct accumulation. With increasing environmental regulations and advancements in renewable energy technologies, emerging methods like electrochemical synthesis are gradually overcoming technical barriers. The electrochemical production of H_2O_2 via the two-electron oxygen reduction reaction ($2e^-$ORR) offers a green, low-energy alternative to the anthraquinone process [103,104]. This approach directly utilizes water and oxygen from air as feedstocks, eliminating reliance on fossil-fuel hydrogen or organic solvents, thereby reducing carbon emissions and pollution risks at the source. Reactions proceed under ambient conditions, circumventing the high energy consumption and safety hazards of high-temperature/pressure processes, and generate no anthraquinone degradation byproducts, minimizing wastewater treatment demands. Additionally, catalyst design can selectively suppress the $4e^-$ORR pathway, markedly potentiating H_2O_2 selectivity.

The design of devices for the electrochemical production of H_2O_2 is critical to augmenting synthesis efficiency and practicality, with current key technological routes including gas diffusion electrode reactors and flow cell systems [105].

A gas diffusion electrode (GDE) reactor is a device that achieves efficient electrochemical reactions through a gas–liquid–solid three-phase interface [106]. Its core structure typically comprises four components: the gas diffusion layer (GDL), microporous layer (MPL), catalyst layer (CL), and current collector. Gas diffusion layer (Figure 10a): Constructed from porous hydrophobic materials (e.g., carbon fiber paper or polytetrafluoroethylene (PTFE) membranes), it ensures efficient gas transport and prevents electrolyte penetration through its pore structure. Microporous layer: A mixture of carbon powder and PTFE forms a hydrophobic and conductive intermediate layer, further refining pore structures, enhancing adhesion between the catalyst layer and GDL, and improving interfacial stability. Catalyst layer: Catalysts are loaded onto carbon supports via spraying, electrodeposition, or electrospinning. Current collector: A conductive metal mesh (e.g., nickel mesh) or carbon cloth provides electrical conduction and mechanical support. The GDE enables high-efficiency reactions by optimizing the three-phase interface (gas–electrolyte–catalyst). In H_2O_2 electrocatalytic synthesis, O_2 is directly conveyed to the catalyst surface, where H_2O_2 is generated via the $2e^-$ORR pathway, significantly reducing mass transfer resistance. Upon reaching the catalyst layer, O_2 forms a gas–liquid–solid three-phase interface with the electrolyte. At this interface, oxygen, ions in the electrolyte, and active sites on the electrode surface interact, driving complex electrochemical reactions. Under the influence of the electric field, ions in the electrolyte migrate to the catalyst layer surface and participate in the reaction with O_2 to produce H_2O_2. Zhang et al. [58] constructed the cathode of a three-phase flow cell device using a gas diffusion layer (GDL) made of carbon paper and a microporous layer (MPL) loaded with Co-N_5/NC catalyst. With 1.0 M KOH as the electrolyte in O_2-saturated alkaline conditions, O_2 was reduced to H_2O_2 via $2e^-$ORR, achieving a production rate of 16.1 mol $g_{cat}^{-1}h^{-1}$ at 150 mA cm^{-2}. Operating at 0.5 V vs. RHE for 20 h, the accumulated H_2O_2 concentration reached 427.7 mM (14,540 ppm) with a Faradaic efficiency (FE) >90%. Liu et al. [76] coated a Co-C/N/O catalyst onto a gas diffusion electrode and assembled a two-compartment flow cell separated by an NR-212 cation-exchange membrane to separate the anode and cathode chambers. In 1 M phosphate-buffered saline (PBS), galvanostatic electrolysis at 20 mA cm^{-2} for 5 h yielded an H_2O_2 production rate of 4.72 mol $g_{cat}^{-1}h^{-1}$. After 5000 cycles of accelerated durability testing, the activity declined by only 20–39 mV, retaining stable H_2O_2 selectivity.

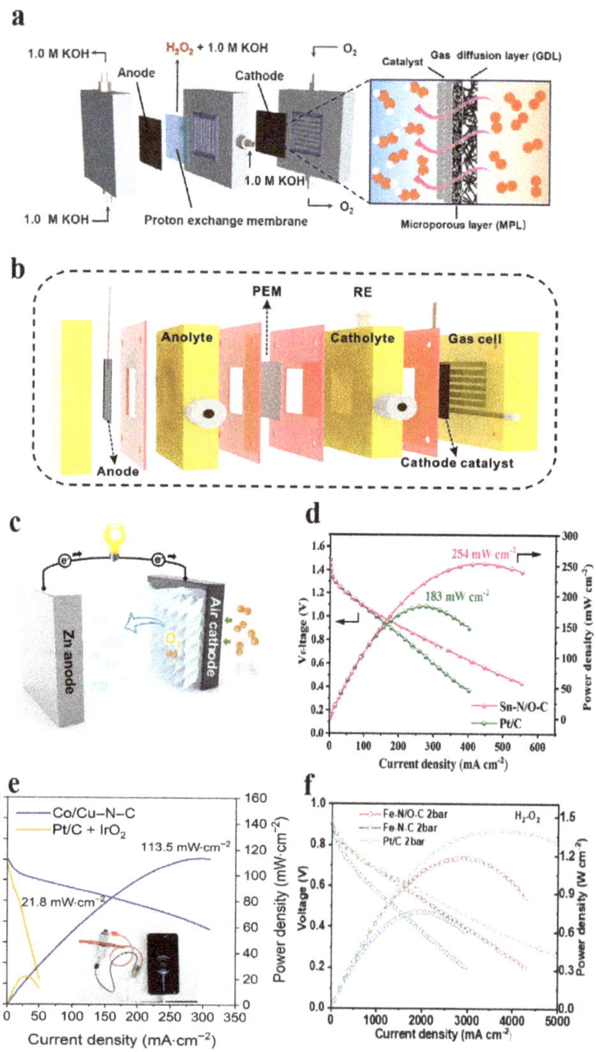

Figure 10. Application of oxygen reduction reaction. (**a**) Schematic diagram of H_2O_2 gas diffusion electrode reactor. (**a**) Reproduced with permission [58]. Copyright 2025, John Wiley and Sons Ltd. (**b**) Schematic diagram of the flow cell for producing H_2O_2. (**b**) Reproduced with permission [81].

Copyright 2024, Wiley-VCH Verlag. (**c**) Schematic diagram of the working principle of zinc–air batteries. (**d**) Polarization curves and corresponding power density diagrams of ZABs containing Sn-N/O-C and Pt/C. (**c**,**d**) Reproduced with permission [88]. Copyright 2024, Wiley-VCH Verlag. (**e**) Discharge polarization curve and power density of a solid-state flexible ZAB. (**e**) Reproduced with permission [107]. Copyright 2025, Tsinghua University Press. (**f**) Polarization curves and corresponding power density graphs of H_2-O_2 PEMFCs assembled with Fe-N/O-C, Fe-N-C, and Pt/C at 2.0 bar. (**f**) Reproduced with permission [89]. Copyright 2024, John Wiley and Sons Ltd.

A flow cell system is an electrochemical device utilizing gas diffusion electrodes (GDEs) and electrocatalytic materials to enable electrocatalytic O_2 reduction at the gas–liquid–solid three-phase interface for H_2O_2 generation [108]. The core structure of the flow cell comprises the GDE (cathode), anode, flowing electrolyte system, membrane materials, and external circulation and storage systems (Figure 10b). The GDE, serving as the cathode, directly supplies gaseous O_2 to the reaction interface, forming a three-phase interface (gas–liquid–solid) and significantly enhancing oxygen mass transfer efficiency. The flowing electrolyte system circulates the electrolyte via pumps to mitigate concentration polarization and rapidly remove generated H_2O_2, preventing its further reduction to water or decomposition. Commonly used membrane materials include proton-exchange membranes (PEMs) or anion-exchange membranes (AEMs). External circulation and storage systems enable continuous electrolyte flow, regulating reactant concentration and flow rate. To enhance H_2O_2 production efficiency, designing superior catalysts remains a critical challenge, particularly in acidic environments where ORR tends to favor complete reduction to H_2O, necessitating precise structural tuning of catalysts to maintain H_2O_2 selectivity. Zhang et al. [63] achieved an H_2O_2 production rate exceeding 2000 mmol $g_{cat}{}^{-1}h^{-1}$ with a Faradaic efficiency (FE) >90% in an acidic electrolyte (0.1 M $HClO_4$) using a flow cell assembled with Co_{SA}-N-C/CNTs, maintaining stable performance for 100 h without decay. In alkaline environments, however, challenges such as insufficient catalyst activity or poor membrane stability necessitate improvements in catalyst performance and membrane durability to enhance H_2O_2 yields. Similarly, Han et al. [55] demonstrated that the Co-N_5-O-C catalyst delivered an H_2O_2 production rate of 11.3 mol $g^{-1}h^{-1}$ at 200 mA cm^{-2} in 1M KOH, retaining ~80% FE after 24 h of continuous operation at 100 mA cm^{-2}.

Future research should focus on multiscale collaborative design, spanning from atomic-level modulation of catalyst electronic structures to optimization of reactor architectures, integrated with computational materials science, to achieve low-cost, high-concentration, and fully green process chains for electrochemical H_2O_2 synthesis, ultimately revolutionizing traditional chemical production paradigms.

4.2. Zinc–Air Battery

Aqueous zinc–air batteries are a novel category of chemical power source that utilize metallic zinc as the anode and oxygen from air as the cathode active material, combining characteristics of both secondary batteries and fuel cells [109]. These batteries typically consist of an air electrode, an alkaline or neutral electrolyte, a separator, and a zinc anode. During discharge, zinc at the anode is oxidized in the alkaline electrolyte to form zincate ions while releasing electrons. At the cathode, oxygen from the air is adsorbed onto the air diffusion electrode and reduced in the electrolyte to generate hydroxide ions (OH^-), which react with zincate ions to produce zinc oxide (ZnO) and H_2O. Electrons flow from the anode to the cathode through an external circuit, generating electrical current. During charging, the anode reaction involves the oxidation of ZnO to Zn^{2+} and OH^- by losing electrons, while the cathode reaction reduces water to produce oxygen and OH^- (Figure 10c). However, aqueous zinc–air batteries suffer from sluggish ORR kinetics, necessitating highly efficient catalysts to reduce overpotentials and enhance reaction rates [110]. When catalysts with metal–nonmetal asymmetric atomic structures are applied, they exhibit superior performance compared to commercial Pt/C catalysts, offering improved stability and catalytic activity for practical applications. Catalysts can be applied in two forms: the first is their direct use as the cathode for catalytic reactions. Zhang et al. [84] employed Co-N_3-C/CNT as the air cathode, where the zinc–air battery demonstrated a maximum power density of 159.3 mW cm^{-2} and a high specific capacity of 892.6 mAh g^{-1}, with no performance decay after 8000 cycles. Yang et al. [86] applied Fe-N_3S/C_{meso} as the cathode material in a zinc–air battery, attaining an open-circuit voltage of 1.45 V, a peak power density of 178 mW cm^{-2}, and stability exceeding 350 h, with 96.17% capacity retention over 100 h. Ren et al. [111] assembled a zinc–air battery using an asymmetric N,P-coordinated Co single-atom catalyst (d-CoN_3P) as the cathode, delivering a specific capacity of 842 mAh g^{-1} and a power density of 175.7 mW cm^{-2}, exceeding Pt/C (662 mAh g^{-1} and 123.8 mW cm^{-2}), and retained stable performance after 1163 charge–discharge cycles. The second approach involves loading catalysts onto the cathode for catalytic reactions, where the loaded catalysts appear to outperform those directly used as cathodes. For example, Ding et al. [83] deposited a Co_1-BNG catalyst onto the anode to form a composite electrode, achieving a zinc–air battery with a maximum power density of 253 mW cm^{-2} and durability exceeding 110 h. Jiang et al. [71] applied S-Cu-ISA/SNC as the air cathode catalyst in a zinc–air battery, demonstrating a maximum power density of 225 mW cm^{-2} and negligible voltage decay during 50 h cycling tests, indicating exceptional discharge stability. Lin et al. [88] employed the Sn-N/O-C catalyst in a ZAB, achieving a maximum power density of 254 mW cm^{-2}, a specific capacity of 781 mAh g^{-1}, and stable performance over 350 h of discharge testing (Figure 10d). Metal–metal asymmetric-atomic-structure catalysts further enhance the performance of aqueous zinc–air batteries, with bimetallic synergistic catalysis exhibiting stronger performance than metal–nonmetal catalysts, yielding higher peak power densities. For instance, Shang et al. [91] applied the SeFe–C_2N catalyst as the cathode in a ZAB, achieving a high power density of 287.2 mW cm^{-2} and a specific capacity of 764.8 mAh g^{-1}, with virtually imperceptible voltage decay over 380 h of charge–discharge cycling. Wang et al. [107] integrated a Co/Cu-N-C catalyst into a ZAB, attaining a peak power density of 256.1 mW cm^{-2} and cycling stability exceeding 500 h.

Flexible solid-state zinc–air batteries (ZABs) are novel energy storage devices characterized by high theoretical energy density, augmented safety, and flexible bendability. Their working principle aligns with aqueous ZABs, with the core innovation being the replacement of traditional liquid electrolytes with solid-state electrolytes and the achievement of flexibility through material and structural design [112]. The key components of a flexible solid-state ZAB include an air cathode, zinc anode, solid-state electrolyte layer, and flexible encapsulation layer. The air cathode facilitates ORR. The zinc anode combines a metallic zinc layer with a flexible substrate to enable bending. The solid-state electrolyte layer conducts ions, prevents short circuits by isolating electrodes, and provides mechanical flexibility and strength. The flexible encapsulation layer is typically made of polymeric materials to block moisture ingress and maintain mechanical flexibility [113]. Cathode catalyst deactivation significantly degrades ORR kinetics and is a primary cause of capacity decay in flexible solid-state ZABs. For instance, Wang et al. [107] synthesized a Co/Cu-N-C catalyst that demonstrated a high specific capacity of 727.1 mAh g^{-1} and stable power density of 113.2 mW cm^{-2} under bending conditions in flexible ZABs (Figure 10e), outperforming Pt/C-based batteries (21.7 mW cm^{-2}). Lv et al. [97] applied the Mn-N-HsGDY catalyst in a 1D solid-state ZAB, achieving an open-circuit voltage of 1.363 V and maintaining stable voltage output during bending to power a timer. In a 2D solid-state ZAB, it delivered a peak power density of 156.6 mW cm^{-2}, surpassing most reported flexible ZABs, and preserved stable voltage differences throughout 1400 min of charge–discharge testing.

4.3. Hydrogen–Oxygen Fuel Cell

Proton-exchange membrane fuel cells (PEMFCs) constitute highly efficient and clean energy conversion devices that directly transform chemical energy into electricity through hydrogen–oxygen electrochemical reactions [114]. Among these, the ORR at the cathode is one of the most critical processes. A PEMFC consists of gas diffusion layers (GDLs), membrane electrode assemblies (MEAs), and bipolar plates. The MEA, which includes catalyst layers (CLs) and a proton-exchange membrane (PEM), serves as the core of the electrochemical conversion process. The performance and efficiency of devices relying on this process depend heavily on the properties of ORR electrocatalysts and PEMs [115]. Thus, developing higher-activity ORR catalysts and cost-effective MEAs with high durability is essential for advancing these energy conversion technologies [116]. Fundamentally, PEMFCs operate as the reverse process of water electrolysis: H_2 and O_2 react in the presence of catalysts to produce H_2O while releasing electrical energy. In the operation of PEMFCs, H_2 enters the anode, where it splits into H^+ and e^- under the action of catalysts. The H^+ ions migrate through the PEM to the cathode, while the e^- are forced through an external circuit, generating an electric current to power electronic devices. At the cathode, O_2 combines with the migrating H^+ and incoming e^- to drive the ORR, producing H_2O. To optimize PEMFC performance, high-performance ORR catalysts are essential. Catalyst particles must be integrated on both sides of the PEM to minimize the transport distances of H^+ and e^-, thereby enhancing reaction kinetics. The chemical stability of the catalysts and their compatibility with the membrane are critical to preventing degradation of both components and maintaining optimal performance. Due to the constrained availability and high cost of noble metal-based catalysts, asymmetric-atomic-structure catalysts have emerged as promising alternatives to Pt-based catalysts. For example, Liu et al. [89] reported a Fe-N/O-C cathode catalyst that achieved a peak power density of 1179 mW cm^{-2} in PEMFCs (Figure 10f). Yin et al. [85] developed a CoN_3S-PC cathode catalyst that delivered a peak power density surpassing 1.32 W cm^{-2} under O_2 at 80 °C. Operating at a constant voltage of 0.6 V, the CoN_3S-PC cathode maintained 53.3% of its current density after 37 h (ultimately stabilizing at approximately 300 mA cm^{-2}). Chang et al. [92] synthesized a (Fe,Co)/N-C catalyst that achieved a peak power density of 505 mW cm^{-2} in PEMFCs (353 K with 0.2 MPa backpressure), reaching 76% of the power density of commercial Pt/C. Additionally, in H_2/air fuel cells, it demonstrated negligible voltage decay when operated continuously for 100 h at constant current densities (600 and 1000 mA cm^{-2}). These advancements highlight the potential of metal–nonmetal asymmetric coordination catalysts to outperform Pt/C in PEMFC applications, driving progress toward cost-effective and durable fuel cell technologies.

5. Conclusions and Prospects

In conclusion, asymmetric atomic structures have demonstrated exceptional performance in the ORR, showing practical applications in high-efficiency H_2O_2 production, metal–air batteries, and hydrogen–oxygen fuel cells, and have materialized as one of the most promising alternatives to traditional noble metal catalysts. Based on recent advancements in theoretical and experimental research, key design principles for developing next-generation high-performance asymmetric-atomic-structure catalysts have been summarized, providing critical guidance for optimizing catalytic activity and selectivity.

(i) Design Principles for Metal Element Doping: In ORR catalysts, transition metal elements are the predominant choice for constructing asymmetric atomic structures owing to their unique electronic configurations and chemical properties. The partially filled d-orbitals of transition metals can hybridize with the antibonding $^*\pi$ orbitals of O_2, modulating the activation energy for O-O bond cleavage. The position of the d-band center critically governs the adsorption strength of oxygen intermediates (e.g., *OOH, *O, *OH). Furthermore, transition metals exhibit multiple oxidation states (e.g., Fe^{2+}/Fe^{3+}, Co^{2+}/Co^{3+}), empowering reversible electron transfer during ORR. This redox flexibility allows their oxidation states to dynamically adapt across reaction steps, thereby facilitating catalytic progression. Additionally, the atomic size and electronegativity of transition metals promote the formation of asymmetric coordination structures with other atoms or ligands. Such configurations induce asymmetric charge distribution and electronic effects, optimizing the adsorption and reactivity toward ORR intermediates. Finally, transition metals possess high chemical and thermal stability, enabling sustained catalytic activity under ORR operating conditions (e.g., acidic/alkaline environments, high potentials) while resisting structural degradation or corrosion.

(ii) Design Principles for Nonmetal Element Doping: In the fabrication of asymmetric-atomic-structure catalysts, the incorporation of high-electronegativity nonmetal elements induces charge polarization through their strong electron-withdrawing capability, creating localized electron density differences. This effectively modulates the electronic structure of the catalyst, enhancing its adsorption capacity for reactants and catalytic activity. Additionally, synergistic co-doping of multiple high-electronegativity nonmetal elements can be considered to achieve more significant catalytic performance improvements. Interactions between different dopants may collectively modulate the electronic and geometric structures of the catalyst, generating synergistic effects. For low-electronegativity nonmetal elements, their weak electron-withdrawing or electron-donating capabilities can modulate the charge distribution, surface reactive sites, and intermediate adsorption behavior of the catalyst structure. Selecting elements with atomic radii matching the substrate can reduce the risk of lattice distortion and preserves structural stability.

(iii) Design Principles for Other Dopants: In asymmetric-atomic-structure ORR catalysts, defect engineering amplifies catalytic activity and selectivity by introducing vacancies, edge sites, and other asymmetric configurations. These defects modulate electronic distribution, expose active sites, and optimize reaction pathways. Defects instigate localized states near the Fermi level, reducing the energy gap between O_2's d-orbitals and the catalyst's electronic states (e.g., carbon vacancies shift the Fermi level downward by 0.2–0.5 eV), thereby enhancing charge transfer efficiency. Finally, defects in porous structures expose additional active sites (e.g., defects in mesoporous carbon walls enhance mass transport and active site density).

(iv) In summary, for the $2e^-$ pathway in the ORR, elements that weakly activate the O-O bond should be selected to reduce O-O bond cleavage capability, stabilize the *OOH intermediate, and retain the O-O bond, thereby enhancing H_2O_2 selectivity. The asymmetric configurations composed of N and O atoms in structures such as Co-N_5, Zn-N_3O and other structures promote the two-electron pathway of the oxygen reduction reaction. For the $4e^-$ pathway, the O-O bond cleavage capability must be strengthened, and active sites should integrate dual functions of O_2 adsorption and O-O bond scission to promote rapid adsorption/desorption equilibrium of oxygen-containing intermediates (*O, *OH). The presence of B and Fe elements in structures such as Co-N_3B and Fe-B-Co/NC promotes the four-electron pathway of the oxygen reduction reaction. Additionally, precise regulation of defect type, location, and concentration can optimize the electronic structure and reaction kinetics of active sites, enabling directional selection of ORR pathways ($2e^-/4e^-$). Finally, the choice of carbon matrix must comprehensively consider the synergistic effects of multi-element doping, defect engineering, hierarchical structures, and coordination modulation to achieve electronic structure optimization, rapid mass transport, and efficient exposure of active sites.

Author Contributions: Resources, H.Q.; conceptualization, S.W.; writing—original draft preparation, H.Q.; writing—review and editing, Q.F. and X.Z.; supervision, X.Z.; funding acquisition, X.Z. All authors have read and agreed to the published version of the manuscript.

Funding: This work was funded by the Colleges and Universities Twenty Terms Foundation of Jinan City (No. 202228053) and the Science, Education, and Industry Integration Innovation Pilot Project of Qilu University of Technology (Shandong Academy of Sciences) (No. 2024RCKY019).

Data Availability Statement: No new data were created or analyzed in this study.

Conflicts of Interest: The authors declare no conflicts of interest.

References

1. Fu, J.; Cano, Z.P.; Park, M.G.; Yu, A.; Fowler, M.; Chen, Z. Electrically Rechargeable Zinc–Air Batteries: Progress, Challenges, and Perspectives. *Adv. Mater.* **2016**, *29*, 1604685. [CrossRef] [PubMed]
2. Fu, J.; Liang, R.; Liu, G.; Yu, A.; Bai, Z.; Yang, L.; Chen, Z. Recent Progress in Electrically Rechargeable Zinc–Air Batteries. *Adv. Mater.* **2018**, *31*, 1805230. [CrossRef] [PubMed]
3. Haider, R.; Wen, Y.; Ma, Z.-F.; Wilkinson, D.P.; Zhang, L.; Yuan, X.; Song, S.; Zhang, J. High temperature proton exchange membrane fuel cells: Progress in advanced materials and key technologies. *Chem. Soc. Rev.* **2021**, *50*, 1138–1187. [CrossRef]
4. Khan, M.A.; Al-Attas, T.; Roy, S.; Rahman, M.M.; Ghaffour, N.; Thangadurai, V.; Larter, S.; Hu, J.; Ajayan, P.M.; Kibria, M.G. Seawater electrolysis for hydrogen production: A solution looking for a problem? *Energy Environ. Sci.* **2021**, *14*, 4831–4839. [CrossRef]
5. Shi, L.; Liu, D.; Lin, X.; Cheng, R.; Liu, F.; Kim, C.; Hu, C.; Qiu, J.; Amal, R.; Dai, L. Stable and High-performance Flow H_2-O_2 Fuel Cells with Coupled Acidic Oxygen Reduction and Alkaline Hydrogen Oxidation Reactions. *Adv. Mater.* **2024**, *36*, 2314077. [CrossRef]
6. Wang, Q.; Kaushik, S.; Xiao, X.; Xu, Q. Sustainable zinc–air battery chemistry: Advances, challenges and prospects. *Chem. Soc. Rev.* **2023**, *52*, 6139–6190. [CrossRef]
7. Xiao, F.; Wang, Y.C.; Wu, Z.P.; Chen, G.; Yang, F.; Zhu, S.; Siddharth, K.; Kong, Z.; Lu, A.; Li, J.C.; et al. Recent Advances in Electrocatalysts for Proton Exchange Membrane Fuel Cells and Alkaline Membrane Fuel Cells. *Adv. Mater.* **2021**, *33*, 2006292. [CrossRef] [PubMed]
8. Yu, Z.; Liu, L. Recent Advances in Hybrid Seawater Electrolysis for Hydrogen Production. *Adv. Mater.* **2023**, *36*, 2308647. [CrossRef]
9. Zhang, W.; Liu, M.; Gu, X.; Shi, Y.; Deng, Z.; Cai, N. Water Electrolysis toward Elevated Temperature: Advances, Challenges and Frontiers. *Chem. Rev.* **2023**, *123*, 7119–7192. [CrossRef]
10. Zhao, X.; Liu, Y. Origin of Selective Production of Hydrogen Peroxide by Electrochemical Oxygen Reduction. *J. Am. Chem. Soc.* **2021**, *143*, 9423–9428. [CrossRef]
11. Xu, S.; Lu, R.; Sun, K.; Tang, J.; Cen, Y.; Luo, L.; Wang, Z.; Tian, S.; Sun, X. Synergistic Effects in N,O-Comodified Carbon Nanotubes Boost Highly Selective Electrochemical Oxygen Reduction to H_2O_2. *Adv. Sci.* **2022**, *9*, 2201421. [CrossRef] [PubMed]
12. Zhou, S.; Huang, S.; Wang, X.; Chen, S.; Wang, L.; Zhang, P.; Zhao, C. Activating the PdN_4 single-atom sites for 4 electron oxygen reduction reaction via axial oxygen ligand modification. *Chem. Eng. J.* **2023**, *472*, 145129. [CrossRef]
13. Tian, X.; Lu, X.F.; Xia, B.Y.; Lou, X.W. Advanced Electrocatalysts for the Oxygen Reduction Reaction in Energy Conversion Technologies. *Joule* **2020**, *4*, 45–68. [CrossRef]
14. Zhai, Z.; Wang, Y.-J.; Pan, L.; Huang, F.; Liu, D.; Wang, B. Accelerating O-O bond dissociation in oxygen reduction reaction on the sp3-hybridized carbon. *Appl. Surf. Sci.* **2025**, *691*, 162668. [CrossRef]

15. Yan, H.-M.; Wang, G.; Lv, X.-M.; Cao, H.; Qin, G.-Q.; Wang, Y.-G. Revealing the Potential-Dependent Rate-Determining Step of Oxygen Reduction Reaction on Single-Atom Catalysts. *J. Am. Chem. Soc.* **2025**, *147*, 3724–3730. [CrossRef]
16. Cheng, J.; Lyu, C.; Li, H.; Wu, J.; Hu, Y.; Han, B.; Wu, K.; Hojamberdiev, M.; Geng, D. Steering the oxygen reduction reaction pathways of N-carbon hollow spheres by heteroatom doping. *Appl. Catal. B Environ.* **2023**, *327*, 122470. [CrossRef]
17. Kumar, K.; Dubau, L.; Jaouen, F.; Maillard, F. Review on the Degradation Mechanisms of Metal-N-C Catalysts for the Oxygen Reduction Reaction in Acid Electrolyte: Current Understanding and Mitigation Approaches. *Chem. Rev.* **2023**, *123*, 9265–9326. [CrossRef]
18. Wolker, T.; Brunnengräber, K.; Martinaiou, I.; Lorenz, N.; Zhang, G.-R.; Kramm, U.I.; Etzold, B.J.M. The effect of temperature on ionic liquid modified Fe-N-C catalysts for alkaline oxygen reduction reaction. *J. Energy Chem.* **2022**, *68*, 324–329. [CrossRef]
19. Du, Y.X.; Yang, Q.; Lu, W.T.; Guan, Q.Y.; Cao, F.F.; Zhang, G. Carbon Black-Supported Single-Atom Co-N-C as an Efficient Oxygen Reduction Electrocatalyst for H_2O_2 Production in Acidic Media and Microbial Fuel Cell in Neutral Media. *Adv. Funct. Mater.* **2023**, *33*, 2300895. [CrossRef]
20. Lim, C.; Fairhurst, A.R.; Ransom, B.J.; Haering, D.; Stamenkovic, V.R. Role of Transition Metals in Pt Alloy Catalysts for the Oxygen Reduction Reaction. *ACS Catal.* **2023**, *13*, 14874–14893. [CrossRef]
21. Valério Neto, E.S.; Almeida, C.V.S.; Colmati, F.; Ciapina, E.G.; Salazar-Banda, G.R.; Eguiluz, K.I.B. Pt nanowires as electrocatalysts for proton-exchange membrane fuel cells applications: A review. *J. Electroanal. Chem.* **2022**, *910*, 116185. [CrossRef]
22. Yin, S.; Song, Y.; Liu, H.; Cui, J.; Liu, Z.; Li, Y.; Xue, T.; Tang, W.; Zhang, D.; Li, H.; et al. Well-Defined PtCo@Pt Core-Shell Nanodendrite Electrocatalyst for Highly Durable Oxygen Reduction Reaction. *Small* **2025**, *21*, 2410080. [CrossRef] [PubMed]
23. Chong, L.; Zhou, H.; Kubal, J.; Tang, Q.; Wen, J.; Yang, Z.; Bloom, I.D.; Abraham, D.; Zhu, H.; Zou, J.; et al. Highly durable fuel cell electrocatalyst with low-loading Pt-Co nanoparticles dispersed over single-atom Pt-Co-N-graphene nanofiber. *Chem Catal.* **2023**, *3*, 100541. [CrossRef]
24. Bhuvanendran, N.; Ravichandran, S.; Lee, S.; Sanij, F.D.; Kandasamy, S.; Pandey, P.; Su, H.; Lee, S.Y. Recent progress in Pt-based electrocatalysts: A comprehensive review of supported and support-free systems for oxygen reduction. *Coord. Chem. Rev.* **2024**, *521*, 216191. [CrossRef]
25. Liu, H.; Zhao, J.; Li, X. Controlled Synthesis of Carbon-Supported Pt-Based Electrocatalysts for Proton Exchange Membrane Fuel Cells. *Electrochem. Energy Rev.* **2022**, *5*, 13. [CrossRef]
26. Sun, Q.; Li, X.-H.; Wang, K.-X.; Ye, T.-N.; Chen, J.-S. Inorganic non-carbon supported Pt catalysts and synergetic effects for oxygen reduction reaction. *Energy Environ. Sci.* **2023**, *16*, 1838–1869. [CrossRef]
27. Peng, J.; Tao, P.; Song, C.; Shang, W.; Deng, T.; Wu, J. Structural evolution of Pt-based oxygen reduction reaction electrocatalysts. *Chin. J. Catal.* **2022**, *43*, 47–58. [CrossRef]
28. Fu, Y.; Zhao, Q.; Wei, Q.; Bowen, C.R.; Wong, W.-Y.; Yang, W. Non-precious metal-based single-atom catalysts for oxygen reduction reaction: Fundamentals and applications. *Mater. Sci. Eng. R Rep.* **2024**, *160*, 100822. [CrossRef]
29. Moghaddam, M.S.; Kafshgari, M.S.; Bahari, A.; Kafshgari, L.A.; Jafari, A. Types, properties, and applications of non-precious oxygen reduction reaction electrocatalyst: A review. *J. Energy Chem.* **2025**, *107*, 305–344. [CrossRef]
30. Ao, X.; Wang, H.; Zhang, X.; Wang, C. Atomically Dispersed Metal–Nitrogen–Carbon Catalysts for Acidic Oxygen Reduction Reaction. *ACS Appl. Mater. Interfaces* **2025**, *17*, 2844–2862. [CrossRef]
31. Yang, Z.; Chen, Y.; Zhang, S.; Zhang, J. Identification and Understanding of Active Sites of Non-Noble Iron-Nitrogen-Carbon Catalysts for Oxygen Reduction Electrocatalysis. *Adv. Funct. Mater.* **2023**, *33*, 2215185. [CrossRef]
32. Xu, W.; Tang, H.; Gu, H.; Xi, H.; Wu, P.; Liang, B.; Liu, Q.; Chen, W. Research progress of asymmetrically coordinated single-atom catalysts for electrocatalytic reactions. *J. Mater. Chem. A* **2022**, *10*, 14732–14746. [CrossRef]
33. Zhao, X.; Hao, Z.; Zhang, X.; Li, L.; Gao, Y.; Liu, L. Alkaline oxygen reduction/evolution reaction electrocatalysis: A critical review focus on orbital structure, non-noble metal catalysts, and descriptors. *Chem. Eng. J.* **2024**, *497*, 155005. [CrossRef]
34. Yu, A.; Yang, Y. Atomically Dispersed Metal Catalysts for Oxygen Reduction Reaction: Two-Electron vs. Four-Electron Pathways. *Angew. Chem. Int. Ed.* **2025**, *64*, e202424161. [CrossRef]
35. Zhu, P.; Xiong, X.; Wang, X.; Ye, C.; Li, J.; Sun, W.; Sun, X.; Jiang, J.; Zhuang, Z.; Wang, D.; et al. Regulating the FeN4 Moiety by Constructing Fe–Mo Dual-Metal Atom Sites for Efficient Electrochemical Oxygen Reduction. *Nano Lett.* **2022**, *22*, 9507–9515. [CrossRef]
36. Buschermöhle, J.G.; Müller-Hülstede, J.; Schmies, H.; Schonvogel, D.; Zierdt, T.; Lucka, R.; Renz, F.; Wagner, P.; Wark, M. Fe–Sn–N–C Catalysts: Advancing Oxygen Reduction Reaction Performance. *ACS Catal.* **2025**, *15*, 4477–4488. [CrossRef]
37. Smith, L.A.; Burrow, J.N.; Eichler, J.E.; Tang, F.; Lauro, S.N.; Zhan, X.; Warner, J.H.; Mullins, C.B. A Deep Dive Into the Study of Nitrogen-Doped Carbons as Electrocatalysts for the Oxygen Reduction Reaction via Design of Experiments. *Small* **2025**, *21*, 2410010. [CrossRef]

38. Zong, L.; Fan, K.; Cui, L.; Lu, F.; Liu, P.; Li, B.; Feng, S.; Wang, L. Constructing Fe-N_4 Sites through Anion Exchange-mediated Transformation of Fe Coordination Environments in Hierarchical Carbon Support for Efficient Oxygen Reduction. *Angew. Chem. Int. Ed.* **2023**, *62*, e202309784. [CrossRef]
39. Wang, X.; Yang, C.; Wang, X.; Zhu, H.; Cao, L.; Chen, A.; Gu, L.; Zhang, Q.; Zheng, L.; Liang, H.-P. Green Synthesis of a Highly Efficient and Stable Single-Atom Iron Catalyst Anchored on Nitrogen-Doped Carbon Nanorods for the Oxygen Reduction Reaction. *ACS Sustain. Chem. Eng.* **2020**, *9*, 137–146. [CrossRef]
40. Zhu, Y.; Deng, F.; Qiu, S.; Ma, F.; Zheng, Y.; Lian, R. Enhanced electro-Fenton degradation of sulfonamides using the N, S co-doped cathode: Mechanism for H2O2 formation and pollutants decay. *J. Hazard. Mater.* **2021**, *403*, 123950. [CrossRef]
41. Wu, L.; Cao, J.; Fu, C.; Chen, Z.; Yin, C.; Lin, Q.; Pan, H.; Wang, K. Lead regulation of electronic properties and local structure of palladium (111) facet for enhanced direct hydrogen peroxide synthesis. *J. Colloid Interface Sci.* **2025**, *682*, 1205–1216. [CrossRef]
42. Liu, J.; Zhang, Z.; Han, C.; Xu, Z.; Zhu, Y.; Attfield, J.P.; Zeng, J.; Yang, M. Ligand-modified nickel nitride for natural seawater H_2O_2 synthesis. *Appl. Catal. B Environ. Energy* **2025**, *373*, 125362. [CrossRef]
43. Musaie, K.; Abbaszadeh, S.; Marais, K.; Nosrati-Siahmazgi, V.; Rezaei, S.; Xiao, B.; Dua, K.; Santos, H.A.; Shahbazi, M.A. H_2O_2-Generating Advanced Nanomaterials for Cancer Treatment. *Adv. Funct. Mater.* **2025**, 2425866. [CrossRef]
44. Yin, J.; Zhang, H.; Luo, M.; Zhao, J.; Zhou, X.; Wang, F.; Zhou, H.; Yuan, Y.; Liu, Y.; Shi, Y.; et al. Electro-Fenton process for wastewater treatment: From selective H_2O_2 generation to efficient *OH conversion. *Chem. Eng. J.* **2025**, *507*, 160709. [CrossRef]
45. Chen, K.-Y.; Huang, Y.-X.; Jin, R.-C.; Huang, B.-C. Single atom catalysts for use in the selective production of hydrogen peroxide via two-electron oxygen reduction reaction: Mechanism, activity, and structure optimization. *Appl. Catal. B Environ.* **2023**, *337*, 122987. [CrossRef]
46. Bu, Y.; Wang, Y.; Han, G.F.; Zhao, Y.; Ge, X.; Li, F.; Zhang, Z.; Zhong, Q.; Baek, J.B. Carbon-Based Electrocatalysts for Efficient Hydrogen Peroxide Production. *Adv. Mater.* **2021**, *33*, 171264. [CrossRef]
47. Zhang, C.; Shen, W.; Guo, K.; Xiong, M.; Zhang, J.; Lu, X. A Pentagonal Defect-Rich Metal-Free Carbon Electrocatalyst for Boosting Acidic O_2 Reduction to H_2O_2 Production. *J. Am. Chem. Soc.* **2023**, *145*, 11589–11598. [CrossRef]
48. Kong, D.; Wang, Y.; Yan, Y.; Li, Z.; Lu, Y.; Chen, G. Breaking planar constraints: Curvature-tailored Co-N-C electrocatalysts for high-efficiency proton exchange membrane H_2O_2 production. *Appl. Catal. B Environ. Energy* **2025**, *376*, 125454. [CrossRef]
49. Yan, L.; Wang, C.; Wang, Y.; Wang, Y.; Wang, Z.; Zheng, L.; Lu, Y.; Wang, R.; Chen, G. Optimizing the binding of the *OOH intermediate via axially coordinated Co-N_5 motif for efficient electrocatalytic H_2O_2 production. *Appl. Catal. B Environ.* **2023**, *338*, 123078. [CrossRef]
50. Yuan, Y.; Ma, J.; Kang, B.; Lee, J.Y. Unraveling the Common Nature of O and S Doping in Improving Electrochemical O_2 Reduction Reaction Performance of FeN_4C. *ACS Catal.* **2025**, *15*, 4039–4050. [CrossRef]
51. Hu, H.; Wang, J.; Liao, K.; Chen, Z.; Zhang, S.; Sun, B.; Wang, X.; Ren, X.; Lin, J.; Han, X. Clarifying the Active Structure and Reaction Mechanism of Atomically Dispersed Metal and Nonmetal Sites with Enhanced Activity for Oxygen Reduction Reaction. *Adv. Mater.* **2024**, *37*, 2416126. [CrossRef]
52. Tang, X.; Zhang, Y.; Tang, S.; Lützenkirchen-Hecht, D.; Yuan, K.; Chen, Y. Structural and Chemical Origin of Dual-Atom Sites for Enhanced Oxygen Electroreduction. *ACS Catal.* **2024**, *14*, 13065–13080. [CrossRef]
53. Li, Y.; Jiang, H.; Lin, L.; Sun, Z.; Sun, G. Single-atomic iron synergistic atom-cluster induce remote enhancement toward oxygen reduction reaction. *J. Energy Chem.* **2025**, *102*, 413–420. [CrossRef]
54. Wu, L.; Wang, Y.; Shao, C.; Wang, L.; Li, B. Copper single atom-modulated functionalization of iron clusters on a porous carbon nanosheet for the oxygen reduction reaction. *J. Mater. Chem. A* **2025**, *13*, 5974–5986. [CrossRef]
55. Lazaridis, T.; Stühmeier, B.M.; Gasteiger, H.A.; El-Sayed, H.A. Capabilities and limitations of rotating disk electrodes versus membrane electrode assemblies in the investigation of electrocatalysts. *Nat. Catal.* **2022**, *5*, 363–373. [CrossRef]
56. Yu, Y.; Jiao, D.; Yi, L.; Zhao, D.; Zou, J.; Cui, X.; Hu, W. Unique electronic regulation in bimetallic MOF for efficient electrochemical synthesis of H_2O_2 at 500 mA cm^{-2}. *Nano Energy* **2025**, *141*, 111096. [CrossRef]
57. Xu, C.; Li, X.; Guo, P.-P.; Yang, K.-Z.; Zhao, Y.-M.; Chi, H.-M.; Xu, Y.; Wei, P.-J.; Wang, Z.-Q.; Xu, Q.; et al. Creating Asymmetric Fe–N_3C–N Sites in Single-Atom Catalysts Boosts Catalytic Performance for Oxygen Reduction Reaction. *ACS Appl. Mater. Interfaces* **2024**, *16*, 37927–37937. [CrossRef]
58. Yu, K.; Guan, S.; Zhang, W.; Zhang, W.; Meng, Y.; Lin, H.; Gao, Q. Engineering Asymmetric Electronic Structure of Co-N-C Single-Atomic Sites Toward Excellent Electrochemical H_2O_2 Production and Biomass Upgrading. *Angew. Chem. Int. Ed.* **2025**, *64*, e202502383. [CrossRef]
59. Liu, C.; Yang, R.; Wang, J.; Liu, B.; Chang, X.; Feng, P.; Zhang, X.; Zhong, L.; Zhao, X.; Niu, L.; et al. Synergistic Catalysts with Fe Single Atoms and Fe_3C Clusters for Accelerated Oxygen Adsorption Kinetics in Oxygen Reduction Reaction. *Angew. Chem. Int. Ed.* **2025**, *64*, e202501266. [CrossRef]

60. Zhao, Y.; Chen, H.C.; Ma, X.; Li, J.; Yuan, Q.; Zhang, P.; Wang, M.; Li, J.; Li, M.; Wang, S.; et al. Vacancy Defects Inductive Effect of Asymmetrically Coordinated Single-Atom Fe-N_3S_1 Active Sites for Robust Electrocatalytic Oxygen Reduction with High Turnover Frequency and Mass Activity. *Adv. Mater.* **2023**, *36*, 2308243. [CrossRef]
61. Zhao, Y.; Gao, Z.; Zhang, S.; Guan, X.; Xu, W.; Liang, Y.; Jiang, H.; Li, Z.; Wu, S.; Cui, Z.; et al. Asymmetric-Charge-Distributed Co-Mn Diatomic Catalyst Enables Efficient Oxygen Reduction Reaction. *Adv. Funct. Mater.* **2025**, *518*, 2504260. [CrossRef]
62. Zhao, Y.; Adiyeri Saseendran, D.P.; Huang, C.; Triana, C.A.; Marks, W.R.; Chen, H.; Zhao, H.; Patzke, G.R. Oxygen Evolution/Reduction Reaction Catalysts: From In Situ Monitoring and Reaction Mechanisms to Rational Design. *Chem. Rev.* **2023**, *123*, 6257–6358. [CrossRef] [PubMed]
63. Zhao, Y.; Huang, M.; Kang, Y.; Fang, Y.; Zhao, T.; Wang, H.; Ou, J.; Liu, J.; Zhong, M.; Wang, T.; et al. Ligand effects enhancing low-temperature oxygen reduction kinetics in neutral conditions. *Energy Environ. Sci.* **2025**, *18*, 5298–5308. [CrossRef]
64. Liao, X.; Lu, R.; Xia, L.; Liu, Q.; Wang, H.; Zhao, K.; Wang, Z.; Zhao, Y. Density Functional Theory for Electrocatalysis. *Energy Environ. Mater.* **2021**, *5*, 157–185. [CrossRef]
65. Xie, L.; Zhou, W.; Huang, Y.; Qu, Z.; Li, L.; Yang, C.; Ding, Y.; Li, J.; Meng, X.; Sun, F.; et al. Elucidating the impact of oxygen functional groups on the catalytic activity of M–N_4–C catalysts for the oxygen reduction reaction: A density functional theory and machine learning approach. *Mater. Horiz.* **2024**, *11*, 1719–1731. [CrossRef]
66. Li, X.; Jiao, D.; Zhao, X. A Comprehensive Descriptor for Understanding High-Shell Heteroatom-Tuned Oxygen Reduction Reaction Activity on Diatomic $FeCoN_6$ Sites. *Renewables* **2024**, *2*, 242–249. [CrossRef]
67. Chen, K.; Huang, J.; Chen, J.; Gao, J.; Lu, Z.; Liu, X.; Lan, S.; Jia, G.; Ci, S.; Wen, Z. Neighboring Iron Single Atomic Sites Boost PtCo Intermetallic for High-Durability ORR Electrocatalysis. *Energy Environ. Sci.* **2025**. [CrossRef]
68. Yuan, M.; Liu, Y.; Du, Y.; Xiao, Z.; Li, H.; Liu, K.; Wang, L. Dual-Shelled Hollow Leafy Carbon Support with Atomically Dispersed (N,S)-Bridged Hydroxy-Coordinated Asymmetric Fe Sites for Oxygen Reduction. *Adv. Funct. Mater.* **2024**, *34*, 2401484. [CrossRef]
69. Genz, N.S.; Kallio, A.J.; Oord, R.; Krumeich, F.; Pokle, A.; Prytz, Ø.; Olsbye, U.; Meirer, F.; Huotari, S.; Weckhuysen, B.M. Operando Laboratory-Based Multi-Edge X-Ray Absorption Near-Edge Spectroscopy of Solid Catalysts. *Angew. Chem. Int. Ed.* **2022**, *61*, e202209334. [CrossRef]
70. Miao, C.; Xu, S.; An, Z.; Pan, X.; Li, Y.; Hu, N.; Li, L.; Zhou, Y.; Zhao, G. Self-Optimized Reconstruction of Metal–Organic Frameworks Introduces Cation Vacancies for Selective Electrosynthesis of Hydrogen Peroxide. *Angew. Chem. Int. Ed.* **2025**, *64*, e202501930. [CrossRef]
71. Shang, H.; Zhou, X.; Dong, J.; Li, A.; Zhao, X.; Liu, Q.; Lin, Y.; Pei, J.; Li, Z.; Jiang, Z.; et al. Engineering unsymmetrically coordinated Cu-S_1N_3 single atom sites with enhanced oxygen reduction activity. *Nat. Commun.* **2020**, *11*, 3049. [CrossRef] [PubMed]
72. Chen, Y.; Rodenbücher, C.; Wippermann, K.; Korte, C. Revealing Interfacial Reactions on Pt Electrodes in Ionic Liquids by In Situ Fourier-Transform Infrared Spectroscopy. *Anal. Chem.* **2023**, *95*, 16618–16624. [CrossRef]
73. Li, Z.M.; Zhang, C.Q.; Liu, C.; Zhang, H.W.; Song, H.; Zhang, Z.Q.; Wei, G.F.; Bao, X.J.; Yu, C.Z.; Yuan, P. High-efficiency Electroreduction of O_2 into H_2O_2 over ZnCo Bimetallic Triazole Frameworks Promoted by Ligand Activation. *Angew. Chem. Int. Ed.* **2023**, *63*, e202314266. [CrossRef]
74. Liu, B.; Guo, P.; Dai, Y.; Liu, B.; Zhang, Z.; Cai, J.; Xia, Y.; Pan, Q.; Shen, L.; Zhang, Y.; et al. Tailoring asymmetric atomic strain of FeN4 sites for enhanced acidic oxygen reduction reaction. *Chem. Eng. J.* **2025**, *507*, 160174. [CrossRef]
75. Du, H.; Li, C.; Liang, Y.; Wu, Z.-S. Key strategies and challenging perspectives of carbon-based electrocatalysts for sustainable H_2O_2 production. *J. Energy Chem.* **2025**, *106*, 864–879. [CrossRef]
76. Liu, L.; Kang, L.; Feng, J.; Hopkinson, D.G.; Allen, C.S.; Tan, Y.; Gu, H.; Mikulska, I.; Celorrio, V.; Gianolio, D.; et al. Atomically dispersed asymmetric cobalt electrocatalyst for efficient hydrogen peroxide production in neutral media. *Nat. Commun.* **2024**, *15*, 4079. [CrossRef]
77. Wei, G.; Liu, X.; Zhao, Z.; Men, C.; Ding, Y.; Gao, S. Constructing ultrahigh-loading unsymmetrically coordinated Zn-N_3O single-atom sites with efficient oxygen reduction for H_2O_2 production. *Chem. Eng. J.* **2023**, *455*, 140721. [CrossRef]
78. Zhang, W.; Choi, J.W.; Kim, S.; Le, T.T.; Nandy, S.; Hwang, C.-K.; Paek, S.Y.; Byeon, A.; Chae, K.H.; Lee, S.Y.; et al. Penta nitrogen coordinated cobalt single atom catalysts with oxygenated carbon black for electrochemical H_2O_2 production. *Appl. Catal. B Environ.* **2023**, *331*, 122712. [CrossRef]
79. Li, J.; Zheng, W.; Meng, S.; Wang, S.; Zhan, S.; Sun, Y.; Hu, W.; Li, Y. Modulated Fe Central Spin State in $FeCu_4$ Alloy Nanoparticles for Efficient and Selective Activation of H_2O_2. *Adv. Funct. Mater.* **2024**, *35*, 2417538. [CrossRef]
80. Yang, X.; Priest, C.; Hou, Y.; Wu, G. Atomically dispersed dual-metal-site PGM-free electrocatalysts for oxygen reduction reaction: Opportunities and challenges. *SusMat* **2022**, *2*, 569–590. [CrossRef]
81. Hu, H.; Zhang, C.; Liu, W.; Qi, H.; Wang, H.; Wang, X.; Zhang, L.; Liu, L.; Bao, L.; Alomar, M.; et al. Coaxial Metal-Nitrogen–Carbon Single-Atom Catalysts Boost Acid Hydrogen Peroxide Production. *Adv. Funct. Mater.* **2024**, *35*, 2419220. [CrossRef]

82. Zhang, S.; Sun, B.; Liao, K.; Wang, X.; Chen, Z.; Wang, J.; Hu, W.; Han, X. Boosting Oxygen Reduction Reaction Performance of Fe Single-Atom Catalysts Via Precise Control of the Coordination Environment. *Adv. Funct. Mater.* **2025**, 2425640. [CrossRef]
83. Guan, G.; Liu, Y.; Li, F.; Shi, X.; Liu, L.; Wang, T.; Xu, X.; Zhao, M.; Ding, J.; Yang, H.B. Atomic Cobalt Metal Centers with Asymmetric N/B-Coordination for Promoting Oxygen Reduction Reaction. *Adv. Funct. Mater.* **2024**, *34*, 2408111. [CrossRef]
84. Tang, C.; Jin, R.; Xia, Y.; Zhang, Q.; Wang, J.; Wu, M.; Cui, X.; Yang, Y. Asymmetric low-coordination tailoring of single-atom cobalt catalysts enabling efficient oxygen reduction reaction. *Nano Energy* **2025**, *137*, 110776. [CrossRef]
85. Chen, L.; Yin, S.; Zeng, H.; Liu, J.; Xiao, X.; Cheng, X.; Huang, H.; Huang, R.; Yang, J.; Lin, W.-F.; et al. Engineering asymmetric electronic structure of cobalt coordination on CoN_3S active sites for high performance oxygen reduction reaction. *J. Energy Chem.* **2024**, *98*, 494–502. [CrossRef]
86. Liu, C.; Yang, X.; Jia, Z.; Pan, J.; Zhao, F.; Ban, Y.; Cui, X.; Bai, Z.; Yang, L. Boosted electron accumulation around Fe atom by asymmetry coordinated FeN_3S_1 for efficient oxygen reduction reaction. *Int. J. Hydrog. Energy* **2024**, *90*, 1003–1011. [CrossRef]
87. Liu, Y.; Pang, H.; Chen, M.; Ji, X.; Meng, F. Hollow polyhedral Fe single-atom catalysts with S/N dual coordination boosting oxygen and iodide redox for high-efficiency zinc–air/iodide hybrid batteries. *J. Mater. Chem. A* **2025**, *13*, 16008–16017. [CrossRef]
88. Lin, X.; Zhang, X.; Liu, D.; Shi, L.; Zhao, L.; Long, Y.; Dai, L. Asymmetric Atomic Tin Catalysts with Tailored p-Orbital Electron Structure for Ultra-Efficient Oxygen Reduction. *Adv. Energy Mater.* **2024**, *14*, 2303740. [CrossRef]
89. Li, Y.; Sun, H.; Ren, L.; Sun, K.; Gao, L.; Jin, X.; Xu, Q.; Liu, W.; Sun, X. Asymmetric Coordination Regulating D-Orbital Spin-Electron Filling in Single-Atom Iron Catalyst for Efficient Oxygen Reduction. *Angew. Chem. Int. Ed.* **2024**, *63*, e202405334. [CrossRef]
90. Brea, C.; Hu, G. Mechanistic Insight into Dual-Metal-Site Catalysts for the Oxygen Reduction Reaction. *ACS Catal.* **2023**, *13*, 4992–4999. [CrossRef]
91. Wang, X.; Zhang, N.; Shang, H.; Duan, H.; Sun, Z.; Zhang, L.; Lei, Y.; Luo, X.; Zhang, L.; Zhang, B.; et al. Precisely designing asymmetrical selenium-based dual-atom sites for efficient oxygen reduction. *Nat. Commun.* **2025**, *16*, 470. [CrossRef] [PubMed]
92. Wang, J.; Huang, Z.; Liu, W.; Chang, C.; Tang, H.; Li, Z.; Chen, W.; Jia, C.; Yao, T.; Wei, S.; et al. Design of N-Coordinated Dual-Metal Sites: A Stable and Active Pt-Free Catalyst for Acidic Oxygen Reduction Reaction. *J. Am. Chem. Soc.* **2017**, *139*, 17281–17284. [CrossRef] [PubMed]
93. Lv, Q.; Li, M.; Li, X.; Yan, X.; Hou, Z.; Huang, C. Introducing hydroxyl groups to tailor the d-band center of Ir atom through side anchoring for boosted ORR and HER. *J. Energy Chem.* **2024**, *90*, 144–151. [CrossRef]
94. Cao, Y.; Zeng, J.; Zheng, X.; Liu, Y.; Lu, J.; Zhang, J.; Wang, Y.; Deng, Y.; Hu, W. Regulating d-orbital spin state of Fe in single-atom electrocatalyst for boosting oxygen reduction activity in neutral electrolyte. *J. Mater. Sci. Technol.* **2025**, *227*, 67–75. [CrossRef]
95. An, Q.; Zhang, X.; Yang, C.; Su, H.; Zhou, W.; Liu, M.; Zhang, X.; Sun, X.; Bo, S.; Yu, F.; et al. Engineering Unsymmetrically Coordinated Fe Sites via Heteroatom Pairs Synergetic Contribution for Efficient Oxygen Reduction. *Small* **2023**, *19*, 2304303. [CrossRef] [PubMed]
96. Yang, M.; Qin, B.; Si, C.; Sun, X.; Li, B. Electrochemical reactions catalyzed by carbon dots from computational investigations: Functional groups, dopants, and defects. *J. Mater. Chem. A* **2024**, *12*, 2520–2560. [CrossRef]
97. Li, M.; Hou, Z.; Li, X.; Gu, X.; Yan, X.; Lv, Q.; Huang, C. Asymmetrical anchor way of manganese atoms on carbon domain edge for enhanced oxygen reduction reaction. *Appl. Catal. B Environ. Energy* **2024**, *356*, 124249. [CrossRef]
98. Bhardwaj, S.; Pathak, A.; Das, S.K.; Das, P.; Thapa, R.; Dey, R.S. Decoding Dual-Functionality in N-doped Defective Carbon: Unveiling Active Sites for Bifunctional Oxygen Electrocatalysis. *Small* **2025**, *21*, 2411035. [CrossRef]
99. Yan, J.; Li, S.; Liu, M.; Zhou, H.; Su, H. Regulating electron effect by interface-induced dislocation in Fe_2P/Fe to accelerate oxygen reduction reactions. *Mater. Today Phys.* **2025**, *54*, 101739. [CrossRef]
100. Tu, H.; Zhang, H.; Song, Y.; Liu, P.; Hou, Y.; Xu, B.; Liao, T.; Guo, J.; Sun, Z. Electronic Asymmetry Engineering of Fe–N–C Electrocatalyst via Adjacent Carbon Vacancy for Boosting Oxygen Reduction Reaction. *Adv. Sci.* **2023**, *10*, 2305194. [CrossRef]
101. Ji, S.; Wang, Y.; Liu, H.; Lu, X.; Guo, C.; Xin, S.; Horton, J.H.; Zhan, F.; Wang, Y.; Li, Z. Regulating the Electronic Synergy of Asymmetric Atomic Fe Sites with Adjacent Defects for Boosting Activity and Durability toward Oxygen Reduction. *Adv. Funct. Mater.* **2024**, *34*, 23146211. [CrossRef]
102. Zhang, Y.; Feng, J.; Shang, Z.; Pang, B.; Zhang, S.; Dong, H.; Yu, L.; Dong, L. Synergistic effects of metal single atoms and clusters on graphene-supported electrocatalysts: Insights into the mechanism of oxygen reduction reaction. *Carbon* **2025**, *233*, 119830. [CrossRef]
103. Tian, Q.; Jing, L.; Wang, W.; Ye, X.; Chai, X.; Zhang, X.; Hu, Q.; Yang, H.; He, C. Hydrogen Peroxide Electrosynthesis via Selective Oxygen Reduction Reactions Through Interfacial Reaction Microenvironment Engineering. *Adv. Mater.* **2024**, *37*, 2414490. [CrossRef] [PubMed]
104. Ma, R.; Han, G.F.; Li, F.; Bu, Y. Rational Design of Carbon-Based Electrocatalysts for H_2O_2 Production by Machine Learning and Structural Engineering. *Adv. Energy Mater.* **2025**, *15*, 2500953. [CrossRef]

105. Wen, Y.; Zhang, T.; Wang, J.; Pan, Z.; Wang, T.; Yamashita, H.; Qian, X.; Zhao, Y. Electrochemical Reactors for Continuous Decentralized H_2O_2 Production. *Angew. Chem. Int. Ed.* **2022**, *61*, e202205972. [CrossRef]
106. Oliveira Silva, T.; Granados-Fernández, R.; Lobato, J.; Lanza, M.R.V.; Rodrigo, M.A. Boosting New Electrochemical Reactor Designs to Improve the Performance in H_2O_2 Production Using Gas Diffusion Electrodes. *ACS Sustain. Chem. Eng.* **2025**, *13*, 3172–3182. [CrossRef] [PubMed]
107. Zhang, L.; Sang, W.; Wang, T.; Wei, X.; Li, Z.; Chen, C.; Kou, Z.; Mu, S. Cu–N–C support confinement stabilizes active Co sites in oxygen reduction reaction. *Nano Res.* **2025**, *18*, 94907345. [CrossRef]
108. Bao, Z.; Cao, Q.; Shao, Y.; Zhang, S.; Peng, X.; Li, Y.; Jiang, C.; Zhong, X.; Wang, J. Carbon and Oxygen-Doped Phosphorus Nitride (COPN) for Continuous Selective and Stable H_2O_2 Production. *ACS Catal.* **2023**, *13*, 14492–14502. [CrossRef]
109. Mohd Shumiri, M.A.I.; Mohd Najib, A.S.; Fadil, N.A. Current status and advances in zinc anodes for rechargeable aqueous zinc-air batteries. *Sci. Technol. Adv. Mater.* **2025**, *26*, 2448418. [CrossRef]
110. Tu, W.B.; Song, L.N.; Liang, S.; Wang, Y.; Sun, Y.; Wang, H.F.; Xu, J.J. Designing Hydrophobic Micro-Three-Phase Reaction Interfaces to Enhance the ORR Kinetics Toward Zinc-Air Battery. *Adv. Energy Mater.* **2025**, 2404946. [CrossRef]
111. Wang, D.; Li, X.; Ren, P.; Wei, Y.; Li, R.; Chen, Y.; Wang, W.; Yang, P.; Chen, Z. Asymmetric N, P-coordinated Co single-atom catalyst for efficient oxygen reduction reaction in Zn-air batteries. *Appl. Catal. B Environ. Energy* **2025**, *375*, 125436. [CrossRef]
112. Zhang, Y.; Deng, Y.-P.; Wang, J.; Jiang, Y.; Cui, G.; Shui, L.; Yu, A.; Wang, X.; Chen, Z. Recent Progress on Flexible Zn-Air Batteries. *Energy Storage Mater.* **2021**, *35*, 538–549. [CrossRef]
113. Wu, J.; Wu, W.-Y.; Wang, S.; Kai, D.; Ye, E.; Thitsartarn, W.; Tan, J.B.H.; Xu, J.; Yan, Q.; Zhu, Q.; et al. Polymer electrolytes for flexible zinc-air batteries: Recent progress and future directions. *Nano Res.* **2024**, *17*, 6058–6079. [CrossRef]
114. Ahmed, S.; Beauger, C.; Zada, A.; Iqbal, W.; Ahmed, N.; Anwar, M.T.; Hassan, M. Recent advancements in designing high-performance proton exchange membrane fuel cells: A comprehensive review. *Appl. Energy* **2025**, *390*, 125753. [CrossRef]
115. Zeng, W.; Guan, B.; Zhuang, Z.; Chen, J.; Zhu, L.; Ma, Z.; Hu, X.; Zhu, C.; Zhao, S.; Shu, K.; et al. Comprehensive review on the advances and comparisons of proton exchange membrane fuel cells (PEMFCs) and anion exchange membrane fuel cells (AFCs): From fundamental principles to key component technologies. *Int. J. Hydrogen Energy* **2025**, *102*, 222–246. [CrossRef]
116. Sharma, R.; Morgen, P.; Chiriaev, S.; Lund, P.B.; Larsen, M.J.; Sieborg, B.; Grahl-Madsen, L.; Andersen, S.M. Insights into Degradation of the Membrane–Electrode Assembly Performance in Low-Temperature PEMFC: The Catalyst, the Ionomer, or the Interface? *ACS Appl. Mater. Interfaces* **2022**, *14*, 49658–49671. [CrossRef]

catalysts

MDPI

Review

Advances in Phosphorus-Based Catalysts for Urea Electrooxidation: A Pathway to Sustainable Waste to Energy Conversion Through Electrocatalysis

Hany M. Abd El-Lateef [1,*], Mai M. Khalaf [1] and Ibrahim M. A. Mohamed [2,*]

1 Department of Chemistry, College of Science, King Faisal University, Al-Ahsa 31982, Saudi Arabia; mmkali@kfu.edu.sa
2 Department of Chemistry, Faculty of Science, Sohag University, Sohag 82524, Egypt
* Correspondence: hmahmed@kfu.edu.sa (H.M.A.E.-L.); ibrahim_mohamed@science.sohag.edu.eg (I.M.A.M.)

Abstract: The electrocatalytic oxidation of urea has gained significant attention as a promising pathway for sustainable energy conversion and wastewater treatment that could address the dual goals of waste remediation and renewable energy generation. Phosphorous function groups-based catalysts have been introduced as potential electrode materials for enhancing the urea electrocatalytic oxidation reaction (UEOR) due to their unique structural properties, high stability, and tunable electronic characteristics. This review presents recent advancements in phosphorous-based catalysts (phosphates/phosphides) for UEOR. It highlights the development of novel phosphorous materials, synthesis approaches, and electrocatalytic insights into urea electrooxidation on phosphorous-based materials surfaces. Key topics include the role of different metal phosphates, surface modifications, and compositional optimizations to improve electrocatalytic efficiency and durability. Through a critical evaluation of current research trends and technological progress, this review underscores the potential of phosphate-based catalysts as environmentally friendly and efficient alternatives for sustainable waste-to-energy conversion via UEOR. The review concludes with a perspective on future directions for optimizing phosphate catalysts, scaling up practical applications, and integrating UEOR systems into renewable energy infrastructures.

Keywords: urea electrooxidation; phosphate-based catalysts; waste-to-energy conversion; electrocatalysis; renewable energy; sustainable catalysis; metal phosphates; catalyst stability; energy-efficient wastewater treatment; green chemistry

1. Introduction

The growing global emphasis on sustainable energy and environmental remediation has placed catalysis points and their applications at the forefront of research, which could drive the development of efficient and environmentally friendly catalytic systems [1,2]. Among various catalytic processes, urea electrooxidation has garnered significant attention as a dual-purpose approach that addresses waste remediation and energy conversion [3,4]. This process offers the potential to transform urea-rich wastewater, which is a persistent pollutant into valuable energy resources, aligning with green chemistry and circular economy [5].

Phosphate-based catalysts have been presented as a promising type of material in this context, owing to their unique structural attributes, high chemical stability, and tunable electronic properties [6,7]. These features make them suitable candidates for enhancing the electrocatalytic systems such as electrooxidation of small molecules [8], as shown in Figure 1, such as hydrazine electrooxidation [9,10], alcohol oxidation [11], and urea electrooxidation reaction (UEOR) [12]. Unlike conventional catalytic systems, phosphate-based materials exhibit advantages such as structural versatility [13], compatibility with surface modifications [14], and the potential for high catalytic efficiency [15,16].

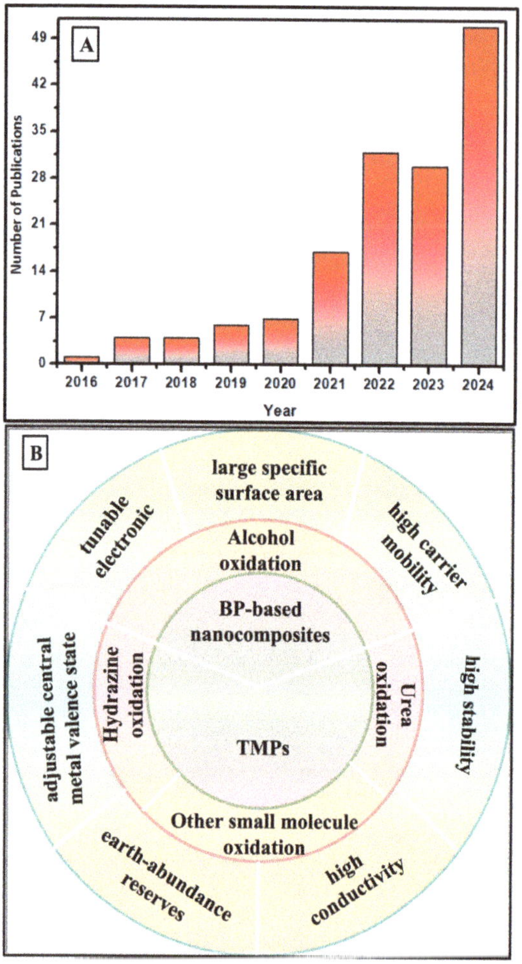

Figure 1. (**A**) Number of publications over the last 9 years, using the "Scopus" database and the following keywords ("urea electrochemical" or "urea oxidation" or "urea electrooxidation" or "urea electrolysis" or "urea electrode") and ("Phosphorus" or "Phosphate" or "Phosphide"). (**B**) Application of black phosphates (BP) and transition metal phosphates (TMPs) towards the electrocatalytic oxidation of small molecules including hydrazine, alcohol, and urea. The figure was reprinted after permission from Elsevier [8], Molecular Catalysis 558, 2024, 114023 and the license number 5926580350419.

This review delves into the advancements in phosphate-based catalysts for UEOR, providing a comprehensive exploration of their synthesis strategies, structural optimizations, and mechanistic insights. Special emphasis is placed on the role of different metal phosphates, surface functionalization techniques, and compositional innovations in improving the catalytic activity, selectivity, and durability of these materials. By critically analyzing current trends and progress, the review underscores the promising role of phosphate-based catalysts in sustainable waste-to-energy conversion technologies. The discussion concludes by addressing challenges in scaling up applications and integrating these catalysts into renewable energy systems, which could offer a perspective on future directions in this exciting catalytic field.

2. Chemistry of Phosphates

Phosphates are a class of chemical materials derived from phosphoric acid (H_3PO_4), which forms various salts and esters when combined with metals or organic molecules [17]. They play an essential role in biological [18], geological [19], and industrial systems [20], contributing to energy transfer [21], mineralization processes [22], 3D printing [23], and fertilizer production [24]. Understanding the types and structures of phosphates offers insight into their diverse applications and natural significance [25]. At the heart of all phosphate compounds is the phosphate ion (PO_4^{3-}), a tetrahedral structure where a phosphorus atom is covalently bonded to four O atoms [26]. In this arrangement, one oxygen atom forms a double bond with phosphorus, and the other three are single-bonded and carry negative charges, which provide more reactivity and the capability of forming various chemical bonds. This flexibility allows phosphates to exist in multiple forms, each with distinct structural and functional characteristics. Phosphates can be categorized into three main types: orthophosphates, polyphosphates, and organophosphates [27]. Orthophosphates (PO_4^{3-}) are the simplest form and serve as the foundation for other phosphate materials. They are prevalent in biological systems, notably in adenosine triphosphate (ATP), where they store and transfer energy within cells [28]. Orthophosphates also exist in minerals such as apatite [29], which is a primary source of phosphorus in nature and an essential component of bones and teeth [30]. The second type of phosphate is polyphosphates, which are chains or rings of phosphate units linked by oxygen bridges [31]. These compounds include pyrophosphate ($P_2O_7^{4-}$) and tripolyphosphate ($P_3O_{10}^{5-}$) and can be synthesized by condensation reactions involving orthophosphates. Polyphosphates are widely used in industrial applications, including detergents, water softeners, and food additives [32]. In living organisms, they act as energy reservoirs and are involved in processes like cell signaling and bone mineralization [33]. The third type of phosphate is organophosphates, which consist of phosphate groups covalently bound to organic molecules [34]. These compounds play a pivotal role in biochemistry, as seen in nucleotides, DNA, RNA, and phospholipids that form cellular membranes [35]. Organophosphates are also the basis for many pesticides and nerve agents [36], exploiting their ability to inhibit essential enzymes like acetylcholinesterase [37]. This duality highlights their importance and the need for careful management in agriculture and medicine [38].

Structurally, phosphates exhibit versatility depending on the phase and chemistry of preparation and saving environment. In crystalline forms (such as minerals), the phosphate ions are coordinated with metal cations like Ca, Na, or K, which lead to complex networks. These structures are stable and contribute to the formation of rocks like phosphorite. In aqueous systems, phosphates exist as hydrated ions or form soluble salts, allowing them to participate in chemical reactions, including but not limited to acid–base equilibria and precipitation. The biological significance of phosphates cannot be overstated. Phosphates are integral to energy metabolism, where ATP molecules release energy through the hydrolysis of phosphate bonds [39]. They also serve as structural components in nucleic acids, linking nucleotides via phosphodiester bonds [40]. Additionally, phosphate groups regulate proteins and enzymes through phosphorylation, a process central to cell signaling and metabolic control [41].

Various synthesis strategies for phosphorus-based catalysts, including techniques such as sol–gel [42], hydrothermal [43], chemical vapor deposition [44], electrochemical methods [45], and MOF-assisted synthesis [46], have been reported. From an environmental perspective, phosphates are a double-edged sword. They are essential for plant growth as fertilizers [24], but excessive phosphate runoff leads to eutrophication, which causes algal blooms and oxygen depletion in aquatic ecosystems [47]. This environmental impact necessitates sustainable phosphate management, including recovery from wastewater and efficient agricultural practices. Geologically, phosphates are key constituents of sedimentary and igneous rocks, forming through biogenic and geochemical processes. Apatite minerals are mined extensively to produce fertilizers, which support global food production. However, phosphate reserves are finite, and their depletion raises concerns about future availability that lead the future research into recycling and alternative sources. Distinct from other P-function groups, phosphides are compounds that contain phosphorus in their negative oxidation state, typically as the anion P^{3-}. These materials are often formed by the reaction of phosphorus with metals or metalloids, resulting in a wide range of metal phosphides, and have gained significant interest in electrocatalysis [48]. Therefore, phosphorous function groups are foundational to life, industry, and the environment. Their diverse types and structures underpin critical biological functions, support agricultural productivity, and pose environmental challenges. It is essential to mention the diverse synthesis strategies employed to develop these catalysts, as these methods directly influence their structure, surface area, electronic properties, and catalytic activity. Common synthesis strategies include sol–gel methods [42], which enable the formation of homogeneously distributed phosphorus within the catalyst matrix, and hydrothermal techniques [43], which are widely used for producing highly crystalline and well-defined nanostructures. Additionally, chemical vapor deposition [44] offers precise control over the deposition of phosphorus onto specific surfaces, making it ideal for creating thin films or layered structures. Electrochemical methods [45], on the other hand, provide a sustainable and scalable approach to phosphorus incorporation while tuning the electrochemical properties of the catalyst. Other innovative approaches, such as MOF-assisted synthesis [46], template-assisted synthesis, pyrolysis of phosphorus-containing precursors, and doping strategies, further expand the design possibilities for these catalysts. The selection of a synthesis method is often dictated by the intended application, as well as the desired properties of the final catalyst, such as high conductivity, stability, and active site availability. A deeper understanding of phosphorous function groups allows us to harness their benefits while mitigating their risks, emphasizing their multifaceted role in shaping the world around us. This review displays the reported application of phosphorous function groups as anode material and electrocatalyst for UEOR towards the commercialization of urea fuel cells (UFC).

3. Electrocatalysis in Urea Fuel Cells

Fuel cells (FCs) are one of the electrochemical devices that change the chemical energy in chemical fuels into electricity with acceptable efficiency [49]. They operate silently and occupy less space compared to conventional energy systems, as well as having minimal environmental impact. Among the various types of fuel cells, UFC holds great promise for sustainable wastewater treatment and energy generation [50]. In these systems, urea is electrochemically oxidized, yielding CO_2, N_2, H_2O, and electricity [51]. UFC systems are notable for their theoretical efficiency (102.9%) and could surpass that of hydrogen fuel cells (83%) because of heat absorption from the surroundings. Additionally, their open-circuit voltage (OCV) is comparable to H_2 fuel cells at approximately 1.21 V. However, enhancing the performance of DUFCs remains a challenge, especially regarding the electrochemical activity of the metal-based catalysts as anode part in UFC commonly used for urea electrocatalytic oxidation. Among the transition metals, nickel (Ni) has been reported as an effective, non-precious electrocatalyst for the UEOR in alkaline conditions [52]. Despite their electrocatalytic activity, Ni electrocatalysts face limitations in electrochemical stability and activity, hindering commercial viability. Efforts to enhance anode performance for fuel cells include incorporating Ni with metals like Pd [53], Cu [54], Co [55], Ce [56], Ti [57,58], Mn [59], Fe [60], Cr [61,62], and Se [63,64] or increasing the catalyst's surface area using supports including carbonaceous support such as C-nanofibers [65,66], graphene oxide [67], 3D carbon sponges [68], and carbon nanotubes [69]. Moreover, employing alternative oxidants like H_2O_2, which exhibits a higher cathodic potential than O_2, or adopting hybrid pH conditions (alkaline anolyte and acidic catholyte) has shown promising results in boosting the open-circuit voltage (OCV) and power output. In situ and ex situ studies reveal that these strategies significantly improve the catalytic efficiency of UFC, opening pathways for their commercialization as dual-purpose devices for energy generation and environmental remediation [49].

The electrocatalytic part in the UFC systems concentrates on two major parts: anodic and cathodic. The anodic part includes electrochemical oxidation of urea, such as the reported Ni-electrocatalysts, which could electrocatalyse urea oxidation via two mechanisms: the indirect pathway, where NiOOH oxidizes urea chemically, and the direct pathway, where urea adsorbs onto NiOOH followed by transferring electrons directly to the electrode [70]. Additionally, both direct and indirect electrocatalytic mechanisms have been reported at the same electrocatalyst, such as Ni_5Sm-P/C [71]. The indirect electrocatalytic mechanism involves competitive adsorption between urea and OH^- on the NiOOH surface, which could be considered an electroactive layer [72]. The reaction kinetics and efficiency are significantly influenced by the alkaline medium's composition, with potassium hydroxide (KOH) emerging as the most effective electrolyte. KOH enhances the oxidation of surface intermediates, facilitates the release of CO_2, and increases the available NiOOH. Increasing the KOH concentration reduces the onset potential (E_{op}) and boosts the oxidation peak current density, improving electrocatalytic activity [73,74]. Furthermore, higher urea concentrations enhance the diffusion-controlled reaction rates, whereas kinetic limitations govern performance at elevated potentials. Therefore, understanding the behavior of electrocatalysts including Ni-based catalysts in alkaline media is crucial for optimizing UFC performance. The redox transitions of Ni between $Ni(OH)_2$ and NiOOH under applied potentials create active sites for urea oxidation, but the formation of intermediates can temporarily block these sites. Techniques like cyclic voltammetry (CV), impedance (EIS), and surface-enhanced Raman spectroscopy have elucidated these mechanisms, confirming the role of β-NiOOH as the primary electrochemical active site for UEOR. Therefore, advances in catalyst design and the use of hybrid conditions are promising steps toward addressing challenges such as catalyst poisoning, stability, and long-term operation. Further research focusing on alternative oxidants, robust electrocatalyst recovery strategies, and mechanistic insights will be pivotal for realizing the full potential of UFC in wastewater treatment and renewable energy applications.

Determining the optimal KOH/urea ratio for large-scale applications is essential for maximizing efficiency and minimizing energy consumption, particularly given the inherent variability in wastewater composition. Empirical studies involving systematic experimentation under different KOH/urea ratios provide valuable data on conditions that optimize urea conversion. Concurrently, real-time wastewater composition analysis allows for dynamic adjustments of the KOH/urea ratio using feedback control systems. Advanced modeling techniques, including computational fluid dynamics (CFD) and reaction kinetic simulations, can predict the optimal ratios under diverse scenarios, significantly reducing the need for trial-and-error experimentation. Pilot-scale testing under representative wastewater conditions further validates the practical applicability of these ratios, ensuring robust performance in real-world scenarios.

Stability tests for both direct and indirect mechanisms of urea oxidation could be conducted through more systematic evaluations under real-world conditions. For direct mechanisms, laboratory studies employing cyclic voltammetry and chronoamperometry under simulated wastewater conditions have provided insights into long-term stability. Indirect mechanisms, which involve reactive intermediates like hydroxyl radicals, are often assessed using advanced spectroscopic methods to monitor catalyst degradation over time. Limited field trials in wastewater treatment facilities have shown promising initial results, but further research is needed to replicate these findings across various operational conditions. To bridge this gap, future studies should focus on accelerated aging tests and continuous-flow setups that closely mimic real-world environments, thereby providing a more comprehensive understanding of catalyst durability.

Certainly, hydroxide ions (^-OH) play a considerable role in urea electrooxidation according to the reported mechanism (as illustrated in Figure 2A), following an electrochemical–chemical (E-C) process. Initially, Ni content reacts with ^-OH to form $Ni(OH)_2$, which is inactive for urea oxidation. $Ni(OH)_2$ is then oxidized by additional ^-OH to form the electroactive NiOOH layer, capable of oxidizing urea into CO_2 and N_2 [62,72]. Urea oxidation involves the attack of six ^-OH groups on specific sites of the urea molecule (the C in the O=C group and the hydrogens in the NH_2 groups). These lead to the cleavage of N–H and C–N bonds, yielding N_2, CO_2, and H_2O. Intermediates formed during this process adsorb onto the NiOOH because of the polarization of O-Ni bonds. Subsequently, ^-OH attacks the N–H bonds, producing H^+, which reacts with ^-OH to form H_2O. The electroactive NiOOH is reduced back to $Ni(OH)_2$ and requires further oxidation to continue the catalytic cycle. However, CO_2 formed during the process can dissolve in the electrolyte and create a carbonate layer, which inhibits the $Ni(OH)_2$/NiOOH conversion, eventually reducing or halting the oxidation reaction [75,76].

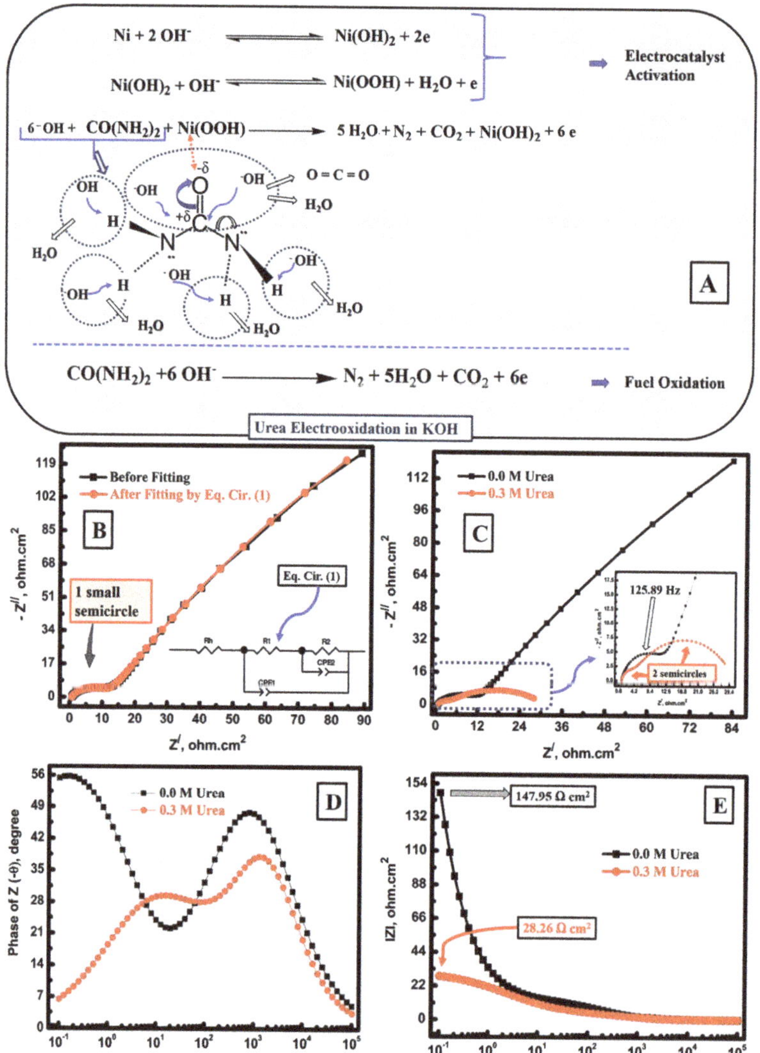

Figure 2. (**A**) Urea electrocatalytic oxidation mechanistic aspects. (**B**) EIS of urea electrocatalytic oxidation using KOH without urea. (**C**) EIS using 0.3 M urea. (**D**) Bode phase with and without urea. (**E**) Total impedance behavior with and without urea. The figure was reprinted after permission from Elsevier [72] and the license number 5926580582366.

From the different reported techniques for studying the electrocatalysis of UEOR, the EIS study is the best one that could provide the charge transfer related to the electrocatalytic mechanism. For example, the EIS with and without urea was displayed (Figure 2B,C, respectively) and confirmed that the relaxation time for urea oxidation is longer than for NiOOH formation, confirming that the rate-determining step (RDS) could be urea degradation (the cleavage of C–N and N–H bonds) [72]. Nyquist plots further illustrate the interaction of γ-NiOOH formation with urea oxidation. For 0.3 M urea, a faster relaxation time (0.797 ms) was observed against what was estimated in the case of urea-free solution (1.264 ms), highlighting the enhanced activity of NiOOH with urea. The large semicircle in the low-frequency region of the Nyquist plot for 0.3 M urea corresponds to urea oxidation, contrasting with the diffusion-controlled process observed in the urea-free solution. The calculated relaxation times and resistance values (R_2 dropping from 1050.3 to 26.55 Ω cm^2 with urea) strongly support the enhanced catalytic activity towards urea electrooxidation. Bode phase plots (Figure 2D) further confirmed the existence of two charge transfer processes: ^{-}OH diffusion to the working electrode surface and the consumption of NiOOH during oxidation. A significant shift in the peaks indicates the active involvement of NiOOH in urea oxidation. Total impedance analysis (Figure 2E) demonstrates that the addition of urea decreases total impedance, indicating improved electrical conductivity due to enhanced charge transfer pathways during urea electrooxidation. Therefore, the study of EIS is a considerable indicator for evaluating the electrocatalysts for UEOR including Nyquist plots, Bode phase plots, and total impedance characteristics.

4. Ni-Phosphate Electrocatalysts for UEOR

Phosphates are promising materials for electrocatalysis due to their robust chemical and thermal stability, tunable electronic properties, and versatility in catalyzing essential reactions such as the oxygen evolution reaction (OER) [77], hydrogen evolution reaction (HER) [78], and carbon dioxide reduction reaction (CO_2RR) [79], as well as UEOR [80]. Their strong P−O bonds ensure durability under harsh reaction conditions, while their structural flexibility permits the incorporation of transition metals, enabling tailored catalytic activity [8]. Phosphates could stabilize key intermediates, lower activation barriers, and promote selective pathways in reactions like UEOR. Their performance can be further enhanced through doping with elements or forming composites with conductive materials such as carbon materials, which will be discussed in detail. Firstly, Ni-phosphate materials could be studied before other bimetallic phosphates, followed by composites containing phosphates.

Song et al. conducted a comprehensive study on the electrochemical catalytic activity of four nickel phosphate catalysts (NiPO-a,b,c,d) synthesized using different methods, focusing on their potential for urea oxidation in alkaline conditions [81]. The structures of these materials varied significantly: NiPO-a featured less-ordered nanotubes, NiPO-b and NiPO-c had a flower-like nanosheet porous structure, and NiPO-d consisted of stacked nanocrystals. Electrochemical analyses, including CV (Figure 3A–D), demonstrated that these phosphates had more negative E_{op} compared to commercial NiO, making them more reactive towards urea. Among them, NiPO-c exhibited the highest oxidation current density (30.0 mA cm^{-2}) and superior stability, retaining 97.47% of its electrochemical activity after a 96-h amperometric i−t test (Figure 3E). This superior performance was attributed to its unique structure and excellent catalytic properties. Additionally, NiPO-c demonstrated favorable electrochemical behavior with clear cathodic/anodic peaks in KOH containing urea, which were absent in NiPO-d and commercial NiO (Figure 3A). The E_{op} for UEOR with NiPO-c was as low as 0.33 V versus Ag/AgCl, lower than other NiPOs and NiO. Mechanistic insights revealed that the reaction involved the conversion of $Ni(OH)_2$ to NiOOH, followed by a chemical reaction with urea, which highlighted the best performance for NiPO-c (Figure 3B). Further investigations into the influence of urea concentration showed that the optimal urea-to-KOH ratio was critical for maximizing current density (Figure 3C), while the ideal KOH concentration for urea oxidation was determined to be 3.0 mol L^{-1} (Figure 3D). Figure 3F has the EIS results, which indicate that NiPO-c had the lowest equivalent series resistance (46 Ω) and charge transfer resistance (R_{ct}), enabling faster reaction kinetics. These findings underscore the potential of NiPO-c as a promising candidate for UEOR.

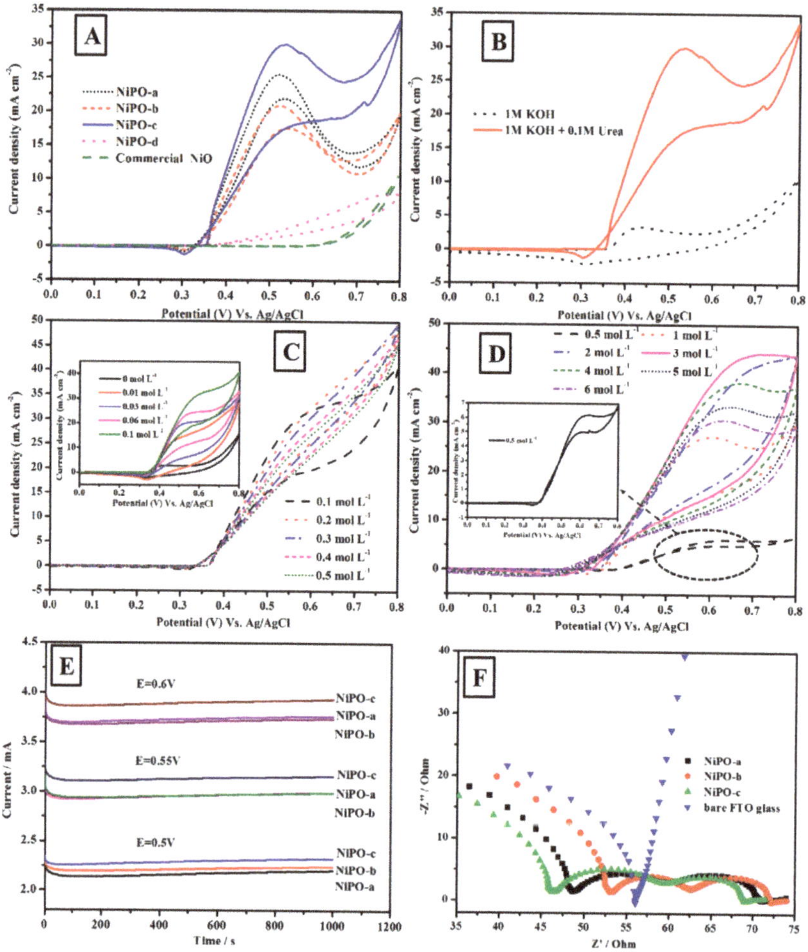

Figure 3. (**A**) CV of Ni-phosphates electrocatalysts in 1 mol/L KOH containing 0.1 mol/L urea at a scan rate of 0.05 V s^{-1}. (**B**) CV of NiPO-c in 1 mol/L KOH and 0.1 M urea. (**C**) CV of NiPO-c at varying urea concentrations; the inset shows CV for lower urea concentrations. (**D**) CV of NiPO-c at different KOH contents with 0.1 mol/L urea. (**E**) Amperometric (i–t) measurements for UEOR on NiPO-a, NiPO-b, and NiPO-c modified fluorine-doped tin oxide (FTO) electrodes at different applied voltages in 0.1 M urea. (**F**) EIS analysis of NiPO-a,b,c modified FTO electrodes at 500 mV after 300 cycles in KOH with 0.1 mol/L urea. The figure was reprinted after permission from Elsevier [81], Electrochimica Acta, 251, 2017, 284–292 and the license number 5926581331801.

Q. Li et al. demonstrated the chemical design of porous rods based on Ni_2P/Ni material composed of nanosize particle subunits through a one-step phosphating method [82]. Density functional theory (DFT) calculations revealed that these assemblies exhibit enhanced conductivity and chemical bond activity compared to NiO/Ni, significantly boosting UOR performance. The optimized Ni_2P/Ni catalyst achieved a low overpotential of

50 mV at 10 $mA·cm^2$, and 87.6 $mV·dec^{-1}$ Tafel slope. Electrochemical characterization indicated that the catalyst possesses a better electrochemical surface area and improved kinetics, as evidenced by its low resistance (R_{ct}) and superior double-layer capacitance. Comparative studies with non-phosphated samples confirmed that phosphating enhances catalytic activity. The Ni_2P/Ni composite, developed by precisely controlling synthesis parameters, displayed superior UEOR activity, achieving higher current density and lower E_{op} than other tested samples. Additionally, the stability of the porous rod-like structure was demonstrated through extended immersion tests, which retained the morphology and performance of the catalyst.

G. Wang et al. synthesized a unique 3D porous Ni_2P nanoflower electrode combined with nickel foam (Ni_2P@NF) using hydrothermal growth strategy and material phosphating [83]. The working electrodes were analyzed by XRD, SEM, EDX, and TEM, and their electrocatalytic performance for UEOR was evaluated via CV, chronoamperometry (CA), and EIS. The phosphating process transformed the $Ni(OH)_2$ precursor into porous nanoflowers (Ni_2P), creating abundant active sites and electron transmission channels that enhanced interfacial reactions. Compared to its precursor, the Ni_2P@NF electrocatalyst displayed superior electrocatalytic activity in KOH (5.00 mol/L) and 0.6 mol/L urea, achieving 750 mA/cm^2 with an E_{op} of 0.24 V, alongside remarkable stability. Electrochemical analyses revealed enhanced redox transitions between Ni^{2+}/Ni^{3+}, a larger electrochemically active surface area (EASA), and improved charge transfer efficiency attributed to the material's porous, flower-like nanosheet structure. The electrode demonstrated optimal catalytic performance at a urea content of 0.60 mol/L, aligning with the theoretical ratio of urea to KOH while maintaining stability under various applied potentials. EIS measurements indicated improved charge transport and distinct conversion processes in the presence of urea. Collectively, these findings establish the Ni_2P@NF electrode as a promising electrocatalyst for alkaline UEOR.

In 2019, Amin Imani et al. explored the chemical design of nanoporous $Ni_3(PO_4)_3·8H_2O$ microspheres (NiP) with an orange peel morphology, which was developed as an advanced catalyst for UEOR [84]. Using a one-pot hydrothermal method, they synthesized NiP materials with tunable morphology and particle size by varying hydrothermal parameters. This material was identified as one-phase $Ni_3(PO_4)_2·8H_2O$, having a monoclinic structure. The optimized NiP110 demonstrated exceptional electrocatalytic activity and stability, obtaining 680 $mA·cm^{-2}·mg^{-1}$ and a low oxidation voltage of 0.33 V in alkaline media. CV analysis revealed that the chemical existence of fuel sharply improved the produced current density, indicating strong catalytic action. The superior performance of NiP110 was attributed to its improved surface area, enabling efficient diffusion of reactants and exposure of active redox sites. The electrocatalytic process involves the electrochemical conversion of fuel into CO_2, water, and N_2 via an electrochemical-catalytic mechanism mediated by NiOOH, generated through the electrooxidation of $Ni(OH)_2$. Additionally, the production of a nickel hydroxide layer on the electrode in the KOH solution facilitated redox reactions. The study also highlighted the impact of the supporting electrolyte-to-urea ratio on oxidation efficiency, emphasizing that an imbalance can lead to reduced current density due to inadequate hydroxide ions or electrode fouling by CO_2. This innovative synthesis of nanoporous NiP microspheres opens avenues for their application in electrochemical sensors for urea detection and underscores the potential to generalize the approach to fabricate other 3D nanoporous phosphate-based materials for diverse electrochemical applications [84].

Phosphorus material based on $Ni_{12}P_5$ nanorods grown in situ on a 3D NF skeleton ($Ni_{12}P_5$/Ni-Pi/NF) has been fabricated, offering a competitive electrocatalyst for UEOR or HER [85]. The hierarchical core–shell architecture of these nanorods provides an improved surface area, enhanced electron transfer, and efficient mass transport, resulting in robust catalytic performance. For UEOR, the catalyst achieves 9 A/cm^2 at 1.378 V in 1.0 mol/L KOH with 0.5 mol/L urea, outperforming many state-of-the-art catalysts. It also demonstrates excellent HER activity, obtaining a promising current (e.g., 500 mA cm^{-2} at −0.312 V) with improved kinetics and durability. Using a two-electrode splitting electrolyzer, the electrode ($Ni_{12}P_5$/Ni-Pi/NF) achieves 0.5 A/cm^{-2} at a cell potential of 1.662 V, which is significantly lower than that required for pure water splitting, highlighting the thermodynamic and kinetic advantages of urea-assisted electrolysis. Durability tests confirm the catalyst's stability under large current densities over extended periods, making it a promising candidate for practical-scale H_2 production. The exceptional catalytic performance can be due to the 3D porous architecture of the core-shell nanorods, which enhances active site exposure, minimizes gas bubble accumulation, and improves electrolyte mass transport. These findings represent a pivotal step toward developing industrial electrolyzers for environmentally friendly hydrogen production through urea decomposition [85].

Y.-C. Tsai et al. studied the electrocatalytic application of zeolitic $Ni_3(PO_4)_2$ nanorods with an open-framework VSB-5 (Versailles-Santa Barbara-5) for UEOR [86]. These nanorods were synthesized hydrothermally on nickel foam using lithium dihydrogen phosphate, forming a highly porous material with nanometer-scale pore diameters (~1 nm) and molecular sieving properties. The VSB-5 nanorods provided micro-, meso-, and macroporous architectures that facilitated electrolyte penetration, electron transport, and gas product evolution. Compared to nickel hydroxide electrodes, the VSB-5 electrode exhibited a larger EASA, enabling enhanced redox cycling of nickel species (Ni^{2+}/Ni^{3+}), which significantly improved its electrocatalytic performance. It achieved a higher current of 160 mA cm^{-2} at 0.6 V (vs. SCE) and an E_{op} of 0.27 V for UEOR, outperforming nickel hydroxide electrodes (75 mA cm^{-2} at 0.28 V). Figure 4 demonstrates the amplified anodic current density in the presence of urea based on CV measurements (Figure 4A) and CA analysis (Figure 4B), confirming the catalytic efficiency of VSB-5 nanorods. CA studies (Figure 4B) revealed the electrode's long-term stability, maintaining high current densities over extended testing periods, while EIS (Figure 4C) highlighted a lower R_{ct} of 5.77 Ω for VSB-5 compared to 21.51 Ω for nickel hydroxide. The open-framework structure of VSB-5 contributed to effective urea adsorption and rapid charge transfer, enhancing the UEOR kinetics. As shown in Figure 4D (current values from staircase wave), the VSB-5 electrode achieved a current density of 123 mA cm^{-2} at 600 mV, significantly surpassing the nickel hydroxide counterpart (35 mA cm^{-2}). The electrode also demonstrated remarkable current efficiency (η) of up to 97% at low potentials, with minimal OER interference, a critical factor for efficient urea electrolysis. These findings underscore the potential of VSB-5 as a promising electrocatalyst for UEOR.

In 2021, nickel phosphate (Ni-P) nanoflakes were introduced via Khalaf et al. for UEOR and represented a promising pathway for morphology modification, particularly when employing cost-effective and high-performance materials such as Ni-P [87]. The synthesis of Ni-P nanoflakes under varying urea proportions has revealed significant impacts on the material's morphology, structure, and electrocatalytic performance. Characterization techniques confirm that Ni-P exhibits a nanoflakes morphology with well-defined structural properties conducive to catalytic activity. Electrochemical studies using CV and EIS in alkaline KOH media demonstrate that Ni-P facilitates efficient UEOR through enhanced redox activity because of Ni(II)/Ni(III) transition and the generation of active NiOOH species on the electrode surface. CV results indicate that urea content during synthesis critically influences the performance of the Ni-P catalyst, with an optimal proportion of 0.8 M yielding the highest current density for UEOR. Beyond this threshold, excessive urea results in larger flake formation, reducing the effective surface area and diminishing catalytic efficiency. The electrooxidation mechanism of UEOC on Ni-P involves the adsorption and activation of urea molecules, leading to bond cleavage and the release of electrons, which contribute to electricity generation. CA experiments confirm the long-term stability and robust performance of the Ni-P electrode, showcasing its ability to sustain high current densities over extended periods with minimal decay. Furthermore, EIS analysis underscores the low ionic and R_{ct} of Ni-P, affirming its suitability as an efficient electrocatalyst in a KOH medium. The proposed mechanism involves the indirect oxidation of fuel on the NiOOH sites, resulting in the production of nitrogen and carbon dioxide as by-products while releasing six electrons per urea molecule. The high activity and stability of Ni-P electrodes highlight their potential for integration into urea-based fuel cells.

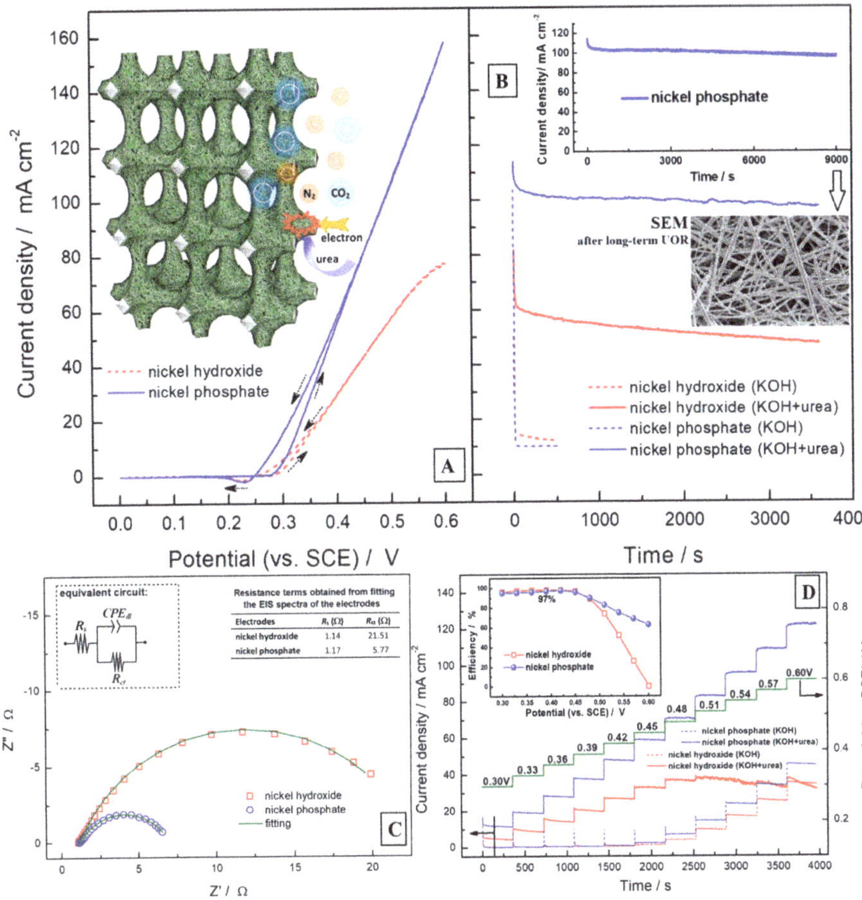

Figure 4. (**A**) CVs of the Ni-hydroxide and Ni-phosphate materials in an alkaline medium with 0.3 mol/L urea at 0.005 V s^{-1}. (**B**) CA of the Ni-hydroxide and Ni-phosphate materials at 500 mV with and without 0.3 mol/L urea. (**C**) EIS analysis of the Ni-hydroxide and Ni-phosphate materials containing urea 0.33 mol/L. (**D**) Voltage-controlled current values from staircase wave for the Ni-hydroxide and Ni-phosphate materials in the presence and absence of 0.3 mol/L urea. The figure was reprinted after permission from Elsevier [86] Applied Surface Science, 529, 2020, 147175 and the license number 5926590137816.

In 2023, Al-Sulmi et al. reported an efficient preparation of mesoporous $Ni_3(PO_4)_2$ (meso-NiPO) through chemical deposition within a template based upon surfactant liquid crystal (Brij®78), achieving a surface area of 43.50 m^2/g compared to 4.26 m^2/g for bare $Ni_3(PO_4)_2$ (bare-NiPO) synthesized without surfactant [88]. Electrochemical studies demonstrated that meso-NiPO exhibits promising electrocatalytic activity for the UEOR in alkaline media, with an oxidation E_{op} of 0.30 V vs. Ag/AgCl, R_{ct} of 3.35 Ω, and a remarkable mass activity of 700.7 $mA/cm^2{\cdot}mg$ at 0.6 V, which is 4.5 times better than that of bare-NiPO. CA tests revealed high stability, retaining 97.5% of the oxidation current after 3 h of fuel electrolysis. Using an H-type electrolyzer, the meso-NiPO catalyst achieved an H_2 production of 415 μmol/h and a Faradic performance of 96.80% at 2.0 V. The mesoporous structure facilitated efficient charge transfer, mass transport, and accessibility of electroactive species, enhancing the UOR [89]. Additionally, the UEOR Tafel slope was found at 20.3 mV/decade indicating faster reaction kinetics compared to bare-NiPO (30.2 mV/decade). Optimization studies showed that the catalytic activity increased with meso-NiPO quantity to 0.2 mg/cm^2 but saturated at higher loadings due to diffusion limitations. Similarly, the UEOR current increased with urea concentrations up to 0.33 M but declined at higher concentrations due to competitive adsorption effects [90]. The process was confirmed to be diffusion-controlled, with the meso-NiPO demonstrating a significant negative shift in the E_{op} and a 15-fold improvement in current with increasing hydroxide ion concentrations from 0.1 to 2.0 M. The catalyst's robustness was underscored by its long-term stability, maintaining higher steady-state currents than bare-NiPO, and showcasing enhanced accessibility of active Ni(II)/Ni(III) redox sites. This innovative synthesis strategy provides a scalable and energy-efficient route for mesoporous catalyst preparation, highlighting the potential of meso-NiPO for UEOR [90].

In 2024, Ghanem et al. introduced an advanced electrocatalyst having $Ni_3(PO_4)_2$-combined with NF (NiPO/NF), which was prepared via anodization for dual applications in UEOR and HER [91]. The preparation steps involved cleaning commercial nickel foam with hydrochloric acid for 3 min to take out surface oxides, followed by rinsing with pure H_2O and organic solvent (acetone), followed by air-drying. Anodization was carried out in a 0.3 mol/L phosphate solution using a 2-electrode setup, with NF as the anode and titanium foil as the cathode, at 5 °C. Direct constant voltage was utilized at 5, 10, or 15 V for 20 min, resulting in electrodes labeled as NiPO/NF@5, NiPO/NF@10, and NiPO/NF@15, respectively. The process induced oxidation at the anode, depositing a nickel phosphate catalyst layer. Electrochemical characterization revealed that the NiPO/NF@10 substrate exhibited the best performance, with optimized redox behavior and electroactive surface area. CV in 2.0 M KOH demonstrated significant enhancements in redox peak currents, with NiPO/NF@10 showing peak current densities of 95.0/−68.5 mA/cm^2 compared to 15.3/−10.3 mA/cm^2 for pristine NF, attributed to the production of Ni(II)/Ni(III) sites and $Ni_3(OH)_3(PO_4)_2$ phase during the hydroxide ion interaction. In 0.3 mol/L urea, the NiPO/NF@10 substrate achieved 20 mA/cm^2 as an oxidation current at an E_{op} of 1.35 V vs. RHE, 100 mV lower than pristine NF, indicating a superior electrocatalytic effect via indirect electrochemical mechanistic aspects. The study underscores the promising potential of the NiPO/NF@10 electrocatalyst for scalable hydrogen production and sustainable UEOR [91].

5. Ni-Free Phosphate Electrocatalysts for UEOR

Despite their promising attributes, the catalytic performance of nickel-free phosphate materials in UEOR remains relatively underexplored compared to their nickel-based counterparts. Challenges such as limited intrinsic electroconductivity, moderate catalytic activity, and optimization of nanostructures to maximize active surface areas persist. Recent advances in nanostructuring techniques, including the fabrication of nanowires, nanosheets, and porous frameworks, have shown considerable potential in overcoming these limitations. Recently, some Ni-free electrocatalysts were reported for electrocatalysis applications away from UEOR, such as efficient CO_2 reduction to multicarbon products via Cu-P [92], glucose detection via Cu-modified Ti-phosphate NPs [93], and highly ethylene-selective electroreduction CO_2 over Cu-P nanostructures with tunable morphology [94]. This part aims to provide an overview of the developments in nickel-free phosphate-based electrocatalysts for UEOR. Hui Shen et al. explored UEOR and proposed a two-step methodology to chemically prepare Cu_3P nanowires on copper foam (Cu_3P-NW/CF) as an electrocatalyst for UEOR in alkaline media [95]. The Cu_3P-NW/CF electrode achieved operational voltage values of 1.43 V and 1.65 V versus RHE at current values of 10 and 100 mA cm^{-2}, respectively, while a two-electrode urea electrolysis system using Cu_3P-NW/CF for both the oxidized and reduced electrodes required a cell voltage of 1.79 V to sustain a current density of 100 mA cm^{-2}. These results surpassed the conventional H_2O electrolysis thresholds without urea, underscoring the energy-saving potential of the urea-assisted process. Structural and kinetic advantages, such as enhanced conductivity, abundant active sites, and low Tafel slopes (33.1 mV dec^{-1} for UEOR), contributed to the catalyst's performance. Electrochemical active surface area analysis revealed a significant improvement in the active area (C_{dl} = 7.5 mF cm^{-2}) compared to a control Cu_3P-P/CF electrode. Furthermore, EIS demonstrated lower charge-transfer resistance, while stability tests showed negligible degradation over 30 h of operation, highlighting the durability of the Cu_3P nanowire architecture. However, limitations remain, as the absence of electroactive metals like Ni, which are often incorporated into UEOR catalysts for enhanced performance, might restrict the catalyst's activity and selectivity. Additionally, while the binder-free design facilitated efficient electron transfer, further investigation into long-term structural integrity under harsh electrochemical conditions is warranted. This study provides a foundation for developing scalable Ni-free UEOR electrocatalysts, yet improvements in material composition and synergy with electroactive metals are needed to maximize catalytic performance for UEOR and related applications [95].

6. Transition Metals Incorporated Ni-Phosphate Electrocatalysts for UEOR

Mono-metal phosphate-based catalysts have demonstrated significant promise in electrocatalysis uses because of their inherent stability, tunable surface chemistry, and ability to mediate redox reactions [96–98]. Despite these advantages, their catalytic activity can be limited by a lack of sufficient active sites, moderate conductivity, and relatively low electrochemical efficiency. To overcome these limitations, the development of bimetal phosphate catalysts has emerged as a transformative approach [99]. Bimetal phosphates introduce a synergistic interplay between two distinct metal centers, offering enhanced catalytic performance compared to their mono-metal counterparts [100–102]. This dual-metal composition allows for a more efficient distribution of electronic density and an optimized local environment, which could facilitate the adsorption and initiate the electrocatalytic process. Moreover, the incorporation of a secondary metal can introduce additional redox-active sites if the second metal is electrochemically active and so improve the overall electrical conductivity [102,103]. Therefore, bimetal phosphates not only enhance the kinetics of the electrocatalysis process but also exhibit higher resistance to electrocatalyst deactivation caused by byproduct accumulation or electrode poisoning. This makes them particularly appealing for practical applications in energy generation and environmental cleanup [104]. By leveraging the unique advantages of bimetallic phosphorus systems, researchers can address the bottlenecks associated with mono-metal phosphates, paving the way for more efficient and sustainable UEOR technologies.

In 2019, Yang et al. synthesized nickel cobalt phosphate (NiCoPO) composites with varying Ni/Co ratios using a co-precipitation method, showcasing their potential as effective electrocatalysts for UEOR [105]. The introduction of cobalt content into nickel phosphate enhanced catalytic performance by regulating electronic levels and texture properties of Ni(III) species, lowering the E_{op}, and increasing the oxidation current. Compared to nickel phosphate (NiPO) without cobalt, the E_{op} of NiCoPO composites with specific Ni/Co ratios decreased significantly, with the NiCoPO (4:6) exhibiting a reduction of 135 mV. Furthermore, this composite achieved a high electrocatalytic current value of 65.4 mA cm^{-2} for UEOR at high pH. The improved activity was attributed to cobalt doping, which increased the chemical defects of Ni-contents, exposed new electroactive sites, regulated the Ni oxidation state, and promoted electron transfer from urea to NiOOH. However, excessive cobalt content (above 70%) hindered performance by decreasing the available active Ni sites due to inactive cobalt species. Electrochemical studies, including CV, amperometric measurements, and EIS confirmed the superior catalytic activity and kinetics of NiCoPO composites, particularly NiCoPO (4:6). The results highlight the potential of cobalt-doped nickel phosphates as cost-effective and efficient catalysts for UEOR [105].

Lulu Qiao et al. demonstrated that the Ni-Co phosphides/amorphous phosphorous combined with the oxide of Mn to form the introduced composite (c-CoNiPx/a-P-MnOy) exhibited exceptional performance as a bifunctional electrocatalyst for the UEOR [106], obtaining 100 mA cm^{-2} as an oxidation current at a remarkably low voltage of 1.35 V vs. RHE. This catalyst outperformed $CoNiP_x$/MnO and MnCoNiO by 3- and 41-fold, respectively, and showed improved kinetics with a Tafel slope of 75 mV dec^{-1} and an R_{ct} of 1.3 Ω. Long-term durability was confirmed with only an 8.4% increase in applied potential over 300 h at 50 mA cm^{-2}, and a mere 40 mV potential increase compared to the fresh electrode. Surface reconstruction during UEOR led to the formation of active NiOOH species from the Ni^{2+}/Ni^{3+} redox couple, with Mn and Co incorporation synergistically enhancing stability and activity by maintaining structural integrity and increasing active sites. The catalyst demonstrated an H_2 production rate of 0.18 mmol h^{-1}, a Faraday performance of 97.2% at 0.02 A cm^{-2}, and maintained promising electrochemical stability. The interplay of Mn and Co elements boosted the electrocatalytic activity, as the UEOR current density of CoNiP-MnO was 4.17-, 1.62-, and 3.44-fold higher than NiP, CoNiP, and MnNiP, respectively. XPS analysis revealed significant surface reconstruction, with a sharp decrease in P content from 19.23% to 3.86%, and enrichment of Ni^{3+} species, which served as key active sites for the indirect oxidation mechanism. Raman and in situ spectroscopic studies confirmed the structural transformation from metal phosphides to NiOOH and CoOOH, while the catalyst retained its hierarchical structure, emphasizing its robustness and efficiency. Lulu Qiao et al.'s work highlights the importance of crystalline/amorphous configurations and Mn/Co incorporation in optimizing bifunctional electrocatalysts for UEOR [106].

In 2023, Abd El-Lateef et al. presented a groundbreaking exploration of bimetallic Co/Ni phosphates (CoNiP-reflux and CoNiP-sol–gel) synthesized via reflux and sol–gel methods, respectively, as cost-effective electrode materials for UEOR in alkaline media, with a focus on their physicochemical properties and electrocatalytic performance [107]. Comprehensive characterization revealed that the sol–gel approach yielded smaller, more uniform particles with enhanced surface area compared to the reflux method, significantly influencing their electrocatalytic behavior. The CV analysis (Figure 5A,B) demonstrated that both CoNiP-reflux and CoNiP-sol–gel exhibited characteristic redox peaks at 350 mV and 430 mV, corresponding to the reversible Ni(II)/Ni(III) transformation, with CoNiP-sol–gel achieving superior current densities due to its higher surface area and improved charge transfer kinetics. The presence of urea further amplified the anodic current density, underscoring the efficiency of both materials, with CoNiP-sol–gel outperforming CoNiP-reflux significantly, achieving the current value of 18.90 mA/cm^2 at 0.46 V compared to 4.71 mA/cm^2 for CoNiP-Reflux under identical conditions.

EIS analysis as shown in Figure 5C,D indicated reduced R_{ct} in CoNiP-sol–gel electrodes, evidenced by the presence of semicircles in Nyquist plots. This could confirm enhanced UEOR efficiency, which could be attributed to the introduction of new charge transfer routes in the presence of urea. In the studied electrochemical cell, three resistances were identified: one associated with the electrolyte, another linked to Ni(II)/Ni(III) redox, and a last one connected to UEOR into N_2 and CO_2, producing six electrons. The last resistance appears only when the electrolyte has urea content, explaining the observed semicircles in the Nyquist plots for both CoNiP-reflux and CoNiP-sol–gel electrodes. Changes in electron transfer pathways were evaluated using equivalent circuit fitting analysis, as previously reported in related studies [72,108,109]. The disparity between experimental and fitted data was illustrated in Figure 5E,F for CoNiP-reflux and CoNiP-sol–gel, respectively. Equivalent circuits provide valuable insights into the electrical behavior of the electrodes by separating their electrical components and determining the effective resistance values from elements coupled in parallel or series. The circuit factors included R_h, which represents series resistance, while R_1/CPE_1 and R_2/CPE_2 correspond to oxidation reactions (Ni and urea, respectively). Notably, R_h went down from 17.36 Ω to 8.89 Ω when the electrode substrate was switched from CoNiP-reflux to CoNiP-sol–gel, indicating improved chemistry between the working electrode and the electrolyte for CoNiP-sol–gel. A similar trend was observed for the R_{ct} elements (R_1 and R_2), highlighting the enhanced electron transfer efficiency of the CoNiP-sol–gel electrode compared to CoNiP-reflux. CA further validated the stability and higher catalytic activity of CoNiP-sol–gel, with current densities reaching 39.98 mA/cm^2 after 1116 s in 0.3 mol/L urea solution, compared to 21.38 mA/cm^2 for CoNiP-reflux, and a deactivation rate of 1.4% for CoNiP-sol–gel, which could highlight its long-term operational stability. Additionally, CV studies across various urea concentrations (0.1–1.0 M) confirmed a direct correlation between urea content and enhanced current densities, with CoNiP-sol–gel maintaining higher peak currents and superior electrochemical performance. The study also examined the impact of scan rates on the CV profiles, showing consistent peak positions and increasing CV areas, indicative of diffusion-controlled processes and robust electrochemical stability. These findings collectively position CoNiP-sol–gel as a promising candidate for UEOR. The results pave the way for further optimization of bimetallic (Co and Ni) phosphate electrodes, leveraging sol–gel synthesis to unlock their full potential in fuel cell technologies [107].

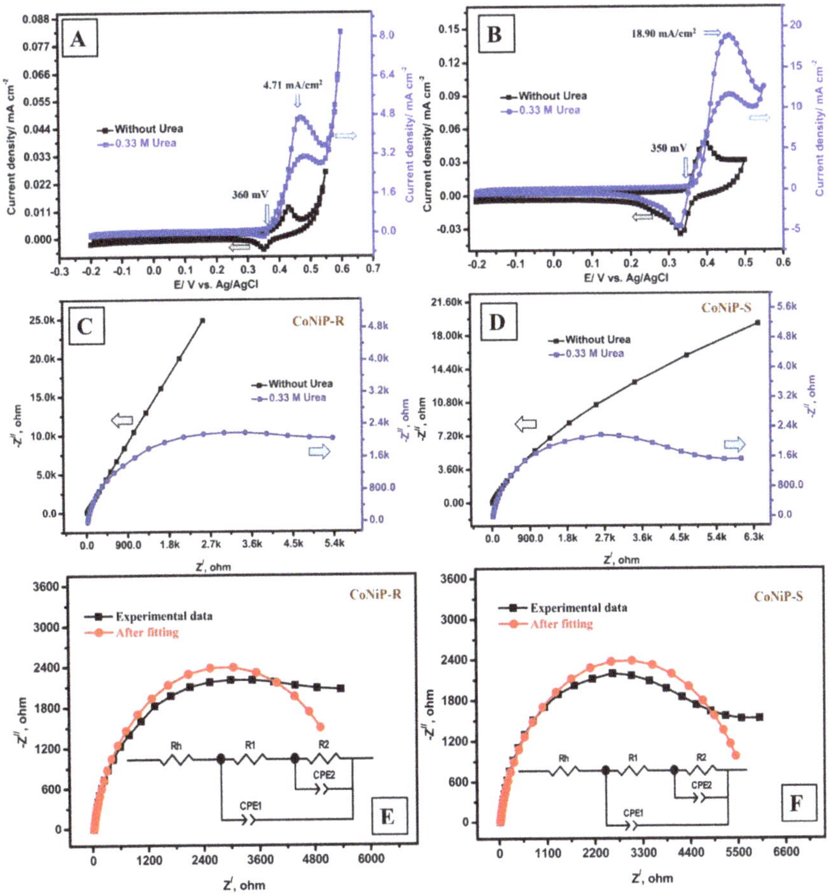

Figure 5. (**A**) CV of the investigated CoNiP-reflux (prepared via reflux) in KOH and/or 0.33 M urea. (**B**) CV of the CoNiP-sol–gel (prepared via sol–gel) substrate in KOH and/or 0.33 mol/L urea. (**C**) The EIS (Z′, −Z″) plots of CoNiP-Reflux. (**D**) EIS (Z′, −Z″) plots of CoNiP-sol–gel in 0.0 mol/L urea and 0.33 mol/L urea. (**E**) The verification of EIS fitting of CoNiP-reflux. (**F**), The verification of EIS fitting of CoNiP-sol–gel. The figure was reprinted after permission from Elsevier [107], Fuel, 334, 2023, 126671 and the license number 5926590385067.

Xiujuan Xu et al. investigated the interfacial charge redistribution of Fe-incorporated $Ni_{12}P_5/Ni_3P$ nanosheets supported on a 3D NiFe foam framework for ultrafast UEOR at industrially relevant current densities [110]. The incorporation of iron not only reinforced the interfacial electric field but also facilitated charge redistribution, effectively reducing the energy barrier during the UEOR. Benefiting from this tailored electronic structure and the ultra-thin nanosheet morphology, the catalyst exhibited remarkable performance and achieved 800 mA cm^{-2} as oxidation current at 1.4 V and an impressive Tafel slope (28.2 mV dec^{-1}). Electrochemical analysis revealed significantly reduced R_{ct} and increased EASA (C_{dl} of 9.36 mF cm^{-2}), which could indicate the efficient charge and mass transfer. Additionally, the Fe-doped electrocatalyst demonstrated excellent hydrophilicity, with a contact angle close to 0° with enhanced gas bubble release properties, mitigating bubble shielding effects and maintaining high UEOR activity. Long-term durability tests confirmed its robust performance, with minimum potential increases (1.1–4.9%) after continuous operation at high current densities for 40 h. Post-UEOR analysis revealed the formation of NiOOH on the electrode surface, further corroborating its high catalytic activity. DFT calculations highlighted the reduced reaction energy barrier on Fe-doped $Ni_{12}P_5$ (202) and Ni_3P (112) facets, elucidating the role of Fe in optimizing the adsorption/desorption of intermediates and boosting intrinsic activity. The turnover frequency of 2.3 s^{-1} is higher than the undoped counterpart (0.7 s^{-1}). Finally, the catalyst demonstrated promising performance in a real-world scenario using urine as a feedstock, which achieved both electrochemical activity and stability for UEOR [110].

Yunfeng Qiu et al. conducted a comprehensive study on the polyoxometalate-mediated design of phosphate-incorporated $NiMoO_{4-x}$ with abundant oxygen vacancies, aiming to develop efficient bifunctional electrocatalysts for hydrogen production via UEOR [80]. The proposed NF/P-$NiMoO_{4-x}$ electrode was synthesized by etching commercial NF with $H_3PMo_{12}O_{40}$, resulting in a phosphate content of 4.46 atomic% and net-like nanostructures. This facile method delivered exceptional HER and UEOR performances, with the electrode achieving an E_{op} of 116 mV at 10 mA/cm^2 for HER and a potential of 1.359 V at 10 mA/cm^2 for UEOR, accompanied by Tafel slopes of 77.5 and 19.3 mV/dec, respectively. The activation energy for UEOR was significantly reduced to 20.6 kJ/mol compared to 37.7 kJ/mol for bare NF, which highlights its enhanced electrocatalytic efficiency. The urea electrolysis setup demonstrated a cell potential of 1.48 V at 10 mA/cm^2, outperforming traditional noble-metal systems such as Ir/C | | Pt/C in terms of both efficiency and stability, with robust operation over 50 h. Remarkably, the etching solution was recyclable for five cycles with minimal loss in catalytic performance, which signifies its potential for large-scale application. Characterization revealed that the etching process optimized the surface area and active site availability, as evidenced by an enhanced BET surface area of 13.7 m^2/g, compared to 3.2 m^2/g for pristine NF, and the identification of mesopores around 3.8 nm. Electrochemical analyses confirmed reduced R_{ct} (37.6 Ω) and increased electrochemical active surface area (ECSA) values, with the electrode demonstrating superior intrinsic activity. The study also elucidated the active role of in situ-formed NiOOH species as the real electrocatalytic sites for UEOR, which facilitate a higher Ni^{3+}/Ni^{2+} ratio and abundant oxygen vacancies. Post-reaction characterization confirmed the structural integrity of the net-like nanostructures and sustained oxygen vacancy content, which ensure long-term catalytic durability. These findings establish NF/P-$NiMoO_{4-x}$ as a cost-effective, scalable, and promising bifunctional electrocatalyst for UEOR [80].

In 2023, Liming Yang et al. developed a feather-like NiCoP electrocatalyst combined with NF for UEOR [111]. The NiCoP/NF was synthesized via a two-step hydrothermal-phosphorization process followed by investigation for UEOR and the found results achieved a 10 mA cm^{-2} current in KOH containing 0.3 mol/L urea at a mere 1.13 V (vs. RHE). DFT calculations revealed that NiCoP's enhanced density of states (DOS) near the Fermi state significantly improves charge transfer and enables efficient catalytic activity. The trifunctional electrocatalyst demonstrated remarkable versatility, requiring a cell voltage of only 1.48 V to drive overall H_2O splitting at 10 mA cm^{-2}, significantly outperforming reference materials such as Co(OH)F/NF, NF, and RuO_2/NF, which required higher voltages. In UEOR, the NiCoP/NF achieved 10 mA cm^{-2} at just 1.36 V, 120 mV lower than the potential required for water splitting, highlighting its efficiency for urea-rich wastewater applications. The electrocatalyst showed excellent reaction kinetics, as evidenced by its low Tafel slope (34.19 mV dec^{-1}) and minimal R_{ct} = 0.67 Ω, coupled with a high double-layer capacitance (C_{dl} = 32.48 mF cm^{-2}), underscoring its superior electrocatalytic characteristics. Long-term stability tests over 30 h affirmed the robust durability of NiCoP/NF with minimal degradation in performance, while post-reaction SEM and XPS analyses revealed the retention of its morphology and conversion of phosphide to catalytically active metal oxo/hydroxides. Moreover, practical applications demonstrated the catalyst's effectiveness in overall water splitting and urea electrolysis using a 1.5 V battery or a solar photovoltaic panel as the power source, with visible hydrogen and oxygen bubbles generated at the electrodes. DFT calculations further elucidated the mechanism underlying its high performance, showing strong adsorption energies for H_2O on the NiCoP (0001) surface, ensuring effective water activation. This study highlights NiCoP/NF as a low-cost, highly active, and stable electrocatalyst, with promising applications in UEOR over noble-metal-free catalysts [111].

In 2024, Ran Zhang et al. investigated the coral-like Co/Zn-Pi@NF (Pi and NF represent phosphate and nickel foam, respectively) electrocatalyst for UEOR [112]. The introduced Co/Zn-Pi@NF was chemically designed via electrodeposition methodology and directly grown on NF to avoid binders and enhance conductivity [113]. The structural and electronic synergy between cobalt and zinc facilitated electrocatalytic performance improvements. Incorporating zinc adjusted the electronic structure of the catalyst. The unique coral-like architecture of Co/Zn-Pi@NF exposed more electroactive sites and better charge transfer efficiency. In the presence of urea (0.5 mol/L), the UEOR performance of Co/Zn-Pi@NF was remarkable, requiring a voltage of 1.40 V while maintaining electrochemical stability over 60 h. The catalyst outperformed comparable materials such as Co-Pi@NF, Zn-Pi@NF, and Co/Zn-OH@NF in terms of lower overpotentials and faster reaction kinetics with the lowest Tafel slope (95.8 mV·dec^{-1}) and the highest exchange current density (j_0) of 7.14 μA·cm^{-2}. EIS revealed the lowest R_{ct} of 0.36 Ω for Co/Zn-Pi@NF, underscoring its superior conductivity. Additionally, the catalyst exhibited the highest C_{dl} of 228.91 mF·cm^{-2}, which indicates a larger EASA. During water-urea electrolysis, employing Co/Zn-Pi@NF as cathode and anode achieved a cell voltage of 1.43 V at 10 mA·cm^{-2}, a reduction of 290 mV compared to H_2O electrolysis. The electrocatalyst demonstrated robust stability under continuous operation at 50 mA·cm^{-2} for 60 h, with only a 1.36% change in voltage, which affirms its potential for practical applications. The study highlights the ability of biphosphate structures (Co/Zn-Pi@NF) to serve as electrocatalysts for UEOR [112].

In 2024, Zhanhong Zhao et al. investigated Mn incorporation on nickel phosphide (Ni_2P) with a self-reconstruction mechanism for the UEOR [114]. By constructing Mn-Ni_2P, the study revealed that manganese doping significantly accelerates surface reconstruction and forms a NiOOH-MnOOH heterostructure characterized by localized electrophilic and nucleophilic regions. This structure was shown to enhance the targeted adsorption and activation of carbon and nitrogen atoms in the fuel, thereby facilitating the cleavage of the C–N bond. Moderate manganese doping not only increased urea adsorption capacity and the affinity for key intermediates but also reduced the energy barrier for gas desorption, which could effectively refresh the electrocatalytic surface. As a result, the Mn-Ni_2P electrode demonstrated promising electrocatalytic performance and achieved an industrial-level current of 1000 mA cm^{-2} at a voltage of 1.46 V (vs. RHE), which is 269 mV lower than the OER under similar environments. Linear-sweep voltammetry (LSV) results (shown in Figure 6A) highlighted that Mn-Ni_2P required only 1.316 V to reach 100 mA cm^{-2}. A significant improvement compared to Ni_2P (Figure 6A), and commercial Pt/C (Figure 6B) could be seen, which required 1.363 V and 1.405 V, respectively. The incorporation of Mn also notably reduced the UEOR E_{op} from 1.334 V (Ni_2P) to 1.27 V (Mn-Ni_2P). Additionally, Tafel slope analysis (Figure 6C) revealed that Mn-Ni_2P possessed a much lower slope (29.61 mV dec^{-1}) compared to Ni_2P (37.46 mV dec^{-1}), Pt/C

(52.12 mV dec^{-1}), and NF (94.72 mV dec^{-1}), indicating significantly enhanced UEOR kinetics. Variations in urea and KOH concentrations were also studied, with results suggesting a shift from diffusion-controlled processes to surface reaction-controlled processes as the urea content exceeded 0.3 M. EIS of the studied Mn-Ni_2P, shown in Figure 6D, further corroborated the enhanced kinetics, showing a pronounced reduction in R_{ct} with increasing applied potential. Mn-Ni_2P exhibited a lower R_{ct} at 1.28 V (vs. RHE) compared to Ni_2P (1.35 V) and confirmed preferential UEOR initiation on Mn-Ni_2P. The C_{dl} measurements (Figure 6E) indicated a higher electrochemical active surface area (ECSA) for Mn-Ni_2P (14.37 mF/cm^2) than Ni_2P (8.57 mF/cm^2). Additionally, the C_{dl} in the case of the Mn-Ni_2P electrode was found at 14.37 mF/cm^2, which is much better than that of the Ni_2P electrode (8.57 mF cm^{-2}), which could suggest the better ESCA for the Mn-Ni_2P electrode. Structural stability studies confirmed that the Mn-Ni_2P electrocatalyst maintained its nanosheet morphology and chemical composition after prolonged operation, with manganese doping effectively mitigating phosphorus leaching and enhancing structural integrity. XPS analysis further revealed the formation of amorphous nickel/manganese oxyhydroxides after extended electrolysis, providing additional evidence for the stability and activity retention of Mn-Ni_2P. The electrocatalyst durability was evaluated via electrolysis experiments as displayed in Figure 3F. The persistent decomposition of urea combined with CO_2 could retard the surface electrocatalytic reaction that leads to the increase in the required potential. The Mn-Ni_2P working electrode has relatively stability for 600 min at 100 mA/cm^2. These findings collectively underscore the position of Mn-Ni_2P as a robust and efficient electrocatalyst for UEOR [114].

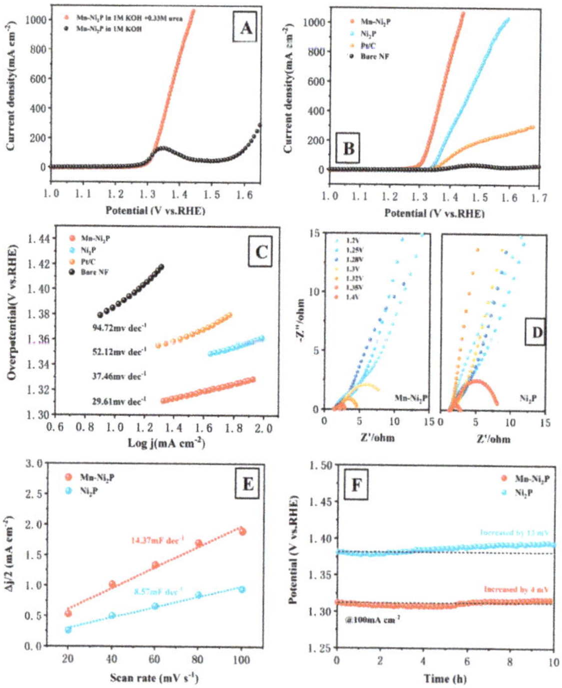

Figure 6. (**A**) LSV of the Mn-Ni_2P and Ni_2P electrodes. (**B**) UEOR polarization curve of Ni_2P, Mn-Ni_2P, Pt/C, and bare NF electrocatalyst. (**C**) Tafel plots for the UEOR. (**D**) Nyquist plots of Ni_2P and Mn-Ni_2P electrodes at variable applied voltages in KOH with 0.3 mol/L urea. (**E**) Linear plots of capacitive current against the applied scan rate for Ni_2P and Mn-Ni_2P. (**F**) Chronopotentiometry measurements of Ni_2P and Mn-Ni_2P for UEOR. The figure was reprinted after permission from Elsevier [114], Water Research, 253, 2024, 121266 and the license number 5926590587949.

7. Heterogeneous Phosphate (Carbon and Metal) Electrocatalysts for UEOR

Heterogeneous phosphate-based materials have emerged as attractive candidates for electrocatalysis applications owing to their excellent stability, tunable electronic properties, and high catalytic activity [2]. For UEOR, these materials provide active sites conducive to the redox transformations of $Ni(OH)_2/NiOOH$ and similar systems, which are pivotal for catalyzing the multi-step oxidation of urea [107]. However, the inherent limitations of phosphate-based materials, such as moderate conductivity and limited surface area, can hinder their overall catalytic performance. Incorporating carbon-based components such as multi-walled carbon nanotubes (MWCNTs) [115], graphene foam [116,117], or activated carbon [118] into phosphate materials offers a synergistic solution to these challenges. Carbonaceous materials could enhance the conductivity and surface area of the heterogeneous composite, facilitating efficient charge transfer and providing additional active sites for the adsorption and oxidation of urea. Furthermore, their robust structural framework ensures mechanical stability and prolonged catalyst life under operational conditions. By integrating heterogeneous phosphate materials with conductive carbonaceous supports, it is possible to significantly lower the E_{op}, improve the electrocatalytic current density, and improve the overall kinetics of UEOR. This hybrid approach not only optimizes catalytic efficiency but also broadens the applicability of these materials in practical energy conversion and wastewater management systems.

Hongmei Zhang et al. studied the UEOR on nickel phosphate-based nanomaterials, including MWCNTs combined with $Ni_3(PO_4)_2$ nanomaterials, which were physically designed via a simple precipitation method [119]. Electrochemical performance was evaluated using CV in alkaline solutions and revealed acceptable catalytic performance for UEOR. The addition of MWCNTs significantly enhanced catalytic current and reduced E_{op} due to their large specific surface area and excellent electrical conductivity. CV studies demonstrated redox behavior attributed to the $Ni(OH)_2/NiOOH$ transformation, with higher current densities observed for NiPO-MWCNT modified electrodes compared to their counterparts without MWCNTs. In the presence of urea, anodic peak currents increased, and peak regions expanded due to overlapping peaks of nickel hydroxide and fuel electrooxidation. Electrochemical reactions involved the oxidation of $Ni(OH)_2$ to NiOOH, then the electrocatalytic oxidation of the fuel, with subsequent regeneration of $Ni(OH)_2$, indicating a quasi-reversible and diffusion-controlled process. Investigations of varying scan rates revealed linear characteristics between current values at the anodic peak and the (scan rate)$^{0.5}$, which could confirm diffusion-controlled reactions. Urea oxidation performance was improved, with increasing urea concentration up to an optimal $OH^-/CO(NH_2)_2$ ratio of 6, beyond which excessive urea negatively impacted performance due to hydrolysis effects, electrolyte depletion, and carbon dioxide accumulation on the electrode surface. Similarly, NaOH concentration influenced urea oxidation, with an increase in peak current and shift in E_{op} to more negative values up to 1.5 M NaOH, after which further increases hindered performance. The study highlights the potential of NiPO-MWCNT electrodes as efficient electrocatalysts for UEOR due to their enhanced electrochemical properties and well-optimized reaction conditions.

In 2022, the innovative synthesis of NiF_3/Ni_2P heterojunctions supported on carbon cloth (NiF_3/Ni_2P@CC) was achieved via a combination of hydrothermal, fluoridation, and phosphorization techniques [120]. Among these, the NiF_3/Ni_2P@CC variant emerges as a standout, offering significant improvements in UEOR performance due to its unique physicochemical properties. This electrode includes abundant high-valence Ni^{3+} sites, optimized charge density, a super hydrophilic surface, and a mesoporous structure, which collectively enhance catalytic activity. Electrochemical evaluations demonstrate that NiF_3/Ni_2P@CC requires a mere voltage of 1.36 V to produce 10 mA cm^{-2} and maintains its activity across extensive cycling, underscoring its durability. Comparative studies reveal that the heterojunction configuration significantly outperforms individual components, NiF_3@CC and Ni_2P@CC, in terms of catalytic efficiency and stability. The superior performance is attributed to the optimized heterointerface density achieved by tuning phosphorization time, which influences the structural and electronic properties of the material. Structural analyses, including XRD and TEM, confirm the coexistence of rhombohedral NiF_3 and hexagonal Ni_2P phases, while surface characterization highlights the remarkable hydrophilicity of the NiF_3/Ni_2P@CC surface, as evidenced by a contact angle of only 11.5°. EIS and Tafel slope analysis further elucidate the rapid charge transfer and favorable reaction kinetics of NiF_3/Ni_2P@CC, with 33 mV dec^{-1} as a Tafel slope indicating a fast UEOR dynamic process. As illustrated in Figure 7A, the NiF_3/Ni_2P@CC electrode demonstrates the smallest semicircle radius at high frequencies, indicative of its low R_{ct}. This reduced R_{ct} can be attributed to the material's abundant heterointerfaces and the conductivity of Ni_2P. The electrochemical surface area, which correlates directly with the C_{dl}, was also assessed. Figure 7B presents the relationship between current density change ($\Delta j = j_a - j_c$ at 0.95 V) and scan rates ranging from 0.02 to 0.12 V/s within the non-Faradaic region. The corresponding C_{dl} was estimated by plotting the slopes of the linear relationship. Notably, NiF_3/Ni_2P@CC exhibited the highest C_{dl} value of 11.7 mF cm^{-2}, which is approximately 10 times greater than those of NiF_3@CC (1.05 mF/cm^2) and Ni_2P@CC (1.14 mF/cm^2). This substantial increase in C_{dl} reflects a significantly larger ECSA, providing a higher density of electroactive sites for the UEOR. A symmetric urea electrolysis system was designed using NiF_3/Ni_2P@CC as the anode as well as the cathode (Figure 7C). As described in Figure 7D, the electrolyzer required a low cell potential of 1.54 V to achieve 10 mA cm^{-2} in KOH aqueous solution containing 0.3 mol/L urea as the fuel content, compared to 1.58 V in KOH without the fuel. The reduced voltage of 1.54 V for 10 mA cm^{-2} surpasses other reported urea electrolyzers employing noble or non-noble electrodes (Figure 7E) [121,122]. Additionally, achieving 50 mA cm^{-2} required 1.83 V voltage in the urea-containing electrolyte, which is significantly lower than the 1.92 V required for H_2O electrolysis. Furthermore, the symmetric urea electrolyzer demonstrated exceptional long-term operational stability and maintained continuous operation for 10 h without degradation (Figure 7F). Additionally, long-term stability assessments reveal minimal morphological and structural changes post-operation, even under extended working conditions, marking this electrocatalyst as robust for practical applications. This research not only establishes NiF_3/Ni_2P@CC as a leading electrocatalyst for UEOR but also sets a precedent for designing multifunctional heterojunction catalysts, with broader implications for scalable and economical energy production technologies. The systematic optimization of synthesis parameters, such as fluorination–phosphorization duration and precursor composition, provides valuable insights into tailoring material and electrocatalytic properties [120].

Xu et al.'s study on phosphorus-induced bottom-up growth of Fe-doped $Ni_{12}P_5$ nanorod arrays for the UEOR presents a significant advancement in the synthesis of self-growing heterogeneous catalysts with robust and regular morphologies. Using 1,1′-bis(diphenylphosphine) ferrocene (DPPF) as both the phosphorus and iron source, the catalysts were synthesized on nitrogen-doped carbon-coated nickel foam (NdC/NF) through high-temperature vapor deposition. This process facilitated the formation of straight Fe-doped $Ni_{12}P_5$ nanorods, with lengths extending to tens of microns, growing via a bottom-up mechanism. DFT calculations suggested that the carburized Ni (111) face on the NF substrate favored phosphorus coordination, enabling uniform incorporation of Fe into the $Ni_{12}P_5$ lattice to form a synergistic catalytic center. An electrochemical evaluation revealed that Fe-$Ni_{12}P_5$/NdC/NF-800 exhibited promising performance in a 1.0 mol/L aqueous KOH + 0.3 mol/L urea solution, achieving 50 and 300 mA cm^{-2} at low potentials of 1.357 V and 1.463 V, respectively. The catalyst demonstrated superior kinetics, with 56.9 mV/dec as a Tafel slope, a maximum C_{dl} of 108.1 mF cm^{-2}, and minimal R_{ct}, reflecting promising conductivity and a large ECSA. Stability tests confirmed its robustness, maintaining 100 mA cm^{-2} for over 12 h with minimal structural degradation. Post-reaction analyses via SEM, XPS, and HRTEM showed the rod morphology remained intact, with surface transformations forming active NiOOH and FeOOH layers, further enhancing electron transfer and providing efficient electroactive sites for urea adsorption and oxidation. Raman spectroscopy indicated increased graphitization post-stability test, with the I_D/I_G

value decreasing from 1.19 to 0.51, signifying improved catalyst durability. Comparative studies underscored the critical role of DPPF loading at 800 °C, as Fe-$Ni_{12}P_5$/NdC/NF-800 outperformed other configurations, demonstrating selective UEOR over OER with a voltage of 1.371 V required for 100 mA/cm^2. The study highlights the synergistic contributions of the NF substrate, N-doped carbon layer, and DPPF, which emphasize the phosphorus-induced self-growth mechanism as a promising approach to designing high-performance, durable electrocatalysts with promising applications in UEOR [123].

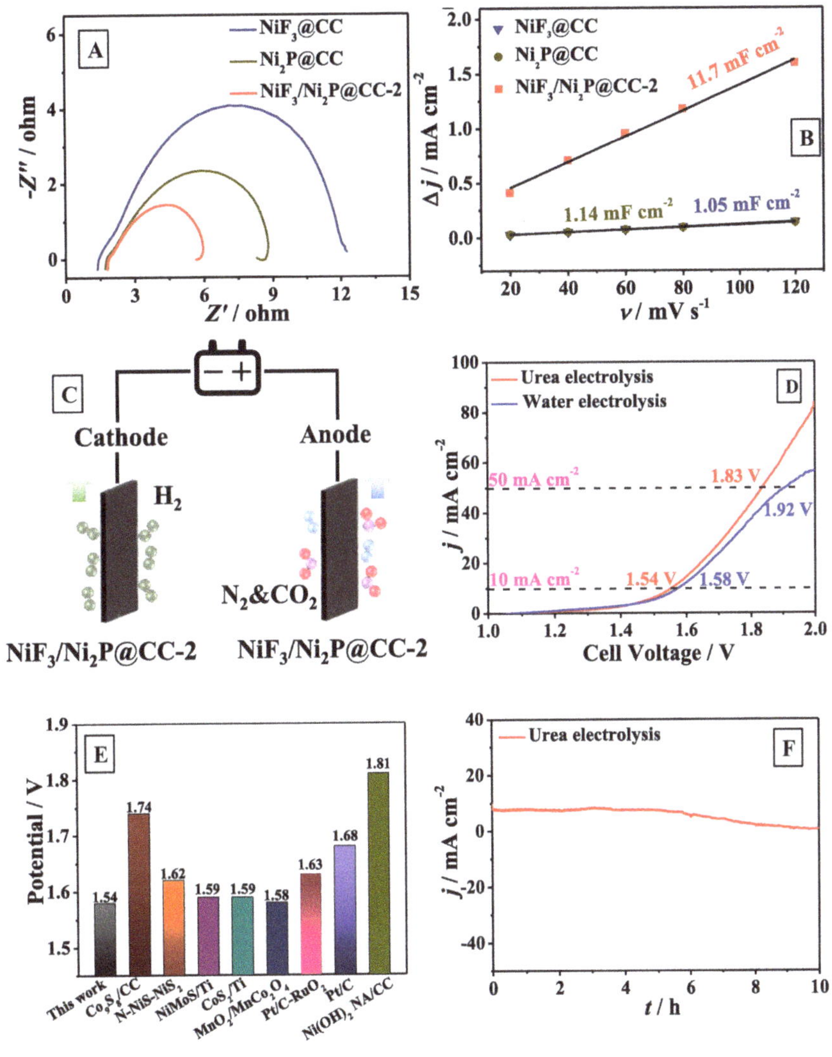

Figure 7. (**A**) Nyquist plots, (**B**) double-layer capacitance study via Δj against scan rates v, (**C**) scheme of urea electrolysis using NiF_3/Ni_2P@CC, (**D**) LSV curves of NiF_3/Ni_2P@CC for overall urea and H_2O electrolysis, (**E**) comparison of voltages at 10 mA cm^{-2} of reported bi-functional electrocatalysts in urea overall splitting, and (**F**) CA curve of NiF_3/Ni_2P@CC in KOH with 0.3 mol/L urea for 600 min. The figure was reprinted after permission from Elsevier [120], Chemical Engineering Journal, 427, 2022, 130865" and the license number 5926590822951.

Finally, Table 1 summarizes various nickel-based phosphates/phosphides as phosphorous-based materials and their performance in UEOR. The materials range from simple nickel phosphates to complex composites involving other metals like cobalt and iron, with distinct crystalline structures including monoclinic, hexagonal, and orthorhombic phases. For example, Ni-phosphate in the monoclinic form exhibits nanotube morphology with a performance of I_a around $600\,mAcm^{-2}mg^{-1}$ and a low R_{ct} $10\,\Omega$ [81]. Other compounds such as Ni_2P@ Ni-foam displayed porous nanoflower structures with promising electrochemical activity (I_a = $750\,mA/cm^2$) and low overpotentials (E_{op} = 0.24 V) [83]. The Ni/Co phosphate demonstrates nanosheet morphology and an impressive I_a = $65.4\,mAcm^{-2}$ at E_{op} = 0.22 V [105], while Ni_2P/Ni features hexagonal Ni_2P and cubic Ni phases with porous rods showing electrochemical activity at low overpotential [82]. The variety of crystalline and morphology correlates to electrochemical performance, with the materials exhibiting promising electrocatalytic properties and potential for energy conversion via UEOR. Additionally, the table highlights the role of dopants (e.g., Mn, Co) and composite structures in enhancing the electrochemical behavior of these materials.

Table 1. A summary of the reported electrocatalytic performance (I_a = anodic current density, E_{op} = onset potential, and R_{ct} = charge-transfer resistance) of the recent literature that introduced phosphorus (phosphate or phosphine) electrocatalysts for UEOR.

No.	Chemistry	Crystallinity	Morphology	Performance	Reference
1	Ni-phosphate	Monoclinic	less-ordered nanotubes of different lengths	I_a = 600 $mAcm^{-2}mg^{-1}$ R_{ct} − 10 Ω	[81]
2	Ni_2P@Nifoam	Hexagonal P-62m	porous nanoflower supported on foam	I_a = 750 $mAcm^2$ with E_{op} = 0.24 V	[83]
3	Ni/Co phosphate	Orthorhombic	nanosheet structure with macropores	I_a = 65.4 mA/cm^2 E_{op} = 0.22 V	[105]
4	nickel (II) phosphate octahydrate	Monoclinic	nanoporous microspheres	I_a = 680 $mA/(cm^2{\cdot}mg)$ & E_{op} = 0.33 V	[84]
5	MWCNTs/Ni-phosphate	---	carbonaceous/ nanoparticles composite	I_a = 9 mA	[119]
6	Ni-phosphate @$Ni_{12}P_5$	Tetragonal I4/m	nanorod arrays	900 $mAcm^{-2}$ at potential of 1.378 V	[85]
7	Ni-phosphate	Hexagonal P63/m	nanorods 80 nm	160 mA/cm^2 at 0.6 V E_{op} = 0.27 V, and R_{ct} = 5.77 Ω	[86]
8	Ni-phosphate	Monoclinic I2/m	nanoflakes aligned in one direction	I_a = 57.45 mA/cm^2 E_{op} = 0.345 V, and R_{ct} = 56.03 Ω	[87]
9	Ni_2P/Ni	Hexagonal Ni_2P & Cubic Ni	porous rods	E_{op} = 0.05 V at 10 $mA{\cdot}cm^{-2}$ and Tafel slope = 87.6 mV/dec	[82]
10	Ni-Phosphate/Ni-foam	---	aggregated nanoparticles with interstitial mesoporous networks	20 and 500 mA/cm^2 at 1.35 and 1.6 V	[91]

Table 1. *Cont.*

No.	Chemistry	Crystallinity	Morphology	Performance	Reference
11	NiF_3/Ni_2P@ carbon cloth	Rhombohedral NiF_3 & hexagonal Ni_2P	irregular nanoparticles with lots of mesopores	I_a = 79.1 mA cm^2 E_{op} = 1.34 V	[120]
12	$CoNiP_x/P\text{-}MnO_y$	Hexagonal P321 &	hierarchical wire-into-flake structure	10 mA cm^{-2} at 1.24 V Faraday efficiency = 97.2%	[106]
13	$Ni_3(PO_4)_2$	Monoclinic P21/a	"berry fruit" like morphology with 25 nm particles	I_a = 135.6 mA/cm^2 E_{op} = 1.34 V	[88]
14	Cu-Phosphide	Hexagonal P63cm	nanowires	Voltage of 1.79 V to achieve 100 $mAcm^{-2}$	[95]
15	Co/Ni-phosphate	monoclinic C2/m	self-assembled flower-like morphology with some distortion	I_a = 105.1 $mAcm^{-2}$ at 0.485 V	[107]
16	$Fe\text{-}Ni_{12}P_5$	Hexagonal	nanorods	1.371 V to get 100 mA cm^{-2}	[123]
17	$Fe\text{-}Ni_{12}P_5/Ni_3P$	Hexagonal	ultrathin nanosheets	800 mA cm^2 at 1.4 V Tafel slope = 28.2 $mV \cdot dec^{-1}$	[110]
18	$NiMoO_{4\text{-}x}$ with O-vacancies	Monoclinic C2/m	net-like nanostructure	1.359 V at 10 mA/cm^2 and Tafel slope = 19.3 mV/dec	[80]
19	Co/Ni-Phosphate/Ni-foam	Hexagonal P6m2	feather-like morphology arranged vertically on the foam	1.13 V to get 10 mA/cm^2 and R_{ct} = 0.67 Ω	[111]
20	Zn/Ni-Phosphate/Ni-foam	Hexagonal	spherical-like flower-shaped nanosheets	1.43 V to get 10 mA/cm^2 and R_{ct} = 0.67 Ω	[112]
21	Mn/Ni-Phosphate	Hexagonal P-62m with Mn-doping	flower-like structure, assembled by smooth interconnecting nanoflakes	1000 mA/cm^2 at 1.46 V	[114]

8. Challenges and Optimization Strategies for Phosphorous-Based Catalysts

Phosphorous-based catalysts such as Ni-phosphates, Ni-free phosphates, NiM phosphates, and heterogeneous phosphate/carbon composites exhibit unique structural properties, electronic tunability, and electrocatalytic stability, making them promising materials for the urea electrocatalytic oxidation reaction (UEOR). However, their practical application is impeded by persistent challenges. Catalyst poisoning by reaction intermediates such as carbonate and urea-derived species significantly hampers catalytic activity and stability. Additionally, achieving optimal catalytic efficiency and selectivity remains difficult due to suboptimal active site exposure, sluggish reaction kinetics, and insufficient electronic conductivity. Addressing these issues requires advanced strategies such as alloying nickel with transition metals like Co, Fe, or Mn, which not only enhance the catalytic activity but also reduce E_{op} by modifying the electronic structure. Incorporation of non-metals, including S or Se, into the phosphorous framework can further alter electronic characteristics and promote synergistic effects, leading to promising solutions for enhanced reaction kinetics. The use of conductive supports such as graphene, carbon nanotubes, or hierarchical porous carbons has demonstrated efficacy in enhancing electron transfer and increasing the specific surface area, thereby improving electrocatalytic performance. Protective coatings, including ultrathin graphene layers, polymer films, or oxide shells, could serve as effective barriers to corrosion and surface poisoning while maintaining active site accessibility. Furthermore, optimizing electrolyte formulations, such as introducing ionic liquids or alkaline media with tailored additives, can suppress undesired side reactions and prolong the catalyst's lifetime. Advanced synthetic techniques such as solvothermal processes, electrodeposition, and in situ growth of catalysts directly onto conductive substrates are pivotal for fabricating highly active, stable, and scalable phosphorous-based electrocatalysts.

Improving selectivity to minimize unwanted by-products like carbonates during urea oxidation involves optimizing both catalyst composition and reaction conditions, which can include four main strategies. The first one is catalyst surface modification, such as introducing bifunctional sites or heteroatom doping (e.g., transition metals like Fe, Ni, or Co) to selectively enhance the breaking of specific urea bonds, suppressing the formation of carbonates. The second one is the tuning electronic structure. Phosphorous-based catalysts with controlled electronic characteristics can facilitate selective oxidation pathways by stabilizing desired intermediates and destabilizing carbonate precursors. The third one is the reaction environment. Operating under optimized pH and temperature conditions reduces carbonate formation. Additionally, maintaining a balance between hydroxyl and urea concentrations suppresses side reactions. The last one is the additives. Introducing co-catalysts or inhibitors that block carbonate-forming pathways without impacting urea oxidation efficiency can further improve selectivity.

Catalyst poisoning by intermediates such as cyanate and ammonia during prolonged urea oxidation is a significant challenge, and addressing this issue requires innovative strategies. Surface regeneration techniques, such as periodic anodic or cathodic cleaning, can effectively remove adsorbed intermediates and restore active sites. Additionally, dynamic surface engineering approaches, such as employing self-regenerating catalysts with surface reconstruction properties, ensure that fresh active sites are continually exposed during the reaction. Introducing promoter elements, such as cerium or manganese, into phosphorous-based catalysts could show promise in enhancing resistance to poisoning by destabilizing adsorbed intermediates. Furthermore, adjusting reaction conditions, including fine-tuning KOH concentrations and applied potentials, can minimize the formation of poisoning intermediates and extend the catalyst's operational lifespan.

Scaling up phosphorous-based catalysts from laboratory to industrial applications poses several challenges. The first one is electrode/catalyst costs. Even though phosphorous-based materials like phosphates and phosphides are relatively abundant, the cost of high-purity precursors and synthesis methods (e.g., high-temperature annealing or chemical vapor deposition) can be prohibitive at scale. Developing cost-effective synthesis routes, such as sol–gel or hydrothermal methods, could address this issue. The second one is sustainability, as energy consumption and waste generation should be minimized. Utilizing green synthesis methods, recycling precursor materials, and integrating renewable energy sources for catalyst production could enhance sustainability. The third one is the scalability of synthesis, which could be achieved by uniformity in catalyst properties (e.g., particle size, surface area, and active sites) during large-scale production. Strategies like continuous flow reactors and scalable templating methods are promising solutions. The last one is the integration with existing systems. Compatibility with current wastewater treatment and energy conversion infrastructures requires extensive pilot studies to optimize operating conditions and ensure seamless integration.

The environmental risks associated with phosphorous-based catalysts, including heavy metal leaching and the generation of toxic by-products, have not been completely addressed in previous studies, and comprehensive evaluations remain necessary. Leaching tests, such as the toxicity characteristic leaching procedure (TCLP), have demonstrated minimal release of heavy metals from well-anchored catalysts under controlled conditions. Analytical techniques like GC-MS and ICP-MS could be employed to detect and quantify potential toxic by-products in treated wastewater, ensuring thorough monitoring of environmental impact. Risk mitigation strategies, including the encapsulation of active sites within inert matrices or the application of protective coatings, further minimize the potential for leaching and enhance environmental safety. Additionally, lifecycle assessments evaluating the environmental impact of catalyst production, usage, and disposal provide a holistic approach to risk management, guiding the development of safer and more sustainable catalytic systems.

9. Conclusions and Future Perspectives

Phosphorous-based catalysts represent a significant advancement in the field of urea electrocatalytic oxidation reaction (UEOR), offering a sustainable solution for wastewater remediation and energy conversion. Their high stability, structural versatility, and environmental compatibility render them promising candidates for future applications. However, the realization of their full potential demands a deeper understanding of their reaction mechanisms, degradation pathways, and structure-activity relationships. In situ and operando techniques, such as X-ray photoelectron spectroscopy (XPS), synchrotron-based X-ray absorption spectroscopy (XAS), and electrochemical impedance spectroscopy (EIS), are crucial for elucidating dynamic surface transformations and intermediate species during the catalytic process. Future research should prioritize the design of hybrid materials that integrate phosphorous-based electrocatalysts with metal oxides, carbides, nitrides, or sulfides to exploit cooperative effects that enhance activity, selectivity, and durability. Scaling up from lab-scale studies to industrially relevant systems necessitates the development of robust reactor designs, optimization of operational parameters, and integration of UEOR technologies into existing renewable energy infrastructures. Incorporating computational methods such as DFT calculations can further accelerate the discovery of novel catalytic materials and guide experimental design. Addressing these challenges through interdisciplinary efforts will establish phosphorous-based electrocatalysts as indispensable components in the global transition toward sustainable energy via UEOR.

Author Contributions: Conceptualization, I.M.A.M., M.M.K. and H.M.A.E.-L.; software, I.M.A.M.; writing—original draft preparation, I.M.A.M., M.M.K. and H.M.A.E.-L.; writing—review and editing, I.M.A.M., M.M.K. and H.M.A.E.-L.; visualization, I.M.A.M., M.M.K. and H.M.A.E.-L.; supervision, I.M.A.M., M.M.K. and H.M.A.E.-L.; project administration, H.M.A.E.-L.; funding acquisition, H.M.A.E.-L. All authors have read and agreed to the published version of the manuscript.

Funding: This work was funded by the Deanship of Scientific Research, Vice Presidency for Graduate Studies and Scientific Research, King Faisal University, Saudi Arabia [Grant No. KFU242725].

Data Availability Statement: Data is contained within the article.

Conflicts of Interest: The authors declare no conflicts of interest.

References

1. Mitchell, S.; Qin, R.; Zheng, N.; Pérez-Ramírez, J. Nanoscale engineering of catalytic materials for sustainable technologies. *Nat. Nanotechnol.* **2021**, *16*, 129–139. [CrossRef] [PubMed]
2. Yang, S.; Yue, K.; Liu, X.; Li, S.; Zheng, H.; Yan, Y.; Cao, R.; Zhang, W. Electrocatalytic water oxidation with manganese phosphates. *Nat. Commun.* **2024**, *15*, 1410. [CrossRef] [PubMed]
3. Raza, S.; Hayat, A.; Bashir, T.; Chen, C.; Shen, L.; Orooji, Y.; Lin, H. Electrochemistry of 2D-materials for the remediation of environmental pollutants and alternative energy storage/conversion materials and devices, a comprehensive review. *Sustain. Mater. Technol.* **2024**, *40*, e00963. [CrossRef]
4. Li, J.; Zhang, J.; Yang, J.-H. Research progress and applications of nickel-based catalysts for electrooxidation of urea. *Int. J. Hydrogen Energy* **2022**, *47*, 7693–7712. [CrossRef]
5. Sun, W.; Zhang, M.; Li, J.; Peng, C. Solar—Driven Catalytic Urea Oxidation for Environmental Remediation and Energy Recovery. *ChemSusChem* **2022**, *15*, e202201263. [CrossRef] [PubMed]
6. Liu, Y.; McCue, A.J.; Li, D. Metal phosphides and sulfides in heterogeneous catalysis: Electronic and geometric effects. *ACS Catal.* **2021**, *11*, 9102–9127. [CrossRef]
7. Zhao, T.; Shen, X.; Wang, Y.; Hocking, R.K.; Li, Y.; Rong, C.; Dastafkan, K.; Su, Z.; Zhao, C. In situ reconstruction of V-doped Ni2P pre-catalysts with tunable electronic structures for water oxidation. *Adv. Funct. Mater.* **2021**, *31*, 2100614. [CrossRef]
8. Fu, Z.; Liu, C.; Bai, M.; Ma, J.; Zhao, P.; Gao, Y.; Zhao, M.; He, Y.; Xiao, H.; Jia, J. Black phosphorus-based nanocomposites and transition metal phosphates for electrocatalytic oxidation of small molecules. *Mol. Catal.* **2024**, *558*, 114023. [CrossRef]

9. Lee, H.; Theerthagiri, J.; Kumari, M.A.; Min, A.; Moon, C.J.; Anbazhagan, V.; Brutchey, R.L.; Choi, M.Y. Leveraging phosphate group in Pd/PdO decorated nickel phosphate microflowers via pulsed laser for robust hydrogen production in hydrazine-assisted electrolyzer. *Int. J. Hydrogen Energy* **2024**, *57*, 176–186. [CrossRef]
10. Zabihinezhad, M.; Shahrabi, T.; Zheng, C.; Shao, T.; Darband, G.B.; Li, J. Pulse-reverse electrodeposited reduced graphene oxide electrode decorated by Ni-P porous nano-layer as a high performance electrocatalyst for H_2 production assisted by hydrazine electrooxidation. *Appl. Catal. A Gen.* **2023**, *666*, 119415. [CrossRef]
11. Yang, X.; Liang, Z.; Chen, S.; Ma, M.; Wang, Q.; Tong, X.; Zhang, Q.; Ye, J.; Gu, L.; Yang, N. A Phosphorus-Doped Ag@Pd Catalyst for Enhanced C—C Bond Cleavage during Ethanol Electrooxidation. *Small* **2020**, *16*, 2004727. [CrossRef] [PubMed]
12. Yang, D.; Li, J.; Li, W.; Jiao, Y.-Z.; Yang, J.-H. The ultrathin regular circular structural Ni-P nanosheet for efficient urea electrooxidation. *J. Environ. Chem. Eng.* **2024**, *12*, 112433. [CrossRef]
13. Feng, X.; Li, G. Versatile phosphate diester-based flame retardant vitrimers via catalyst-free mixed transesterification. *ACS Appl. Mater. Interfaces* **2020**, *12*, 57486–57496. [CrossRef] [PubMed]
14. Zhao, H.; Yuan, Z.Y. Insights into transition metal phosphate materials for efficient electrocatalysis. *ChemCatChem* **2020**, *12*, 3797–3810. [CrossRef]
15. Khiar, H.; Barka, N.; Puga, A. Metal phosphates for the design of advanced heterogeneous photocatalysts. *Coord. Chem. Rev.* **2024**, *510*, 215814. [CrossRef]
16. Zhang, X.; Hou, Q.; Cao, S.; Lin, X.; Chen, X.; Wang, Z.; Wei, S.; Liu, S.; Dai, F.; Lu, X. Research status, opportunities, and challenges of cobalt phosphate based materials as OER electrocatalysts. *Green Chem.* **2023**, *25*, 7883–7903. [CrossRef]
17. Liu, H.; Yang, J.; Yang, X.; Hu, C.; Yang, L. A non-aqueous phase extraction system using tributyl phosphate for H_3PO_4 separation from wet-process superphosphoric acid: Extraction equilibrium and mechanism. *Korean J. Chem. Eng.* **2022**, *39*, 1659–1672. [CrossRef]
18. Tian, J.; Ge, F.; Zhang, D.; Deng, S.; Liu, X. Roles of phosphate solubilizing microorganisms from managing soil phosphorus deficiency to mediating biogeochemical P cycle. *Biology* **2021**, *10*, 158. [CrossRef]
19. Idrissi, H.; Taha, Y.; Elghali, A.; El Khessaimi, Y.; Aboulayt, A.; Amalik, J.; Hakkou, R.; Benzaazoua, M. Sustainable use of phosphate waste rocks: From characterization to potential applications. *Mater. Chem. Phys.* **2021**, *260*, 124119. [CrossRef]
20. Villalba, G.; Liu, Y.; Schroder, H.; Ayres, R.U. Global phosphorus flows in the industrial economy from a production perspective. *J. Ind. Ecol.* **2008**, *12*, 557–569. [CrossRef]
21. Li, W.; Jin, L.; Gao, F.; Wan, H.; Pu, Y.; Wei, X.; Chen, C.; Zou, W.; Zhu, C.; Dong, L. Advantageous roles of phosphate decorated octahedral CeO_2 {111}/g-C_3N_4 in boosting photocatalytic CO_2 reduction: Charge transfer bridge and Lewis basic site. *Appl. Catal. B* **2021**, *294*, 120257. [CrossRef]
22. Salama, A. Recent progress in preparation and applications of chitosan/calcium phosphate composite materials. *Int. J. Biol. Macromol.* **2021**, *178*, 240–252. [CrossRef] [PubMed]
23. Fang, B.; Hu, Z.; Shi, T.; Liu, Y.; Wang, X.; Yang, D.; Zhu, K.; Zhao, X.; Zhao, Z. Research progress on the properties and applications of magnesium phosphate cement. *Ceram. Int.* **2023**, *49*, 4001–4016. [CrossRef]
24. Timofeeva, A.; Galyamova, M.; Sedykh, S. Prospects for using phosphate-solubilizing microorganisms as natural fertilizers in agriculture. *Plants* **2022**, *11*, 2119. [CrossRef]
25. Pal, S.; Ghosh, T.K.; Ghosh, R.; Mondal, S.; Ghosh, P. Recent advances in recognition, sensing and extraction of phosphates: 2015 onwards. *Coord. Chem. Rev.* **2020**, *405*, 213128. [CrossRef]
26. Tankov, I.; Yankova, R. Crystal structure, chemical bonds nature and thermodynamic functions for the new ionic liquid pyridinium bis (dihydrogen phosphate). *J. Mol. Struct.* **2022**, *1247*, 131363. [CrossRef]
27. Lin, H.; Wang, Y.; Dong, Y. A review of methods, influencing factors and mechanisms for phosphorus recovery from sewage and sludge from municipal wastewater treatment plants. *J. Environ. Chem. Eng.* **2023**, *12*, 111657. [CrossRef]
28. Deng, J.; Walther, A. ATP-responsive and ATP-fueled self-assembling systems and materials. *Adv. Mater.* **2020**, *32*, 2002629. [CrossRef] [PubMed]
29. Shen, J.; Gaines, D.; Shahabfar, S.; Li, Z.; Kang, D.; Griesemer, S.; Salgado-Casanova, A.; Liu, T.-C.; Chou, C.-T.; Xia, Y. Phase Stability of Lead Phosphate Apatite $Pb_{10-x}Cu_x(PO_4)_6O$, $Pb_{10-x}Cu_x(PO_4)_6(OH)_2 (x = 0, 1)$, and $Pb_8Cu_2(PO_4)_6$. *Chem. Mater.* **2023**, *36*, 275–285. [CrossRef]
30. Kono, T.; Sakae, T.; Nakada, H.; Kaneda, T.; Okada, H. Confusion between carbonate apatite and biological apatite (carbonated hydroxyapatite) in bone and teeth. *Minerals* **2022**, *12*, 170. [CrossRef]
31. Peresypkina, E.; Virovets, A.; Scheer, M. Organometallic polyphosphorus complexes as diversified building blocks in coordination chemistry. *Coord. Chem. Rev.* **2021**, *446*, 213995. [CrossRef]
32. Panseri, S.; Arioli, F.; Biolatti, C.; Mosconi, G.; Pavlovic, R.; Chiesa, L.M. Detection of polyphosphates in seafood and its relevance toward food safety. *Food Chem.* **2020**, *332*, 127397. [CrossRef]
33. Cao, J.; Wang, A.; Huang, N.; Wu, F.; Chen, P.; Li, C.; Wang, X. Role and mechanism of polyphosphate in bone tissue regeneration. *Chin. J. Tissue Eng. Res.* **2023**, *27*, 3375. [CrossRef]
34. Sidhu, G.K.; Singh, S.; Kumar, V.; Dhanjal, D.S.; Datta, S.; Singh, J. Toxicity, monitoring and biodegradation of organophosphate pesticides: A review. *Crit. Rev. Environ. Sci. Technol.* **2019**, *49*, 1135–1187. [CrossRef]
35. Kolodiazhnyi, O.I. Phosphorus compounds of natural origin: Prebiotic, stereochemistry, application. *Symmetry* **2021**, *13*, 889. [CrossRef]

36. Mukherjee, S.; Gupta, R.D. Organophosphorus nerve agents: Types, toxicity, and treatments. *J. Toxicol.* **2020**, *2020*, 3007984. [CrossRef]
37. Aroniadou-Anderjaska, V.; Figueiredo, T.H.; de Araujo Furtado, M.; Pidoplichko, V.I.; Braga, M.F. Mechanisms of organophosphate toxicity and the role of acetylcholinesterase inhibition. *Toxics* **2023**, *11*, 866. [CrossRef]
38. Alozi, M.; Rawas-Qalaji, M. Treating organophosphates poisoning: Management challenges and potential solutions. *Crit. Rev. Toxicol.* **2020**, *50*, 764–779. [CrossRef]
39. Nath, S. Phosphorus chemistry at the roots of bioenergetics: Ligand permutation as the molecular basis of the mechanism of ATP synthesis/hydrolysis by F_0F_1-ATP synthase. *Molecules* **2023**, *28*, 7486. [CrossRef] [PubMed]
40. Nardi, A.N.; Olivieri, A.; D'Abramo, M.; Amadei, A. A Theoretical-Computational Study of Phosphodiester Bond Cleavage Kinetics as a Function of the Temperature. *ChemPhysChem* **2024**, *25*, e202300952. [CrossRef] [PubMed]
41. Seok, S.-H. Structural insights into protein regulation by phosphorylation and substrate recognition of protein kinases/phosphatases. *Life* **2021**, *11*, 957. [CrossRef] [PubMed]
42. Foroutan, F.; Abrahams, I.; Smales, G.J.; Kanwal, N.; di Pasquale, R.; Knowles, J.C.; Smith, A.J.; Carta, D. A sol-gel templating route for the synthesis of hierarchical porous calcium phosphate glasses containing zinc. *Ceram. Int.* **2024**, *50*, 38174–38182. [CrossRef]
43. Ou, J.; Kang, S.; Chen, J.; Jiaqiang, E.; Liao, G.; Zhang, F.; Leng, E. Study on the selective recovery of metals from lithium iron phosphate cathode materials based on hydrothermal oxidation. *J. Energy Storage* **2024**, *101*, 113832. [CrossRef]
44. Katsui, H.; Kondo, N. Preferred orientations and microstructures of lanthanum phosphate films prepared via laser chemical vapor deposition. *J. Cryst. Growth* **2019**, *519*, 46–53. [CrossRef]
45. Lee, S.-J.; Toan, D.L.H.; Chen, Y.-H.; Peng, H.-C. Performance Improvement of Phosphate Coating on Carbon Steel by Cathodic Electrochemical Method. *Int. J. Electrochem. Sci.* **2016**, *11*, 2306–2316. [CrossRef]
46. Moumen, E.; El Hankari, S. MOF-assisted synthesis of mesoporous zirconium phosphate for enhanced cadmium removal from water. *Colloids Interface Sci. Commun.* **2023**, *52*, 100691. [CrossRef]
47. Jin, Z.; Liao, P.; Jaisi, D.P.; Wang, D.; Wang, J.; Wang, H.; Jiang, S.; Yang, J.; Qiu, S.; Chen, J. Suspended phosphorus sustains algal blooms in a dissolved phosphorus-depleted lake. *Water Res.* **2023**, *241*, 120134. [CrossRef]
48. Kumar, S.; Bhanuse, G.B.; Fu, Y.P. Phosphide-Based Electrocatalysts for Urea Electrolysis: Recent Trends and Progress. *ChemPhysChem* **2024**, *25*, e202300924. [CrossRef] [PubMed]
49. Sayed, E.T.; Eisa, T.; Mohamed, H.O.; Abdelkareem, M.A.; Allagui, A.; Alawadhi, H.; Chae, K.-J. Direct urea fuel cells: Challenges and opportunities. *J. Power Sources* **2019**, *417*, 159–175. [CrossRef]
50. Yang, K.; Hao, L.; Hou, Y.; Zhang, J.; Yang, J.-H. Summary and application of Ni-based catalysts for electrocatalytic urea oxidation. *Int. J. Hydrogen Energy* **2024**, *51*, 966–981. [CrossRef]
51. Gao, X.; Zhang, S.; Wang, P.; Jaroniec, M.; Zheng, Y.; Qiao, S.-Z. Urea catalytic oxidation for energy and environmental applications. *Chem. Soc. Rev.* **2024**, *53*, 1552–1591. [CrossRef] [PubMed]
52. Geng, S.-K.; Zheng, Y.; Li, S.-Q.; Su, H.; Zhao, X.; Hu, J.; Shu, H.-B.; Jaroniec, M.; Chen, P.; Liu, Q.-H. Nickel ferrocyanide as a high-performance urea oxidation electrocatalyst. *Nat. Energy* **2021**, *6*, 904–912. [CrossRef]
53. Mohamed, I.M.; Yasin, A.S.; Barakat, N.A.; Song, S.A.; Lee, H.E.; Kim, S.S. Electrocatalytic behavior of a nanocomposite of Ni/Pd supported by carbonized PVA nanofibers towards formic acid, ethanol and urea oxidation: A physicochemical and electro-analysis study. *Appl. Surf. Sci.* **2018**, *435*, 122–129. [CrossRef]
54. Yang, Y.; Yuwono, J.A.; Whittaker, T.; Ibáñez, M.M.; Wang, B.; Kim, C.; Borisevich, A.Y.; Chua, S.; Prada, J.P.; Wang, X. Double Hydroxide Nanocatalysts for Urea Electrooxidation Engineered toward Environmentally Benign Products. *Adv. Mater.* **2024**, *36*, 2403187. [CrossRef]
55. Mohamed, I.M.A.; Motlak, M.; Fouad, H.; Barakat, N.A.M. Cobalt/Chromium Nanoparticles-Incorporated Carbon Nanofibers as Effective Nonprecious Catalyst for Methanol Electrooxidation in Alkaline Medium. *Nano* **2016**, *11*, 1650049. [CrossRef]
56. Abd El-Lateef, H.M.; Almulhim, N.F.; Mohamed, I.M. Physicochemical and electrochemical investigations of an electrodeposited $CeNi_2$@ NiO nanomaterial as a novel anode electrocatalyst material for urea oxidation in alkaline media. *J. Mol. Liq.* **2020**, *297*, 111737. [CrossRef]
57. Zhan, G.; Hu, L.; Li, H.; Dai, J.; Zhao, L.; Zheng, Q.; Zou, X.; Shi, Y.; Wang, J.; Hou, W.; et al. Highly selective urea electrooxidation coupled with efficient hydrogen evolution. *Nat. Commun.* **2024**, *15*, 5918. [CrossRef]
58. Mohamed, I.M.; Shehata, A.G.; Al-Anazi, A.; Khairy, M.; Newair, E.F. A template-assisted method for synthesizing TiO_2 nanoparticles and Ni/TiO_2 nanocomposites for urea electrooxidation. *Mater. Chem. Phys.* **2024**, *316*, 129112. [CrossRef]
59. Yang, M.; Ge, W.; Liu, Z.; Lv, Y.; Wu, H.; Zhang, J. Selectively etching manganese from nickel-manganese hydroxides via thiourea corrosion for urea electro-oxidation. *J. Alloys Compd.* **2024**, *981*, 173731. [CrossRef]
60. Chen, H.; Zhang, K.; Xia, Y.; Li, J.; Hui, B. (B, P, Co, Fe)-Ni Modified on Nanowood for Boosting Seawater Urea Electro-Oxidation. *Green Chem.* **2025**, *27*, 133–143. [CrossRef]
61. Yan, X.; Zhang, W.-D.; Xu, H.; Xiang, L.; Liu, J.; Yang, J.; Zhu, H.; Gu, Z.-G. Hierarchical NiCr hydroxide nanospheres with tunable domain boundaries for highly efficient urea electro-oxidation. *Electrochim. Acta* **2021**, *388*, 138633. [CrossRef]
62. Mohamed, I.M.; Liu, C. Chemical design of novel electrospun CoNi/Cr nanoparticles encapsulated in C-nanofibers as highly efficient material for urea oxidation in alkaline media. *Appl. Surf. Sci.* **2019**, *475*, 532–541. [CrossRef]

63. Ojha, S.K.; Kamlesh; Tiwari, S.; Mudgal, M.; Raizada, P.; Singh, P.; Singh, A. Electrodeposited Ni-Se on recycled steel substrate as eco-friendly electrodes with enhanced electrocatalytic activity towards urea oxidation reaction. *J. Appl. Electrochem.* **2024**, *54*, 1–17. [CrossRef]
64. Gao, X.; Bai, X.; Wang, P.; Jiao, Y.; Davey, K.; Zheng, Y.; Qiao, S.-Z. Boosting urea electrooxidation on oxyanion-engineered nickel sites via inhibited water oxidation. *Nat. Commun.* **2023**, *14*, 5842. [CrossRef] [PubMed]
65. Hameed, R.A.; Mohamed, I.M.; Al-Enizi, A.M.; Abutaleb, A.; Shaikh, S.F.; Yousef, A. Fabrication of electrospun nickel sulphide nanoparticles onto carbon nanofibers for efficient urea electro-oxidation in alkaline medium. *Int. J. Hydrogen Energy* **2021**, *46*, 12944–12960. [CrossRef]
66. Abd El-Lateef, H.M.; Khalaf, M.M.; Mohamed, I.M. An efficient and non-precious anode electrocatalyst of NiO-modified carbon nanofibers towards electrochemical urea oxidation in alkaline media. *Ceram. Int.* **2020**, *46*, 20376–20384. [CrossRef]
67. Alamro, F.S.; Medany, S.S.; Al-Kadhi, N.S.; Ahmed, H.A.; Hefnawy, M.A. Modified $NiFe_2O_4$-Supported Graphene Oxide for Effective Urea Electrochemical Oxidation and Water Splitting Applications. *Molecules* **2024**, *29*, 1215. [CrossRef] [PubMed]
68. Li, B.; Song, C.; Rong, J.; Zhao, J.; Wang, H.-E.; Yang, P.; Ye, K.; Cheng, K.; Zhu, K.; Yan, J. A new catalyst for urea oxidation: $NiCo_2S_4$ nanowires modified 3D carbon sponge. *J. Energy Chem.* **2020**, *50*, 195–205. [CrossRef]
69. Li, P.; Li, W.; Huang, Y.; Huang, Q.; Tian, S. 3D Hierarchical-Architectured Nanoarray Electrode for Boosted and Sustained Urea Electro-Oxidation. *Small* **2023**, *19*, 2300725. [CrossRef] [PubMed]
70. Guo, X.; Qiu, L.; Li, M.; Tian, F.; Ren, X.; Jie, S.; Geng, S.; Han, G.; Huang, Y.; Song, Y. Accelerating the generation of NiOOH by in-situ surface phosphating nickel sulfide for promoting the proton-coupled electron transfer kinetics of urea electrolysis. *Chem. Eng. J.* **2024**, *483*, 149264. [CrossRef]
71. Wang, W.; Chai, D.; Zhang, J.; Xue, S.; Wang, Y.; Lei, Z. Ni_5Sm-P/C ternary alloyed catalyst as highly efficient electrocatalyst for urea electrooxidation. *J. Taiwan Inst. Chem. Eng.* **2017**, *80*, 326–332. [CrossRef]
72. Mohamed, I.M.A.; Kanagaraj, P.; Yasin, A.S.; Iqbal, W.; Liu, C. Electrochemical impedance investigation of urea oxidation in alkaline media based on electrospun nanofibers towards the technology of direct-urea fuel cells. *J. Alloys Compd.* **2020**, *816*, 152513. [CrossRef]
73. Bezerra, Â.C.S.; De Sa, E.L.; Nart, F.C. In situ vibrational study of the initial steps during urea electrochemical oxidation. *J. Phys. Chem. B* **1997**, *101*, 6443–6449. [CrossRef]
74. Vedharathinam, V.; Botte, G.G. Understanding the electro-catalytic oxidation mechanism of urea on nickel electrodes in alkaline medium. *Electrochim. Acta* **2012**, *81*, 292–300. [CrossRef]
75. Singh, R.K.; Rajavelu, K.; Montag, M.; Schechter, A. Advances in catalytic electrooxidation of urea: A review. *Energy Technol.* **2021**, *9*, 2100017. [CrossRef]
76. Alex, C.; Shukla, G.; John, N.S. Introduction of surface defects in NiO with effective removal of adsorbed catalyst poisons for improved electrochemical urea oxidation. *Electrochim. Acta* **2021**, *385*, 138425. [CrossRef]
77. Qiao, H.; Wang, X.; Dong, Q.; Zheng, H.; Chen, G.; Hong, M.; Yang, C.-P.; Wu, M.; He, K.; Hu, L. A high-entropy phosphate catalyst for oxygen evolution reaction. *Nano Energy* **2021**, *86*, 106029. [CrossRef]
78. Majhi, K.C.; Yadav, M. Facile hydrothermal synthesis of rare earth phosphate for boosting hydrogen evolution reaction. *Int. J. Hydrogen Energy* **2022**, *47*, 14092–14103. [CrossRef]
79. Zhai, R.; Zhang, L.; Gu, M.; Zhao, X.; Zhang, B.; Cheng, Y.; Zhang, J. A review of phosphorus structures as CO_2 reduction photocatalysts. *Small* **2023**, *19*, 2207840. [CrossRef]
80. Qiu, Y.; Dai, X.; Wang, Y.; Ji, X.; Ma, Z.; Liu, S. The polyoxometalates mediated preparation of phosphate-modified $NiMoO_{4-x}$ with abundant O-vacancies for H_2 production via urea electrolysis. *J. Colloid Interface Sci.* **2023**, *629*, 297–309. [CrossRef]
81. Song, X.; Gao, L.; Li, Y.; Chen, W.; Mao, L.; Yang, J.-H. Nickel phosphate-based materials with excellent durability for urea electro-oxidation. *Electrochim. Acta* **2017**, *251*, 284–292. [CrossRef]
82. Li, Q.; Li, X.; Gu, J.; Li, Y.; Tian, Z.; Pang, H. Porous rod-like Ni_2P/Ni assemblies for enhanced urea electrooxidation. *Nano Res.* **2021**, *14*, 1405–1412. [CrossRef]
83. Wang, G.; Ye, K.; Shao, J.; Zhang, Y.; Zhu, K.; Cheng, K.; Yan, J.; Wang, G.; Cao, D. Porous Ni_2P nanoflower supported on nickel foam as an efficient three-dimensional electrode for urea electro-oxidation in alkaline medium. *Int. J. Hydrogen Energy* **2018**, *43*, 9316–9325. [CrossRef]
84. Imani, A.; Oveisi, H. Self-Assembly Assisted Fabrication of Nanoporous Nickel (II) Phosphate Octahydrate Microspheres Catalyst with Orange Peel Surface toward Urea Oxidation. *ChemistrySelect* **2019**, *4*, 7338–7342. [CrossRef]
85. Xu, X.; Du, P.; Guo, T.; Zhao, B.; Wang, H.; Huang, M. In Situ Grown Ni phosphate@$Ni_{12}P_5$ Nanorod Arrays as a Unique Core–Shell Architecture: Competitive Bifunctional Electrocatalysts for Urea Electrolysis at Large Current Densities. *ACS Sustain. Chem. Eng.* **2020**, *8*, 7463–7471. [CrossRef]
86. Tsai, Y.-C.; Wu, M.-S. Zeolitic nickel phosphate nanorods with open-framework structure (VSB-5) for catalytic application in electro-oxidation of urea. *Appl. Surf. Sci.* **2020**, *529*, 147175. [CrossRef]
87. Khalaf, M.M.; Abd El-Lateef, H.M.; Touny, A.H.; Saleh, M.M.; Mohamed, I.M.A. Electrocatalytic performance of inorganic nanoflakes nickel phosphates under adjusted synthetic parameters towards urea and methanol oxidation in alkaline media. *Microchem. J.* **2021**, *163*, 105901. [CrossRef]

88. Al-Sulmi, W.A.; Ghanem, M.A.; Aladeemy, S.A.; Al-Mayouf, A.M.; Alotaibi, N.H. Chemical deposition from a liquid crystal template: A highly active mesoporous nickel phosphate electrocatalyst for hydrogen green production via urea electro-oxidation in an alkaline solution. *Arab. J. Chem.* **2023**, *16*, 105266. [CrossRef]
89. Rasal, A.S.; Chen, H.M.; Yu, W.-Y. Electronic structure engineering of electrocatalyst for efficient urea oxidation reaction. *Nano Energy* **2024**, *121*, 109183. [CrossRef]
90. Lu, J.; Jiang, W.; Deng, R.; Feng, B.; Yin, S.; Tsiakaras, P. Tailoring competitive adsorption sites of hydroxide ion to enhance urea oxidation-assisted hydrogen production. *J. Colloid Interface Sci.* **2024**, *667*, 249–258. [CrossRef] [PubMed]
91. Ghanem, M.A.; Al-Mayouf, A.M.; Abdulkader, M.O.; Alfudhayli, K.A. Modification of nickel foam with nickel phosphate catalyst layer via anodizing for boosting the electrocatalytic urea oxidation and hydrogen evolution reactions. *J. Saudi Chem. Soc.* **2024**, *28*, 101913. [CrossRef]
92. Gao, T.; Gu, Y.; Wu, M.; Liu, Y.; Yang, L.; Han, T.; Zhang, S.; Li, S.; Wei, W.; Chen, W.; et al. Phosphate-Derived Copper-Based Catalysts for Efficient CO_2 Reduction to Multicarbon Products. *Energy Fuels* **2023**, *37*, 19053–19062. [CrossRef]
93. Martín-Yerga, D.; Carrasco-Rodríguez, J.; Fierro, J.L.G.; García Alonso, F.J.; Costa-García, A. Copper-modified titanium phosphate nanoparticles as electrocatalyst for glucose detection. *Electrochim. Acta* **2017**, *229*, 102–111. [CrossRef]
94. Zhang, L.; Men, Y.; Wu, B.; Feng, Y.; Song, C.; Liu, S.; Wang, J.; An, W.; Magkoev, T.T. Highly Ethylene-Selective Electroreduction CO_2 Over Cu Phosphate Nanostructures with Tunable Morphology. *Top. Catal.* **2023**, *66*, 1527–1538. [CrossRef]
95. Shen, H.; Wei, T.; Ding, J.; Liu, X. Copper Phosphide Nanowires as High-Performance Catalysts for Urea-Assisted Hydrogen Evolution in Alkaline Medium. *Materials* **2023**, *16*, 4169. [CrossRef]
96. Zhang, T.; Zhang, S.; Cao, S.; Yao, Q.; Lee, J.Y. A Self-Templating Redox-Mediated Synthesis of Hollow Phosphated Manganese Oxide Nanospheres as Noble-Metal-like Oxygen Electrocatalysts. *Chem. Mater.* **2018**, *30*, 8270–8279. [CrossRef]
97. Wang, J.; Ping, Y.; Chen, Y.; Liu, S.; Dong, J.; Ruan, Z.; Liang, X.; Lin, J. Improvement of electrocatalytic water oxidation activity of novel copper complex by modulating the axial coordination of phosphate on metal center. *Dalton Trans.* **2024**, *53*, 5222–5229. [CrossRef]
98. Yang, S.; Li, X.; Li, Y.; Wang, Y.; Jin, X.; Qin, L.; Zhang, W.; Cao, R. Effect of Proton Transfer on Electrocatalytic Water Oxidation by Manganese Phosphates. *Angew. Chem. Int. Ed.* **2023**, *62*, e202215594. [CrossRef] [PubMed]
99. Li, J.; Chen, Y.; Bai, R.; Chen, C.; Wang, W.; Pan, Y.; Liu, Y. Construction of Pd/Ni_2P-Ni foam nanosheet array electrode by in-situ phosphatization-electrodeposition strategy for synergistic electrocatalytic hydrodechlorination. *Chem. Eng. J.* **2022**, *435*, 134932. [CrossRef]
100. Zhao, S.; Qu, G.; Wang, C.; Zhang, X.; Wang, X.; Hou, P.; Zhao, G.; Leng, J.; Xu, X.; Zhai, T. Synergetic modulation on structure and interface enables bimetallic phosphide nanoarrays for advanced aqueous zinc batteries. *Chem. Eng. J.* **2023**, *461*, 141790. [CrossRef]
101. Liu, L.; Meng, F.; Wang, H.; Ma, S.; Lukyanov, D.; Zhu, W.; Li, Y.; Ren, P.; Kondratiev, V.; Yang, P.; et al. Insight into the electronic modulation on nickel-cobalt bimetallic phosphates towards high-efficiency electrocatalytic hydrogen evolution. *J. Alloys Compd.* **2024**, *1002*, 175259. [CrossRef]
102. Wang, J.; Yan, M.; Wang, S.; Liu, Z.; Zhang, L.; Xu, Y.; Zhao, X.; Jiao, T. Electron density regulation of bimetal phosphates via heterojunction interface engineering for electrocatalytic reaction. *Colloids Surf. A Physicochem. Eng. Asp.* **2024**, *683*, 132989. [CrossRef]
103. Colodrero, R.M.P.; Olivera-Pastor, P.; Cabeza, A.; Bazaga-García, M. Properties and Applications of Metal Phosphates and Pyrophosphates as Proton Conductors. *Materials* **2022**, *15*, 1292. [CrossRef]
104. Abd El-Lateef, H.M.; Khalaf, M.M.; Mohamed, I.M.A. XPS analysis, voltammetric, and impedance characteristics of novel heterogeneous biphosphates based on Cu/Ni for tri(ammonium) phosphate oxidation: A new direction for material processing in fuel technology. *Fuel* **2024**, *356*, 129618. [CrossRef]
105. Yang, J.-H.; Song, X.; Zhao, X.; Wang, Y.; Yang, Y.; Gao, L. Nickel phosphate materials regulated by doping cobalt for urea and methanol electro-oxidation. *Int. J. Hydrogen Energy* **2019**, *44*, 16305–16314. [CrossRef]
106. Qiao, L.; Zhu, A.; Liu, D.; Feng, J.; Chen, Y.; Chen, M.; Zhou, P.; Yin, L.; Wu, R.; Ng, K.W.; et al. Crystalline phosphides/amorphous oxides composite for energy-saving hydrogen production assisted by efficient urea oxidation reaction. *Chem. Eng. J.* **2023**, *454*, 140380. [CrossRef]
107. Abd El-Lateef, H.M.; Khalaf, M.M.; Alnajjar, A.O.; Mohamed, I.M.A. Facile synthesis of Co/Ni bimetallic phosphate as electrode material for urea fuel cells: Effect of synthetic strategy on the physicochemical and electrocatalytic behavior. *Fuel* **2023**, *334*, 126671. [CrossRef]
108. Guo, F.; Ye, K.; Du, M.; Huang, X.; Cheng, K.; Wang, G.; Cao, D. Electrochemical impedance analysis of urea electro-oxidation mechanism on nickel catalyst in alkaline medium. *Electrochim. Acta* **2016**, *210*, 474–482. [CrossRef]
109. Panigrahi, K.; Mal, S.; Bhattacharyya, S. Deciphering interfacial charge transfer mechanisms in electrochemical energy systems through impedance spectroscopy. *J. Mater. Chem. A* **2024**, *12*, 14334–14353. [CrossRef]
110. Xu, X.; Zhang, C.; Li, J.; Liu, H.; Su, G.; Shi, Z.; Huang, M. Redistributing interfacial charge density of $Ni_{12}P_5$/Ni_3P via Fe doping for ultrafast urea oxidation catalysis at large current densities. *Chem. Eng. J.* **2023**, *452*, 139362. [CrossRef]
111. Yang, L.; Ru, F.; Shi, J.; Yang, T.; Guo, C.; Chen, Y.; Wang, E.; Du, Z.; Chou, K.-C.; Hou, X. Trifunctional electrocatalysts based on feather-like NiCoP 3D architecture for hydrogen evolution, oxygen evolution, and urea oxidation reactions. *Ceram. Int.* **2023**, *49*, 659–668. [CrossRef]

112. Zhang, R.; Wang, Y.; Gao, X.; Duan, D.; Wang, J.; Liu, S. Electrodeposited zinc cobalt bimetallic phosphate as a bifunctional catalyst for hydrogen evolution and urea oxidation. *J. Alloys Compd.* **2024**, *1009*, 176818. [CrossRef]
113. Zhang, L.; Wang, X.; Huang, Y.; Cheng, W.; Zhang, Y.; Guo, Y.; Sun, Y.; Tang, X. Rational design of bamboo mat-like 3D Ni foam@NiO@PPy nanoarray electrode for binder-free, high-loading and high-areal capacity lithium storage. *J. Alloys Compd.* **2021**, *874*, 159957. [CrossRef]
114. Zhao, Z.; Dong, Y.; Ding, H.; Li, X.; Chang, X. Manganese-facilitated dynamic active-site generation on Ni_2P with self-termination of surface reconstruction for urea oxidation at high current density. *Water Res.* **2024**, *253*, 121266. [CrossRef] [PubMed]
115. Sharmila, V.; Packiaraj, R.; Nallamuthu, N.; Parthibavarman, M. Fabrication of MWCNTs wrapped nickel manganese phosphate asymmetric capacitor as a supercapattery electrode for energy storage applications. *Inorg. Chem. Commun.* **2020**, *121*, 108194. [CrossRef]
116. Mirghni, A.A.; Oyedotun, K.O.; Fasakin, O.; Mahmoud, B.A.; Tarimo, D.J.; Manyala, N. High-performance bimetallic Ni-Mn phosphate hybridized with 3-D graphene foam for novel hybrid supercapacitors. *J. Energy Storage* **2020**, *31*, 101584. [CrossRef]
117. Mahmoud, B.A.; Mirghni, A.A.; Fasakin, O.; Oyedotun, K.O.; Manyala, N. Bullet-like microstructured nickel ammonium phosphate/graphene foam composite as positive electrode for asymmetric supercapacitors. *RSC Adv.* **2020**, *10*, 16349–16360. [CrossRef] [PubMed]
118. Ruangudomsakul, M.; Osakoo, N.; Keawkumay, C.; Kongmanklang, C.; Butburee, T.; Kiatphuengporn, S.; Faungnawakij, K.; Chanlek, N.; Wittayakun, J.; Khemthong, P. Influential properties of activated carbon on dispersion of nickel phosphides and catalytic performance in hydrodeoxygenation of palm oil. *Catal. Today* **2021**, *367*, 153–164. [CrossRef]
119. Zhang, H.; Zhang, Y.; Wang, M.; Shen, Y.; Liu, B. Electro-oxidation of Urea on the Nickel Phosphate-based Nanomaterials. *Int. J. Electrochem. Sci.* **2020**, *15*, 3400–3409. [CrossRef]
120. Wang, K.; Huang, W.; Cao, Q.; Zhao, Y.; Sun, X.; Ding, R.; Lin, W.; Liu, E.; Gao, P. Engineering NiF_3/Ni_2P heterojunction as efficient electrocatalysts for urea oxidation and splitting. *Chem. Eng. J.* **2022**, *427*, 130865. [CrossRef]
121. Liu, H.; Liu, Z.; Wang, F.; Feng, L. Efficient catalysis of N doped NiS/NiS_2 heterogeneous structure. *Chem. Eng. J.* **2020**, *397*, 125507. [CrossRef]
122. Hao, P.; Zhu, W.; Li, L.; Tian, J.; Xie, J.; Lei, F.; Cui, G.; Zhang, Y.; Tang, B. Nickel incorporated Co_9S_8 nanosheet arrays on carbon cloth boosting overall urea electrolysis. *Electrochim. Acta* **2020**, *338*, 135883. [CrossRef]
123. Xu, W.; Feng, Y.; Sun, Z.; Guo, L.; Li, C.; Li, H.; Wang, Y.; Sun, H.-B. P-induced bottom-up growth of Fe-doped $Ni_{12}P_5$ nanorod arrays for urea oxidation reaction. *J. Colloid Interface Sci.* **2023**, *633*, 746–753. [CrossRef] [PubMed]

MDPI AG
Grosspeteranlage 5
4052 Basel
Switzerland
Tel.: +41 61 683 77 34

Catalysts Editorial Office
E-mail: catalysts@mdpi.com
www.mdpi.com/journal/catalysts

www.ingramcontent.com/pod-product-compliance
Lightning Source LLC
Chambersburg PA
CBHW061827210926
51850CB00006B/828